AF356555

Mitogen Activated Protein Kinases

Mitogen Activated Protein Kinases

Functions in Signal Transduction and Human Diseases

Special Issue Editors

Ritva Tikkanen
David Nikolic-Paterson

MDPI • Basel • Beijing • Wuhan • Barcelona • Belgrade • Manchester • Tokyo • Cluj • Tianjin

Special Issue Editors
Ritva Tikkanen
Justus-Liebig University of Giessen
Germany

David Nikolic-Paterson
Monash University
Australia

Editorial Office
MDPI
St. Alban-Anlage 66
4052 Basel, Switzerland

This is a reprint of articles from the Special Issue published online in the open access journal *International Journal of Molecular Sciences* (ISSN 1422-0067) (available at: https://www.mdpi.com/journal/ijms/special_issues/MAPK).

For citation purposes, cite each article independently as indicated on the article page online and as indicated below:

LastName, A.A.; LastName, B.B.; LastName, C.C. Article Title. *Journal Name* **Year**, *Article Number*, Page Range.

ISBN 978-3-03928-070-4 (Hbk)
ISBN 978-3-03928-071-1 (PDF)

Cover image courtesy of David Nikolic-Paterson.

Contents

About the Special Issue Editors

Ritva Tikkanen is Professor of Biochemistry and Molecular Biology at the Medical Faculty of the University of Giessen, Germany. The main focus of her research is rare diseases, including lysosomal storage disorders, metabolic diseases, and autoimmune diseases. Her laboratory is specialized in developing personalized therapies for genetic diseases such as aspartylglucosaminuria, SSADH deficiency, and neuronal ceroid lipofuscinoses. In addition, she is interested in the regulation of MAP kinase signaling in cancer cells and in autoimmune diseases.

David Nikolic-Paterson is a basic scientist with degrees in Biochemistry and in Cellular Immunology. He is Head of Laboratory Research in the Department of Nephrology at Monash Medical Centre in Melbourne, Australia. Dr. Nikolic-Paterson's research focuses on mechanisms of inflammation and fibrosis in acute and chronic kidney disease and kidney transplant rejection, with a particular interest in the contribution of mitogen-activated protein kinase signaling pathways in disease pathogenesis.

International Journal of
Molecular Sciences

Editorial

Mitogen-Activated Protein Kinases: Functions in Signal Transduction and Human Diseases

Ritva Tikkanen [1,*] and David J. Nikolic-Paterson [2]

[1] Institute of Biochemistry, Medical Faculty, Justus-Liebig University of Giessen, Friedrichstrasse 24, D-35392 Giessen, Germany
[2] Department of Nephrology, Monash Health and Monash University Centre for Inflammatory Diseases, 246 Clayton Road, Clayton, Victoria 3168, Australia; david.nikolic-paterson@monash.edu
*   Correspondence: Ritva.Tikkanen@biochemie.med.uni-giessen.de

Received: 26 September 2019; Accepted: 26 September 2019; Published: 29 September 2019

Mitogen-activated protein kinases (MAPKs) are involved in signaling processes induced by various stimuli, such as growth factors, stress, or even autoantibodies. Sequential activation and phosphorylation of kinases upstream of the MAPKs result in activation of the MAPKs and phosphorylation of their downstream substrates that can be either nuclear (e.g., transcription factors) or cytoplasmic (e.g., proteins involved in cell migration). MAPKs comprise three functional families, namely, extracellularly regulated kinases (ERKs), c-Jun N-terminal kinases (JNKs) and p38. MAPK signaling thus provides a versatile signaling module that regulates cell proliferation, differentiation, homeostasis and even cell death.

Individual proteins associated with the MAPK signaling cascades are regulated not only by phosphorylation but also by scaffolding proteins that can associate with several signaling proteins simultaneously and thus facilitate their regulation and interaction. Numerous scaffolding proteins are known to exist (for a review, see [1]). In this Special Issue, Dietel et al. summarize the role of one of these scaffolders, protein tyrosine phosphatase interacting protein 51 (PTPIP51), in the regulation of MAPK signaling [2]. The authors provide an overview about the function of this interesting protein in physiological signaling, but also in various diseases associated with dysregulation of MAPK signaling.

While MAPK pathways have homeostatic roles, the pathological mechanisms underpinning many disease processes involve signaling via one or more of the p38, JNK or ERK pathways. This Special Issue provides new insights into how MAPK signaling contributes to specific pathological processes across a range of conditions, including disorders of lung development, type 2 diabetes, cancer, proliferative skin diseases, cardiovascular disease and neurological disease.

Bronchopulmonary dysplasia (BPD) is a chronic lung disease of premature infants characterized by interrupted lung development. Supplemental oxygen is critical for preterm infants with respiratory failure. However, this hyperoxic environment contributes to BPD pathogenesis. Using a newborn mouse model, Menon et al. [3] found that hyperoxia induced a transient increase in endothelial ERK1/2 signaling during the period of lung vascularization at day 7—possibly as an adaptive response to mitigate hyperoxic injury—but this signaling was not sustained at day 14 when endothelial apoptosis and development of BPD occurred. This finding was replicated in cultured fetal human pulmonary artery endothelial cells where hyperoxia induced ERK-dependent proliferation.

Excessive fibroblast growth factor receptor (FGFR) signaling has been shown to be involved in human diseases such as cancer and skeletal disorders [4]. Thus, negative feedback regulation of FGFR signaling (e.g., by ERK and p38 MAPK) is required to ensure that this signaling cascade is precisely controlled. FGFR substrate 2 (FRS2) is an adapter protein that becomes Tyr-phosphorylated during FGFR signaling and feeds the signal forward towards MAPK activation. A negative feedback regulation loop that involves ERK-mediated Thr phosphorylation of FRS2 has been shown to be a major downregulation mode for FGFR signaling [5]. In this Special Issue, Zakrzewska et al. provide a deeper insight into this important feedback mechanism by showing that not only ERK, but also p38 kinase can also phosphorylate FRS2 and thus contributes to the negative regulation of FGFR [6]. This regulation mechanism appears to require the activity of both ERK and p38 kinases, since inhibition of both kinases resulted in prolonged FGFR signaling.

Activation of MAPKs, in particular JNKs, has been shown to inhibit insulin signaling and contribute to inflammation in type 2 diabetes [7]. In this issue, Cui et al. [8] show that the traditional Chinese medicines *Scutellaria baicalensis* Georgi and *Coptis chinensis* Franch, individually and in combination gave significant protection against hyperglycemia, hyperinsulinemia and lipid abnormalities in a rat model of type 2 diabetes. This was associated with reduced systemic and liver inflammation, with data suggestive of reduced JNK, NF-κB, and Akt signaling as the underlying protective mechanisms. Interestingly, the effects of these traditional medicines are comparable to that of metformin that acts via AMP-activated protein kinase signaling which inhibits JNK signaling [7,8].

Activation of JNK signaling pathways is also associated with other pathological processes such as liver diseases. In this Special Issue, Win et al. discuss the implications of JNK signaling mode as a modulator of cell death [9]. The authors stress that not only the level of signaling activation, but also its duration is important for the signaling outcome. In their review article, Win et al. mainly focus on the interplay of JNK signaling with the mitochondrial SAB (SH3 domain binding protein 5) protein that promotes the release of reactive oxygen species (ROS) and cell death. Therefore, modulation of SAB function and thus JNK signaling may provide means for therapy in conditions such as cancer, in which the balance between survival and cell death frequently involves the JNK-SAB-ROS axis.

In addition to MAPK, death-associated protein kinases (DAPKs) have been shown to be important regulators of cell death. In this Special Issue, Elbadawy et al. provide an overview of the functions of DAPKs in various diseases such as cancer, neuronal injury and cardiovascular disease [10]. A special focus of this review is the crosstalk of DAPK and MAPK signaling pathways and its implication for diseases, as well as the potential of DAPKs as therapeutic targets. As small molecules that inhibit DAPKs have become available, it will be interesting to see if these can be used for the treatment of the diseases associated with DAPK activity.

Infantile myofibromatosis is a rare disorder of mesenchymal proliferation characterized by nonmetastatic tumors and which is associated with specific point mutations in the platelet-derived growth factor receptor B (PDGFRB) gene. Using primary tumor samples and the tumor-derived NSTS-47 cell line, Sramek et al. [11] demonstrated that a PDGFRB mutation results in prominent PDGFRB/ERK signaling. Interestingly, while tumor cell proliferation was resistant to ERK inhibitors, the non-selective tyrosine kinase inhibitor sunitinib inhibited tumor cell proliferation via reducing PDGFRB and Akt phosphorylation, explaining the beneficial effects seen with this drug in some patients [12].

Another chronic proliferative disease is psoriasis, a multisystemic disease characterized by abnormal keratinocyte proliferation resulting in erythematous lesions on the skin. Dermal fibroblasts can regulate the function of adjacent keratinocytes, and studies by Becatti et al. [13] identified a marked activation of p38 and JNK signaling pathways in fibroblasts from psoriatic lesions versus control fibroblasts. This enhanced kinase signaling was associated with increased mitochondrial superoxide production and apoptosis of psoriatic fibroblasts. Mechanistically, psoriatic fibroblasts exhibited reduced expression of the NAD$^+$-dependent protein deacetylase, SIRT1. Addition of the SIRT1 activator, SRT1720, diminished p38 and JNK signaling and normalized function in psoriatic fibroblasts, suggesting therapeutic potential for this approach.

Inflammatory bowel disease (IBD) is characterized by intestinal barrier dysfunction that is strongly associated with gastrointestinal infections caused by Gram-negative bacteria. Cecropin A is an antimicrobial peptide (AMP) that has been shown to exhibit antimicrobial activity, but its potential role in intestinal barrier function has remained unknown. In this Special Issue, Zhai et al. now show that cecropin A can increase the expression of tight junction proteins zonula occludens 1 (ZO-1), claudin 1 and occludin, thus increasing intestinal cell barrier function [14]. Cecropin A treatment also resulted in downregulation of ERK phosphorylation, and chemical inhibition of ERK activity showed a synergistic effect on barrier function when applied together with cecropin A. These data are highly interesting for the treatment of IBD, since resistance to antibiotics that are used for the treatment of IBD-associated bacterial infections is a major problem. Therefore, AMPs such as cecropin A may provide an alternative as antibacterial therapy in IBD or other diseases associated with compromised intestinal barrier function.

MAPK signaling is also involved in other gastrointestinal inflammatory diseases such as gastric ulcer. In this issue, Akanda et al. show that an extract from the perennial shrub *Rabdosia inflexa* (RI) exhibits potent anti-inflammatory and gastro-protective effects by affecting proinflammatory cytokine release and MAPK signaling [15]. RI was able to protect RAW 264.7 macrophage cells against various toxic stimuli such as lipopolysaccharide, NO and ROS. The authors showed that RI treatment resulted in reduced inflammatory signs, such as cyclooxygenase 2 expression or nuclear factor kappa B (NF-κB) activation. Importantly, gastric damage in HCl- and ethanol-treated mice was efficiently reduced by RI. Thus, RI, which is traditionally used in Chinese medicine for gastrointestinal problems, may exert its beneficial effects against inflammatory diseases of the gut by regulating MAPK/NF-κB signaling and cytokine release.

Atherosclerosis and aortic valve sclerosis are major cardiovascular diseases in Western societies. Reustle and Torzewski [16] discuss how p38 MAPK signaling in endothelial cells, vascular smooth muscle cells and (myo)fibroblasts may contribute to the pathogenesis of these conditions and make the argument that p38 MAPK inhibition may be useful in the treatment of these diseases.

Bones are subject to continuous remodeling that involves osteoclasts that degrade the bone and the counteracting osteoblasts that reform the bone. Physiologically, it is important that the activity of these cells is kept in a delicate balance, and tipping of this balance in one or the other direction is involved in numerous diseases. In this issue, Lee et al. review the roles of MAPKs in osteoclast biology [17]. Osteoclast differentiation and activation are regulated by MAPK signaling, including ERK, JNK and p38. In their review, Lee et al. discuss the upstream regulation of MAPK signaling in osteoclasts and provide insights into the differential kinetics and crosstalk of the three MAPK signaling modes during osteoclast function.

ERK5, an atypical member of the ERK family, is thought to contribute to neuron differentiation and survival. In this context, Kashino et al. [18] used the PC-12 cell line to describe a mechanism whereby ERK5 phosphorylates the cytoplasmic domain of the Kv4.2 voltage-gated K^+ channel to inhibit the A-type current inactivation, leading to rapid repolarization towards resting potential, thus causing an increase in firing frequency, which may contribute to the neuronal differentiation process.

Parkinson's disease (PD) is a neurodegenerative disorder caused by insufficient dopamine production due to the loss of dopaminergic neurons. Both oxidative stress and endoplasmic reticulum stress are directly implicated in the pathogenesis of PD. As reviewed by Bohush et al. [19], both of these stressors are potent activators of JNK and p38 MAPK signaling which induce both apoptosis of neurons and microglial activation, leading to chronic neuroinflammation, identifying these kinases as potential therapeutic targets in this disease.

Organometallic drugs such as naphthalimides and their derivatives have shown promise in the treatment of cancers, but their clinical use is limited by their toxic side effects [20]. Thus, there is great interest in developing less toxic and efficient anticancer agents based on organometallic derivatives. In this issue, Dabiri et al. characterize the molecular mechanisms of rhodium(I) and ruthenium(II) containing naphthalimides conjugated with an *N*-heterocyclic carbene (NHC) moiety [21]. They show that these compounds induce a profound activation of p38 MAPK and elevated generation of reactive oxygen species, without affecting other MAPK signaling modules. These findings thus suggest that the effect of organometallic naphthalimides may be exerted through p38 signaling cascade.

There have been numerous setbacks in the clinical use of MAPK inhibitors, in particular due to acute liver toxicity with p38 inhibitors [22]. However, there is renewed interest in targeting MAPKs in disease. Recently, the ERK1/2 pathway inhibitor trametinib was granted an accelerated approval by the

US Food and Drug Administration for the treatment of unresectable metastatic melanoma. Apoptosis signal-regulating kinase 1 (ASK1/MAP3K5) is an upstream activator of p38 and JNK pathways, and an ASK1 inhibitor, selonsertib, is currently in phase 3 trials in liver fibrosis and diabetic kidney disease. Finally, CC90001 is a JNK inhibitor currently in phase 2 trials of idiopathic pulmonary fibrosis and in liver fibrosis. Thus, there is reason for optimism that targeting the pathological role of MAPK signaling may provide therapeutic benefit across a range of human diseases.

Funding: The work in the Tikkanen laboratory was supported by the German Research Council DFG (Grant TI 291/10-1 to R.T.). David J. Nikolic-Paterson is a Senior Research Fellow of the National Health and Medical Research Council of Australia (GNT1122073).

Conflicts of Interest: R.T. declares no conflict of interest. D.J.N.-P. received funding for research involving ASK1 inhibitors (Gilead Sciences) and JNK inhibitors (Celgene).

References

1. Meister, M.; Tomasovic, A.; Banning, A.; Tikkanen, R. Mitogen-Activated Protein (MAP) kinase scaffolding proteins: A recount. *Int. J. Mol. Sci.* **2013**, *14*, 4854–4884. [CrossRef] [PubMed]
2. Dietel, E.; Brobeil, A.; Gattenlohner, S.; Wimmer, M. The importance of the right framework: Mitogen-activated protein kinase pathway and the scaffolding protein PTPIP51. *Int. J. Mol. Sci.* **2018**, *19*, 3282. [CrossRef] [PubMed]
3. Menon, R.T.; Shrestha, A.K.; Barrios, R.; Shivanna, B. Hyperoxia disrupts extracellular signal-regulated kinases 1/2-induced angiogenesis in the developing lungs. *Int. J. Mol. Sci.* **2018**, *19*, 1525. [CrossRef] [PubMed]
4. Ornitz, D.M.; Itoh, N. The fibroblast growth factor signaling pathway. *Wiley Interdiscip. Rev. Dev. Biol.* **2015**, *4*, 215–266. [CrossRef] [PubMed]
5. Lax, I.; Wong, A.; Lamothe, B.; Lee, A.; Frost, A.; Hawes, J.; Schlessinger, J. The docking protein FRS2alpha controls a MAP kinase-mediated negative feedback mechanism for signaling by FGF receptors. *Mol. Cell.* **2002**, *10*, 709–719. [CrossRef]
6. Zakrzewska, M.; Opalinski, L.; Haugsten, E.M.; Otlewski, J.; Wiedlocha, A. Crosstalk between p38 and Erk 1/2 in Downregulation of FGF1-induced signaling. *Int. J. Mol. Sci.* **2019**, *20*, 1826. [CrossRef] [PubMed]
7. Engin, A. Human protein kinases and obesity. *Adv. Exp. Med. Biol.* **2017**, *960*, 111–134.
8. Cui, X.; Qian, D.W.; Jiang, S.; Shang, E.X.; Zhu, Z.H.; Duan, J.A. Scutellariae radix and coptidis rhizoma improve glucose and lipid metabolism in T2DM rats via regulation of the metabolic profiling and MAPK/PI3K/Akt signaling pathway. *Int. J. Mol. Sci.* **2018**, *19*, 3634. [CrossRef] [PubMed]
9. Win, S.; Than, T.A.; Kaplowitz, N. The regulation of JNK signaling pathways in cell death through the Interplay with mitochondrial SAB and upstream post-translational effects. *Int. J. Mol. Sci.* **2018**, *19*, 3657. [CrossRef]
10. Elbadawy, M.; Usui, T.; Yamawaki, H.; Sasaki, K. Novel functions of death-associated protein kinases through mitogen-activated protein kinase-related signals. *Int. J. Mol. Sci.* **2018**, *19*, 3031. [CrossRef]
11. Sramek, M.; Neradil, J.; Macigova, P.; Mudry, P.; Polaskova, K.; Slaby, O.; Noskova, H.; Sterba, J.; Veselska, R. Effects of sunitinib and other kinase inhibitors on cells harboring a PDGFRB mutation associated with infantile myofibromatosis. *Int. J. Mol. Sci.* **2018**, *19*, 2599. [CrossRef] [PubMed]
12. Mudry, P.; Slaby, O.; Neradil, J.; Soukalova, J.; Melicharkova, K.; Rohleder, O.; Jezova, M.; Seehofnerova, A.; Michu, E.; Veselska, R.; et al. Case report: Rapid and durable response to PDGFR targeted therapy in a child with refractory multiple infantile myofibromatosis and a heterozygous germline mutation of the PDGFRB gene. *BMC Cancer* **2017**, *17*, 119. [CrossRef] [PubMed]
13. Becatti, M.; Barygina, V.; Mannucci, A.; Emmi, G.; Prisco, D.; Lotti, T.; Fiorillo, C.; Taddei, N. Sirt1 protects against oxidative stress-induced apoptosis in fibroblasts from psoriatic patients: A new insight into the pathogenetic mechanisms of psoriasis. *Int. J. Mol. Sci.* **2018**, *19*, 1572. [CrossRef] [PubMed]
14. Zhai, Z.; Ni, X.; Jin, C.; Ren, W.; Li, J.; Deng, J.; Deng, B.; Yin, Y. Cecropin A modulates tight junction-related protein expression and enhances the barrier function of porcine intestinal epithelial cells by suppressing the MEK/ERK pathway. *Int. J. Mol. Sci.* **2018**, *19*, 1941. [CrossRef] [PubMed]

15. Akanda, M.R.; Kim, I.S.; Ahn, D.; Tae, H.J.; Nam, H.H.; Choo, B.K.; Kim, K.; Park, B.Y. Anti-inflammatory and gastroprotective roles of rabdosia inflexa through downregulation of pro-inflammatory cytokines and MAPK/NF-kappaB signaling pathways. *Int. J. Mol. Sci.* **2018**, *19*, 584. [CrossRef] [PubMed]

16. Reustle, A.; Torzewski, M. Role of p38 MAPK in atherosclerosis and aortic valve sclerosis. *Int. J. Mol. Sci.* **2018**, *19*, 3761. [CrossRef] [PubMed]

17. Lee, K.; Seo, I.; Choi, M.H.; Jeong, D. Roles of mitogen-activated protein kinases in osteoclast biology. *Int. J. Mol. Sci.* **2018**, *19*, 3004. [CrossRef] [PubMed]

18. Kashino, Y.; Obara, Y.; Okamoto, Y.; Saneyoshi, T.; Hayashi, Y.; Ishii, K. ERK5 phosphorylates Kv4.2 and inhibits inactivation of the A-type current in PC12 cells. *Int. J. Mol. Sci.* **2018**, *19*, 2008. [CrossRef] [PubMed]

19. Bohush, A.; Niewiadomska, G.; Filipek, A. Role of mitogen activated protein kinase signaling in Parkinson's disease. *Int. J. Mol. Sci.* **2018**, *19*, 2973. [CrossRef]

20. Tandon, R.; Luxami, V.; Kaur, H.; Tandon, N.; Paul, K. 1,8-naphthalimide: A potent DNA intercalator and target for cancer therapy. *Chem. Rec.* **2017**, *17*, 956–993. [CrossRef]

21. Dabiri, Y.; Schmid, A.; Theobald, J.; Blagojevic, B.; Streciwilk, W.; Ott, I.; Wolfl, S.; Cheng, X. A Ruthenium(II) N-Heterocyclic Carbene (NHC) complex with naphthalimide ligand triggers apoptosis in colorectal cancer cells via activating the ROS-p38 MAPK pathway. *Int. J. Mol. Sci.* **2018**, *19*, 3964. [CrossRef]

22. Hammaker, D.; Firestein, G.S. "Go upstream, young man": Lessons learned from the p38 saga. *Ann. Rheum. Dis.* **2010**, *69*, i77–i82. [CrossRef]

International Journal of
Molecular Sciences

Review

The Importance of the Right Framework: Mitogen-Activated Protein Kinase Pathway and the Scaffolding Protein PTPIP51

Eric Dietel [1,*], Alexander Brobeil [2], Stefan Gattenlöhner [2] and Monika Wimmer [1,3]

[1]   Institute of Anatomy and Cell Biology, Justus-Liebig-University, 35392 Giessen, Germany;
      monika.wimmer@anatomie.med.uni-giessen.de
[2]   Institute of Pathology, Justus-Liebig-University, 35392 Giessen, Germany;
      alexander.brobeil@patho.med.uni-giessen.de (A.B.); stefan.gattenloehner@patho.med.uni-giessen.de (S.G.)
[3]   Institute of Anatomy, Johannes-Kepler-University Linz, 4040 Linz, Austria
*   Correspondence: eric.dietel@med.uni-giessen.de; Tel.: +49-641-99-47012

Received: 26 August 2018; Accepted: 19 October 2018; Published: 22 October 2018

Abstract: The protein tyrosine phosphatase interacting protein 51 (PTPIP51) regulates and interconnects signaling pathways, such as the mitogen-activated protein kinase (MAPK) pathway and an abundance of different others, e.g., Akt signaling, NF-κB signaling, and the communication between different cell organelles. PTPIP51 acts as a scaffold protein for signaling proteins, e.g., Raf-1, epidermal growth factor receptor (EGFR), human epidermal growth factor receptor 2 (Her2), as well as for other scaffold proteins, e.g., 14-3-3 proteins. These interactions are governed by the phosphorylation of serine and tyrosine residues of PTPIP51. The phosphorylation status is finely tuned by receptor tyrosine kinases (EGFR, Her2), non-receptor tyrosine kinases (c-Src) and the phosphatase protein tyrosine phosphatase 1B (PTP1B). This review addresses various diseases which display at least one alteration in these enzymes regulating PTPIP51-interactions. The objective of this review is to summarize the knowledge of the MAPK-related interactome of PTPIP51 for several tumor entities and metabolic disorders.

Keywords: mitogen-activated protein kinase pathway (MAPK pathway); protein tyrosine phosphatase interacting protein 51 (PTPIP51); protein-protein interaction (PPI); cancer signaling

---

1. Background

The mitogen-activated protein kinase (MAPK) pathway is one of the best described signaling system in cancer. Almost one third of all human cancers have reported alterations in MAPK signaling, indicating the high relevance of the precise understanding of this pathway [1]. The basic role of the MAPK pathway is to transduce extracellular signals into the cell to regulate fundamental cellular functions including growth, cell migration, differentiation, and apoptosis [2]. To achieve a correct regulation of these diverging functions several distinct pathways are necessary [2]. The MAPK signaling consists of three different signaling systems, the extracellular signal-regulated kinase (ERK) pathway, the C-Jun N-terminal kinase/stress-activated protein kinase (JNK/SAPK) pathway and the p38 kinase pathway [3]. Each of these different signaling systems is strictly hierarchically structured and consists of a MAPK kinase kinase (MAPKKK), which is superior to a MAPK kinase (MAPKK), which controls a MAPK [3]. Of these different systems, the ERK pathway is the best studied MAPK pathway. The ERK signaling can be activated by numerous extracellular stimuli, e.g., growth factors or mitogens. One example for a classical activation path is represented by the activation of the epidermal growth factor receptor (EGFR). Its hetero- or homodimerization induced by binding of epidermal growth factors leads to an autophosphorylation of the receptor [4]. Consequently, a signaling cascade consisting of growth factor receptor-bound protein 2 (GRB2), son of sevenless (SOS), and the small GTPase Ras is activated. GTP-bound Ras recruits Raf kinases to the cell membrane for activation [4]. The Raf kinases represent the MAPKKK in the ERK pathway. Subsequently, Raf kinases activate MEK1/2 (MAPKK) and ERK1/2 (MAPK) [2]. The targets of ERK1/2 are diverse and include p90^{RSK}, mitogen-activated protein kinase interacting protein kinases 1 and 2 (MNK1/2), Ets, Ets domain-containing protein (Elk1), Myc, signal transducer and activator of transcription 1/3 (STAT1/3) and estrogen receptor (ER), to name some of them [2]. These many targets are necessary for the precise regulation of the various aforementioned cellular functions, e.g., differentiation, growth, apoptosis, and migration.

Since, the MAPK pathway controls these essential functions a precise regulation and titration of the signaling activity is needed. A perfect example for such a strict regulation is the upstream positioned Raf kinases. The Raf kinase family consists of three different Raf proteins, Raf-1, B-Raf and A-Raf. Although their structures are almost similar, their activation modes are extremely different. After recruitment of Raf kinases to GTP-bound Ras, a complex series of phosphorylations is induced for activation. These phosphorylations are needed for the activation of the kinase domain and reduction of the autoinhibition. The activation of Raf-1 and A-Raf requires phosphorylation of the N-region, dephosphorylation of the S259 inhibitory site, and phosphorylation of the activation loop. B-Raf is already in a preactivated state and can be fully activated by Ras alone, whereas the activation of Raf-1 and A-Raf requires other factors [1]. Due to this, only small aberrations in the structure of B-Raf, such as the V600E exchange, are needed to induce a constitutive activation. Alterations such as these are found in almost two thirds of malignant melanoma and in glioblastoma, where of more than 40 different mutations in the B-Raf gene 90% are at residue 600 in exon 15 [5].

Another regulating mode of the Raf kinases is the binding to scaffolding proteins such as the 14-3-3 protein family [6]. Such adaptor and scaffolding proteins facilitate the correct subcellular localization, provide a proximity of different signaling partners and support the formation of multiprotein complexes [3]. Moreover, scaffolding proteins can shield activated signaling molecules from deactivating phosphatases to allow an adequate signaling strength [3]. Additionally, scaffolds provide crosstalks between different signaling pathways.

Of note, the protein tyrosine phosphatase interacting protein 51 (PTPIP51) represents another scaffold protein, which regulates MAPK activation on Raf-1 level [7]. PTPIP51 exerts its regulating effect on the MAPK pathway on Raf-1 level via the scaffold protein 14-3-3β [7]. The recruitment of PTPIP51 into the MAPK signaling leads to an activation of the MAPK pathway. A well-titrated signal is a prerequisite for an optimal cellular function. Therefore, the formation of the PTPIP51/14-3-3β/Raf-1 complex is tightly regulated by kinases and phosphatases [8,9]. One of the crucial spots for this regulation is the tyrosine 176 residue of PTPIP51 [9–11]. Its phosphorylation results in a break-up of the PTPIP51/14-3-3β/Raf-1 complex and hence the stimulation of the MAPK signaling is omitted [8,9,11,12]. The phosphorylation of the tyrosine 176 residue is under the control of the EGFR and other kinases, such as the cellular sarcoma kinase (c-Src) [8,10–12]. Dephosphorylation is mainly performed by PTP1B [9,11,12]. Another important phosphorylation site of PTPIP51 is the serine 212 residue. Computational models of the PTPIP51 molecule show a cleft in its tertiary structure, which is surrounded by the aforementioned tyrosine 176 residue and serine 212 residue, respectively [9]. Up to now, we assume, that the cleft represents a binding site for the Raf kinases [9]. Contrary to the interaction inhibiting tyrosine 176 residue, phosphorylation of the serine 212 residue leads to an augmentation of the interaction with Raf-1 via 14-3-3β [7–9,12]. Besides the cleft, PTPIP51 protein structure contains tetratricopeptide domains, which are known to serve as binding sites for protein-protein interactions [9]. Additionally, in the structure of PTPIP51 two conserved regions are found. These sites facilitate the interaction with the scaffolding protein 14-3-3β [7,9]. In summary, PTPIP51 possesses the perfect scaffolding protein equipment, encompassing several binding sites for protein-protein interactions and the capability of modulating these bindings via phosphorylation and dephosphorylation of tyrosine and serine residues (Figure 1A).

Besides the direct regulation of the MAPK pathway, PTPIP51 is involved in a broad range of cellular functions and signaling systems. The panel of interaction partners ranges from NF-κB signaling proteins (RelA, I-κB) over mitochondrial associated ER membrane-related proteins (VAPB, ORP5/ORP8), autophagy-related signaling, and mitosis associated proteins (CGI-99, Nuf2) [13–20]. These interactions of PTPIP51 are already reviewed and analyzed by studies of our group and other scientists. Therefore, the focus of this review is to highlight the regulation of PTPIP51 and its functional consequences affecting the MAPK signaling in diseases associated with an aberrant MAPK signaling.

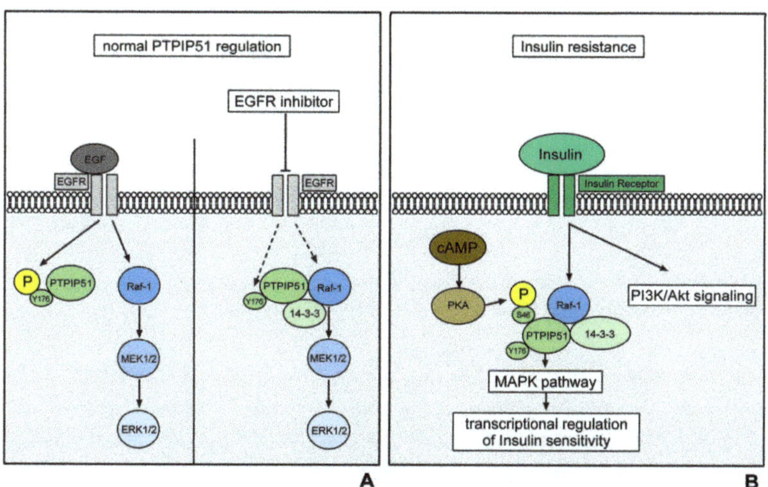

Figure 1. Regulation of PTPIP51 interactions in normal cells (represented by the HaCat cell line). Activation of the EGFR via the binding of EGF induces an activation of Raf-1 via several signaling molecules. Raf-1 depicts the MAPKKK of the ERK signaling. Its activation triggers a signaling cascade via MEK1/2 and ERK1/2, which ultimately initiates transcription. The EGFR also phosphorylates the Tyr176 residue of PTPIP51 and thereby inhibits its interaction with Raf-1. This mechanism prevents an overshooting activation of the MAPK pathway. The right side of the figure represents the interactions when EGFR is inhibited. The inhibition of EGFR leads to an omission of Tyr176 phosphorylation of PTPIP51 via the EGFR. The dephosphorylation of PTPIP51 at Tyr176 induces the formation of the Raf-1/14-3-3β/PTPIP51 complex and thus a stimulation of the MAPK pathway. This mechanism partially compensates for the EGFR inhibition (black arrows indicate a phosphorylation/activation; dotted black arrows indicate a reduced phosphorylation/activation) (**A**); regulation of PTPIP51 interactions in insulin resistance. Activation of the insulin receptor induces the activation of the PI3K-Akt-mTOR signaling and the MAPK pathway, especially the ERK signaling. Here, PTPIP51 stimulates the signaling on Raf-1 level and potentially modulates the insulin sensitivity on transcriptional level. Protein kinase A (PKA) phosphorylates the Ser46 residue of PTPIP51 and thereby stimulates the binding of PTPIP51 and Raf-1 via 14-3-3β (black arrows indicate a phosphorylation/activation) (**B**).

2. Regulation of PTPIP51 in Metabolic Signaling

2.1. Insulin Resistance and Obesity

Obesity and insulin resistance and their secondary diseases have reached epidemic proportions. More than one third of the US population has a diagnosed obesity and 10% of the adults are affected of diabetes [21]. The annual medical costs to manage these diseases are estimated to exceed $245 billion showing not only the medical but also the economical relevance of the precise understanding of the underlying mechanisms of obesity and insulin resistance to develop an effective therapy [21].

The most important signaling pathways for the regulation of lipid storages in adipocytes are the insulin receptor and the PKA signaling, which act as antagonists [22]. Whereas insulin receptor activity leads to lipogenesis, the PKA activation induces lipolysis [22–24]. Insulin receptor activation is mediated by binding of insulin and subsequently the autophosphorylation of the receptor molecules [24]. This results in the binding of insulin receptor substrates, which ultimately induce the activation of the two most important downstream signaling pathways, the PI3K-Akt-mTOR pathway and the ERK pathway [24]. While PI3K-Akt-mTOR signaling controls the lipid synthesis, glycogen synthesis, expression of metabolism-related proteins and glucose transport, the ERK pathway is needed for proliferation signaling [24]. The activation of ERK signaling also provides a mechanism for the adjustment of insulin sensitivity [25,26]. Zhang and coworkers showed, that the MAPK pathway controls the expression of the insulin-like receptor via the ETS-1 transcription factor pointed in Drosophila, thus ensuring the correct insulin sensitivity for the maintenance of adequate glucose levels [25].

The activity level of the insulin receptor is not only determined by the presence of insulin, but also by the phosphorylation status of the receptor. The phosphatase PTP1B represents a crucial downregulator of insulin receptor activity [24,27]. The impact of PTP1B on insulin-related signaling is mirrored by the results of PTP1B knockout models [28,29]. PTP1B knockout mice do not develop obesity or insulin resistance if fed a high-fat diet [28,29]. Thus, the inhibition of PTP1B via small molecule inhibitors seems a promising therapeutic strategy, but a safe and selective PTP1B inhibitor has yet to be identified [27].

PTPIP51 in Insulin Resistance and Obesity

The expression of PTPIP51 in adipocytes is tightly regulated by the diet and the submission to training in mice [22]. Bobrich and coworkers showed, that the expression of PTPIP51 protein correlates with the grade of insulin sensitivity, whereas the adipose tissue of normal mice contains the highest amount of PTPIP51 protein and that of high-fat diet-fed mice the lowest amount [22]. Mice fed a high-fat diet and subjected to training expressed levels of PTPIP51 protein lying in between the two aforementioned groups. Interestingly, also the interaction of PTP1B and PTPIP51 is correlated with the training and to the diet [22]. High interaction levels of PTPIP51 and PTP1B as seen in normal and trained mice ensure the stimulating effect of PTPIP51 on MAPK pathway via dephosphorylation of the PTPIP51 Tyr176 residue and the formation of the Raf-1/14-3-3β/PTPIP51 complex [22]. Subsequently, a similar insulin sensitizing effect via transcriptional control of the insulin receptor as cited above for Drosophila could be possible [25]. This hypothesis is supported by the upregulated interaction of 14-3-3β and PTPIP51 in high-fat diet, trained mice [22]. Here, the stimulation of the MAPK pathway by PTPIP51 could be a bypassing mechanism for the high-fat diet induced insulin receptor resistance.

PTPIP51 also interacts with the insulin receptor signaling antagonist, the PKA [8,9]. PKA is a serine/threonine kinase, which is activated by high cAMP levels in consequence of extracellular signals [30]. PKA phosphorylates PTPIP51 at serine 46 and serine 212 in vitro [10]. Additionally, using the Group-based Prediction System 3.0 (http://gps.biocuckoo.org/), PKA was identified to phosphorylate the serine residue 149 of PTPIP51 [9]. Interestingly, the serine residues 46 and 149 are in direct vicinity of the two conserved regions of PTPIP51 [9]. The serine residue 212 is located close to the cleft of PTPIP51, which presumably depicts the binding pocket for Raf-1 [9]. Thus, PKA can modulate the binding of PTPIP51 to 14-3-3 proteins and Raf-1 [8,9,11]. Here, PTPIP51 represent a possible mediator between the katabolic PKA signaling and the anabolic insulin signaling.

The central position in metabolic signaling mirrors the important role of PTPIP51 in the genesis of insulin resistance and obesity. Up to now, PTPIP51 seems to maintain a level of insulin sensitivity via stimulation of MAPK signaling. As mentioned above this stimulating effect is active if PTPIP51 is dephosphorylated at the Tyr176 residue. This conflicts with the observed outstanding findings of PTP1B knockout mice, since PTP1B is essential for the dephosphorylation of PTPIP51 [8–12]. The functional and interactional consequences of PTP1B knockout or inhibition must be subject to further studies, as the aforementioned medical and economic implications strongly highlight the importance of the precise understanding of PTPIP51 in these signaling systems (Figure 1B).

3. Regulation of PTPIP51 in Cancer

3.1. Breast Cancer

Breast cancer is the most common neoplasm in women, accounting for about 25% of all diagnosed tumors. Although early diagnosis and enhanced therapies of breast carcinomas greatly improved the overall survival time, breast cancer is still the third most cause of cancer deaths in the US [31,32].

About one third of the breast tumors exhibit an overexpression of the HER2/ErbB2 receptor, leading to a more aggressive and invasive growth of the cancer cells and thus an impaired overall survival time [33]. The amplification of the Her2 receptor induces an over-activation of mainly two different signaling pathways, the MAPK pathway and the Akt signaling [34]. The activation of these signaling pathways is mediated by the enhanced formation of homodimers and heterodimers of the Her2 receptor and other members of the Her family, namely EGFR, Her2, and Her4 [34]. Subsequently, the signaling is channeled via the aforementioned signaling molecules into the ERK signaling, resulting in an enhanced growth and proliferation of the cancer cells [34,35].

The knowledge about Her2 receptor amplification and its functional consequences in breast tumors led to the development of several Her2 targeted therapies encompassing small molecule tyrosine kinase inhibitors and monoclonal antibodies [33,36]. Introduction of these substances prompted a great progress in the clinical management of Her2 amplified breast cancer [33,36]. For example, targeting the HER2 receptor with the monoclonal antibody trastuzumab improved the disease-free survival rates at 5 years from 75% to 81–84% in HER2-positive early stage breast cancer [37].

Despite the good clinical results, the management of anti Her2 therapy resistances are still challenging. Several signaling pathways have been identified to mediate these resistances, e.g., the PI3K-Akt-mTOR signaling, c-MET signaling pathway or low immune response [38–40].

The regulation of the non-receptor tyrosine kinase c-Src depicts another crucial resistance mechanism [41,42]. c-Src is involved in many cellular processes regulating cell proliferation and cell survival [43]. These functions are exerted via the interaction of c-Src with several receptor tyrosine kinases and other signaling hubs, e.g., the MAPK pathway and PI3K-Akt signaling [43]. Activation of c-Src is observed in tumors of the colon, liver, lung, and pancreas [41]. In breast tumors high levels of activated c-Src correlate with poor prognosis, lower overall survival time and trastuzumab resistance [40,41]. In fact, the activation of c-Src alone is capable of conferring trastuzumab resistance [44]. Thus, the combination of anti-Her2 and anti-c-Src therapies seems a promising concept for the treatment of resistant tumors. Actually, recent studies verified this approach. Combination of Saracatinib, a small molecule inhibitor of c-Src and Lapatinib, a small molecule inhibitor of Her2 resulted in prolonged survival in a xenograft mouse model [45]. Furthermore, a direct interaction of c-Src and Her family members is pivotal in Her2 amplified breast cancer cells for the exertion of mitogenesis upon EGF stimulation and the correct transduction of growth promoting effects of heregulin [46]. The physical interaction of Her2 and c-Src seems to be crucial for the transduction of survival and growth signals of Her2 heterodimers [46]. Furthermore, Her2/c-Src interaction promotes the anchorage-independent growth of Her2 amplified breast cancer cells [46].

For the precise titration of phosphorylation levels of receptor tyrosine kinases (RTKs), signaling kinases and scaffolding proteins not only kinases, such as c-Src, are needed but also phosphatases. One of the best described phosphatases is the protein tyrosine phosphatase 1B (PTP1B) [47]. PTP1B is involved in the modulation of several RTK, e.g., EGFR and Her2 [28]. The relevance of PTP1B in breast cancer is mirrored in its interaction with the Her2 receptor. Usually, PTP1B is involved in the dephosphorylation and thereby deactivation of RTKs. In contrast, the Her2 receptor is activated by PTP1B via an up to now unknown mode of action [28]. The combination of Her2 overexpression and PTP1B knockdown in a mouse model resulted in a delayed tumor development of about 85 days compared to mice with normal PTP1B expression [48]. Furthermore, inhibition of PTP1B in breast gland cells leads to a reduced proliferation, it also affects the epithelial-mesenchymal-transition, which is a hallmark in the formation of metastasis [49]. All these findings stress the importance of PTP1B in the development and the progression of breast tumors, especially in Her2 amplified breast cancer cells. A precise understanding of the PTP1B affected signaling pathways is of the utmost interest to provide a platform for the development of novel therapeutic strategies.

PTPIP51 in Breast Carcinoma

The scaffolding protein PTPIP51 represents a crucial crossing point of all the aforementioned tumor promoting and resistance inducing mechanisms. PTPIP51 is expressed in normal breast glands as well as in breast cancer cells, whereby the expression of PTPIP51 protein is diminished in the cancer cells (ongoing studies of our group). Further analysis of the crucial Tyr 176 phosphorylation site of PTPIP51 showed a strong upregulation of the phosphorylation in breast cancer cells. As shown by Brobeil et al. the Tyr 176 residue phosphorylation regulates the binding of PTPIP51 to Raf-1 via 14-3-3β and thereby exerts its MAPK stimulating effect [8]. The downregulation of PTPIP51 in combination with the high phosphorylation of the Tyr176 residue depicts a potential inhibition of the MAPK stimulating effect of PTPIP51 in breast cancer cells. Thus, the regulation of PTPIP51 seems to counteract the activation of the tumor promoting MAPK signaling.

In Her2 amplified breast cancer cells, the phosphorylation of PTPIP51 at Tyr176 is to a great extend performed by the EGFR. Inhibition of the EGFR in Her2 amplified breast cancer cells induces a reduction of PTPIP51 phosphorylation at the Tyr176 residue accompanied by a formation of the Raf-1/14-3-3β/PTPIP51 interactome, thus proofing a normal regulation of MAPK-related interactions of PTPIP51 [50].

Interestingly, PTPIP51 also interacts with the Her2 receptor, but it is not clear if the Her2 receptor phosphorylates PTPIP51 or if PTPIP51 forms a scaffold for the interaction of the Her2 receptor with other signaling molecules [50]. Of note, selective inhibition of the Her2 receptor with the TKI Mubritinib induces a formation of a potential ternary interactome consisting of the Her2 receptor, PTPIP51 and c-Src [50]. As mentioned above, c-Src plays a crucial role in Her2 targeted therapy resistance and the transduction of growth and survival signals [41,42,44,46]. The formation of the ternary complex Her2/c-Src/PTPIP51 stresses a pivotal role of PTPIP51 in the mediation of these resistance mechanisms [50].

In addition, own studies showed the relevance of the PTPIP51/c-Src interaction in correlation with the sensitivity of Her2 positive breast cancer cells to EGFR/Her2 targeted TKIs. Application of several EGFR/Her2 targeted TKIs to the Her2 amplified breast cancer cell line SK-BR3 led to a highly significant augmentation of PTPIP51/c-Src interaction, whereas the same treatment of BT474 cells, also a Her2 amplified breast cancer cell line, did not alter or even reduce the PTPIP51/c-Src interaction. Interestingly, the enhanced interaction of PTPIP51 and c-Src was accompanied by a less sensitivity to TKI treatment, representing another potential resistance mechanism [50].

Of note, as aforementioned the interaction of PTPIP51 and c-Src is severely altered in Her2 amplified breast cancer cells when treated with EGFR/Her2 targeted TKIs, but the interaction of PTPIP51 and its crucial phosphatase PTP1B remains nearly unaffected [50]. Solely long-term application of Gefitinib (EGFR TKI) and Lapatinib (EGFR/Her2 TKI) induced an upregulation of PTPIP51/PTP1B interactions [50], thus, portraying a potential adaption mechanism of the cancer cell to the applied TKI treatment and not a direct effect of the TKI. Interestingly, this adaption seems contradictory since the inhibition of EGFR results in a reduced Tyr176 phosphorylation and a compensation would be a downregulation of PTPIP51/PTP1B interaction. Moreover, the upregulation of PTPIP51/PTP1B interaction upon EGFR inhibition is accompanied by an increase in the sensitivity of SK-BR3 cells (Her2 amplified breast cancer cell line) to the treatment, which is not seen after short term application. Interestingly, these regulations of PTPIP51/PTP1B interaction do not occur under selective Her2 inhibition [50]. Up to now it is not clear but seems probable whether there is a link between the aforementioned differing effects of PTP1B on EGFR and Her2 activity and the diverging effects of EGFR inhibition and selective Her2 inhibition on the PTPIP51/PTP1B interaction. Interestingly, the application of the EGFR/Her2 TKI Neratinib did not reduce the Tyr176 phosphorylation of PTPIP51 to the same extend as Gefitinib and Lapatinib. This regulation is accompanied by a reduced interaction of

PTPIP51 and PTP1B. Thus, the reduced phosphorylation by EGFR is counter regulated via a reduced dephosphorylation of PTPIP51 through PTP1B under Neratinib treatment.

Additionally, the importance of the PTPIP51/PTP1B interaction is underlined by the fact, that this interaction is directly correlated with the grading of breast carcinomas. Ongoing studies of our group examined the interaction of PTPIP51 and PTP1B in breast carcinomas of no special type and showed a significant enhancement of the interaction in grade 3 carcinomas compared to grade 1 and 2 carcinomas. Moreover, Her2 amplified carcinomas also displayed a significantly upregulated PTPIP51/PTP1B interaction compared to Her2 negative breast cancer samples. This may be supported by the upregulation of PTP1B in Her2 amplified breast cancer.

To sum up, PTPIP51 is needed for the normal function of healthy mammary glands and its expression is altered in the development of breast tumors. Furthermore, the interaction of PTPIP51 and PTP1B correlates with the grading and the Her2 amplification, indicating an alteration of PTPIP51 phosphorylation during the progression of breast carcinoma. We were able to unveil a potential role of PTPIP51 in tumor promoting signaling and therapy resistance against EGFR/Her2 targeted TKIs mediated through the non-receptor kinase c-Src and the phosphatase PTP1B [50]. In consequence, PTPIP51 plays a pivotal role in the oncogenesis of breast carcinoma and it is of the utmost interest to unveil the regulations of PTPIP51 in respect of therapy resistance and growth signaling (Figure 2A).

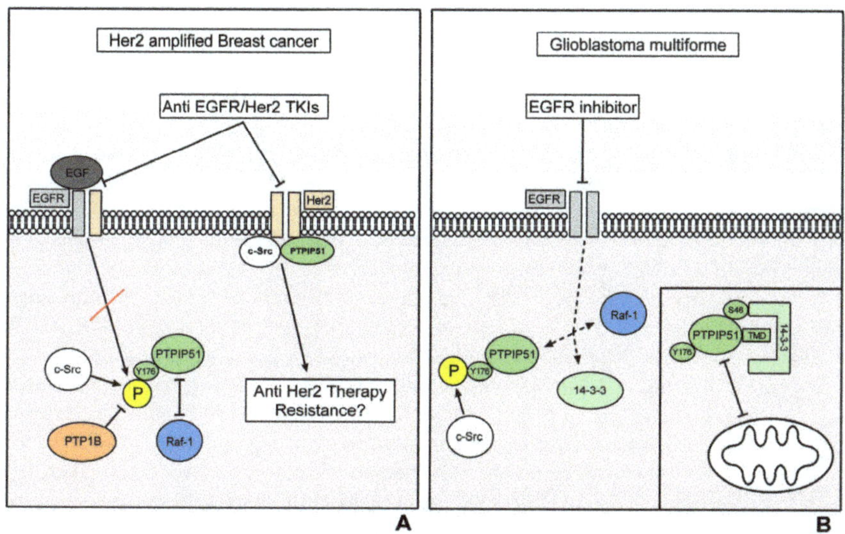

Figure 2. Regulation of PTPIP51 interactions in breast cancer. The inhibition of EGFR in Her2 amplified breast cancer cells induces the same effects as observed in the HaCat cell line upon EGFR inhibition regarding the formation of the Raf-1/14-3-3β/PTPIP51 complex. The sensitivity of Her2 amplified breast cancer cells towards EGFR-targeted TKIs correlates with the regulation of the interaction of PTPIP51 with c-Src. The selective inhibition of Her2 induces a formation of a PTPIP51/Her2/c-Src complex, which depicts a potential resistance mechanism against anti-Her2 therapies (black arrows indicate a phosphorylation/activation, arrows with vertical bar as arrow head indicate an inhibition of interaction/activation) (**A**); regulation of PTPIP51 interactions in Glioblastoma multiforme. The left side of the figures depicts the regulation of PTPIP51 interactome under EGFR inhibition. Contrary to the expectations, the inhibition of the EGFR induces a disruption of the Raf-1/14-3-3β/PTPIP51 complex. The right side of the figure shows that the upregulated 14-3-3 protein levels in gliomas of high malignancy potentially inhibit the translocation of PTPIP51 to the mitochondrion and thus its apoptosis-inducing effects (black arrows indicate a phosphorylation/activation, arrows with vertical bar as arrow head indicate an inhibition of interaction/activation, dotted black arrows indicate a dissolution of the Raf-1/14-3-3β/PTPIP51 complex via EGFR inhibition) (**B**).

3.2. Glioblastoma Multiforme

One of the most malignant tumors with an overall survival time of about one year, despite significant advances in the therapeutic options, is the glioblastoma multiforme (GBM) [51–53]. The GBM represents the most common primary brain tumor in adults, accounting for almost 80% of all primary brain tumors [52,53]. The clinical presentation depends on the location and the size of the GBM and ranges from headache over neurological deficits to seizures, which are present in about 25% of the patients at the time of diagnosis [52]. The multimodal treatment of GBM includes radical surgical resection, radiotherapy, and chemotherapy. Despite the extensive clinical management, the prognosis of GBM is poor and a curation is not possible [52,54]. Due to this the Cancer Genome Atlas project and the genomic profiling of more than 600 genes of 200 human tumor samples were performed to get an insight in the most frequent signaling aberrations in GBM. The three most commonly activated signaling pathways were the p53 signaling, the retinoblastoma pathway and the receptor tyrosine kinase-Ras-PI3K pathway [55]. One special receptor tyrosine kinase in GBM is the EGFR, respectively, its constitutively activated mutant, the EGFRvIII [53,56,57]. This mutation occurs in about 30% of all GBM [58]. The constitutive activation of the EGFRvIII leads to an over-activation of the MAPK pathway and thereby an exaggerated growth and survival signaling [58]. Targeting the aberrant EGFR and thereby blocking the aggressive growth of GBM seems a reasonable therapy strategy, but several clinical trials demonstrated only discrete improvements of a small percentage of GBM patients, when treated with EGFR inhibitors [57]. The most recent strategy is a vaccination against the EGFRvIII, which seemed a promising approach. However, again the theory did not approve in the clinical setting [52].

Consequently, many GBMs must have an intrinsic or a rapidly acquired resistance against EGFR inhibitors or the signaling is bypassed via alternative growth and survival promoting pathways [59]. Therefore, it is of the utmost interest to unveil the downstream pathways of the EGFR in GBM.

The most commonly activated signaling pathway downstream of the EGFRvIII is the ERK signaling [60]. The activated EGFRvIII induces an activation of Raf-1 by signaling molecules mentioned in the background section. Subsequently, the activation of Raf-1 facilitates the activation p90^{RSK}, MNK1/2, Ets, Elk1, Myc, STAT1/3, which mediate the proliferation and growth inducing effects of the EGFRvIII activation [2]. A crucial regulator of Raf-1 is the scaffolding protein 14-3-3β [6,7]. 14-3-3 proteins belong to a highly conserved protein family and are expressed in all human cells. Due to the lack of a kinase domain, 14-3-3 proteins exert their function via binding of serine phosphorylated proteins [61]. Thereby, 14-3-3β ensures proximity of signaling substrates and provides a corresponding reaction matrix [61]. Furthermore, for the activation of some signaling molecules 14-3-3 proteins are of utmost need as seen for the activation of Raf-1 [6]. Thorson and coworkers reported, that an activation of Raf-1 in the absence of 14-3-3β is not possible. The addition of 14-3-3β restored the activation of Raf-1 [6].

In the setting of GBM 14-3-3 proteins also seem to play a crucial role. Yang and coworkers showed a direct correlation between the amount of expressed 14-3-3β and 14-3-3η and the grade of malignancy of glioma [62]. The importance of 14-3-3 proteins in GBM is further mirrored by their mediation of radio- and chemotherapy resistance. Park and coworkers found, that the depletion of 14-3-3η enhances the radiosensitivity of GBM cells [63]. Similar results were found in earlier studies concerning the sensitivity of GBM to chemotherapeutics, such as microtubule agents [63].

The examination of the aforementioned cellular signaling and many more, e.g., platelet-derived growth factor receptor (PDGFR) signaling, neurofibromatosis type 1 (NF1) and the tumor microenvironment, lead to crucial insights in the aberrant tumor signaling, but up to now a linkage between these structures is still missing.

PTPIP51 in Glioblastoma

PTPIP51 depicts a possible functional linkage between the constitutively activated EGFRvIII and the scaffolding protein 14-3-3β. PTPIP51 is expressed in glioma of low malignancy as well as in GBM [64]. The expression of PTPIP51 protein and mRNA directly correlates with the grade of malignancy and thus mirrors the expression of 14-3-3 proteins in primary brain tumors [64]. The upregulation of 14-3-3β and 14-3-3η is associated with reduced apoptosis [63]. Of note, PTPIP51 includes an N-terminal transmembrane domain (TMD), when expressed in its full form. The TMD is responsible for the translocation of PTPIP51 towards the mitochondrion. Lv and coworkers showed, that the overexpression of PTPIP51 leads to an accumulation at the mitochondrion and subsequently to the induction of apoptosis [65]. Besides the apoptosis-related functions, mitochondrion located PTPIP51 is involved in the correct formation of mitochondrion-related endoplasmic reticulum membranes (MAM) and the calcium signaling between mitochondrion and endoplasmic reticulum, but these functions are not in the scope of this review [13–16].

Interestingly, the PTPIP51 protein structure includes two conserved regions, which facilitate the binding to 14-3-3 proteins [7,9]. Moreover, these sites are accompanied by serine and tyrosine residues, whose phosphorylation control the binding of PTPIP51 and 14-3-3β [9]. The conserved region 1, which spans from aas 43 to aas 48 is near the TMD [9]. Binding of 14-3-3β at the conserved region 1 leads to a capping of the TMD and thus the translocation of PTPIP51 towards the mitochondrion is abrogated [9,12]. In consequence, the hindrance of the translocation of PTPIP51 to the mitochondrion via binding of the upregulated 14-3-3 proteins in gliomas of high malignancy may depict a potential apoptosis resistance mechanism.

As aforementioned, the inhibition of the constitutively activated EGFRvIII seemed a promising therapeutic strategy, but up to now the clinical results lack a sufficient improvement of the patients' outcome [52]. Interestingly, the inhibition of EGFR with TKIs in glioblastoma cells leads to completely unexpected regulations in the MAPK-related interactome of PTPIP51. Usually, the inhibition of the EGFR induces a reduction of Tyr176 phosphorylation of PTPIP51 and consequently to a formation of the Raf-1/14-3-3β/PTPIP51 complex, which at least partially compensates for the reduced MAPK pathway activation [8–12]. In glioblastoma cells the regulations upon EGFR inhibition differ. Duolink proximity ligation assays revealed a decrease in the interaction of 14-3-3β/PTPIP51 and Raf-1/PTPIP51, indicating an omission of the PTPIP51 mediated MAPK pathway stimulating effect. Accordingly, the phosphorylation of Tyr176 of PTPIP51 was only slightly altered [66]. A possible explanation of these regulations is an overshooting compensation for the loss of the EGFR activity by c-Src activation [66]. c-Src is frequently overexpressed and over-activated in gliomas of high malignancy [67]. Likewise, mechanisms were seen in Her2 amplified breast cancer cells. Here, c-Src/PTPIP51 interaction directly correlated with the resistance towards EGFR-targeted therapies [50]. Thus, in tumors with an altered RTK signaling the interaction of PTPIP51 and c-Src may constitute a tumor-interspecies mechanism of anti EGFR therapy resistance (Figure 2B).

3.3. Melanoma

Melanomas belong to a cancer entity arising from melanocytes originally derived from neural crest cells found in skin and uvea. Among the three main types of skin cancer, beside melanoma there are basal cell carcinoma and squamous cell carcinoma. The melanomas account for about 2% of skin cancers, but for 75% to 90% of deaths [68]. Melanomas display the highest mutation frequency of all cancers [69]. About 2.3% of people will develop a melanoma during life time. In 2015 an estimated 1.2 million Americans were living with a melanoma (NIH National Cancer Institute, Surveillance, Epidemiology, and End Results https://seer.cancer.gov/statfacts/html/melan.html). From all cancers melanomas have the highest probability to metastasize to the brain [70]. In 50% of patients dying from melanoma the reasons are brain metastasis [71].

For the pathogenesis of melanomas, the activation of the MAPK pathway is essential [72]. This is reflected by genetic alterations in melanomas affecting molecules linked to the MAPK signaling. Possible candidates are the BRAF kinase, the small GTPase NRAS and the c-KIT receptor. The BRAF mutations are present in about 40–50%, mutations of NRAS in 20% and of c-KIT in 1–3% of the melanomas [69,73]. Metastatic melanomas are subdivided by their mutation profile into four subtypes

BRAF driven (about 52%), NRAS driven (28%), *Neurofibromin 1 (NF1)* mutated (14%) and in "triple wild type" [74]. The knowledge of these specific mutations within the MAPK pathway is essential for therapeutic options.

CRAF/Raf-1 is expressed in all cells, but BRAF is associated with cells of neuronal origin such as melanocytes. CRAF mutations are rare in contrast to BRAF mutations [75]. BRAF mutations seen in melanomas are MAPK pathway activating mutations. Commonly 80 to 90% of the BRAF mutations are a single amino acid substitution in the BRAF protein where valine is substituted by glutamic acid at position 600 (BRAF V600E). In 15% of BRAF mutated melanomas valine is substituted by lysine (BRAF V600K) and in 3% of the cases by arginine (V600R) or by aspartic acid (V600D). Yet, any of these mutations leads to a constitutively activated BRAF kinase as the mutation is in the activation loop of the kinase [76]. Due to this modification within the activation segment of BRAF no extracellular signal is needed for upregulated MAPK signaling by increased MEK phosphorylation which in turn phosphorylates ERK stimulating downstream signaling and resulting in proliferation [77]. The knowledge of these mutations prompted new therapy options by targeting the overactive BRAF kinase through specific inhibitors such as vemurafinib and dabrafenib, which bind to the active site of BRAF kinase, encorafenib an ATP-competitive Raf kinase inhibitor and sorafenib, a multi-kinase inhibitor [76].

For example, vemurafinib treated patients showed an overall response rate of 52.2% and the median progression-free survival was 8.3 months; the median overall survival was 13.5 months according to a study of Si and coworkers [78]. This form of therapy is also challenged by acquired resistance, despite initial good results.

Mechanisms promoting resistances can be manifold: a reactivation of MAPK signaling (amplification or activation of target kinases); bypass via different signaling pathways, e.g., PI3K-Akt-mTOR signaling; or by the surrounding stromal cells, which secrete HGF in BRAFi-resistant melanomas, activating MAPK and PI3K-Akt-mTOR.

Next to the more common BRAF oncogene, mutations in the NRAS gene result in more aggressive melanomas [76]. BRAF mutations can effectively be targeted by specific small molecule inhibitors, whereas NRAS-mutated subtypes are more susceptible to immunotherapy [79]. NRAS mutations are seen in all types of melanomas but seem to be slightly more numerous in melanomas of sun-damaged skin [80]. The NRAS gene belongs to the Ras oncogene family coding for the NRAS protein a GTPase, which is activated by bound GTP. When activated, NRAS binds to Raf and changes its conformation, thus activating the Raf kinase stimulating the MAPK pathway.

NRAS mutations lead to a substitution of the amino acids at positions G12 or G13 making NRAS insensitive to the inactivation by the Ras GTPase-activating proteins, or more common at position Q61, where Q61R is predominant in melanoma cells [81]. Ras mutations at position Q61 are associated with impaired GTPase activity, thus NRAS is constitutively activated due to its GTP-associated conformation [82].

As there exist no specific NRAS/Ras inhibitors, different strategies for treatment of NRAS-mutated melanomas were administered for therapy. The use of farnesyl transferase inhibitors to prevent posttranslational modification of Ras and its insertion into the plasma membrane failed in clinical studies [83]. Third generation MEK1/2 inhibitors such as binimetinib (MEK162)-a potent allosteric inhibitor, improved the response rate and progression-free survival of patients, but failed to prolong overall survival of the patients [84].

The first immunotherapies successfully approved interleukin 2 and anti-CTLA4 antibody both stimulating the immune response as treatment options [84]. As seen in clinical trials 25 to 50% of patients respond to the inhibition of the immune checkpoint PD1 [84]. In 2016 Johnson and coworkers already stated, that NRAS mutations displayed a superior clinical outcome in immune therapy compared to BRAF mutations [85]. This better response of NRAS-mutated melanoma to immune therapy is due to a higher level of immunosuppression in the tumor microenvironment compared to BRAF-mutant melanoma [86]. Immune therapies of NRAS-mutated melanomas are now the first-line treatment for NRAS and WT melanoma [84]. Immune checkpoint blockade inhibitors significantly improved overall survival rates [87]. Yet, eventually any of these treatments leads to a resistant behavior of the tumor cells. Echevarri'a-Vargas et al. addressed this problem by a new therapeutic strategy blocking two different pathways via inhibition of bromodomain and extraterminal domain (BET) and MEK pathways treating successfully NRAS-mutant and immune therapy-resistant melanoma [88].

PTPIP51 in Melanoma

PTPIP51 protein and its interactome in melanoma cells in relation to normal melanocytes derived from common nevi, as well as to melanocytes from dysplastic nevi revealed a characteristic profile within the three different entities. Dysplastic nevi represent an intermediate state between nevi and melanomas as they are morphologically and biologically intermediate between these two entities [89].

Despite massive changes in the molecular characteristics of melanocytes from normal nevi to dysplastic nevi and further on to malignant melanocytes, there is no significant change in the amount of PTPIP51 protein in these different stages. Yet, if the grade of Tyrosine 176 phosphorylation is analyzed, melanocytes from dysplastic nevi displayed the highest level in Tyr 176 phosphorylated PTPIP51 compared to the Tyr176 phosphorylation status of melanocytes derived from normal nevi, exhibiting the lowest phosphorylation level and when compared to the phosphorylation state in melanocytes from malignant melanomas displaying a tyrosine phosphorylation level laying in between. These findings are corroborated by the interaction profile of PTPIP51 and PTP1B, the phosphatase which is responsible for dephosphorylation of tyrosine 176, with the highest number of interactions in normal melanocytes and lowest number of interactions in melanocytes from dysplastic nevi and an interaction profile laying in between the aforementioned for malignant melanocytes. As described above, Tyr176 phosphorylation regulates the interaction of PTPIP51 with Raf. Solely when dephosphorylated at Tyr176 PTPIP51 can bind to Raf or 14-3-3β protein promoting MAPK signaling.

PTPIP51 affects the MAPK pathway by interacting with Raf-1 to stimulate downstream signaling. In melanocytes the interaction with the Raf-1/CRAF protein is relatively low. Nevertheless, there are quantitative differences for the three entities with lowest interaction numbers in malignant melanocytes and highest numbers in melanocytes from dysplastic nevi. Noteworthy, the interaction with BRAF is up to 100 times more numerous in the three different entities. PTPIP51/BRAF interaction is lowest in dysplastic nevi melanocytes, and somewhat higher in melanoma cells and highest in healthy melanocytes, where the number of interactions was five times higher compared to that of dysplastic nevi cells. This pattern corresponds well to that seen for PTPIP51 and PTP1B interaction and the measured phosphorylation status of PTPIP51. In addition, as indicated by the interaction profile of PTPIP51 and PKA, serine 212 phosphorylation which enhances the interaction of PTPIP51 and Raf, is much more reduced in dysplastic nevi and to a lesser degree in malignant melanoma cells as indicated by their lower interaction levels of PTPIP51/PKA in comparison to an almost doubled interaction seen in normal melanocytes. Noteworthy, serine212 phosphorylation promotes the activation of MAPK signaling by augmenting the interactome of PTPIP51, 14-3-3β and Raf-1.

These data argue for a counter-regulatory function of PTPIP51 in dedifferentiating melanocytes trying to reduce the stimulation by PTPIP51 interaction on BRAF level displaying lowest levels in dysplastic nevi and somewhat higher levels in malignant melanomas. The relative rise in the number of interactions in malignant melanocytes probably reflects the ongoing further dedifferentiation of the cells, thus leading to an increased MAPK signaling compared to dysplastic nevi and the highest stimulation to proliferate.

BRAF inhibitor resistance of melanoma can be a consequence of bypassing MAPK signaling via the PI3K-Akt pathway. Noteworthy, PTPIP51 is also involved in Akt signaling via a direct interaction with Akt protein. In untreated samples either from healthy control nevi or dysplastic nevi or from melanoma, melanocytes of the control display a high PTPIP51/Akt interaction, which is strongly reduced in samples from dysplastic nevi or melanoma. Akt stabilizes the communication site between mitochondria and endoplasmic reticulum. PTPIP51 likewise interferes in this mitochondria-ER relation via its VAPB interaction. The communication between both organelles is among other factors necessary for apoptosis [50]. Presumably, the reduced PTPIP51/Akt interaction is linked to the reduced apoptosic rate in dedifferentiated melanoma cells. An alternative mechanism for resistance is the activation of NF-κB [68]. NF-κB/PTPIP51 interaction is enhanced both in dysplastic nevi and melanoma cells with highest levels in the dysplastic nevi compared to the level in healthy melanocytes. The expression of PTPIP51 mRNA and protein is negatively regulated by RelA, thus affecting the apoptotic function of PTPIP51 [19].

To overcome therapy resistance in melanoma MEK inhibitors are tested in combination with a variety of drugs that use different approaches: inhibition of upstream Ras effectors, inhibition of PI3K-Akt-mTOR, inhibition of cell cycle regulators and activation of anti-tumor immunity but all seem to fail according to the existing cross-resistances [76].

This strongly emphasizes the need for a better understanding of the MAPK/PTPIP51 interactome in melanoma (Figure 3A).

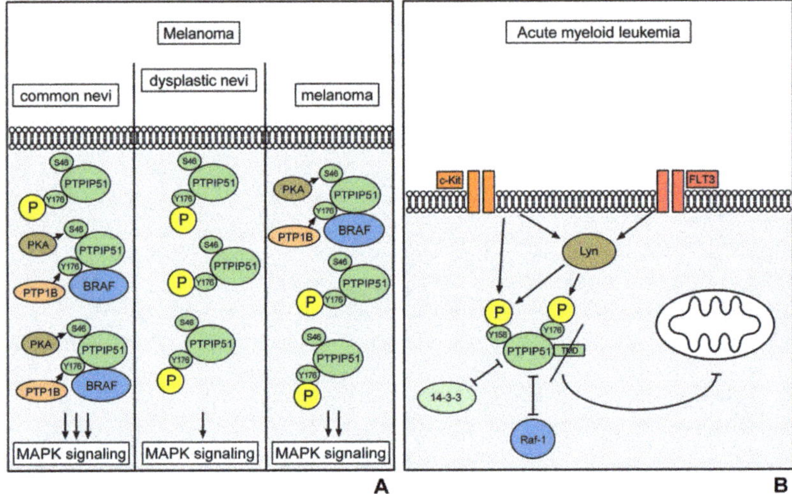

Figure 3. Regulation of PTPIP51 in melanoma. The left panel depicts the regulations of PTPIP51 in normal nevi. The phosphorylation level of the Tyr176 residue is low due to the high interaction with PTP1B. The phosphorylation of the Ser46 residue of PTPIP51 via PKA and the low phosphorylation level of the Tyr176 residue induce the formation of the PTPIP51/14-3-3β (not depicted)/BRAF complex and thereby a stimulation of the MAPK pathway. In the progression of the dysregulation of signaling as represented by the dysplastic nevi, the phosphorylation of PTPIP51 at Tyr176 is upregulated and thus inhibits the MAPK pathway stimulation of PTPIP51. This potentially mirrors a counter-regulation against the dysregulated growth and proliferation signaling in dysplastic nevi (middle panel). In melanoma cells the regulation of the phosphorylation and the interactions of PTPIP51 lie in between the normal nevi and the dysplastic nevi. This depicts the complete dysregulation of signaling since the counter-regulation of MAPK pathway via PTPIP51 phosphorylation is also deregulated (black arrows indicate a phosphorylation/interaction) (**A**); regulation of PTPIP51 interactions in acute myeloid leukemia. Activating mutations of the FMS like tyrosine kinase 3 (FLT3) receptor induce an activation of the Src family kinase Lyn. Lyn phosphorylates the Tyr158 and the Tyr176 residue of PTPIP51, which inhibit the formation of the Raf-1/14-3-3β/PTPIP51 complex and thus the MAPK pathway stimulation of PTPIP51. In acute myeloid leukemia blasts the N-terminus of PTPIP51 is missing. Therefore, the PTPIP51 protein does not contain the TMD. Due to the loss of the TMD a translocation to the mitochondrion is not possible and the apoptosis-inducing function of PTPIP51 is omitted (black arrows indicate a phosphorylation/activation, arrows with vertical bar as arrow head indicate an inhibition of interaction/activation) (**B**).

3.4. Acute Myeloid Leukemia

Leukemias are a group of heterogenous diseases with highly different malignant potential. In this context, two entities must be discriminated: acute leukemias and chronic leukemias. The later encompasses the chronic lymphocytic leukemia, which is classified as a non-Hodgkin lymphoma, and the chronic myelogenous leukemia. Both diseases are not curable but exhibit a prolonged course, e.g., with an eight years survival time of almost 87% for the chronic myeloid leukemia [90].

Acute leukemias are also subdivided into the two cell lineages, lymphocytic and myelogenous. Here, the group of acute myeloid leukemia (AML) is a highly malignant neoplasm exhibiting a near stable incidence over the last years with 3.7 affections per 100,000 persons and an age-dependent mortality of 2.7 to nearly 18 per 100,000 persons. The disease continuously shows two peaks in early childhood and later adulthood [91]. Moreover, despite advances in the therapeutic regimens of AML, the prognosis in the elderly who account for most new cases remains poor.

Interestingly, the prognosis as well as therapeutic decisions are tightly linked to specific cytogenetic and molecular alterations within the malignant transformed cells [92], encompassing microscopically detectable chromosome aberration, submicroscopic gene mutations and changes in gene expression. According to the underlying cytogenetic and mutational burden, patients can be classified into three prognostic categories: favorable, intermediate, and adverse [93]. Two types of cooperating mutations lead to alterations in self-renewal capacity, cellular differentiation, and cell survival of the AML blasts. So-called class II mutations affect transcription factors and lead to impaired differentiation. The class I mutations occur in RTKs, such as FLT3, c-KIT, or downstream effectors, such as Ras [94]. In consequence, they enhance cell survival and proliferation [95]. Interestingly, most of the alterations involve RTK, namely c-Kit and FLT3 [96].

The FMS-like tyrosine kinase 3 (FLT3) is a receptor tyrosine kinase which is expressed on the surface of CD34+ hematopoietic stem cells and other immature hematopoietic progenitors.

The receptor is classified as a type-1 transmembrane receptor tyrosine kinase encompassing several functional domains, e.g., an extracellular domain with Ig domains, TMD, and an intracellular tyrosine kinase domain with two kinase domains [97]. FLT3 is classified as a class III receptor tyrosine kinase besides the platelet-derived growth factor receptor (PDGFR), macrophage colony-stimulating factor receptor, and stem cell factor receptor (c-KIT). Upon activation with the cognate ligands the FLT3 tyrosine kinase couples to distinct downstream pathways, namely phosphatidylinositol-3 kinase (PI3K)-Akt pathway, the janus kinase (JAK)/signal transducer and activator of transcription (STAT) pathway and the MAPK pathway [97].

Approximately 30% of AML patients harbor some form of FLT3 mutation, which can be divided itself into two mutational classes: internal tandem duplications (FLT3/ITD mutations) in or near the juxtamembrane domain of the receptor and point mutations resulting in single amino acid substitutions involving the activation loop of the tyrosine kinase domain (FLT3/TKD mutations) [94]. Yet, these mutations are clinically and therapeutically challenging because of the nature of the mutation and the context in which it occurs [98]. Notably, FLT3/ITD mutation leads to uncontrolled cellular proliferation, survival, and differentiation through constitutive activation of FLT3 and the coupled downstream pathways, e.g., the consecutive activated MAPK pathway [97].

Besides the activation of whole signaling cascades, FLT3 is also able to phosphorylate specific signaling molecules, e.g., the Src family kinase Lyn. Compared to the wild type FLT3 receptor, FLT3/ITD displayed a higher affinity to bind to Lyn and the affinity was relative to the intensity of tyrosin phosphorylation of the receptor [99]. Lyn is known to play a critical role in leukemogenesis [95].

The other class III receptor tyrosine kinase, c-KIT, which also plays a pivotal role in AML, is expressed by myeloblasts in about 60% to 80% of patients: Therefore, next to the described FLT3 aberrations c-KIT harbors the most frequently observed activating RTK mutations in AML with an overall incidence of 17%. The mutations also encompass, for example, internal tandem duplications in c-Kit [100]. The coupled downstream pathways are identical to the FLT3 coupled pathways. Of note, Lyn activation can also be exerted by c-KIT [101].

PTPIP51 in AML

PTPIP51 is expressed in malignant transformed blasts of AML in an isotype specific manner. PTPIP51 protein isoforms could be traced with molecular weights of 13 kDa, 25 kDa, 38 and 52 kDa, respectively [102]. Using peptide sequence specific antibodies, only the peptide sequence for the C-terminal portion of PTPIP51 could be traced, whereas no staining of the N-terminal or the aa 114–129 protein sequence could be observed [102]. As reviewed by Brobeil et al. alternative splicing and the leaky scanning mechanism may build the base for the (disease) specific isoform expression of PTPIP51 [9]. The resulting protein sequence can lack distinct functional domains. In the case of AML, PTPIP51 lost the TMD located at the N-terminal portion of PTPIP51 [102], which is crucial for the mitochondrial binding and the apoptotic function of PTPIP51 [7]. In silico experiments using the Group-based Prediction System 2.1 (GPS 2.1) with the full-length protein sequence as input (SwissProt acc. no. Q96TC7) disclosed that PTPIP51 can be phosphorylated by Lyn, c-Src and c-Kit, respectively [9]. These interactions could be verified in AML blasts by using an in situ approach (DuoLink proximity ligation assay) [102]. Tyrosine 158 and 176 of PTPIP51 serve as phosphorylation residues for Lyn and c-Src, while c-Kit phosphorylation is limited to tyrosine 158 [9]. Interestingly, two functional domains of PTPIP51 are near these phosphorylation sites arbitrating the 14-3-3β binding and, therefore, the Raf-1 and MAPK signaling modulation [7]. The conserved region 1 spans aas 45 to 48 but is completely missing on the protein sequence of PTPIP51 in AML blast [7,102]. The conserved region 2, spanning aas 146–154, is near the tyrosine 158 and 176 residue [9]. As shown by HaCaT cell line experiments exposing the cells to EGF, the phosphorylation of these tyrosine residues leads to the disassembly of the PTPIP51/14-3-3β/Raf-1 complex resulting in a decreased stimulation of the MAPK pathway by PTPIP51 on Raf-1 level [11]. As FLT3 and c-KIT, both RTKs, are constitutive active in most AML cases, the physiological mechanism of regulating the MAPK stimulatory function of PTPIP51 may be still intact in AML blasts. This is also resembled by the high tyrosine 176 phosphorylation levels of PTPIP51 in AML blasts [102]. In none of the AML samples an interaction of PTPIP51 with Raf-1 could be traced due to the high tyrosine 176 phosphorylation levels, despite regions with small residues of normal hematopoiesis [102]. The probable apoptotic function of hyperphosphorylated PTPIP51 is omitted as the TMD is missing leading to uncontrolled proliferation by the constitutive activation of the MAPK by FLT3 and c-KIT [7,8,102]. Using the GPS 2.1 algorithm an enzyme-substrate relation between FLT3 and PTPIP51 could not be verified in contrast to the results gained for c-Kit [102]. PTPIP51 is co-located with Lyn and the interaction of both could be proved in situ by the DuoLink proximity ligation assay in AMLs blast [102]. Thus, c-Kit activity probably leads to the phosphorylation of tyrosine 158 residue and FLT3 mediated Lyn activation leads to phosphorylation of tyrosine 176 residue preventing PTPIP51 to bind 14-3-3β and in consequence modulating the MAPK activity in AML blasts. This inhibited interaction cannot be reversed by dephosphorylation of PTPIP51, as the main dephosphorylating enzyme PTP1B is absent in AML blasts [8,10,11,102].

As Dasatinib blocks Lyn activity with consecutive apoptosis in imatinib-resistant CML cells [103], the interaction blockage of PTPIP51 with the MAPK could probably be abolished with the initiation of apoptosis by administrating Dasatinib. Moreover, Dasatinib also binds to the c-KIT receptor suppressing its activity and promotes cellular apoptosis via activation of the caspase-dependent apoptotic pathway in AML blasts [104].

In summary, PTPIP51 displays a disease-related isoform expression in AML with loss of functional domains. Yet, the physiological regulation of the MAPK binding capacity of PTPIP51 seems to be intact. Thus, the involved signaling molecules of PTPIP51 can be directly targeted by small molecules to induce apoptosis in AML blasts (Figure 3B).

4. Summary

Protein-protein interactions are the foundation of all signaling events in normal cells as well as in dysregulated tumor cells. The framework for the correct procedure of signaling, the interconnection of different pathways and the appropriate subcellular localization is provided by scaffold proteins [105].

In this review, we highlight the central position of PTPIP51 within the dysregulated MAPK pathway signaling of several disparate diseases. The interactions and thus the functions of PTPIP51 are regulated by the serine and tyrosine phosphorylation status of PTPIP51 [8–12]. It exerts its main function in the MAPK pathway via the binding and stimulation of Raf proteins [7]. As seen in melanoma, the inhibition of the Raf stimulating effect of PTPIP51 can be used as a counter-regulatory mechanism against the BRAF deregulation within the sequence from normal nevi over dysplastic nevi to melanoma. Likewise, mechanisms were seen in breast cancer samples. Here, the observed high Tyr176 phosphorylation prevents the binding and thus the stimulation of Raf-1 via PTPIP51. In these tumors PTPIP51-related signaling represents a counter-regulation against the dysregulated growth and proliferation signaling (ongoing studies of our group). In insulin signaling, not the inhibition but the stimulation of the MAPK signaling via PTPIP51 represents the counter-regulatory mechanism against insulin resistance, showing the importance of the cellular setting in which the signaling takes place.

During the progression of tumorous diseases, the PTPIP51 regulating kinases and phosphatases often succumb alterations, in the form of activating mutations, overexpression or over-activation. C-Src represents a perfect example for tumor promoting and therapy resistance inducing kinases, which also regulates the phosphorylation of PTPIP51 [8,11,41,46,67]. The interaction of c-Src and PTPIP51 determines the sensitivity of Her2 amplified breast cancer cells towards EGFR-targeted TKIs [50]. The same interaction might be the reason for the anti-EGFR-therapy resistance of glioblastoma cells. A summary of the relevance of PTPIP51 as a diagnostic biomarker and as a therapeutic target in the various described diseases is presented in Table 1.

To sum up, PTPIP51 modulates the upmost position of the ERK pathway, the MAPKKK Raf-1. Furthermore, PTPIP51 crosslinks this signaling node with several tumor relevant RTKs, non-receptor tyrosine kinases and protein tyrosine phosphatases. PTPIP51 connects the MAPK pathway with several other signaling systems, e.g., the Akt and NF-κB signaling, which exceed the scope of this review. The precise understanding of the PTPIP51-related interactome in tumors is of the utmost interest and offers the possibility to understand dysregulated signaling systems and potential targetable signaling molecules. In particular, the interaction of PTPIP51 and c-Src seems to be of great relevance for therapy resistance mechanisms in several tumor entities and needs further investigation.

Table 1. Summary table of the PTPIP51-related mechanisms of the various diseases and their implications on the role of PTPIP51 as a potential biomarker and targetable molecule.

Disease	PTPIP51-Related Mechanisms	Role of PTPIP51 as Diagnostic Biomarker	Targetable Molecule
Insulin Resistance	Transcriptional regulation of the IR via MAPK activation through the formation of the PTPIP51/14-3-3β/Raf-1 complex	expression of PTPIP51 negatively correlates with the grade of insulin sensitivity in mice	shifting PTPIP51 into MAPK signaling could enhance the transcription of IR and thus the insulin sensitivity
Breast Cancer	Sensitivity to EGFR/Her2 targeted TKIs depends on the formation of the Her2/c-Src/PTPIP51 complex	PTPIP51/PTP1B interaction positively correlates with the grading	Targeting the formation of the Her2/c-Src/PTPIP51 complex could overcome anti-EGFR/Her2 therapy resistances
Glioblastoma Multiforme	EGFR-targeted therapies are potentially bypassed via an enhanced interaction of c-Src and PTPIP51	PTPIP51 mRNA expression positively correlates with the grading of glioma	Targeting the PTPIP51/c-Src interaction could overcome anti-EGFR therapy resistances. Inhibition of the PTPIP51/14-3-3β interaction could unveil the TMD of PTPIP51 and promote the apoptosis-inducing functions of PTPIP51
Melanoma	Modulation of the serine and tyrosine phosphorylation of PTPIP51 via PKA and PTP1B induces a disease-stage-dependent alteration of the formation of the PTPIP51/14-3-3β/Raf-1 complex and thus the MAPK pathway activation	Phosphorylation and interaction profile of PTPIP51 is altered stage-dependently	Inhibition of PKA and PTP1B could reduce the interaction of PTPIP51 and BRAF and thus the MAPK stimulating effect of PTPIP51
Acute Myeloid Leukemia	Loss of the TMD of PTPIP51 inhibits the apoptosis-inducing functions of PTPIP51. Phosphorylation of PTPIP51 via Lyn and c-Kit prevents the PTPIP51-induced MAPK pathway activation	PTPIP51 is expressed in a disease-related isoform without TMD	

Author Contributions: E.D. wrote the manuscript. A.B. wrote and corrected the manuscript. S.G. corrected the manuscript. M.W. wrote and corrected the manuscript.

Funding: This research received no external funding.

Conflicts of Interest: The authors declare no conflict of interest.

References

1. Dhillon, A.S.; Hagan, S.; Rath, O.; Kolch, W. MAP kinase signalling pathways in cancer. *Oncogene* **2007**, *26*, 3279–3290. [CrossRef] [PubMed]
2. Zhang, W.; Liu, H.T. MAPK signal pathways in the regulation of cell proliferation in mammalian cells. *Cell Res.* **2002**, *12*, 9–18. [CrossRef] [PubMed]
3. Meister, M.; Tomasovic, A.; Banning, A.; Tikkanen, R. Mitogen-Activated Protein (MAP) Kinase Scaffolding Proteins: A Recount. *Int. J. Mol. Sci.* **2013**, *14*, 4854–4884. [CrossRef] [PubMed]
4. Wee, P.; Wang, Z. Epidermal Growth Factor Receptor Cell Proliferation Signaling Pathways. *Cancers* **2017**, *9*, 52. [CrossRef]
5. Tosuner, Z.; Geçer, M.Ö.; Hatiboğlu, M.A.; Abdallah, A.; Turna, S. BRAF V600E mutation and BRAF VE1 immunoexpression profiles in different types of glioblastoma. *Oncol. Lett.* **2018**, *16*, 2402–2408. [CrossRef] [PubMed]
6. Thorson, J.A.; Yu, L.W.; Hsu, A.L.; Shih, N.Y.; Graves, P.R.; Tanner, J.W.; Allen, P.M.; Piwnica-Worms, H.; Shaw, A.S. 14-3-3 proteins are required for maintenance of Raf-1 phosphorylation and kinase activity. *Mol. Cell. Biol.* **1998**, *18*, 5229–5238. [CrossRef] [PubMed]
7. Yu, C.; Han, W.; Shi, T.; Lv, B.; He, Q.; Zhang, Y.; Li, T.; Zhang, Y.; Song, Q.; Wang, L.; et al. PTPIP51, a novel 14-3-3 binding protein, regulates cell morphology and motility via Raf-ERK pathway. *Cell Signal* **2008**, *20*, 2208–2220. [CrossRef] [PubMed]
8. Brobeil, A.; Bobrich, M.; Tag, C.; Wimmer, M. PTPIP51 in protein interactions: Regulation and in situ interacting partners. *Cell Biochem. Biophys.* **2012**, *63*, 211–222. [CrossRef] [PubMed]
9. Brobeil, A.; Bobrich, M.; Wimmer, M. Protein tyrosine phosphatase interacting protein 51–a jack-of-all-trades protein. *Cell Tissue Res.* **2011**, *344*, 189–205. [CrossRef] [PubMed]
10. Stenzinger, A.; Schreiner, D.; Koch, P.; Hofer, H.-W.; Wimmer, M. *Chapter 6 Cell and Molecular Biology of the Novel Protein Tyrosine-Phosphatase-Interacting Protein 51*; Elsevier: Amsterdam, The Netherlands, 2009; pp. 183–246.
11. Brobeil, A.; Koch, P.; Eiber, M.; Tag, C.; Wimmer, M. The known interactome of PTPIP51 in HaCaT cells—Inhibition of kinases and receptors. *Int. J. Biochem. Cell Biol.* **2014**, *46*, 19–31. [CrossRef] [PubMed]
12. Brobeil, A.; Dietel, E.; Gattenlöhner, S.; Wimmer, M. Orchestrating cellular signaling pathways-the cellular "conductor" protein tyrosine phosphatase interacting protein 51 (PTPIP51). *Cell Tissue Res.* **2017**, *368*, 411–423. [CrossRef] [PubMed]
13. Stoica, R.; de Vos, K.J.; Paillusson, S.; Mueller, S.; Sancho, R.M.; Lau, K.-F.; Vizcay-Barrena, G.; Lin, W.-L.; Xu, Y.-F.; Lewis, J.; et al. ER-mitochondria associations are regulated by the VAPB-PTPIP51 interaction and are disrupted by ALS/FTD-associated TDP-43. *Nat. Commun.* **2014**, *5*, 3996. [CrossRef] [PubMed]
14. Stoica, R.; Paillusson, S.; Gomez-Suaga, P.; Mitchell, J.C.; Lau, D.H.; Gray, E.H.; Sancho, R.M.; Vizcay-Barrena, G.; de Vos, K.J.; Shaw, C.E.; et al. ALS/FTD-associated FUS activates GSK-3β to disrupt the VAPB-PTPIP51 interaction and ER-mitochondria associations. *EMBO Rep.* **2016**, *17*, 1326–1342. [CrossRef] [PubMed]
15. De Vos, K.J.; Morotz, G.M.; Stoica, R.; Tudor, E.L.; Lau, K.-F.; Ackerley, S.; Warley, A.; Shaw, C.E.; Miller, C.C.J. VAPB interacts with the mitochondrial protein PTPIP51 to regulate calcium homeostasis. *Hum. Mol. Genet.* **2012**, *21*, 1299–1311. [CrossRef] [PubMed]
16. Gomez-Suaga, P.; Paillusson, S.; Stoica, R.; Noble, W.; Hanger, D.P.; Miller, C.C.J. The ER-Mitochondria Tethering Complex VAPB-PTPIP51 Regulates Autophagy. *Curr. Biol.* **2017**, *27*, 371–385. [CrossRef] [PubMed]
17. Gomez-Suaga, P.; Paillusson, S.; Miller, C.C.J. ER-mitochondria signaling regulates autophagy. *Autophagy* **2017**, *13*, 1250–1251. [CrossRef] [PubMed]
18. Galmes, R.; Houcine, A.; van Vliet, A.R.; Agostinis, P.; Jackson, C.L.; Giordano, F. ORP5/ORP8 localize to endoplasmic reticulum-mitochondria contacts and are involved in mitochondrial function. *EMBO Rep.* **2016**, *17*, 800–810. [CrossRef] [PubMed]

19. Brobeil, A.; Kämmerer, F.; Tag, C.; Steger, K.; Gattenlöhner, S.; Wimmer, M. PTPIP51-A New RelA-tionship with the NFκB Signaling Pathway. *Biomolecules* **2015**, *5*, 485–504. [CrossRef] [PubMed]

20. Brobeil, A.; Chehab, R.; Dietel, E.; Gattenlöhner, S.; Wimmer, M. Altered Protein Interactions of the Endogenous Interactome of PTPIP51 towards MAPK Signaling. *Biomolecules* **2017**, *7*, 55. [CrossRef] [PubMed]

21. Bhupathiraju, S.N.; Hu, F.B. Epidemiology of Obesity and Diabetes and Their Cardiovascular Complications. *Circ. Res.* **2016**, *118*, 1723–1735. [CrossRef] [PubMed]

22. Bobrich, M.; Brobeil, A.; Mooren, F.C.; Krüger, K.; Steger, K.; Tag, C.; Wimmer, M. PTPIP51 interaction with PTP1B and 14-3-3β in adipose tissue of insulin-resistant mice. *Int. J. Obes.* **2011**, *35*, 1385–1394. [CrossRef] [PubMed]

23. Pessin, J.E.; Saltiel, A.R. Signaling pathways in insulin action: Molecular targets of insulin resistance. *J. Clin. Investig.* **2000**, *106*, 165–169. [CrossRef] [PubMed]

24. Boucher, J.; Kleinridders, A.; Kahn, C.R. Insulin receptor signaling in normal and insulin-resistant states. *Cold Spring Harb. Perspect. Biol.* **2014**, *6*, a009191. [CrossRef] [PubMed]

25. Zhang, W.; Thompson, B.J.; Hietakangas, V.; Cohen, S.M. MAPK/ERK signaling regulates insulin sensitivity to control glucose metabolism in Drosophila. *PLoS Genet.* **2011**, *7*, e1002429. [CrossRef] [PubMed]

26. Ozaki, K.-I.; Awazu, M.; Tamiya, M.; Iwasaki, Y.; Harada, A.; Kugisaki, S.; Tanimura, S.; Kohno, M. Targeting the ERK signaling pathway as a potential treatment for insulin resistance and type 2 diabetes. *Am. J. Physiol. Endocrinol. Metab.* **2016**, *310*, E643–E651. [CrossRef] [PubMed]

27. Koren, S.; Fantus, I.G. Inhibition of the protein tyrosine phosphatase PTP1B: Potential therapy for obesity, insulin resistance and type-2 diabetes mellitus. *Best Pract. Res. Clin. Endocrinol. Metab.* **2007**, *21*, 621–640. [CrossRef] [PubMed]

28. Tonks, N.K.; Muthuswamy, S.K. A brake becomes an accelerator: PTP1B—A new therapeutic target for breast cancer. *Cancer Cell* **2007**, *11*, 214–216. [CrossRef] [PubMed]

29. Tonks, N.K. PTP1B: From the sidelines to the front lines! *FEBS Lett.* **2003**, *546*, 140–148. [CrossRef]

30. Das, R.; Esposito, V.; Abu-Abed, M.; Anand, G.S.; Taylor, S.S.; Melacini, G. cAMP activation of PKA defines an ancient signaling mechanism. *Proc. Natl. Acad. Sci. USA* **2007**, *104*, 93–98. [CrossRef] [PubMed]

31. Tao, Z.; Shi, A.; Lu, C.; Song, T.; Zhang, Z.; Zhao, J. Breast Cancer: Epidemiology and Etiology. *Cell Biochem. Biophys.* **2015**, *72*, 333–338. [CrossRef] [PubMed]

32. Slamon, D.; Clark, G.; Wong, S.; Levin, W.; Ullrich, A.; McGuire, W. Human breast cancer: Correlation of relapse and survival with amplification of the HER-2/neu oncogene. *Science* **1987**, *235*, 177–182. [CrossRef] [PubMed]

33. Slamon, D.; Eiermann, W.; Robert, N.; Pienkowski, T.; Martin, M.; Press, M.; Mackey, J.; Glaspy, J.; Chan, A.; Pawlicki, M.; et al. Adjuvant Trastuzumab in HER2-Positive Breast Cancer. *N. Engl. J. Med.* **2011**, *365*, 1273–1283. [CrossRef] [PubMed]

34. Moasser, M.M. The oncogene HER2: Its signaling and transforming functions and its role in human cancer pathogenesis. *Oncogene* **2007**, *26*, 6469–6487. [CrossRef] [PubMed]

35. Moasser, M.M. Targeting the function of the HER2 oncogene in human cancer therapeutics. *Oncogene* **2007**, *26*, 6577–6592. [CrossRef] [PubMed]

36. Geyer, C.E.; Forster, J.; Lindquist, D.; Chan, S.; Romieu, C.G.; Pienkowski, T.; Jagiello-Gruszfeld, A.; Crown, J.; Chan, A.; Kaufman, B.; et al. Lapatinib plus Capecitabine for HER2-Positive Advanced Breast Cancer. *N. Engl. J. Med.* **2006**, *355*, 2733–2743. [CrossRef] [PubMed]

37. Marty, M.; Cognetti, F.; Maraninchi, D.; Snyder, R.; Mauriac, L.; Tubiana-Hulin, M.; Chan, S.; Grimes, D.; Anton, A.; Lluch, A.; et al. Randomized phase II trial of the efficacy and safety of trastuzumab combined with docetaxel in patients with human epidermal growth factor receptor 2-positive metastatic breast cancer administered as first-line treatment: The M77001 study group. *J. Clin. Oncol.* **2005**, *23*, 4265–4274. [CrossRef] [PubMed]

38. Gajria, D.; Chandarlapaty, S. HER2-amplified breast cancer: Mechanisms of trastuzumab resistance and novel targeted therapies. *Expert. Rev. Anticancer Ther.* **2011**, *11*, 263–275. [CrossRef] [PubMed]

39. Canonici, A.; Gijsen, M.; Mullooly, M.; Bennett, R.; Bouguern, N.; Pedersen, K.; O'Brien, N.A.; Roxanis, I.; Li, J.-L.; Bridge, E.; et al. Neratinib overcomes trastuzumab resistance in HER2 amplified breast cancer. *Oncotarget* **2013**, *4*, 1592–1605. [CrossRef] [PubMed]

40. Gagliato, D.d.M.; Jardim, D.L.F.; Marchesi, M.S.P.; Hortobagyi, G.N. Mechanisms of resistance and sensitivity to anti-HER2 therapies in HER2+ breast cancer. *Oncotarget* **2016**, *27*, 64431–64446. [CrossRef] [PubMed]

41. Peiró, G.; Ortiz-Martínez, F.; Gallardo, A.; Pérez-Balaguer, A.; Sánchez-Payá, J.; Ponce, J.J.; Tibau, A.; López-Vilaro, L.; Escuin, D.; Adrover, E.; et al. Src, a potential target for overcoming trastuzumab resistance in HER2-positive breast carcinoma. *Br. J. Cancer* **2014**, *111*, 689–695. [CrossRef] [PubMed]

42. Mueller, K.L.; Hunter, L.A.; Ethier, S.P.; Boerner, J.L. Met and c-Src cooperate to compensate for loss of epidermal growth factor receptor kinase activity in breast cancer cells. *Cancer Res.* **2008**, *68*, 3314–3322. [CrossRef] [PubMed]

43. Martin, G.S. The hunting of the Src. *Nat. Rev. Mol. Cell Biol.* **2001**, *2*, 467–475.

44. Zhang, S.; Huang, W.-C.; Li, P.; Guo, H.; Poh, S.-B.; Brady, S.W.; Xiong, Y.; Tseng, L.-M.; Li, S.-H.; Ding, Z.; et al. Combating trastuzumab resistance by targeting SRC, a common node downstream of multiple resistance pathways. *Nat. Med.* **2011**, *17*, 461–469. [CrossRef] [PubMed]

45. Formisano, L.; Nappi, L.; Rosa, R.; Marciano, R.; D'Amato, C.; D'Amato, V.; Damiano, V.; Raimondo, L.; Iommelli, F.; Scorziello, A.; et al. Epidermal growth factor-receptor activation modulates Src-dependent resistance to lapatinib in breast cancer models. *Breast Cancer Res.* **2014**, *16*, R45. [CrossRef] [PubMed]

46. Belsches-Jablonski, A.P.; Biscardi, J.S.; Peavy, D.R.; Tice, D.A.; Romney, D.A.; Parsons, S.J. Src family kinases and HER2 interactions in human breast cancer cell growth and survival. *Oncogene* **2001**, *20*, 1465–1475. [CrossRef] [PubMed]

47. Feldhammer, M.; Uetani, N.; Miranda-Saavedra, D.; Tremblay, M.L. PTP1B: A simple enzyme for a complex world. *Crit. Rev. Biochem. Mol. Biol.* **2013**, *48*, 430–445. [CrossRef] [PubMed]

48. Julien, S.G.; Dubé, N.; Read, M.; Penney, J.; Paquet, M.; Han, Y.; Kennedy, B.P.; Muller, W.J.; Tremblay, M.L. Protein tyrosine phosphatase 1B deficiency or inhibition delays ErbB2-induced mammary tumorigenesis and protects from lung metastasis. *Nat. Genet.* **2007**, *39*, 338–346. [CrossRef] [PubMed]

49. Hilmarsdottir, B.; Briem, E.; Halldorsson, S.; Kricker, J.; Ingthorsson, S.; Gustafsdottir, S.; Mælandsmo, G.M.; Magnusson, M.K.; Gudjonsson, T. Inhibition of PTP1B disrupts cell-cell adhesion and induces anoikis in breast epithelial cells. *Cell Death Dis.* **2017**, *8*, e2769. [CrossRef] [PubMed]

50. Dietel, E.; Brobeil, A.; Tag, C.; Gattenloehner, S.; Wimmer, M. Effectiveness of EGFR/HER2-targeted drugs is influenced by the downstream interaction shifts of PTPIP51 in HER2-amplified breast cancer cells. *Oncogenesis* **2018**, *7*, 64. [CrossRef] [PubMed]

51. McLendon, R.E.; Turner, K.; Perkinson, K.; Rich, J. Second messenger systems in human gliomas. *Arch. Pathol. Lab. Med.* **2007**, *131*, 1585–1590. [PubMed]

52. Davis, M.E. Glioblastoma: Overview of Disease and Treatment. *Clin. J. Oncol. Nurs.* **2016**, *20*, S2–S8. [CrossRef] [PubMed]

53. Jovčevska, I.; Kočevar, N.; Komel, R. Glioma and glioblastoma—How much do we (not) know? *Mol. Clin. Oncol.* **2013**, *1*, 935–941. [CrossRef] [PubMed]

54. Kelly, C.; Majewska, P.; Ioannidis, S.; Raza, M.H.; Williams, M. Estimating progression-free survival in patients with glioblastoma using routinely collected data. *J. Neurooncol.* **2017**, *135*, 621–627. [CrossRef] [PubMed]

55. Verhaak, R.G.W.; Hoadley, K.A.; Purdom, E.; Wang, V.; Qi, Y.; Wilkerson, M.D.; Miller, C.R.; Ding, L.; Golub, T.; Mesirov, J.P.; et al. Integrated genomic analysis identifies clinically relevant subtypes of glioblastoma characterized by abnormalities in PDGFRA, IDH1, EGFR, and NF1. *Cancer Cell* **2010**, *17*, 98–110. [CrossRef] [PubMed]

56. Vleeschouwer, S.D. *Glioblastoma*; Codon Publications: Brisbane, Australia, 2017.

57. Yang, J.; Yan, J.; Liu, B. Targeting EGFRvIII for glioblastoma multiforme. *Cancer Lett.* **2017**, *403*, 224–230. [CrossRef] [PubMed]

58. Loew, S.; Schmidt, U.; Unterberg, A.; Halatsch, M.-E. The epidermal growth factor receptor as a therapeutic target in glioblastoma multiforme and other malignant neoplasms. *Anticancer Agents Med. Chem.* **2009**, *9*, 703–715. [CrossRef] [PubMed]

59. Clark, P.A.; Iida, M.; Treisman, D.M.; Kalluri, H.; Ezhilan, S.; Zorniak, M.; Wheeler, D.L.; Kuo, J.S. Activation of multiple ERBB family receptors mediates glioblastoma cancer stem-like cell resistance to EGFR-targeted inhibition. *Neoplasia (N. Y.)* **2012**, *14*, 420–428. [CrossRef]

60. Golding, S.E.; Morgan, R.N.; Adams, B.R.; Hawkins, A.J.; Povirk, L.F.; Valerie, K. Pro-survival AKT and ERK signaling from EGFR and mutant EGFRvIII enhances DNA double-strand break repair in human glioma cells. *Cancer Biol. Ther.* **2009**, *8*, 730–738. [CrossRef] [PubMed]

61. Yaffe, M.B. How do 14-3-3 proteins work?—Gatekeeper phosphorylation and the molecular anvil hypothesis. *FEBS Lett.* **2002**, *513*, 53–57. [CrossRef]

62. Yang, X.; Cao, W.; Lin, H.; Zhang, W.; Lin, W.; Cao, L.; Zhen, H.; Huo, J.; Zhang, X. Isoform-specific expression of 14-3-3 proteins in human astrocytoma. *J. Neurol. Sci.* **2009**, *276*, 54–59. [CrossRef] [PubMed]

63. Park, G.-Y.; Han, J.Y.; Han, Y.K.; Kim, S.D.; Kim, J.S.; Jo, W.S.; Chun, S.H.; Jeong, D.H.; Lee, C.-W.; Yang, K.; et al. 14-3-3 eta depletion sensitizes glioblastoma cells to irradiation due to enhanced mitotic cell death. *Cancer Gene Ther.* **2014**, *21*, 158–163. [CrossRef] [PubMed]

64. Petri, M.K.; Koch, P.; Stenzinger, A.; Kuchelmeister, K.; Nestler, U.; Paradowska, A.; Steger, K.; Brobeil, A.; Viard, M.; Wimmer, M. PTPIP51, a positive modulator of the MAPK/Erk pathway, is upregulated in glioblastoma and interacts with 14-3-3β and PTP1B in situ. *Histol. Histopathol.* **2011**, *26*, 1531–1543. [PubMed]

65. Lv, B.F.; Yu, C.F.; Chen, Y.Y.; Lu, Y.; Guo, J.H.; Song, Q.S.; Ma, D.L.; Shi, T.P.; Wang, L. Protein tyrosine phosphatase interacting protein 51 (PTPIP51) is a novel mitochondria protein with an N-terminal mitochondrial targeting sequence and induces apoptosis. *Apoptosis* **2006**, *11*, 1489–1501. [CrossRef] [PubMed]

66. Petri, M.K.; Brobeil, A.; Planz, J.; Bräuninger, A.; Gattenlöhner, S.; Nestler, U.; Stenzinger, A.; Paradowska, A.; Wimmer, M. PTPIP51 levels in glioblastoma cells depend on inhibition of the EGF-receptor. *J. Neurooncol.* **2015**, *123*, 15–25. [CrossRef] [PubMed]

67. Lewis-Tuffin, L.J.; Feathers, R.; Hari, P.; Durand, N.; Li, Z.; Rodriguez, F.J.; Bakken, K.; Carlson, B.; Schroeder, M.; Sarkaria, J.N.; et al. Src family kinases differentially influence glioma growth and motility. *Mol. Oncol.* **2015**, *9*, 1783–1798. [CrossRef] [PubMed]

68. Chan, X.Y.; Singh, A.; Osman, N.; Piva, T.J. Role Played by Signalling Pathways in Overcoming BRAF Inhibitor Resistance in Melanoma. *Int. J. Mol. Sci.* **2017**, *18*, E1527. [CrossRef] [PubMed]

69. Davis, E.J.; Johnson, D.B.; Sosman, J.A.; Chandra, S. Melanoma: What do all the mutations mean? *Cancer* **2018**, *124*, 3490–3499. [CrossRef] [PubMed]

70. Suki, D. The Epidemiology of Brain Metastasis. In *Intracranial Metastases: Current Management Strategies*; Sawaya, R., Ed.; Blackwell Futura: Malden, MA, USA, 2004; pp. 20–34.

71. Siu, T.L.; Huang, S. Cerebral metastases from malignant melanoma: Current treatment strategies, advances in novel therapeutics and future directions. *Cancers* **2010**, *2*, 364–375. [CrossRef] [PubMed]

72. Amaral, T.; Sinnberg, T.; Meier, F.; Krepler, C.; Levesque, M.; Niessner, H.; Garbe, C. The mitogen-activated protein kinase pathway in melanoma part I-Activation and primary resistance mechanisms to BRAF inhibition. *Eur. J. Cancer* **2017**, *73*, 85–92. [CrossRef] [PubMed]

73. Sperduto, P.W.; Jiang, W.; Brown, P.D.; Braunstein, S.; Sneed, P.; Wattson, D.A.; Shih, H.A.; Bangdiwala, A.; Shanley, R.; Lockney, N.A.; et al. The Prognostic Value of BRAF, C-KIT, and NRAS Mutations in Melanoma Patients with Brain Metastases. *Int. J. Radiat. Oncol. Biol. Phys.* **2017**, *98*, 1069–1077. [CrossRef] [PubMed]

74. Sullivan, R.J. The role of mitogen-activated protein targeting in melanoma beyond BRAFV600. *Curr. Opin. Oncol.* **2016**, *28*, 185–191. [CrossRef] [PubMed]

75. Emuss, V.; Garnett, M.; Mason, C.; Marais, R. Mutations of C-RAF are rare in human cancer because C-RAF has a low basal kinase activity compared with B-RAF. *Cancer Res.* **2005**, *65*, 9719–9726. [CrossRef] [PubMed]

76. Lim, S.Y.; Menzies, A.M.; Rizos, H. Mechanisms and strategies to overcome resistance to molecularly targeted therapy for melanoma. *Cancer* **2017**, *123*, 2118–2129. [CrossRef] [PubMed]

77. Pilat, M.J.; Bumma, N.; Back, J.B.; Pandit, T.; Weise, A.M. BRAF: From Discovery to Drug Resistance. *J. Sarcoma Res.* **2018**, *2*, 1007.

78. Si, L.; Zhang, X.; Xu, Z.; Jiang, Q.; Bu, L.; Wang, X.; Mao, L.; Zhang, W.; Richie, N.; Guo, J. Vemurafenib in Chinese patients with BRAFV600 mutation-positive unresectable or metastatic melanoma: An open-label, multicenter phase I study. *BMC Cancer* **2018**, *18*, 520. [CrossRef] [PubMed]

79. Kirchberger, M.C.; Ugurel, S.; Mangana, J.; Heppt, M.V.; Eigentler, T.K.; Berking, C.; Schadendorf, D.; Schuler, G.; Dummer, R.; Heinzerling, L. MEK inhibition may increase survival of NRAS-mutated melanoma patients treated with checkpoint blockade: Results of a retrospective multicentre analysis of 364 patients. *Eur. J. Cancer* **2018**, *98*, 10–16. [CrossRef] [PubMed]

80. Lee, J.-H.; Choi, J.-W.; Kim, Y.-S. Frequencies of BRAF and NRAS mutations are different in histological types and sites of origin of cutaneous melanoma: A meta-analysis. *Br. J. Dermatol.* **2011**, *164*, 776–784. [CrossRef] [PubMed]

81. Burd, C.E.; Liu, W.; Huynh, M.V.; Waqas, M.A.; Gillahan, J.E.; Clark, K.S.; Fu, K.; Martin, B.L.; Jeck, W.R.; Souroullas, G.P.; et al. Mutation-specific RAS oncogenicity explains NRAS codon 61 selection in melanoma. *Cancer Discov.* **2014**, *4*, 1418–1429. [CrossRef] [PubMed]

82. Hobbs, G.A.; Der, C.J.; Rossman, K.L. RAS isoforms and mutations in cancer at a glance. *J. Cell Sci.* **2016**, *129*, 1287–1292. [CrossRef] [PubMed]

83. Muñoz-Couselo, E.; Adelantado, E.Z.; Ortiz, C.; García, J.S.; Perez-Garcia, J. NRAS-mutant melanoma: Current challenges and future prospect. *Oncol. Targets Ther.* **2017**, *10*, 3941–3947.

84. Boespflug, A.; Caramel, J.; Dalle, S.; Thomas, L. Treatment of NRAS-mutated advanced or metastatic melanoma: Rationale, current trials and evidence to date. *Ther. Adv. Med. Oncol.* **2017**, *9*, 481–492. [CrossRef] [PubMed]

85. Johnson, D.B.; Lovly, C.M.; Sullivan, R.J.; Carvajal, R.D.; Sosman, J.A. Melanoma driver mutations and immune therapy. *Oncoimmunology* **2016**, *5*, e1051299. [CrossRef] [PubMed]

86. Thomas, N.E.; Edmiston, S.N.; Alexander, A.; Groben, P.A.; Parrish, E.; Kricker, A.; Armstrong, B.K.; Anton-Culver, H.; Gruber, S.B.; From, L.; et al. Association Between NRAS and BRAF Mutational Status and Melanoma-Specific Survival Among Patients with Higher-Risk Primary Melanoma. *JAMA Oncol.* **2015**, *1*, 359–368. [CrossRef] [PubMed]

87. Iorgulescu, J.B.; Harary, M.; Zogg, C.K.; Ligon, K.L.; Reardon, D.A.; Hodi, F.S.; Aizer, A.A.; Smith, T.R. Improved Risk-Adjusted Survival for Melanoma Brain Metastases in the Era of Checkpoint Blockade Immunotherapies: Results from a National Cohort. *Cancer Immunol. Res.* **2018**, *6*, 1039–1045. [CrossRef] [PubMed]

88. Echevarría-Vargas, I.M.; Reyes-Uribe, P.I.; Guterres, A.N.; Yin, X.; Kossenkov, A.V.; Liu, Q.; Zhang, G.; Krepler, C.; Cheng, C.; Wei, Z.; et al. Co-targeting BET and MEK as salvage therapy for MAPK and checkpoint inhibitor-resistant melanoma. *EMBO Mol. Med.* **2018**, *10*, e8446. [CrossRef] [PubMed]

89. Goldstein, A.M.; Tucker, M.A. Dysplastic nevi and melanoma. *Cancer Epidemiol. Biomark. Prev.* **2013**, *22*, 528–532. [CrossRef] [PubMed]

90. Kantarjian, H.; O'Brien, S.; Jabbour, E.; Garcia-Manero, G.; Quintas-Cardama, A.; Shan, J.; Rios, M.B.; Ravandi, F.; Faderl, S.; Kadia, T.; et al. Improved survival in chronic myeloid leukemia since the introduction of imatinib therapy: A single-institution historical experience. *Blood* **2012**, *119*, 1981–1987. [CrossRef] [PubMed]

91. Deschler, B.; Lübbert, M. Acute myeloid leukemia: Epidemiology and etiology. *Cancer* **2006**, *107*, 2099–2107. [CrossRef] [PubMed]

92. Bacher, U.; Schnittger, S.; Haferlach, C.; Haferlach, T. Molecular diagnostics in acute leukemias. *Clin. Chem. Lab. Med.* **2009**, *47*, 1333–1341. [CrossRef] [PubMed]

93. Mrózek, K.; Bloomfield, C.D. Chromosome aberrations, gene mutations and expression changes, and prognosis in adult acute myeloid leukemia. *Hematol. Am. Soc. Hematol. Educ. Program.* **2006**, 169–177. [CrossRef] [PubMed]

94. Grafone, T.; Palmisano, M.; Nicci, C.; Storti, S. An overview on the role of FLT3-tyrosine kinase receptor in acute myeloid leukemia: Biology and treatment. *Oncol. Rev.* **2012**, *6*, e8. [CrossRef] [PubMed]

95. Dos Santos, C.; Demur, C.; Bardet, V.; Prade-Houdellier, N.; Payrastre, B.; Récher, C. A critical role for Lyn in acute myeloid leukemia. *Blood* **2008**, *111*, 2269–2279. [CrossRef] [PubMed]

96. Stirewalt, D.L.; Meshinchi, S.; Radich, J.P. Molecular targets in acute myelogenous leukemia. *Blood Rev.* **2003**, *17*, 15–23. [CrossRef]

97. El Fakih, R.; Rasheed, W.; Hawsawi, Y.; Alsermani, M.; Hassanein, M. Targeting FLT3 Mutations in Acute Myeloid Leukemia. *Cells* **2018**, *7*, 4. [CrossRef] [PubMed]

98. Levis, M. FLT3 mutations in acute myeloid leukemia: What is the best approach in 2013? *Hematol. Am. Soc. Hematol. Educ. Program.* **2013**, *2013*, 220–226. [CrossRef] [PubMed]

99. Okamoto, M.; Hayakawa, F.; Miyata, Y.; Watamoto, K.; Emi, N.; Abe, A.; Kiyoi, H.; Towatari, M.; Naoe, T. Lyn is an important component of the signal transduction pathway specific to FLT3/ITD and can be a therapeutic target in the treatment of AML with FLT3/ITD. *Leukemia* **2007**, *21*, 403–410. [CrossRef] [PubMed]

100. Malaise, M.; Steinbach, D.; Corbacioglu, S. Clinical implications of c-Kit mutations in acute myelogenous leukemia. *Curr. Hematol. Malig. Rep.* **2009**, *4*, 77–82. [CrossRef] [PubMed]

101. Linnekin, D.; DeBerry, C.S.; Mou, S. Lyn associates with the juxtamembrane region of c-Kit and is activated by stem cell factor in hematopoietic cell lines and normal progenitor cells. *J. Biol. Chem.* **1997**, *272*, 27450–27455. [CrossRef] [PubMed]

102. Brobeil, A.; Bobrich, M.; Graf, M.; Kruchten, A.; Blau, W.; Rummel, M.; Oeschger, S.; Steger, K.; Wimmer, M. PTPIP51 is phosphorylated by Lyn and c-Src kinases lacking dephosphorylation by PTP1B in acute myeloid leukemia. *Leukemia Res.* **2011**, *35*, 1367–1375. [CrossRef] [PubMed]

103. Okabe, S.; Tauchi, T.; Tanaka, Y.; Ohyashiki, K. Dasatinib preferentially induces apoptosis by inhibiting Lyn kinase in nilotinib-resistant chronic myeloid leukemia cell line. *J. Hematol. Oncol.* **2011**, *4*, 32. [CrossRef] [PubMed]

104. Heo, S.-K.; Noh, E.-K.; Kim, J.Y.; Jeong, Y.K.; Jo, J.-C.; Choi, Y.; Koh, S.; Baek, J.H.; Min, Y.J.; Kim, H. Targeting c-KIT (CD117) by dasatinib and radotinib promotes acute myeloid leukemia cell death. *Sci. Rep.* **2017**, *7*, 15278. [CrossRef] [PubMed]

105. Garbett, D.; Bretscher, A. The surprising dynamics of scaffolding proteins. *Mol. Biol. Cell* **2014**, *25*, 2315–2319. [CrossRef] [PubMed]

International Journal of
Molecular Sciences

Article

Hyperoxia Disrupts Extracellular Signal-Regulated Kinases 1/2-Induced Angiogenesis in the Developing Lungs

Renuka T. Menon [1], Amrit Kumar Shrestha [1], Roberto Barrios [2] and Binoy Shivanna [1,*]

[1] Section of Neonatology, Department of Pediatrics, Baylor College of Medicine, Houston, TX 77030, USA; Renuka.Menon@bcm.edu (R.T.M.); Amrit.Shrestha@bcm.edu (A.K.S.)

[2] Department of Pathology and Genomic Medicine, Houston Methodist Hospital, Houston, TX 77030, USA; rbarrios@houstonmethodist.org

* Correspondence: shivanna@bcm.edu; Tel.: +1-832-824-6474; Fax: +1-832-825-3204

Received: 24 April 2018; Accepted: 18 May 2018; Published: 20 May 2018

Abstract: Hyperoxia contributes to the pathogenesis of bronchopulmonary dysplasia (BPD), a chronic lung disease of infants that is characterized by interrupted alveologenesis. Disrupted angiogenesis inhibits alveologenesis, but the mechanisms of disrupted angiogenesis in the developing lungs are poorly understood. In pre-clinical BPD models, hyperoxia increases the expression of extracellular signal-regulated kinases (ERK) 1/2; however, its effects on the lung endothelial ERK1/2 signaling are unclear. Further, whether ERK1/2 activation promotes lung angiogenesis in infants is unknown. Hence, we tested the following hypotheses: (1) hyperoxia exposure will increase lung endothelial ERK1/2 signaling in neonatal C57BL/6J (WT) mice and in fetal human pulmonary artery endothelial cells (HPAECs); (2) ERK1/2 inhibition will disrupt angiogenesis in vitro by repressing cell cycle progression. In mice, hyperoxia exposure transiently increased lung endothelial ERK1/2 activation at one week of life, before inhibiting it at two weeks of life. Interestingly, hyperoxia-mediated decrease in ERK1/2 activation in mice was associated with decreased angiogenesis and increased endothelial cell apoptosis. Hyperoxia also transiently activated ERK1/2 in HPAECs. ERK1/2 inhibition disrupted angiogenesis in vitro, and these effects were associated with altered levels of proteins that modulate cell cycle progression. Collectively, these findings support our hypotheses, emphasizing that the ERK1/2 pathway is a potential therapeutic target for BPD infants with decreased lung vascularization.

Keywords: extracellular signal-regulated kinases 1/2; hyperoxia; bronchopulmonary dysplasia; HPAECs; angiogenesis; cell cycle

---

1. Introduction

Bronchopulmonary dysplasia (BPD) is a chronic lung disease of premature infants that is characterized by interrupted lung development [1]. The incidence of BPD has remained unchanged over the past few decades, and BPD is still the most common long-term morbidity of preterm infants [2]. Importantly, there are no specific therapies for BPD. In addition, BPD is the second most expensive childhood disease after asthma. Therefore, there is a need for improved therapies to prevent and treat BPD.

Decreased alveolarization or alveolar simplification and dysmorphic lung vascularization are histopathological hallmarks of BPD [3,4]. Lung blood vessels are crucial for healthy lungs. Abnormal lung angiogenesis is a characteristic feature of BPD [5]. Lung angiogenesis facilitates alveolarization (lung development), and disrupted angiogenesis can interrupt alveolarization in the developing lungs [6]. Therefore, understanding the mechanisms that promote the development and function of the lung blood vessels is vital to prevent and treat this human disease. Toward this end, vascular endothelial growth factor (VEGF) and nitric oxide (NO) signaling pathways have been extensively investigated and have been shown to be necessary for lung development in health and disease in neonatal animals [7–11]. VEGF restores the alveolar and pulmonary vascular structure and function via the endothelial nitric oxide synthase pathway in experimental BPD and pulmonary hypertension (PH) [12–14]. However, these results were not replicated in clinical studies [15,16]. Recent evidence suggests that inhaled NO combined with vitamin A can decrease the incidence of BPD better than NO therapy alone [17]. Thus, there is a need to identify additional druggable molecular targets that can complement the inhaled NO therapy to promote the development and function of the lung vascular system.

Lung development is orchestrated by a complex process involving signaling by growth factors [18], which mediate their effects mostly via the activation of mitogen-activated protein (MAP) and phosphatidylinositol 3-OH kinases. Among the four major families of MAP kinases, the extracellular signal-regulated kinases (ERK)1/2 were shown to primarily mediate proliferation and differentiation of many cell types, whereas c-Jun NH_2-terminal kinases and p38 kinase mainly induce cell apoptosis [19]. In fact, ERK1/2 are activated during development in many organisms [20,21] and regulate morphogenesis in several organs including the lungs [22–24]. Thus, it seems logical that disruption of these signaling pathways may mechanistically contribute to a developmental lung disease such as BPD.

Supplemental oxygen is frequently used as a life-saving therapy in preterm infants with respiratory failure; however, excessive oxygen exposure or hyperoxia contributes to BPD pathogenesis. We [25] and others [26,27] have demonstrated that hyperoxia-induced lung parenchymal and vascular injury in newborn mice leads to a phenotype that is similar to that of human BPD. So, we used this model to investigate the effects of hyperoxia on the expression and activation of ERK1/2 proteins in the developing lungs. ERK1/2 activation is shown to protect alveolar epithelial cells against hyperoxic injury [28,29]. However, there are several knowledge gaps, including: (1) the effects of hyperoxia on the expression and activation of ERK1/2 in neonatal mouse and fetal human lung endothelial cells; (2) the effects of ERK1/2 signaling on lung angiogenesis in preterm infants. Therefore, using neonatal C57BL/6J wild type (WT) and fetal human lung cells, we tested the following hypotheses: (1) hyperoxia exposure will increase endothelial ERK1/2 signaling in neonatal C57BL/6J (WT) mouse lungs and fetal human lung endothelial cells; (2) inhibition of ERK1/2 signaling will disrupt angiogenesis in vitro by repressing cell cycle progression.

2. Results

2.1. Hyperoxia Exposure Transiently Activates ERK1/2 in Neonatal Mouse Lungs

The level of protein phosphorylation correlates strongly with the activity of a protein. Therefore, we quantified phosphorylated (p) ERK1/2 protein levels in whole lung homogenates to determine if hyperoxia activates ERK1/2. Western blot analyses (Figure 1) showed that hyperoxia exposure for one week increased ERK1/2 phosphorylation (0.6 ± 0.07 vs. 0.38 ± 0.02). However, prolonged hyperoxia exposure (14 d) decreased ERK1/2 activation compared to age-matched controls (0.31 ± 0.06 vs. 0.43 ± 0.08), suggesting that hyperoxia transiently activates ERK1/2 in the developing lungs.

2.2. Hyperoxia Exposure Transiently Activates ERK1/2 in Neonatal Mouse Lung Endothelial Cells

To investigate if hyperoxia activates ERK1/2 in mouse lung endothelial cells, we performed immunofluorescence colocalization experiments using lung sections from neonatal mice exposed to normoxia or hyperoxia for one or two weeks. We localized pERK1/2 protein expression in endothelial cells by immunofluorescence labelling using anti-pERK1/2 and anti-von Willebrand factor (vWF) antibodies. Figure 2B shows a clear overlap between the green (pERK1/2) and red (vWF) signals, indicating that pERK1/2 is expressed in lung endothelial cells. Similar to our results of immunoblotting experiments with whole lung homogenates, hyperoxia increased ERK1/2 activation in lung endothelial cells compared with the normoxia group at one week of life (Figure 2B). However, analyses of the time-dependent effects of hyperoxia revealed that ERK1/2 activation significantly decreased in the hyperoxia group at two weeks of life (Figure 2D), a time point at which lung development is still occurring.

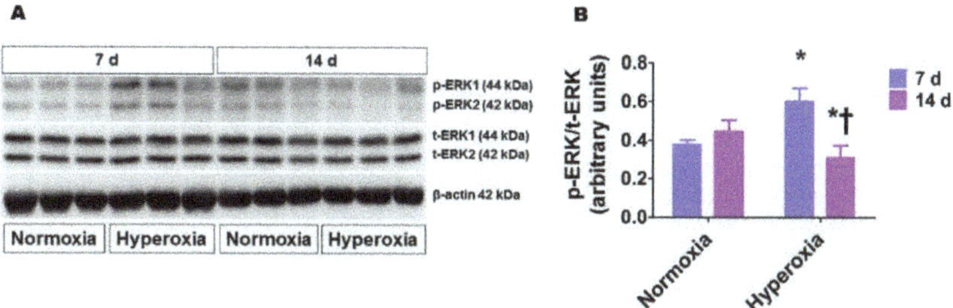

Figure 1. Lung phosphorylated extracellular signal-regulated kinases (ERK)1/2 protein levels in neonatal wild-type (WT) mice exposed to hyperoxia. Lung proteins obtained from neonatal WT mice exposed to 21% O_2 (normoxia) or 70% O_2 (hyperoxia) for up to two weeks (n = 6/exposure) were subjected to immunoblotting using antibodies against total ERK1/2, phosphorylated ERK1/2, or β-actin. Representative immunoblot showing total ERK1/2 and phosphorylated ERK1/2 protein expression (**A**). Densitometric analyses wherein the phosphorylated ERK1/2 band intensities were quantified and normalized to those of total ERK1/2 (**B**). The values are presented as mean ± SD. Significant differences between the normoxia and hyperoxia groups are indicated by * p < 0.05. Significant differences between the hyperoxia groups are indicated by † p < 0.001 (Two-way ANOVA).

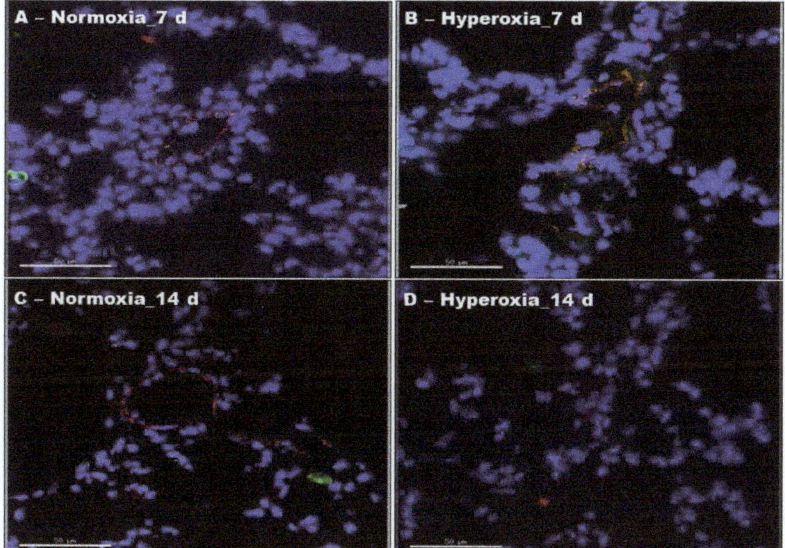

Figure 2. Phosphorylated ERK1/2 protein levels in lung endothelial cells of neonatal WT mice exposed to hyperoxia. One-day-old WT mice were exposed to either 21% O_2 (normoxia) or 70% O_2 (hyperoxia) for one or two weeks (n = 6/exposure/time-point), after which lung sections were processed for colocalization studies. (**A–D**) Representative merged images of lung sections stained with anti-pERK1/2 (green) and anti-vWF (red) antibodies, and DAPI (blue). Scale bar = 50 μM.

2.3. Prolonged Hyperoxia Exposure Interrupts Pulmonary Vascularization in Neonatal Mice

We next determined pulmonary vascularization by quantifying the vWF-stained lung blood vessels. Interestingly, the changes in pulmonary vascularization followed an pattern identical to that of lung endothelial cell ERK1/2 activation in hyperoxia-exposed animals. One week of hyperoxia exposure (Figure 3A,B,E) significantly increased the number of vWF-stained lung blood vessels (7.3 ± 3.4 vs. 4.6 ± 2.7), whereas prolonged (two weeks) hyperoxia exposure (Figure 3C–E) decreased the number of blood vessels (4.9 ± 2.4 vs. 7.4 ± 2.4) in comparison with normoxia-exposed mice.

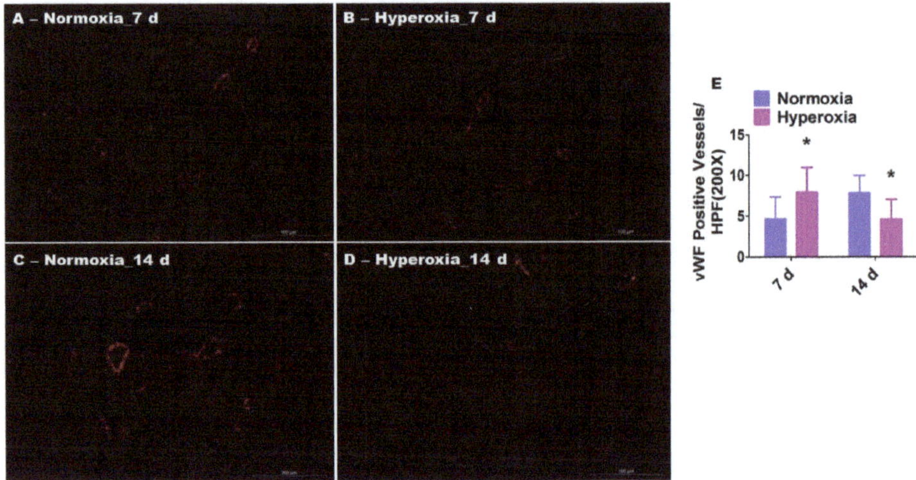

Figure 3. Pulmonary vascularization in neonatal WT mice exposed to hyperoxia. One-day-old WT mice were exposed to either 21% O_2 (normoxia) or 70% O_2 (hyperoxia) for one or two weeks (n = 6/exposure/time-point), following which the lung sections were stained with anti-von Willebrand factor (vWF) antibodies. (**A–D**) Representative vWF-stained lung blood vessels (red). (**E**) Quantitative analysis of vWF-stained lung blood vessels per high-power field (HPF). The values are presented as the mean ± SD. Two-way ANOVA analysis showed an effect of hyperoxia and duration of exposure and an interaction between them for the dependent variable, vWF-stained vessels, in this figure. Significant differences between normoxia- and hyperoxia-exposed mice are indicated by * p < 0.01 (Two-way ANOVA). Scale bar = 100 µM.

2.4. Hyperoxia Exposure Increases Pulmonary Endothelial Cell Apoptosis in Neonatal Mice

To identify the mechanisms through which hyperoxia interrupts lung angiogenesis, we performed immunofluorescence colocalization experiments using lung sections from neonatal mice exposed to normoxia or hyperoxia for one or two weeks. We determined apoptosis in lung endothelial cells by immunofluorescence labelling using an indirect TUNEL assay and anti-vWF antibodies. Figure 4 shows an increased intensity of apoptotic stain (green) in vWF-stained endothelial cells (red) in animals exposed to hyperoxia (Figure 4B,D), indicating that hyperoxia causes lung endothelial cell apoptosis.

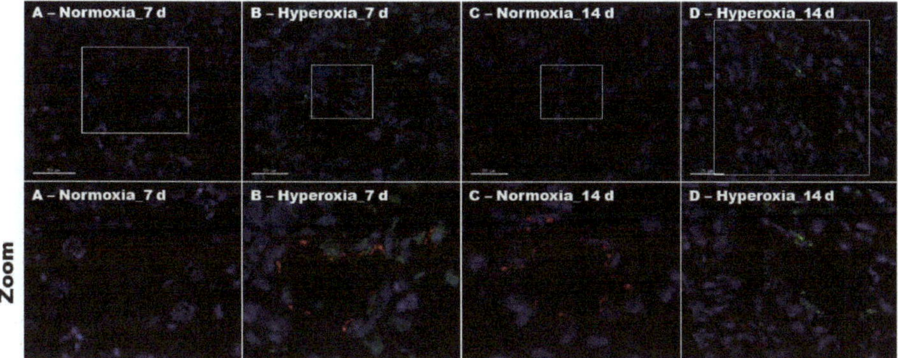

Figure 4. Lung endothelial cell apoptosis in neonatal WT mice exposed to hyperoxia. One-day-old WT mice were exposed to either 21% O_2 (normoxia) or 70% O_2 (hyperoxia) for one or two weeks ($n = 6$/exposure/time-point), after which the lung sections were processed for colocalization studies. (**A–D**) Representative merged images of lung sections stained with TUNEL (green), anti-vWF (red) antibody, and DAPI (blue). The frames in the original magnification figures represent the zoomed regions. Scale bar = 50 μM.

2.5. Hyperoxia Exposure Activates ERK1/2 in HPAECs

To examine the clinical significance of our findings, we used fetal HPAECs to determine if hyperoxia similarly activates ERK1/2 in the developing lungs of preterm infants. Similar to our findings in neonatal mouse lungs, time-dependent studies showed that hyperoxia increased ERK1/2 phosphorylation (0.35 ± 0.01 vs. 0.22 ± 0.03), and the extent of phosphorylation declined (0.28 ± 0.01 vs. 0.21 ± 0.02) as the duration of hyperoxia was prolonged (Figure 5).

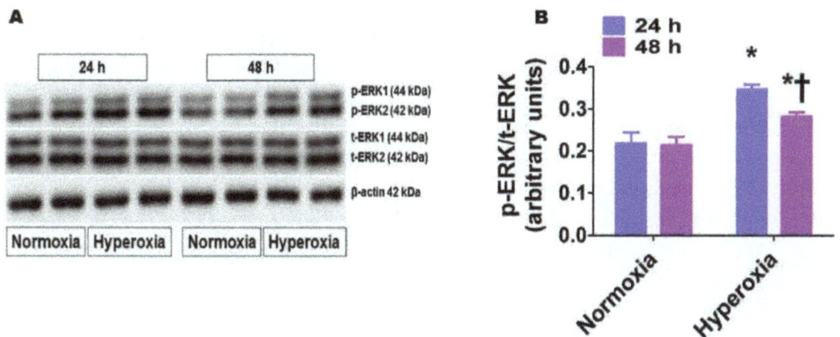

Figure 5. Phosphorylated ERK1/2 protein levels in human pulmonary artery endothelial cells (HPAECs) exposed to hyperoxia. HPAECs were exposed to normoxia or hyperoxia for 24 or 48 h, following which whole-cell proteins were extracted, and immunoblotting was performed using antibodies against total ERK1/2, phosphorylated ERK1/2, or β-actin. Representative immunoblot showing total ERK1/2 and phosphorylated ERK1/2 protein expression (**A**). Densitometric analyses wherein the phosphorylated ERK1/2 band intensities were quantified and normalized to those of total ERK1/2 (**B**). The values are presented as mean $\pm$ SD ($n = 6$/group). Two-way ANOVA analysis showed an effect of hyperoxia and duration of exposure and an interaction between them for the dependent variable, p-ERK1/2, in this figure. Significant differences between normoxia- and hyperoxia-exposed cells are indicated by * $p < 0.001$. Significant differences between hyperoxia-exposed cells are indicated by † $p < 0.01$ (Two-way ANOVA).

2.6. PD98059 Efficiently Inhibits ERK1/2 Activation in HPAECs

To investigate if ERK1/2 plays a direct role in angiogenesis in the developing lungs, we performed in vitro angiogenesis assays after inhibiting ERK1/2 activity by PD98059 in HPAECs. To this end, we first validated ERK1/2 inhibition by determining the expression of phosphorylated ERK1/2 protein levels in PD98059-treated cells. As expected, PD98059 decreased phosphorylated ERK1/2 protein expression in a dose-dependent manner (Figure 6), indicating that PD98059 efficiently inhibits ERK1/2 activation also in these cells.

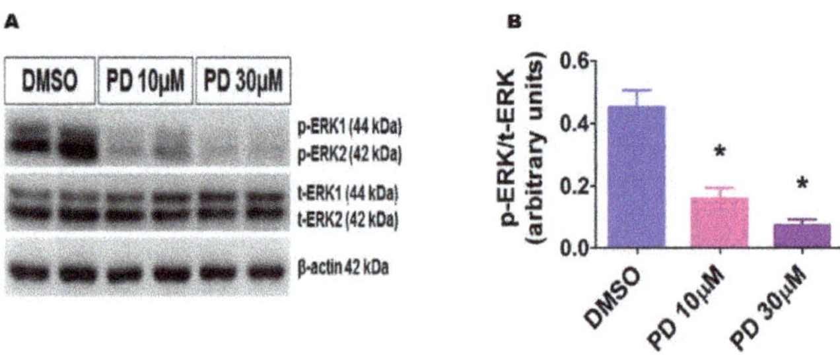

Figure 6. Decreased phosphorylated ERK1/2 protein levels in HPAECs treated with PD98059. HPAECs were treated with dimethylsulfoxide (DMSO) or PD98059 at concentrations of 10 (PD 10) or 30 (PD 30) μM for 30 min, after which whole-cell proteins were extracted, and immunoblotting was performed using antibodies against total ERK1/2, phosphorylated ERK1/2, or β-actin. Representative immunoblot showing total ERK1/2 and phosphorylated ERK1/2 protein expression (**A**). Densitometric analyses wherein the phosphorylated ERK1/2 band intensities were quantified and normalized to those of total ERK1/2 (**B**). The values are presented as mean ± SD (n = 6/group). Significant differences between DMSO- and PD-treated cells are indicated by * p < 0.001 (One-way ANOVA).

2.7. ERK1/2 Inhibition Decreases HPAEC Migration

Endothelial cell migration, proliferation, and tubule formation are essential steps in angiogenesis. To determine the effects of ERK1/2 inhibition on these steps, we performed scratch, proliferation, and tubule formation assays in cells treated with varying concentrations of PD98059. We initially performed the scratch assay and quantified cell migration by assessing the extent of wound closure in mitomycin-treated cell monolayers exposed to the vehicle or PD98059. The extent of wound closure was significantly decreased by PD98059 in a dose-dependent manner (Figure 7), indicating that ERK1/2 inhibition decreases cell migration. When compared with vehicle-treated cells, 10, 20, and 30 μM of PD98059 inhibited cell migration by 29.6, 55.5, and 91.8%, respectively (Figure 7I).

2.8. ERK1/2 Inhibition Decreases HPAEC Proliferation

The MTT activity reflects the cell number, and, thus, the measured absorbance positively correlates with cell proliferation. As expected, there was a time-dependent effect on cell proliferation in vehicle-treated cells. Their proliferation rate increased by 38% in 24 h (Figure 8). However, PD98059 decreased cell proliferation in a dose-dependent manner (Figure 8). When compared with vehicle-treated cells, 10, 20, and 30 μM of PD98059 inhibited cell proliferation by 6.7, 26.4, and 29.1%, respectively, at 24 h and by 25, 29, and 30.8%, respectively, at 48 h.

2.9. ERK1/2 Inhibition Decreases HPAEC Tubule and Mesh Formation

A Matrigel assay was performed to determine the extent of tubule and mesh formation in cells treated with the vehicle or with 30 μM PD98059. Consistent with its effects on cell proliferation and migration, PD98059 decreased HPAEC tubule (1.9 ± 1.3 vs. 5.1 ± 2.3) (Figure 9A–C) and mesh (0.4 ± 0.6 vs. 1.7 ± 0.9) (Figure 9A,B,D) formation, in comparison with vehicle-treated cells.

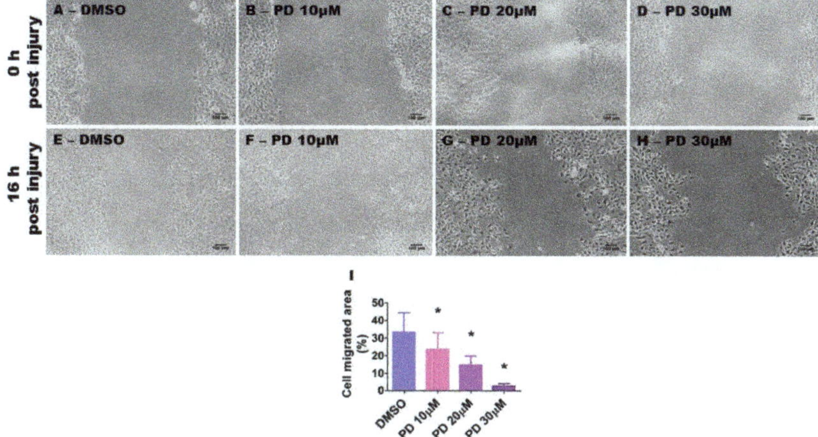

Figure 7. Suppression of ERK1/2 activity decreases HPAEC migration. HPAECs grown as monolayers in six-well plates were treated with 10 μg/mL of mitomycin for 2 h and scratched with a 200 μL pipette tip. The cells were then treated with dimethylsulfoxide (DMSO) or PD98059 at concentrations of 10 (PD 10), 20 (PD 20), or 30 (PD 30) μM. The wound closure area was analyzed using Image J software after 16 h of treatment. (**A–H**) Representative photographs showing cell migration. (**I**) Quantitative analysis of cell migration. The values are presented as mean ± SD (n = 6/group). Significant differences between DMSO- and PD-treated cells are indicated by *, (DMSO vs. PD 10 [$p < 0.05$]; DMSO vs. PD 20 and PD 30 [$p < 0.001$]) (One-way ANOVA). Scale bar = 100 μM.

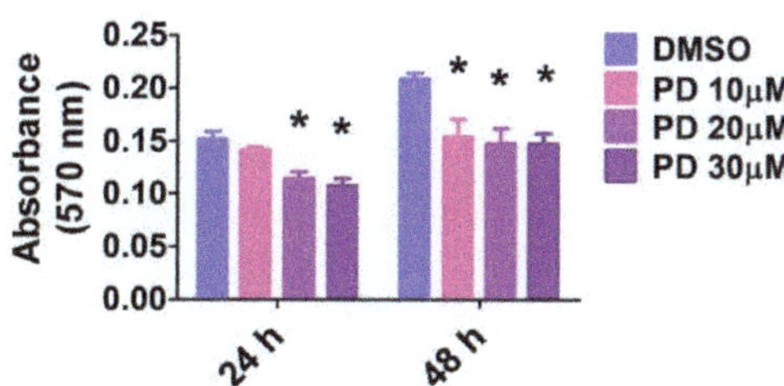

Figure 8. Suppression of ERK1/2 activity decreases HPAEC proliferation. HPAECs were treated with dimethylsulfoxide (DMSO) or PD98059 at concentrations of 10 (PD 10), 20 (PD 20), or 30 (PD 30) μM for 24 or 48 h, following which cell proliferation was assessed by the MTT (3-(4,5-dimethylthiazolyl-2)-2,5-diphenyltetrazolium bromide) assay. The values are presented as mean ± SD (n = 10/group). Two-way ANOVA analysis showed an effect of PD and duration of exposure and an interaction between them for the dependent variable, absorbance at 570 nm, in this figure. Significant differences between DMSO- and PD-treated cells are indicated by * $p < 0.001$ (Two-way ANOVA).

2.10. ERK1/2 Inhibition Alters the Level of Proteins That Regulate Cell Cycle Progression in HPAECs

We finally investigated the mechanisms by which ERK1/2 inhibition disrupts angiogenesis in vitro. Because cell cycle regulation is the major biological pathway affected by ERK1/2 signaling, we determined the effects of ERK1/2 inhibition on the expression of proteins that modulate the crucial G1/S phase of cell cycle. Cyclins A and D and Cdk4 promote G1/S phase transition, while the cyclin-dependent kinase inhibitor p27 prevents this transition. Exposure of cells to the ERK1/2 inhibitor for 24 h decreased the protein levels of cyclin A and Cdk4 and increased the protein levels of p27. Cyclin D expression was similar between the vehicle- and PD-treated cells (Figure 10). These findings indicate that repression of cell cycle progression is one of the mechanisms by which ERK1/2 inhibition interrupts angiogenesis.

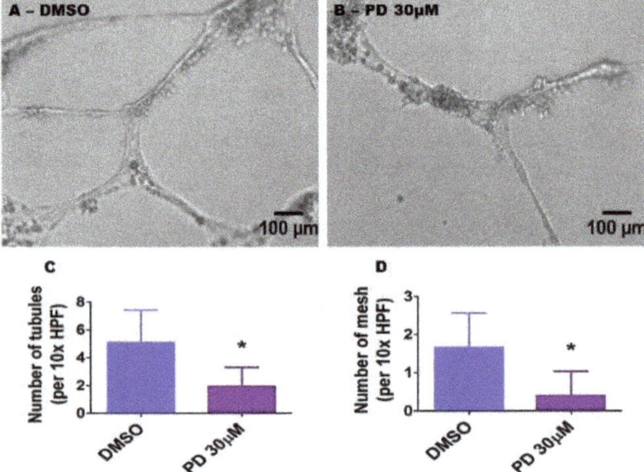

Figure 9. Suppression of ERK1/2 activity decreases HPAEC tubule and mesh formation. HPAECs were pre-treated with dimethylsulfoxide (DMSO) or 30 µM PD98059 (PD 30) for 30 min before being loaded on growth factor-reduced Matrigel (BD Bioscience) in 96-well plates. Following an incubation period of 18 h, tubule formation was quantified. (**A,B**) Representative photographs showing tubule formation in growth factor-reduced Matrigel. (**C,D**) Quantitative analysis of tubule (**C**) and mesh (**D**) formation. The values are presented as mean $\pm$ SD (n = 9/group). Significant differences between DMSO- and PD-treated cells are indicated by * $p < 0.001$ (*t*-test). Scale bar = 100 µM.

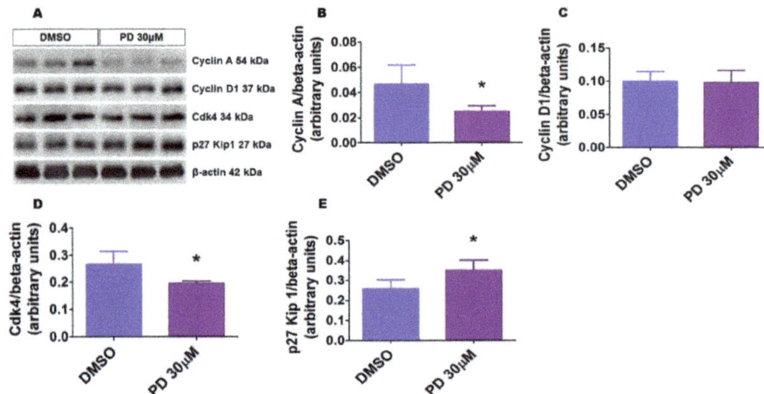

Figure 10. ERK1/2 inhibition affects the expression of cell cycle regulatory proteins. HPAECs were treated with dimethylsulfoxide (DMSO) or 30 µM PD98059 (PD 30) for 24 h, after which whole-cell protein were extracted, and immunoblotting was performed using antibodies against the following proteins: cyclin A, cyclin D, Cdk4, p27, and β-actin. Representative immunoblots showing the expression of the above proteins (**A**). Densitometric analyses wherein cyclin A (**B**), cyclin D (**C**), Cdk4 (**D**), and p27 Kip 1 (**E**) band intensities were quantified and normalized to those of total β-actin. The values are presented as mean $\pm$ SD (n = 6/group). Significant differences between DMSO- and PD-treated cells are indicated by * $p < 0.05$ (*t*-test).

3. Discussion

In this study, we investigated the interaction between hyperoxia and ERK1/2 activation in mouse lungs and fetal HPAECs. Previously, we demonstrated that moderate hyperoxia exposure leads to increased lung oxidative stress and inflammation and causes alveolar and pulmonary vascular simplification, pulmonary vascular remodeling, and PH [25]. Further, we showed in the same model that exposure of mice to neonatal hyperoxia causes lung developmental abnormalities that persist into adolescence [30]. These findings indicate that the phenotype of our mouse model closely aligns with that of preterm infants with BPD and PH. Therefore, we chose the same model for this study and found that hyperoxia exposure transiently increases ERK1/2 activation before decreasing it to below-baseline levels. Further, we show that hyperoxia exposure affects mouse lung vascularization in an identical pattern. Finally, using fetal HPAECs, we demonstrate that inhibition of ERK1/2 pathway disrupts angiogenesis by repressing cell cycle progression.

Lung angiogenesis actively contributes to alveologenesis during development, and healthy lung blood vessels are necessary to maintain the structural and functional integrity of alveolar structures later in life. Numerous in vitro studies have demonstrated that ERK1/2 promote angiogenesis. For example, Mavria et al. [31] demonstrated that these kinases promote endothelial cell survival and sprouting by repressing Rho kinase signaling. Similarly, Murphy et al. [32] showed that sorafenib exerts tumoricidal anti-angiogenic effects by inhibiting of ERK1/2 signaling. Further, an elegant in vivo study using mice lacking endothelial ERK1/2 genes demonstrated that ERK1/2 genes are necessary for embryonic angiogenesis [33]. Therefore, we initially determined the effects of hyperoxia on ERK1/2 activation in the lungs. Hyperoxia exposure was shown to increase ERK1/2 activation in whole lung homogenates and lung epithelial cells by several investigators [34–39]; however, the effects of hyperoxia on lung endothelial cell ERK1/2 activation are poorly understood. Our study demonstrates that hyperoxia exposure increases lung endothelial cell ERK1/2 activation at postnatal day (PND) 7, but decreases their activation at PND14. Importantly, PND14 is still a critical time period for lung development in mice. Lung development occurs at an accelerated rate between PND5 and PND14, and the maximal alveolar number is reached by PND39 [40]. These findings indicate that there may be a mechanistic link between endothelial ERK1/2 signaling and hyperoxia-induced developmental lung injury. To this end, we determined the time-dependent effects of hyperoxia on pulmonary vascularization in our mouse model.

Brief exposures to a high inspired O_2 concentration (>95% O_2) or prolonged exposures to a moderate oxygen concentration was shown to inhibit lung angiogenesis. Here, we show that exposure to a moderate oxygen concentration (70% O_2) initially increased and later induced a significant decline in angiogenesis. Interestingly, hyperoxia-induced changes in lung vascularization paralleled those of ERK1/2 activation. Further, the decreased ERK1/2 activation correlated with increased apoptosis in lung endothelial cells. It is possible that upon exposure to moderate hyperoxia, several angiogenic molecules, such as ERK1/2, are activated to promote or maintain angiogenesis and facilitate lung repair. However, with prolonged hyperoxia exposure, these changes are not sustained to promote healing and prevent further damage from hyperoxic injury. Therefore, we hypothesized that the early ERK1/2 activation upon hyperoxia exposure is an adaptive response to mitigate rather than to potentiate hyperoxic lung injury. To test this hypothesis and to examine the clinical significance of our animal studies, we investigated the effects of hyperoxia exposure on ERK1/2 activation and the effects of ERK1/2 inhibition on angiogenesis using fetal HPAECs. HPAECs were selected because: (1) their proliferation and maturation are crucial for alveolarization and lung growth; (2) their dysfunction contributes to BPD pathogenesis; (3) arterial endothelial cells are enriched in ERK1/2 proteins. Similar to our findings in mouse lungs, hyperoxia exposure transiently increased ERK1/2 activation before decreasing it, when compared with normoxia exposure. Activation of ERK1/2 promotes cell proliferation and differentiation in health [41] and protects against cell death in pathological states [42–44]. However, it is unclear if ERK1/2 signaling attenuates or potentiates hyperoxia-mediated cytotoxicity. Several investigators have demonstrated that ERK1/2 activation protects lung epithelial cells against hyperoxia-induced cell death [28,29,45,46]. Similarly, Ahmad et al. [47] showed that ERK1/2 activation protects adult human lung pulmonary microvascular endothelial cells against hyperoxia-induced cell death. On the other hand, Zhang et al. [36] have demonstrated that inhibition of ERK1/2 signaling decreases cytochrome c release, caspase-9 and -3 activation, and poly (ADP-ribosyl) polymerase cleavage and attenuates lung epithelial cell death in hyperoxic conditions. Similarly, Carnesecchi et al. [46] demonstrated that NADPH oxidase 1 inhibition decreases oxidative stress-mediated ERK1/2 activation and attenuates acute hyperoxic lung injury in adult mice. Further, it was also shown that ERK1/2 activation potentiates hyperoxia-induced developmental lung injury, primarily by regulating the proliferation and differentiation of fibroblasts [35]. The fate of alveolar interstitial fibroblasts influences lung epithelial proliferation and differentiation, i.e., lung development. Differentiation of alveolar interstitial fibroblasts into lipofibroblast promotes lung epithelial proliferation and differentiation [48,49], whereas their differentiation into myofibroblasts interrupts lung development [49]. Under hyperoxic conditions, ERK1/2 activation was shown to be associated with increased myofibroblast differentiation [35]. It is possible that the biological response of ERK1/2 activation may be dependent upon several factors, including the cell and tissue types, magnitude and duration of ERK1/2 activation, and the interactions between ERK1/2 and other activated pathways. Further studies using ERK1/2 transgenic mice are needed to address these knowledge gaps.

The proangiogenic effects of ERK1/2 signaling is well established in the field of cancer biology. However, it is important to note that endothelial cells display substantial organ- and tissue-specific diversity [50,51]. Further, whether ERK1/2 signaling regulates angiogenesis in the developing lungs of humans needs to be determined. Consequently, we inhibited ERK1/2 activation by PD98035 and determined the resulting effects on in vitro angiogenesis using fetal HPAECs. Angiogenesis is a highly coordinated multistep process that includes cell migration, proliferation, and tubule formation [52]. Inhibition of any of these processes disrupts angiogenesis [53]. We observed that ERK1/2 inhibition decreased fetal lung endothelial cell migration, proliferation, and tubule and mesh formation, indicating that ERK1/2 signaling regulates angiogenesis in the developing lungs. Others have shown that ERK1/2 activation is associated with pulmonary vascular development [54]; however, to the best of our knowledge, ours is the first study to indicate that ERK1/2 signaling is necessary for pulmonary vascular development. The angiogenic molecule VEGF promotes lung angiogenesis via ERK1/2 activation [54], but the downstream effectors of ERK1/2 activation are poorly characterized. Hence, we finally investigated the mechanisms by which ERK1/2 signaling regulates lung angiogenesis. A large body of evidence indicate that cell cycle regulation is the predominant biological pathway regulated by ERK1/2 signaling. G1/S phase transition is critical for cell cycle progression. Cyclins A and D and Cdk4 promote this transition, whereas cyclin-dependent kinase inhibitor p27 prevents this transition [55]. Importantly, cyclin A regulates cell cycle progression at multiple levels and therefore can be considered a master regulator of the cell cycle [56]. The Cdks determine the biological activity of cyclin A. For example, cyclin A-mediated activation of Cdk2 promotes G1/S transition, whereas activation of Cdk1 promotes G2/M phase transition. We demonstrate that ERK1/2 inhibition prevented cell cycle progression by decreasing the expression of cyclin A and Cdk4 and increasing that of p27, a conclusion supported by other investigators [33,57,58].

The major limitation of this study is that the interactions between ERK1/2 signaling and lung angiogenesis were determined by in vitro studies. However, our studies in a clinically relevant model of hyperoxia-induced developmental lung injury suggest that there may be similar interactions in vivo. Our future studies will address this limitation by performing in vivo studies using ERK1/2 transgenic mice.

In summary, we demonstrate that exposure of neonatal mice to hyperoxia causes parallel changes in lung endothelial cell ERK1/2 activation and lung vascularization, wherein it initially increases ERK1/2 activation and lung vascularization and later induces a significant decline of these biological processes. Further, our in vitro studies using primary fetal human lung endothelial cells show that

ERK1/2 signaling may be necessary for pulmonary vascular development. Our findings signify that targeting ERK1/2 signaling may be beneficial for BPD infants who have decreased lung vascularization.

4. Materials and Methods

4.1. In Vivo Experiments

4.1.1. Animals

This study was approved and conducted in strict accordance with the federal guidelines for the humane care and use of laboratory animals by the Institutional Animal Care and Use Committee of Baylor College of Medicine (AN-5631, 12/12/2016). C57BL/6J wild-type (WT) mice were obtained from The Jackson Laboratory (Bar Harbor, ME, USA). Timed-pregnant mice raised in our animal facility were used for the experiments. The dams were fed standard mice food and water ad libitum, and all the experimental animals were maintained in 12 h day–night cycles.

4.1.2. Hyperoxia Experiments

Within 24 h of birth, WT dams and their male and female pups were exposed to 21% O_2 (normoxia, $n = 18$) or 70% O_2 (hyperoxia, $n = 18$) for up to two weeks. The dams were rotated between normoxia- and hyperoxia-exposed litters every 24 h during the experiment to prevent oxygen toxicity in the dams and to control for the maternal effects between the groups. Oxygen exposures were conducted in plexiglass chambers, and the animals were monitored as described previously [59].

4.1.3. Lung Tissue Harvest and Protein Extraction

The lungs from a subset of study animals ($n = 6$/exposure) were snap-frozen in liquid nitrogen and stored at −80 °C for the subsequent isolation of total proteins. A mortar and pestle were used to homogenize the lung tissue in a buffer containing 50 mM Tris-HCL (pH 7.5), 0.5 M KCL, 1 M MgCL, and 0.5 M EDTA. The homogenates were centrifuged at 2400× g for 5 min at 4 °C. The supernatants (protein lysate) were stored at −80 °C.

4.1.4. Western Blot Assays

The protein lysates from the experimental animals were separated by 10% SDS-polyacrylamide gel electrophoresis and transferred to polyvinylidene difluoride membranes. The membranes were incubated overnight at 4 °C with the following primary antibodies: anti-β-actin (Santa Cruz Biotechnologies, Santa Cruz, CA, USA; sc-47778, dilution 1:1000), anti-total ERK1/2 (Cell Signaling, Danvers, MA, USA; 4695, dilution 1:1000), and anti-phospho-ERK1/2 (Cell Signaling, Danvers, MA, USA; 9106, dilution 1:1000). The primary antibodies were detected by incubation with the appropriate horseradish peroxidase-conjugated secondary antibodies. The immunoreactive bands were detected by chemiluminescence methods, and the band densities were quantified by Image lab 5.2.1 software (Bio-Rad Laboratories, Inc., Hercules, CA, USA).

4.1.5. Tissue Preparation for Immunofluorescence and Lung Vascular Morphometry Studies

A separate group of pups were euthanized at one and two weeks of life ($n = 6$/exposure/time-point), and their lungs were inflated and fixed via the trachea with 10% formalin at 25 cm H_2O pressure for at least 10 min. Sections of the paraffin-embedded lungs were obtained for immunofluorescence studies and for the analysis of lung vascularization, as described previously [59].

4.1.6. Immunofluorescence Studies

We performed double immunostaining with anti-pERK1/2 and anti-von Willebrand factor (vWF; endothelial specific marker) antibodies to localize ERK1/2 activation in the lung endothelial cells. Fresh frozen lung tissues were incubated with 7.5% normal donkey serum for 1 h to block nonspecific protein binding, after which they were incubated overnight at 4 °C with the following primary antibodies: anti-phospho-ERK1/2 (Cell Signaling, Danvers, MA, USA; 4370, dilution 1:150) and anti-vWF (Abcam, Cambridge, MA, USA; ab11713, dilution 1:50). The primary anti-pERK1/2 and anti-vWF antibodies were detected by incubation with fluorescein-conjugated donkey anti-rabbit (Alexa Fluor 488, dilution 1:200) and donkey anti-sheep (Alexa Fluor 633, dilution 1:200) secondary antibodies, respectively. An indirect TUNEL assay was used to detect apoptosis using the ApopTag Fluorescein In Situ Apoptosis detection kit (MilliporeSigma, St. Louis, MO, USA; S7110), as per the manufacturer's recommendations. The localization of the apoptotic process in the endothelial cells was determined using vWF antibodies, as described above. All the slides were counterstained with 4',6-diamidino-2-phenylindole (DAPI) and analyzed by confocal microscopy. The observers analyzing these slides were masked to the experimental conditions.

4.1.7. Analyses of Pulmonary Vascularization

Pulmonary vessel density was determined in these animals on the basis of the immunofluorescence staining with anti-vWF antibody (Abcam, Cambridge, MA, USA; ab11713, dilution 1:50), which is an endothelial-specific marker [60]. At least 10 counts from 10 random non-overlapping fields (original magnification, 200×) were performed for each animal (n = 6/exposure/time-point). The observers performing these measurements were masked to the slide identity.

4.2. In Vitro Experiments

4.2.1. Cell Culture

The human pulmonary artery endothelial cells (HPAECs) derived from the lungs of human fetus (26 weeks gestational age) were obtained from ScienCell research laboratories (San Diego, CA, USA; 3100). HPAECs were grown in 95% air and 5% CO_2 at 37 °C in specific endothelial cell medium, according to the manufacturer's protocol. Briefly, the cells were grown in fibronectin-coated plates containing basal endothelial cell medium supplemented with fetal bovine serum, antibiotics, and an endothelial cell growth supplement in a humidifier containing 5% CO_2 at 37 °C. When the cell culture reached >90% confluence, the cells were subcultured with a split ratio of 1:3. Cells between passages 5–7 were used for all our experiments.

4.2.2. Hyperoxia Exposure

Hyperoxia experiments were conducted in a plexiglass, sealed chamber into which a mixture of 95% O_2 and 5% CO_2 was circulated continuously. The chamber was placed in a Forma Scientific water-jacketed incubator at 37 °C. Once the O_2 level inside the chamber reached 95%, the cells were placed inside the chamber for up to 48 h [61]. The cells were harvested at 24 h and 48 h of exposure to determine the effects of hyperoxia on ERK1/2 activation.

4.2.3. Cell Treatment

HPAECs were treated with either 0.01% *v*/*v* dimethyl sulfoxide (DMSO) (Sigma-Aldrich, St. Louis, MO, USA; 276855) or the ERK1/2 inhibitor PD98059 (Sigma-Aldrich, St. Louis, MO, USA; P215) at varying concentrations up to 30 μM. The cells were then harvested to determine the effects of PD98059 on ERK1/2 activation, cell proliferation, migration, tubule formation, and expression of cell cycle regulatory proteins.

4.2.4. Western Blot Assays

The cells were grown in complete medium on six-well plates to 70–80% confluence, after which they were exposed to normoxia (95% air and 5% CO_2) or hyperoxia (95% O_2 and 5% CO_2) for up to 48 h. In a separate set of experiments, the cells grown on six-well plates were treated with DMSO or 30 μM PD98059 for 24 h. Following these treatments, whole-cell protein extracts were obtained by using the radio immunoprecipitation assay lysis buffer system (Santa Cruz Biotechnologies, Santa Cruz, CA, USA; sc-24948) and subjected to western blotting with the following antibodies: anti-β-actin (Santa Cruz Biotechnologies, Santa Cruz, CA, USA; sc-47778, dilution 1:1000), anti-cyclin A (Santa Cruz Biotechnologies, Santa Cruz, CA, USA; sc-751, dilution 1:1000), anti-cyclin D1 (Santa Cruz Biotechnologies, Santa Cruz, CA, USA; sc-8396, dilution 1:250), anti-cyclin-dependent kinase (Cdk) 4 (Santa Cruz Biotechnologies, Santa Cruz, CA, USA; sc-23896, dilution 1:250), anti-p27 Kip 1 (Abcam, Cambridge, MA, USA ; ab32034, dilution 1:1000), anti-total ERK1/2 (Cell Signaling, Danvers, MA, USA; 4695, dilution 1:1000), anti-phospho-ERK1/2 (Cell Signaling, Danvers, MA, USA; 9106, dilution 1:1000). The immunoreactive bands were detected and quantified as described in the "in vivo experiments" section.

4.2.5. Cell Proliferation Assay

Cell proliferation was determined by a colorimetric assay based on the ability of viable cells to reduce the tetrazolium salt MTT (3-(4,5-dimethylthiazolyl-2)-2,5-diphenyltetrazolium bromide) (American Type Culture Collection, Manassas, VA, USA; ATCC 30-1010K) to formazan. HPAECs were grown in 96-well microplates overnight at a density of 5×10^4 cells per well in 100 µL of complete medium, followed by an additional period of growth for 24 h in basal medium containing 0.5% fetal bovine serum (FBS). The cells were then treated with varying concentrations of PD98059 and grown under reduced serum conditions for up to 48 h, after which cell proliferation was assessed by the MTT reduction assay, as outlined in the MTT Assay protocol (American Type Culture Collection, Manassas, VA, USA). Briefly, at the end of the experiments, 10 µL of MTT reagent was added to each well, and the cells were incubated in a humidifier containing 5% CO_2 at 37 °C for 2 h, at the end of which precipitates were visible in all wells. Following the incubation, 100 µL of detergent was added to each well, and the cells were incubated at room temperature in the dark for additional 2 h; the absorbance was measured at 570 nm. The absorbance readings are directly proportional to the number of cells.

4.2.6. Scratch Assay

A scratch assay was used to quantify cell migration [62]. HPAECs were grown in complete medium in six-well plates to 70–80% confluence, followed by a period of growth for 24 h in basal medium containing 0.5% fetal bovine serum. To mitigate the effects of cell proliferation on wound closure, the cells were pre-treated with mitomycin (10 µg/mL) (MilliporeSigma, St. Louis, MO, USA; M4287) for 2 h. The cells were then scratched with a 200 µL pipette tip before they were treated with PD98059. The wound closure or cell migration area was estimated using Image J software (National Institutes of Health, Bethesda, MD, USA) by comparing the wounded areas at 0 h and 16 h.

4.2.7. Tubule and Mesh Formation Assays

A Matrigel assay was used to determine tubule and mesh formation, as described previously [63,64]. HPAECs were grown in 96-well microplates at a density of 2×10^4 cells per well in 100 µL of basal medium containing 0.5% FBS. The cells were pretreated with 30 µM PD98059 for 30 min and loaded on top of growth factor-reduced Matrigel (Corning, NY, USA; 356230). Following an incubation period of 18 h, tubule and mesh formation were quantified.

4.3. Statistical Analyses

The results were analyzed by the GraphPad Prism 5 software (La Jolla, CA, USA). The data were expressed as mean ± SD. In vivo experiments: At least three separate experiments were performed for each measurement (n = total animals from the three experiments). The effects of hyperoxia exposure on ERK1/2 activation were assessed by t-test, whereas the effects of time-point and exposure on pulmonary vascularization were assessed using two-way ANOVA. In vitro experiments: At least three separate experiments were performed for each measurement. One-way ANOVA was used to determine the dose-dependent effects of PD98059 on cell proliferation and ERK1/2 phosphorylation, while a nonparametric test (Kruskal–Wallis test) was used to determine the dose-dependent effects of PD98059 on cell migration. The effects of PD98059 on tubule and mesh formation were determined by *t*-test. A *p* value of <0.05 was considered significant.

Author Contributions: R.T.M., A.K.S., R.B., and B.S. participated in the research design, performed data analysis, and contributed to the writing of the manuscript. R.T.M., A.K.S., and B.S. conducted the experiments.

Acknowledgments: We thank Pamela Parsons for her timely processing of histopathology slides. This work was supported by National Institutes of Health grants: HD-073323 to Binoy Shivanna and P30DK056338 to the Digestive Disease Center Core at Baylor College of Medicine, and grants from the American Heart Association BGIA-20190008, American Lung Association RG-349917, and Texas Children's Hospital Pediatric Pilot Award program to Binoy Shivanna.

Conflicts of Interest: The authors report no conflicts of interest, financial or otherwise in this work.

Abbreviations

BPD	bronchopulmonary dysplasia
Cdk	cyclin-dependent kinase
ERK	extracellular signal-regulated kinases
DAPI	4′,6-diamidino-2-phenylindole
HPAECs	human pulmonary artery endothelial cells
MAP kinase	mitogen-activated protein kinase
MTT	3-(4,5-dimethylthiazolyl-2)-2,5-diphenyltetrazolium bromide
PD	PD98059
PH	pulmonary hypertension
NO	nitric oxide
PND	postnatal day
WT	wild type
VEGF	vascular endothelial growth factor
vWF	von Willebrand factor

References

1. Jobe, A.H. Animal models, learning lessons to prevent and treat neonatal chronic lung disease. *Front. Med.* **2015**, *2*, 49. [CrossRef] [PubMed]

2. Fanaroff, A.A.; Stoll, B.J.; Wright, L.L.; Carlo, W.A.; Ehrenkranz, R.A.; Stark, A.R.; Bauer, C.R.; Donovan, E.F.; Korones, S.B.; Laptook, A.R.; et al. Trends in neonatal morbidity and mortality for very low birthweight infants. *Am. J. Obstet. Gynecol.* **2007**, *196*, 147.e1–147.e8. [CrossRef] [PubMed]

3. Husain, A.N.; Siddiqui, N.H.; Stocker, J.T. Pathology of arrested acinar development in postsurfactant bronchopulmonary dysplasia. *Hum. Pathol.* **1998**, *29*, 710–717. [CrossRef]

4. Coalson, J.J. Pathology of new bronchopulmonary dysplasia. *Semin. Neonatol.* **2003**, *8*, 73–81. [CrossRef]

5. Bhatt, A.J.; Pryhuber, G.S.; Huyck, H.; Watkins, R.H.; Metlay, L.A.; Maniscalco, W.M. Disrupted pulmonary vasculature and decreased vascular endothelial growth factor, Flt-1, and TIE-2 in human infants dying with bronchopulmonary dysplasia. *Am. J. Respir. Crit. Care Med.* **2001**, *164*, 1971–1980. [CrossRef] [PubMed]

6. Thebaud, B.; Abman, S.H. Bronchopulmonary dysplasia: Where have all the vessels gone? Roles of angiogenic growth factors in chronic lung disease. *Am. J. Respir. Crit. Care Med.* **2007**, *175*, 978–985. [CrossRef] [PubMed]

7. Jakkula, M.; Le Cras, T.D.; Gebb, S.; Hirth, K.P.; Tuder, R.M.; Voelkel, N.F.; Abman, S.H. Inhibition of angiogenesis decreases alveolarization in the developing rat lung. *Am. J. Physiol. Lung Cell. Mol. Physiol.* **2000**, *279*, L600–L607. [CrossRef] [PubMed]

8. Le Cras, T.D.; Markham, N.E.; Tuder, R.M.; Voelkel, N.F.; Abman, S.H. Treatment of newborn rats with a vegf receptor inhibitor causes pulmonary hypertension and abnormal lung structure. *Am. J. Physiol. Lung Cell. Mol. Physiol.* **2002**, *283*, L555–L562. [CrossRef] [PubMed]

9. Thebaud, B.; Ladha, F.; Michelakis, E.D.; Sawicka, M.; Thurston, G.; Eaton, F.; Hashimoto, K.; Harry, G.; Haromy, A.; Korbutt, G.; et al. Vascular endothelial growth factor gene therapy increases survival, promotes lung angiogenesis, and prevents alveolar damage in hyperoxia-induced lung injury: Evidence that angiogenesis participates in alveolarization. *Circulation* **2005**, *112*, 2477–2486. [CrossRef] [PubMed]

10. Stenmark, K.R.; Abman, S.H. Lung vascular development: Implications for the pathogenesis of bronchopulmonary dysplasia. *Annu. Rev. Physiol.* **2005**, *67*, 623–661. [CrossRef] [PubMed]

11. Yamamoto, H.; Yun, E.J.; Gerber, H.P.; Ferrara, N.; Whitsett, J.A.; Vu, T.H. Epithelial-vascular cross talk mediated by VEGF-A and HGF signaling directs primary septae formation during distal lung morphogenesis. *Dev. Biol.* **2007**, *308*, 44–53. [CrossRef] [PubMed]

12. Lin, Y.J.; Markham, N.E.; Balasubramaniam, V.; Tang, J.R.; Maxey, A.; Kinsella, J.P.; Abman, S.H. Inhaled nitric oxide enhances distal lung growth after exposure to hyperoxia in neonatal rats. *Pediatr. Res.* **2005**, *58*, 22–29. [CrossRef] [PubMed]

13. Tang, J.R.; Markham, N.E.; Lin, Y.J.; McMurtry, I.F.; Maxey, A.; Kinsella, J.P.; Abman, S.H. Inhaled nitric oxide attenuates pulmonary hypertension and improves lung growth in infant rats after neonatal treatment with a VEGF receptor inhibitor. *Am. J. Physiol. Lung Cell. Mol. Physiol.* **2004**, *287*, L344–L351. [CrossRef] [PubMed]

14. Kunig, A.M.; Balasubramaniam, V.; Markham, N.E.; Seedorf, G.; Gien, J.; Abman, S.H. Recombinant human vegf treatment transiently increases lung edema but enhances lung structure after neonatal hyperoxia. *Am. J. Physiol. Lung Cell. Mol. Physiol.* **2006**, *291*, L1068–L1078. [CrossRef] [PubMed]

15. Allen, M.C.; Donohue, P.; Gilmore, M.; Cristofalo, E.; Wilson, R.F.; Weiner, J.Z.; Robinson, K. Inhaled nitric oxide in preterm infants. *Evid. Rep./Technol. Assess.* **2010**, *195*, 1–315.

16. Donohue, P.K.; Gilmore, M.M.; Cristofalo, E.; Wilson, R.F.; Weiner, J.Z.; Lau, B.D.; Robinson, K.A.; Allen, M.C. Inhaled nitric oxide in preterm infants: A systematic review. *Pediatrics* **2011**, *127*, e414–e422. [CrossRef] [PubMed]

17. Gadhia, M.M.; Cutter, G.R.; Abman, S.H.; Kinsella, J.P. Effects of early inhaled nitric oxide therapy and vitamin a supplementation on the risk for bronchopulmonary dysplasia in premature newborns with respiratory failure. *J. Pediatr.* **2014**, *164*, 744–748. [CrossRef] [PubMed]

18. Madurga, A.; Mizikova, I.; Ruiz-Camp, J.; Morty, R.E. Recent advances in late lung development and the pathogenesis of bronchopulmonary dysplasia. *Am. J. Physiol. Lung Cell. Mol. Physiol.* **2013**, *305*, L893–L905. [CrossRef] [PubMed]

19. English, J.; Pearson, G.; Wilsbacher, J.; Swantek, J.; Karandikar, M.; Xu, S.; Cobb, M.H. New insights into the control of map kinase pathways. *Exp. Cell Res.* **1999**, *253*, 255–270. [CrossRef] [PubMed]

20. Gabay, L.; Seger, R.; Shilo, B.Z. Map kinase in situ activation atlas during *Drosophila embryogenesis*. *Development* **1997**, *124*, 3535–3541.

21. Kashimata, M.; Sayeed, S.; Ka, A.; Onetti-Muda, A.; Sakagami, H.; Faraggiana, T.; Gresik, E.W. The ERK-1/2 signaling pathway is involved in the stimulation of branching morphogenesis of fetal mouse submandibular glands by EGF. *Dev. Biol.* **2000**, *220*, 183–196. [CrossRef] [PubMed]

22. Karihaloo, A.; O'Rourke, D.A.; Nickel, C.; Spokes, K.; Cantley, L.G. Differential mapk pathways utilized for hgf- and egf-dependent renal epithelial morphogenesis. *J. Biol. Chem.* **2001**, *276*, 9166–9173. [CrossRef] [PubMed]

23. Niemann, C.; Brinkmann, V.; Birchmeier, W. Hepatocyte growth factor and neuregulin in mammary gland cell morphogenesis. *Adv. Exp. Med. Biol.* **2000**, *480*, 9–18. [PubMed]

24. Kling, D.E.; Lorenzo, H.K.; Trbovich, A.M.; Kinane, T.B.; Donahoe, P.K.; Schnitzer, J.J. MEK-1/2 inhibition reduces branching morphogenesis and causes mesenchymal cell apoptosis in fetal rat lungs. *Am. J. Physiol. Lung Cell. Mol. Physiol.* **2002**, *282*, L370–L378. [CrossRef] [PubMed]

25. Reynolds, C.L.; Zhang, S.; Shrestha, A.K.; Barrios, R.; Shivanna, B. Phenotypic assessment of pulmonary hypertension using high-resolution echocardiography is feasible in neonatal mice with experimental bronchopulmonary dysplasia and pulmonary hypertension: A step toward preventing chronic obstructive pulmonary disease. *Int. J. Chron. Obstr. Pulm. Dis.* **2016**, *11*, 1597–1605.

26. Aslam, M.; Baveja, R.; Liang, O.D.; Fernandez-Gonzalez, A.; Lee, C.; Mitsialis, S.A.; Kourembanas, S. Bone marrow stromal cells attenuate lung injury in a murine model of neonatal chronic lung disease. *Am. J. Respir. Crit. Care Med.* **2009**, *180*, 1122–1130. [CrossRef] [PubMed]

27. Lee, K.J.; Berkelhamer, S.K.; Kim, G.A.; Taylor, J.M.; O'Shea, K.M.; Steinhorn, R.H.; Farrow, K.N. Disrupted pulmonary artery cgmp signaling in mice with hyperoxia-induced pulmonary hypertension. *Am. J. Respir. Cell Mol. Biol.* **2014**, *50*, 369–378. [CrossRef] [PubMed]

28. Xu, D.; Guthrie, J.R.; Mabry, S.; Sack, T.M.; Truog, W.E. Mitochondrial aldehyde dehydrogenase attenuates hyperoxia-induced cell death through activation of ERK/MAPK and PI3K-Akt pathways in lung epithelial cells. *Am. J. Physiol. Lung Cell. Mol. Physiol.* **2006**, *291*, L966–L975. [CrossRef] [PubMed]

29. Buckley, S.; Driscoll, B.; Barsky, L.; Weinberg, K.; Anderson, K.; Warburton, D. ERK activation protects against DNA damage and apoptosis in hyperoxic rat AEC2. *Am. J. Physiol.* **1999**, *277*, L159–L166. [CrossRef] [PubMed]

30. Menon, R.T.; Shrestha, A.K.; Shivanna, B. Hyperoxia exposure disrupts adrenomedullin signaling in newborn mice: Implications for lung development in premature infants. *Biochem. Biophys. Res. Commun.* **2017**, *487*, 666–671. [CrossRef] [PubMed]

31. Mavria, G.; Vercoulen, Y.; Yeo, M.; Paterson, H.; Karasarides, M.; Marais, R.; Bird, D.; Marshall, C.J. ERK-MAPK signaling opposes Rho-kinase to promote endothelial cell survival and sprouting during angiogenesis. *Cancer Cell* **2006**, *9*, 33–44. [CrossRef] [PubMed]

32. Murphy, D.A.; Makonnen, S.; Lassoued, W.; Feldman, M.D.; Carter, C.; Lee, W.M. Inhibition of tumor endothelial ERK activation, angiogenesis, and tumor growth by sorafenib (bay43-9006). *Am. J. Pathol.* **2006**, *169*, 1875–1885. [CrossRef] [PubMed]

33. Srinivasan, R.; Zabuawala, T.; Huang, H.; Zhang, J.; Gulati, P.; Fernandez, S.; Karlo, J.C.; Landreth, G.E.; Leone, G.; Ostrowski, M.C. Erk1 and Erk2 regulate endothelial cell proliferation and migration during mouse embryonic angiogenesis. *PLoS ONE* **2009**, *4*, e8283. [CrossRef] [PubMed]

34. Konsavage, W.M.; Zhang, L.; Wu, Y.; Shenberger, J.S. Hyperoxia-induced activation of the integrated stress response in the newborn rat lung. *Am. J. Physiol. Lung Cell. Mol. Physiol.* **2012**, *302*, L27–L35. [CrossRef] [PubMed]

35. Sakurai, R.; Villarreal, P.; Husain, S.; Liu, J.; Sakurai, T.; Tou, E.; Torday, J.S.; Rehan, V.K. Curcumin protects the developing lung against long-term hyperoxic injury. *Am. J. Physiol. Lung Cell. Mol. Physiol.* **2013**, *305*, L301–L311. [CrossRef] [PubMed]

36. Zhang, X.; Shan, P.; Sasidhar, M.; Chupp, G.L.; Flavell, R.A.; Choi, A.M.; Lee, P.J. Reactive oxygen species and extracellular signal-regulated kinase 1/2 mitogen-activated protein kinase mediate hyperoxia-induced cell death in lung epithelium. *Am. J. Respir. Cell Mol. Biol.* **2003**, *28*, 305–315. [CrossRef] [PubMed]

37. Buckley, S.; Barsky, L.; Weinberg, K.; Warburton, D. In vivo inosine protects alveolar epithelial type 2 cells against hyperoxia-induced DNA damage through map kinase signaling. *Am. J. Physiol. Lung Cell. Mol. Physiol.* **2005**, *288*, L569–L575. [CrossRef] [PubMed]

38. Papaiahgari, S.; Zhang, Q.; Kleeberger, S.R.; Cho, H.Y.; Reddy, S.P. Hyperoxia stimulates an Nrf2-ARE transcriptional response via ROS-EGFR-PI3K-Akt/ERK map kinase signaling in pulmonary epithelial cells. *Antioxid. Redox Signal.* **2006**, *8*, 43–52. [CrossRef] [PubMed]

39. Mao, Q.; Gundavarapu, S.; Patel, C.; Tsai, A.; Luks, F.I.; De Paepe, M.E. The fas system confers protection against alveolar disruption in hyperoxia-exposed newborn mice. *Am. J. Respir. Cell Mol. Biol.* **2008**, *39*, 717–729. [CrossRef] [PubMed]

40. Pozarska, A.; Rodriguez-Castillo, J.A.; Surate Solaligue, D.E.; Ntokou, A.; Rath, P.; Mizikova, I.; Madurga, A.; Mayer, K.; Vadasz, I.; Herold, S.; et al. Stereological monitoring of mouse lung alveolarization from the early postnatal period to adulthood. *Am. J. Physiol. Lung Cell. Mol. Physiol.* **2017**, *312*, L882–L895. [CrossRef] [PubMed]

41. Marshall, C.J. Specificity of receptor tyrosine kinase signaling: Transient versus sustained extracellular signal-regulated kinase activation. *Cell* **1995**, *80*, 179–185. [CrossRef]

42. Nguyen, D.T.; Kebache, S.; Fazel, A.; Wong, H.N.; Jenna, S.; Emadali, A.; Lee, E.H.; Bergeron, J.J.; Kaufman, R.J.; Larose, L.; et al. Nck-dependent activation of extracellular signal-regulated kinase-1 and regulation of cell survival during endoplasmic reticulum stress. *Mol. Biol. Cell* **2004**, *15*, 4248–4260. [CrossRef] [PubMed]

43. Wada, T.; Penninger, J.M. Mitogen-activated protein kinases in apoptosis regulation. *Oncogene* **2004**, *23*, 2838–2849. [CrossRef] [PubMed]

44. Hung, C.C.; Ichimura, T.; Stevens, J.L.; Bonventre, J.V. Protection of renal epithelial cells against oxidative injury by endoplasmic reticulum stress preconditioning is mediated by ERK1/2 activation. *J. Biol. Chem.* **2003**, *278*, 29317–29326. [CrossRef] [PubMed]

45. Truong, S.V.; Monick, M.M.; Yarovinsky, T.O.; Powers, L.S.; Nyunoya, T.; Hunninghake, G.W. Extracellular signal-regulated kinase activation delays hyperoxia-induced epithelial cell death in conditions of akt downregulation. *Am. J. Respir. Cell Mol. Biol.* **2004**, *31*, 611–618. [CrossRef] [PubMed]

46. Carnesecchi, S.; Deffert, C.; Pagano, A.; Garrido-Urbani, S.; Metrailler-Ruchonnet, I.; Schappi, M.; Donati, Y.; Matthay, M.A.; Krause, K.H.; Barazzone Argiroffo, C. Nadph oxidase-1 plays a crucial role in hyperoxia-induced acute lung injury in mice. *Am. J. Respir. Crit. Care Med.* **2009**, *180*, 972–981. [CrossRef] [PubMed]

47. Ahmad, S.; Ahmad, A.; Ghosh, M.; Leslie, C.C.; White, C.W. Extracellular ATP-mediated signaling for survival in hyperoxia-induced oxidative stress. *J. Biol. Chem.* **2004**, *279*, 16317–16325. [CrossRef] [PubMed]

48. McGowan, S.E.; Torday, J.S. The pulmonary lipofibroblast (lipid interstitial cell) and its contributions to alveolar development. *Annu. Rev. Physiol.* **1997**, *59*, 43–62. [CrossRef] [PubMed]

49. Torday, J.S.; Torres, E.; Rehan, V.K. The role of fibroblast transdifferentiation in lung epithelial cell proliferation, differentiation, and repair in vitro. *Pediatr. Pathol. Mol. Med.* **2003**, *22*, 189–207. [CrossRef] [PubMed]

50. Plendl, J.; Neumuller, C.; Vollmar, A.; Auerbach, R.; Sinowatz, F. Isolation and characterization of endothelial cells from different organs of fetal pigs. *Anat. Embryol.* **1996**, *194*, 445–456. [CrossRef] [PubMed]

51. Gumkowski, F.; Kaminska, G.; Kaminski, M.; Morrissey, L.W.; Auerbach, R. Heterogeneity of mouse vascular endothelium. In vitro studies of lymphatic, large blood vessel and microvascular endothelial cells. *Blood Vessels* **1987**, *24*, 11–23. [PubMed]

52. Goodwin, A.M. In vitro assays of angiogenesis for assessment of angiogenic and anti-angiogenic agents. *Microvasc. Res.* **2007**, *74*, 172–183. [CrossRef] [PubMed]

53. Carmeliet, P. Mechanisms of angiogenesis and arteriogenesis. *Nat. Med.* **2000**, *6*, 389–395. [CrossRef] [PubMed]

54. Pang, J.; Hoefen, R.; Pryhuber, G.S.; Wang, J.; Yin, G.; White, R.J.; Xu, X.; O'Dell, M.R.; Mohan, A.; Michaloski, H.; et al. G-protein-coupled receptor kinase interacting protein-1 is required for pulmonary vascular development. *Circulation* **2009**, *119*, 1524–1532. [CrossRef] [PubMed]

55. Sherr, C.J.; Roberts, J.M. CDK inhibitors: Positive and negative regulators of G1-phase progression. *Genes Dev.* **1999**, *13*, 1501–1512. [CrossRef] [PubMed]

56. Yam, C.H.; Fung, T.K.; Poon, R.Y. Cyclin A in cell cycle control and cancer. *Cell. Mol. Life Sci. CMLS* **2002**, *59*, 1317–1326. [CrossRef] [PubMed]

57. Shen, H.; Zhou, E.; Wei, X.; Fu, Z.; Niu, C.; Li, Y.; Pan, B.; Mathew, A.V.; Wang, X.; Pennathur, S.; et al. High density lipoprotein promotes proliferation of adipose-derived stem cells via S1P1 receptor and Akt, ERK1/2 signal pathways. *Stem Cell Res. Ther.* **2015**, *6*, 95. [CrossRef] [PubMed]

58. Ling, L.; Wei, T.; He, L.; Wang, Y.; Wang, Y.; Feng, X.; Zhang, W.; Xiong, Z. Low-intensity pulsed ultrasound activates ERK1/2 and PI3K-Akt signalling pathways and promotes the proliferation of human amnion-derived mesenchymal stem cells. *Cell Prolif.* **2017**, *50*. [CrossRef] [PubMed]

59. Shivanna, B.; Zhang, W.; Jiang, W.; Welty, S.E.; Couroucli, X.I.; Wang, L.; Moorthy, B. Functional deficiency of aryl hydrocarbon receptor augments oxygen toxicity-induced alveolar simplification in newborn mice. *Toxicol. Appl. Pharmacol.* **2013**, *267*, 209–217. [CrossRef] [PubMed]

60. Shivanna, B.; Zhang, S.; Patel, A.; Jiang, W.; Wang, L.; Welty, S.E.; Moorthy, B. Omeprazole attenuates pulmonary aryl hydrocarbon receptor activation and potentiates hyperoxia-induced developmental lung injury in newborn mice. *Toxicol. Sci.* **2015**, *148*, 276–287. [CrossRef] [PubMed]

61. Shivanna, B.; Chu, C.; Welty, S.E.; Jiang, W.; Wang, L.; Couroucli, X.I.; Moorthy, B. Omeprazole attenuates hyperoxic injury in H441 cells via the aryl hydrocarbon receptor. *Free Radic. Biol. Med.* **2011**, *51*, 1910–1917. [CrossRef] [PubMed]

62. Liang, C.C.; Park, A.Y.; Guan, J.L. In vitro scratch assay: A convenient and inexpensive method for analysis of cell migration in vitro. *Nat. Protoc.* **2007**, *2*, 329–333. [CrossRef] [PubMed]

63. Arnaoutova, I.; Kleinman, H.K. In vitro angiogenesis: Endothelial cell tube formation on gelled basement membrane extract. *Nat. Protoc.* **2010**, *5*, 628–635. [CrossRef] [PubMed]

64. Ribatti, D.; Guidolin, D.; Conconi, M.T.; Nico, B.; Baiguera, S.; Parnigotto, P.P.; Vacca, A.; Nussdorfer, G.G. Vinblastine inhibits the angiogenic response induced by adrenomedullin in vitro and in vivo. *Oncogene* **2003**, *22*, 6458–6461. [CrossRef] [PubMed]

 International Journal of
Molecular Sciences

Communication

Crosstalk between p38 and Erk 1/2 in Downregulation of FGF1-Induced Signaling

Malgorzata Zakrzewska [1,*], Lukasz Opalinski [1], Ellen M. Haugsten [2,3], Jacek Otlewski [1] and Antoni Wiedlocha [3,4]

[1] Department of Protein Engineering, Faculty of Biotechnology, University of Wroclaw, 50-383 Wroclaw, Poland; lukasz.opalinski@uwr.edu.pl (L.O.); jacek.otlewski@uwr.edu.pl (J.O.)
[2] Department of Tumor Biology, Institute for Cancer Research, Oslo University Hospital, Montebello, 0379 Oslo, Norway; ellen.m.haugsten@rr-research.no
[3] Centre for Cancer Cell Reprogramming, Institute of Clinical Medicine, Faculty of Medicine, University of Oslo, Montebello, 0379 Oslo, Norway; antoni.wiedlocha@rr-research.no
[4] Department of Molecular Cell Biology, Institute for Cancer Research, Oslo University Hospital, Montebello, 0379 Oslo, Norway
* Correspondence: malgorzata.zakrzewska@uwr.edu.pl; Tel.: +48-71-375-2889

Received: 14 January 2019; Accepted: 10 April 2019; Published: 12 April 2019

Abstract: Mitogen-activated protein kinases (MAPK): Erk1 and Erk2 are key players in negative-feedback regulation of fibroblast growth factor (FGF) signaling. Upon activation, Erk1 and Erk2 directly phosphorylate FGF receptor 1 (FGFR1) at a specific serine residue in the C-terminal part of the receptor, substantially reducing the tyrosine phosphorylation in the receptor kinase domain and its signaling. Similarly, active Erks can also phosphorylate multiple threonine residues in the docking protein FGF receptor substrate 2 (FRS2), a major mediator of FGFR signaling. Here, we demonstrate that in NIH3T3 mouse fibroblasts and human osteosarcoma U2OS cells stably expressing FGFR1, in addition to Erk1 and Erk2, p38 kinase is able to phosphorylate FRS2. Simultaneous inhibition of Erk1/2 and p38 kinase led to a significant change in the phosphorylation pattern of FRS2 that in turn resulted in prolonged tyrosine phosphorylation of FGFR1 and FRS2 and in sustained signaling, as compared to the selective inhibition of Erks. Furthermore, excessive activation of p38 with anisomycin partially compensated the lack of Erks activity. These experiments reveal a novel crosstalk between p38 and Erk1/2 in downregulation of FGF-induced signaling.

Keywords: FGF-induced signaling; FRS2; phosphorylation; downregulation; p38; MAPK

---

1. Introduction

Fibroblast growth factor (FGF) receptor (FGFR) -dependent signaling plays a crucial role during embryonic development, as well as in adult life. It stimulates growth, differentiation, survival, injury repair, regeneration, and metabolism. Consequently, excessive activation of FGFRs may result in severe abnormalities, such as cancer development and progression, and skeletal disorders [1].

Downregulation of FGFR signaling consists of a number of mechanisms that enable precise control of cellular outcome. They include regulation of FGFR synthesis, receptor internalization, followed by degradation and modulation of FGFR tyrosine phosphorylation via the activity of several negative regulators and feedbacks [1–6].

Previously we showed that phosphorylation of Ser777 of FGFR1 by activated MAPK Erk1 and Erk2 significantly reduces the tyrosine phosphorylation in the kinase domain of the receptor. We suggested that the phosphorylated Ser777 could act as a binding site for tyrosine phosphatases responsible for receptor inactivation [7]. FGFR1 and FGFR2 are also phosphorylated on Ser779 in response to FGF. Upon phosphorylation, this residue serves as a binding site for 14-3-3 family of phosphoserine/threonine-binding adaptor/scaffold proteins and tunes up Ras/MAPK signaling [8,9]. In addition, ribosomal S6 kinase 2 (RSK2) phosphorylates Ser789 of FGFR1. Inhibition of RSK2 activity leads to prolonged tyrosine phosphorylation of FGFR1 resulting from reduced FGFR1 ubiquitination and endocytosis [10].

Furthermore, FGFR downstream signaling molecules are also affected by inhibitory loops and negative regulators, including Sprouty proteins (SPRY1–SPRY4), MAPK phosphatase 3 (MKP3), SEF (similar expression to FGF), and protein tyrosine phosphatase receptor type G (PTPRG). MKP3 directly dephosphorylates MAPK turning off the Ras cascade [11]. Sprouty is phosphorylated in response to FGFR activation and competes for Grb2 (growth factor receptor-bound protein 2), binding to FRS2 and Shp2 (SH2 domain-containing protein tyrosine phosphatase-2), preventing Ras activation. Phosphorylated Sprouty also binds to Raf and blocks subsequent MAPK signaling [12,13]. SEF protein, in addition to its role in inhibiting MAPK signaling, is capable of interacting directly with FGFRs and blocking receptor phosphorylation [14,15]. The protein tyrosine phosphatase receptor type G, PTPRG, directly dephosphorylates the receptor itself, thereby turning off the signaling [16].

However, one of the most effective downregulation mechanism described so far is based on Erk1/2-FRS2 negative feedback loop [17]. FRS2, a lipid anchored adaptor protein, is associated constitutively through its PTB domain with the juxtamembrane domain of FGFR and is phosphorylated on multiple tyrosines upon receptor activation [18,19]. Phosphorylated FRS2 forms two specific binding sites for Shp2 and four binding sites for Grb2. Grb2 is constitutively bound via its SH3 domains to Sos and Gab1, and these proteins constitute a signaling complex activating the RAS/MAPK and PI3K/Akt pathways [1]. Activated extracellular signal-regulated protein kinases (Erk1/2), irrespective of the upstream receptor tyrosine kinase (RTK), phosphorylate FRS2 at eight threonines (PXTP motifs) [17]. Such phosphorylation reduces FRS2 tyrosine phosphorylation, decreases recruitment of Grb2, and attenuates downstream signaling response, providing a control mechanism to regulate FGFR activity [17].

Besides Erks, the MAPK family includes also p38 and c-Jun N-terminal kinases (JNKs). Erks are activated by mitogenic factors and are associated with growth, differentiation, and proliferation, whereas p38 and JNKs take part mainly in the cell response to stress conditions and inflammatory cytokines, and are activated only weakly by proliferative stimuli [20]. There are four isoforms of p38 identified so far in mammals, among which p38α and p38β are the best characterized and most widely expressed [21].

Interestingly, all MAPKs share specificity for a common consensus phosphorylation site [22]. We showed previously that two different MAPKs, Erk1/2 and p38, are able to phosphorylate FGFR1 at the same serine residue [7,23]. Phosphorylation of FGFR1 by p38 seems to be necessary for FGF1 to be translocated into the cytosol and nucleus upon internalization [23], while phosphorylation by Erk1/2 provides a negative-feedback mechanism that controls FGF signaling, and thereby protects the cell against excessive activation of FGFR [7].

Here, we found that in addition to Erks, p38 can also phosphorylate FRS2 and modulate its phosphorylation at tyrosine residues. Simultaneous inhibition of both types of MAPKs resulted in prolonged activation of FGFR1, FRS2, and downstream signaling, as compared to the elimination of Erks activity alone. Our results reveal a novel regulatory feedback mechanism, where p38 is able to partially substitute Erk1 and Erk2 in the regulation of FGFR activity.

2. Results

2.1. The p38 Activity Influences Electrophoretic Mobility Shift of FRS2

We analyzed by immunoblotting the activation of FGF1-induced signaling in NIH3T3 mouse fibroblast cells upon 15-min stimulation with FGF1 in the presence of different combinations of MAP kinases inhibitors (MEK1/2 inhibitors, p38 inhibitor) and p38 kinase activator (anisomycin) (Figure 1a). As expected, in the presence of specific MEK1/2 inhibitors (U0126 or SL327, lanes 3, 4, 5, 6) that prevent

Erk1/2 activation, we observed augmented tyrosine phosphorylation of FGFR1 (Tyr653/Tyr654) and lack of phosphorylation at Ser777. In addition, MEK1/2 inhibition led to increased tyrosine phosphorylation of FRS2. This is in accordance with findings by Lax and co-workers [17]. The band for FRS2, visualized by antibodies recognizing total and phoshorylated (Tyr196) forms, displayed faster migration (lanes 3 and 4) as compared to the cells untreated with inhibitor (lane 2). Interestingly, we also found that in the simultaneous presence of MEK1/2 inhibitor (U0126) and p38 inhibitor (SB203580) (lane 5), the FRS2 band migrated slightly faster than in the presence of U0126 or SL327 alone (lanes 3 and 4). Addition of anisomycin (p38 activator) in the presence of U0126 resulted in a small up-shift of the FRS2 band (lane 6). We confirmed these findings using another model cell line, human osteosarcoma U2OS cells that stably expresses FGFR1 (U2OS-R1) (Figure 1b).

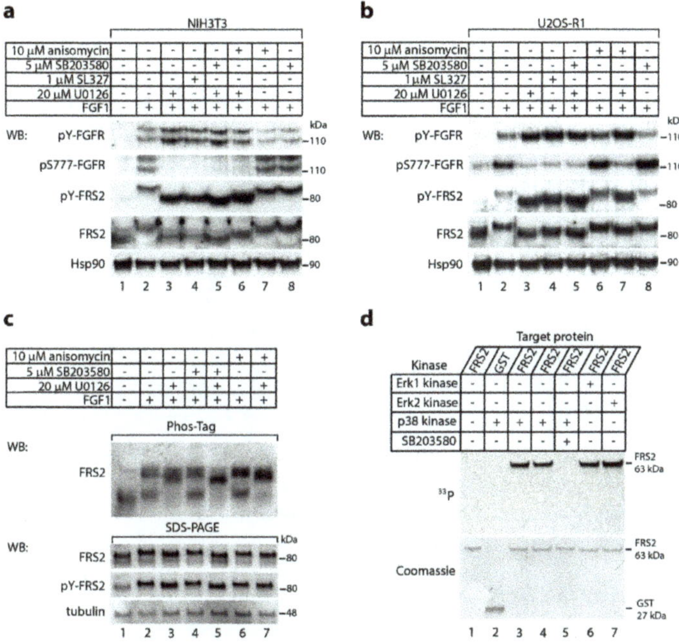

Figure 1. The effect of p38 activity on FRS2. (**a**–**c**) FGF1-induced electrophoretic mobility shift of FRS2. Serum-starved (**a**,**c**) NIH3T3 and (**b**) U2OS-R1 were pretreated for 20 min with or without MEK1/2 inhibitors (20 μM U0126, 1 μM SL327), p38 inhibitor (5 μM SB203580), and p38 activator (10 μM anisomycin), and then stimulated with the growth factor in the presence of heparin (10 U/mL) for 15 min. Cells were lysed, and the cellular material was analyzed by sodium dodecyl sulfate polyacrylamide gel electrophoresis (SDS-PAGE) (**a**,**b**) or Phos-Tag SDS-PAGE (**c**) and immunoblotting using the following antibodies: anti-phospho-FGFR (Tyr653/Tyr654) (pY-FGFR), anti-phospho-FGFR1 (Ser777) (pS777-FGFR1), anti-phospho-FRS2 (Tyr196) (pY-FRS), anti-FRS2, and anti-Hsp90 or anti-tubulin as a loading control. (**d**) In vitro phosphorylation of FRS2 by p38α kinase. Recombinant, active kinase p38α and partial recombinant fusion protein FRS2α with GST tag were incubated with (γ-^{33}P) ATP in reaction buffer at 30 °C for 30 min in the presence or absence of 5 μM SB203580. Erk1 and Erk2 kinases served as positive controls. The proteins were analyzed by SDS-PAGE, electroblotting, and autoradiography (upper panel), and then the membrane was stained with Coomassie (lower panel). Representative experiments are shown, for (**a**) and (**b**) n = 4, and for (**c**) and (**d**) n = 2.

As conventional SDS-PAGE is not optimal for separation of phosphorylated populations of a protein from a non-modified fraction, we studied FRS2 migration upon treatment with various kinases inhibitors using a Phos-Tag technology that is well suited for phosphorylation profiling of proteins [24,25]. Cell lysates were separated on Phos-Tag gels followed by immunoblotting with anti-FRS2 antibody (Figure 1c).

In the serum starved NIH3T3 cells, FRS2 was observed mainly as a single, fast migrating band on Phos-Tag gel, likely representing non-phosphorylated FRS2 (Figure 1c, lane 1). Stimulation of cells with FGF1 drastically changed the migration pattern of FRS2 in the Phos-Tag gel. FRS2 was detected in three slowly migrating bands that represent differentially phosphorylated FRS2 (Figure 1c, lane 2). Treatment of cells with MEK1/2 inhibitor caused clear changes in the phosphorylation status of FRS2 (Figure 1c, lane 3). Interestingly, concomitant inhibition of p38 and MEK1/2 altered gel migration of FRS2 in relation to the inhibition of either MEK1/2 or p38 alone (Figure 1c, lane 5 vs. lanes 3 and 4). In agreement with these findings, the over-activation of p38 by anisomycin, either in the presence or absence of MEK1/2 inhibitor, altered the migration of FRS2 on Phos-Tag gels (Figure 1c, lanes 6 and 7 vs lanes 2 and 3). These data suggest that p38 kinase affects the phosphorylation state of FRS2.

2.2. In Vitro Phosphorylation of FRS2 by p38α Kinase

Since the addition of either p38 inhibitor or p38 activator resulted in FRS2 mobility changes in SDS-PAGE and Phos-Tag PAGE, we hypothesized that p38, in addition to Erks, is also able to phosphorylate FRS2. To verify such possibility, we performed in vitro phosphorylation using recombinant FRS2 and recombinant active form of p38α kinase. Using autoradiography we observed a clear band corresponding to phosphorylated FRS2 when active p38α was present during the reaction (Figure 1d, lanes 3 and 4, two parallel samples). FRS2 phosphorylation was dependent on p38 activity, since in the presence of the specific p38 inhibitor, SB203580, there was no trace of radioactivity (Figure 1d, lane 5). Interestingly, we did not observe any differences in the efficiency of phosphorylation (assessed by intensity of bands) of FRS2 in the case of p38α and both Erks (Erk1/2) (Figure 1d). These data indicate that p38 can phosphorylate FRS2.

2.3. Synergistic Effect of MEK1/2 and p38 Inhibitors on Kinetics of FGF1-Induced Signaling

Next, we studied the impact of p38 activity on the kinetics of signaling cascades activated by FGF1 in NIH3T3 (Figure 2a). Experiments were performed in the presence of brefeldin A (2 µg/mL) to prevent the appearance of newly synthesized receptors [7]. In the presence of 5 µM of SB203580, p38 inhibitor (Figure 2a, lanes 9–13), or 10 µM anisomycin, p38 activator (Figure 2a, lanes 16–20), we did not observe significant differences in the intensity of FRS2 phosphorylation and FRS2 mobility, as compared to untreated cells. However, when the cells were simultaneously treated with 20 µM of MEK1/2 inhibitor (U0126) and either p38 inhibitor (SB203580) or p38 activator (anisomycin), the phosphorylation pattern of FRS2 varied, as well as the duration of FGFR1 tyrosine phosphorylation. We observed a synergistic effect of SB203580 and U0126. When activity of both kinds of MAP kinases (Erks and p38) were blocked, not only was the intensity of the band corresponding to tyrosine-phoshorylated FGFR1 stronger and lasted longer (Figure 2a, lanes 30–34), but also the electrophoretic mobility shift of FRS2 was prolonged, as compared to the inhibition of MEK1/2 with U0126 alone (Figure 2a, lanes 23–27). Fifteen minutes after FGF1 stimulation, FRS2 migrated faster in the presence of U0126, as well as in the presence of U0126 and SB203580, than in the absence of inhibitors. In the course of the experiment, the FRS2 band shifted upwards. When Erks and p38 were blocked at the same time (Figure 2a, lanes 30–34), we observed a more gradual change in the position of the FRS2 band than in the presence of U0126 alone (Figure 2a, lanes 23–27). In contrast to the effect of p38 inhibition, hyper-activation of p38 with 10 µM anisomycin used in the combination with 20 µM U0126 compensated, to some extent, the lack of Erk activity (Figure 2a, lanes 37–41). In the presence of anisomycin and U0126, the FRS2 band was less down-shifted after 15-min treatment with FGF1 and moved to the direction of higher molecular masses much faster over time than in the case of cells treated only with U0126. We confirmed all these findings in U2OS-R1 cells (Figure 2b). These data demonstrate that Erks and p38 regulate phosphorylation status of FRS2, which in turn modulates kinetics of FGFR1 signaling (Figure 2b).

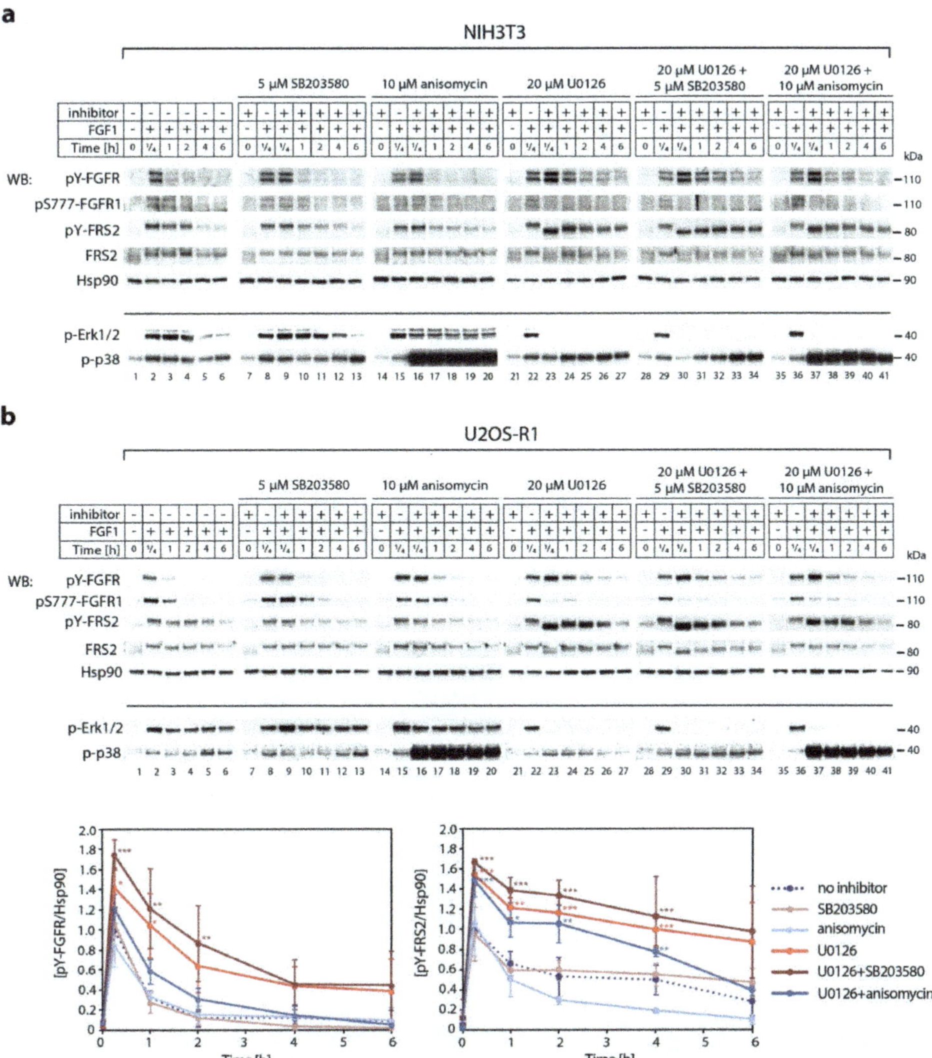

Figure 2. The crosstalk between p38 and Erk1/2 in downregulation of FGF1-induced signaling. Serum-starved (**a**) NIH3T3 and (**b**) U2OS-R1 cells were pretreated for 30 min with or without 20 μM U0126, 5 μM SB203580 and 10 μM anisomycin, and then stimulated with the growth factor in the presence of heparin (10 U/mL) and brefeldin A (2 μg/mL) for different time points. Cells were lysed, and the cellular material was analyzed by SDS-PAGE and immunoblotting using the following antibodies: anti-phospho-FGFR (Tyr653/Tyr654) (pY-FGFR), anti-phospho-FGFR1 (Ser777) (pS777-FGFR1), anti-phospho-FRS2 (Tyr196) (pY-FRS), anti-FRS2, and anti-Hsp90 as a loading control. Anti-phospho-Erk1/2 (p-Erk1/2) and anti-phospho-p38 MAPK (Thr180/Tyr182) (p-p38) were used to control the effect of U0126 (MEK inhibitor) and anisomycin (p38 activator). Representative experiments are shown, n = 3. The graphs present quantification of bands from panel b corresponding to phospho-FGFR1 (Tyr653/Tyr654) and phospho-FRS2 (Tyr196) normalized to loading control (Hsp90) and expressed as a fraction of the maximal response in the absence of inhibitor. Data are means ± SD of three independent experiments; * $p < 0.05$, ** $p < 0.01$, *** $p < 0.001$.

2.4. The Effect of p38 Kinase on Erk1/2 Activity

Finding a synergistic effect of the inhibition of both types of MAP kinases, Erks and p38, we asked if they were able to compensate for each other. Indeed, we observed that FGF1-dependent Erk1/2 phosphorylation in the presence of p38 inhibitor was slightly stronger (Figure 2a, lanes 8 and 9), whereas in the presence of p38 activator (anisomycin), was notably weaker (Figure 2a, lanes 15 and 16) as compared to the untreated cells (Figure 1a, lane 2). Analysis of signaling kinetics revealed that the inhibition of Erks increased the phosphorylation of p38 (Figure 2a, lanes 23–27 and lanes 30–34).

Next, we verified the impact of p38 inhibition on the activity of Erks without FGF1 stimulation. We found that in the presence of increasing concentrations of the specific p38 kinase inhibitor, SB203580, the basal level of phosphorylated Erk1/2 in serum-starved NIH3T3 was augmented (Figure 3a), showing that reduced activity of one MAP kinase (p38) extorted increased activity of the other (Erk1/2). We hypothesize that this crosstalk functions as a backup system in cells. If one of the two kinases is turned off, the other one can take over.

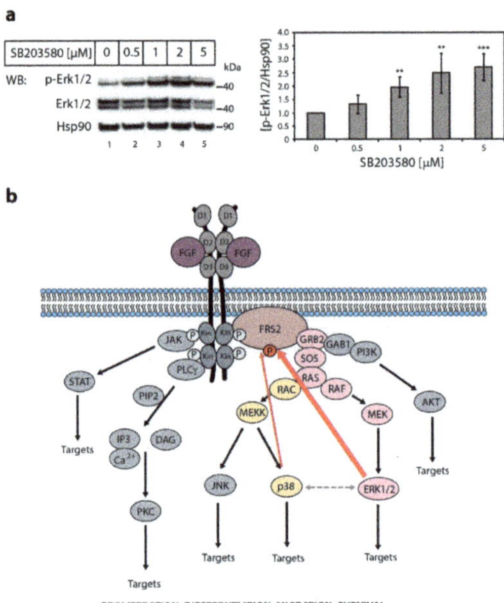

Figure 3. The effect of p38 kinase activity on Erk1/2 activity and FGF1-induced signaling. (**a**) The effect of p38 specific inhibitor on Erk1/2 activity. Serum-starved NIH3T3 cells were treated with increasing concentration of the specific p38 kinase inhibitor SB203580 for 30 min. Then, the cells were lysed and the cellular material was analyzed by SDS-PAGE and immunoblotting using the following antibodies: anti-phospho-Erk1/2 (p-Erk1/2), anti-Erk1/2, and anti-Hsp90 antibodies as a loading control. A representative experiment is shown, n = 4. The graph presents quantification of bands corresponding to phospho-Erk1/2 (p-Erk1/2) normalized to loading control (Hsp90) and expressed as a fold of change in comparison with untreated control. Data are means ± SD of four independent experiments; ** $p < 0.01$, *** $p < 0.001$. (**b**) Schematic representation of synergistic effect of p38 and Erk1/2 in the downregulation of FGF1-induced signaling through FRS2. FGF1-induced tyrosine phosphorylation of FGFR1 leads to the activation of FRS2 followed by GRB2/SOS-mediated activation of RAS and MAP kinases (Erk1/2 and p38). Activated Erks are supported by p38 in phosphorylation of FRS2 (red arrows), constituting a negative feedback loop that results in reduced tyrosine phosphorylation of FRS2 and consequent attenuation of FGFR signaling. Grey dashed line represents functional cross-talk between Erks and p38.

3. Discussion

Several mechanisms to attenuate FGFR activity have been reported, however, the overall picture of how the FGFR signaling is controlled has not been yet fully described [6]. FRS2 definitely plays an important role in the regulation of FGFR activity, constituting a key component in the Ras/MAPK pathway and being at the same time under control of its effectors (Erks) in a negative feedback loop [17].

Our results show that p38 kinase phosphorylates FRS2 in vitro, and together with Erk1 and Erk2 takes part in modulation of the duration and the intensity of FGFR signaling. Experiments with anisomycin, a p38 activator, revealed that in specific conditions (e.g., stress), p38 can compensate for Erks' action. Despite the similar activity in vitro, the effect of p38 on FRS2 phosphorylation in cells is much weaker than the effect of Erk1 and Erk2. It is probably caused by the fact that growth factors, including FGF1, evoke stronger activation of pro-proliferative MAPKs, i.e., Erks, than stress-response MAPKs, such as p38 kinase [20]. We also found that inhibition of p38 slightly increases the basal activity of Erks. It seems that p38 provides a kind of fine-tuning of the Erks-driven feedback system, taking part in the complicated and multilevel network of downregulation mechanisms of FGFR signaling (Figure 3B). However, to elucidate how one MAP kinase can substitute for another in the absence of FGF ligand, further studies are required. We speculate that inhibition of p38 by SB203580, reducing the degree of FRS2 and FGFR basal serine/threonine phosphorylation, evokes minor upregulation of their tyrosine phosphorylation. A slight increase in FGFR and FRS2 activity results in amplification of Erks phosphorylation.

All MAPK family members share a common phosphorylation site motif [22]. Their substrate specificity is achieved through the docking interactions occurring between the regions outside the phosphorylation site in the substrate and fragments distal from the active site in the kinase [22]. The best characterized docking site involved in interaction with MAPK is the D-site, which consists of a basic region, followed by an LXL motif and a hydrophobic region [22,26]. Interestingly, while Erks require three parts of the D-site, p38 can phosphorylate a substrate without the LXL motif [27]. Another conserved motif, known as the DEF site or the FXFP motif, was identified in several substrates of Erks, including transcription factors, phosphatases, scaffolding, and focal adhesion proteins [28]. It was also shown that p38α has a DEF site motif similar to Erk2, being highly selective for aromatic residues at P1 and P3 position with high preference to Trp [22]. Therefore, it is not surprising that Erk1/2 substrate, FRS2, can also be phosphorylated by p38 kinase.

MAPK signaling seems to be a dynamic map of non-linear interactions rather than static connections. Different MAPK cascades operate in parallel, but at the same time they constitute a network and modulate each other's activity [29]. Here, we have described a cross-talk between two classes of MAPKs in the regulation of FGFR-dependent signaling, but there are other cases of cooperation between different MAPKs. High specificity is observed among upstream kinases, but at the level of MAPK substrate a cross-talk is not unusual [29]. A good example is Elk-1, a member of Ets family of transcription factors controlling the expression of various signaling molecules, which is a target of all three MAPK (Erk1/2, p38, and JNK). Activation of MAPK pathways results in phosphorylation and positive regulation of Elk1, leading to the transcription of early response genes [30]. Another example is MSK1 protein, mitogen-, and stress-activated kinase 1, which is a substrate of both Erks and p38-MAPK during oxidative stress in skeletal myoblasts. Phosphorylation of MSK1 was suggested to influence NF-κB signal specificity in stress response [31]. Another group showed that Erk1/2 and p38 pathways cooperate to promote p21CIP1 expression in order to ensure a sustained G1 cell cycle arrest [32]. The activation of both Erks and p38 was also demonstrated to be essential in regulating delayed STAT3 phosphorylation, as well as in ANF (atrial natriuretic factor) expression profile in response to IL-1β treatment, suggesting their simultaneous role in the development of IL-1β-induced hypertrophy in cardiac myocytes [33]. Furthermore, it was suggested that the function of p38α kinase can be modulated by the cross talk between JNK and p38 kinases [34].

Therapeutic targeting of components of the MAPK cascades is an urgent medical need. Inhibitors of Erks and p38 signaling pathways are in many clinical trials dedicated to treating different types of cancer, inflammation, pain, rheumatoid arthritis, asthma, and neurodegenerative diseases, including Alzheimer's [35,36]. However, it is under question whether targeting of a single signaling molecule can provide an effective therapy. A growing number of drugs are used in combination. Especially in the case of targeting the Erk pathway, vertical inhibition to block two subsequent steps in the cascade is now standard care in a few types of cancer [36,37]. Moreover, to overcome different compensatory mechanisms generated by signaling feedback loops and cross-talks resulting in resistance, efforts are focused to target multiple pathways simultaneously (horizontal inhibition) by combining selective agents [36,38]. Thus, understanding of interplay between different cascades and signaling components is of great importance. Cross-talk between Erk1/2 and p38 and their potential compensation effect should be taken into account during biochemical studies, and might have implications in the design of effective targeted therapies.

4. Materials and Methods

The following primary antibodies were used: rabbit anti-MAPK (Erk1/2, p44/p42; #9102), mouse anti-phospho-MAPK (Erk1/2, p44/p42) (Thr202/Tyr204; #9106), mouse anti-phospho-FGFR (Tyr653/Tyr654; #3476), and rabbit anti-phospho-FRS2α (Tyr196; #3864) from Cell Signaling Technology (Danvers, MA, USA), rabbit anti-FRS2α (H-91; sc-8318), from Santa Cruz Biotechnology (Dallas, TX, USA), mouse anti-phospho-p38 MAPK antibody (Thr180/Tyr182; 612280) and mouse anti-Hsp90 (610418) from BD Transduction Laboratory (San Jose, CA, USA), mouse anti-tubulin (T6557) from Sigma-Aldrich (St Louis, MO, USA). Specific anti-phospho-FGFR1 (Ser777) (pS777-FGFR1) antibody was made by GenScript (Piscataway, NJ, USA) using the following phospho-specific peptide CSMPLDQYpSPSFPDTR. The antibody was purified using the phosphopeptide and by cross-adsorption to the corresponding non-phosphopeptide. HRP-conjugated secondary antibodies were from Jackson Immuno Research Laboratories (West Grove, PA, USA). Heparin-Sepharose CL-6B affinity resin was from GE Healthcare (Piscataway, NJ, USA). Brefeldin A, SB203580, anisomycin, and SL327 were from Calbiochem (San Diego, CA, USA). Heparin, sodium orthovanadate, and U0126 were from Sigma-Aldrich. Phos-tag gels were from Wako Chemicals (Osaka, Japan). All other chemicals were from Sigma-Aldrich.

4.1. Cell Lines and Bacterial Strain

NIH3T3 cells were grown in DMEM from Thermo Fisher Scientific (Waltham, MA, USA) or (Biowest, Nuaille, France) supplemented with 10% bovine serum (Thermo Fisher Scientific) and 100 U/mL penicillin and 100 μg/mL streptomycin (Thermo Fisher Scientific). The U2OS cells stably expressing FGFR1 (U2OS-R1) have been described previously [39]. The cells were propagated in DMEM supplemented with 10% fetal bovine serum (Gibco), 100 U/mL penicillin, and 100 μg/mL streptomycin and 0.2 mg/mL geneticin (Invitrogen). For expression of FGF1 wild-type *Escherichia coli* strain Bl21(DE3)pLysS from New England Biolabs (Ipswich, MA, USA) was used.

4.2. Recombinant Proteins

GST-tagged FRS2 recombinant protein (Q01) was purchased from Anova (Taipei, Taiwan), active human recombinant p38α MAPK was from R&D Systems (Minneapolis, MN, USA) and active human recombinant MAP kinases, Erk1 (p44) and Erk2 (p42), from Calbiochem. Recombinant FGF1 was produced in *E. coli*, as described previously [40].

4.3. Analysis of Signaling Cascades

Serum-starved cells were treated with 20 ng/mL FGF1 in the presence of 10 U/mL heparin and in the presence or absence of indicated inhibitors for 15 min. Signaling kinetics were carried out in the presence or absence of 2 μg/mL brefeldin A. The cells were lysed with SDS sample buffer, scraped, and sonicated. Total cell lysates were separated by SDS-PAGE or by Phos-tag SDS-PAGE, transferred onto Immobilon-P membrane and subjected to immunoblotting. The sectioned membrane was stripped with Restore Western Blot Stripping Buffer (Thermo Fisher Scientific) and re-probed maximally four times with different antibodies. ImageLab software (version 6.0.1) from Bio-Rad (Hercules, CA, USA) was used to quantify the intensity of bands. The intensity of bands corresponding to phospho-proteins (pY-FGFR, pY-FRS2, p-Erk1/2) was normalized to the intensity of bands corresponding to Hsp90 and then expressed as a fraction of control.

4.4. In Vitro Phosphorylation of Recombinant FRS2

In vitro phosphorylation experiments were performed with recombinant proteins. One microgram of a fusion protein was incubated with kinases and 40 μCi/mL [γ-33P] ATP in reaction buffer (25 mM HEPES, pH 7.5, 20 mM MgCl2, 1 mM Na2MO3, 20 mM sodium β-glycerophosphate, 1 mM DTT, 5 mM EGTA) at 30 °C for 30 min. The proteins were separated by SDS-PAGE, electroblotted, and analyzed by autoradiography. Equal loading was ensured by membrane staining with Coomassie Blue.

4.5. Statistical Analysis

For statistical analysis, one-way analysis of variance (ANOVA) with Tukey's posttest was applied using SigmaPlot 12 software from (Systat Software, San Jose, CA, USA); $p < 0.05$ was considered statistically significant.

Author Contributions: M.Z., E.M.H., L.O. and A.W. designed the research. M.Z. and L.O. performed the research. M.Z., L.O., E.M.H., J.O. and A.W. analyzed the data. M.Z. wrote the paper. M.Z., L.O., E.M.H., A.W. and J.O. reviewed and edited the paper.

Funding: This research was funded by grant Sonata Bis 2015/18/E/NZ3/00501 from the National Science Centre. L.O.'s work was supported by the First TEAM programme (POIR.04.04.00-00-43B2/17-00) of the Foundation for Polish Science, co-financed by the European Union under the European Regional Development Fund.

Acknowledgments: The skilful technical assistance of Anne Engen and Agnieszka Kubiak is gratefully acknowledged. We thank Daniel Krowarsch for help in statistical analysis.

Conflicts of Interest: The authors declare no conflict of interest.

Abbreviations

FGF	Fibroblast Growth Factor
FGFR	Fibroblast Growth Factor Receptor
FRS2	FGF receptor substrate 2
MAPK	mitogen-activated protein kinase

References

1. Ornitz, D.M.; Itoh, N. The Fibroblast Growth Factor signaling pathway. *Wiley Interdiscip Rev. Dev. Biol.* **2015**, *4*, 215–266. [CrossRef] [PubMed]

2. Wong, A.; Lamothe, B.; Lee, A.; Schlessinger, J.; Lax, I. FRS2 alpha attenuates FGF receptor signaling by Grb2-mediated recruitment of the ubiquitin ligase Cbl. *Proc. Natl. Acad. Sci. USA* **2002**, *99*, 6684–6689. [CrossRef]

3. Cho, J.Y.; Guo, C.; Torello, M.; Lunstrum, G.P.; Iwata, T.; Deng, C.; Horton, W.A. Defective lysosomal targeting of activated fibroblast growth factor receptor 3 in achondroplasia. *Proc. Natl. Acad. Sci. USA* **2004**, *101*, 609–614. [CrossRef] [PubMed]

4. Haugsten, E.M.; Sorensen, V.; Brech, A.; Olsnes, S.; Wesche, J. Different intracellular trafficking of FGF1 endocytosed by the four homologous FGF receptors. *J. Cell Sci.* **2005**, *118*, 3869–3881. [CrossRef] [PubMed]

5. Dufour, C.; Guenou, H.; Kaabeche, K.; Bouvard, D.; Sanjay, A.; Marie, P.J. FGFR2-Cbl interaction in lipid rafts triggers attenuation of PI3K/Akt signaling and osteoblast survival. *Bone* **2008**, *42*, 1032–1039. [CrossRef] [PubMed]

6. Turner, N.; Grose, R. Fibroblast growth factor signalling: From development to cancer. *Nat. Rev. Cancer* **2010**, *10*, 116–129. [CrossRef]

7. Zakrzewska, M.; Haugsten, E.M.; Nadratowska-Wesolowska, B.; Oppelt, A.; Hausott, B.; Jin, Y.; Otlewski, J.; Wesche, J.; Wiedlocha, A. ERK-mediated phosphorylation of fibroblast growth factor receptor 1 on Ser777 inhibits signaling. *Sci. Signal.* **2013**, *6*, ra11. [CrossRef]

8. Lonic, A.; Barry, E.F.; Quach, C.; Kobe, B.; Saunders, N.; Guthridge, M.A. Fibroblast growth factor receptor 2 phosphorylation on serine 779 couples to 14-3-3 and regulates cell survival and proliferation. *Mol. Cell. Biol.* **2008**, *28*, 3372–3385. [CrossRef]

9. Lonic, A.; Powell, J.A.; Kong, Y.; Thomas, D.; Holien, J.K.; Truong, N.; Parker, M.W.; Guthridge, M.A. Phosphorylation of serine 779 in fibroblast growth factor receptor 1 and 2 by protein kinase C(epsilon) regulates Ras/mitogen-activated protein kinase signaling and neuronal differentiation. *J. Biol. Chem.* **2013**, *288*, 14874–14885. [CrossRef]

10. Nadratowska-Wesolowska, B.; Haugsten, E.M.; Zakrzewska, M.; Jakimowicz, P.; Zhen, Y.; Pajdzik, D.; Wesche, J.; Wiedlocha, A. RSK2 regulates endocytosis of FGF receptor 1 by phosphorylation on serine 789. *Oncogene* **2014**, *33*, 4823–4836. [CrossRef]

11. Zhao, Y.; Zhang, Z.Y. The mechanism of dephosphorylation of extracellular signal-regulated kinase 2 by mitogen-activated protein kinase phosphatase 3. *J. Biol. Chem.* **2001**, *276*, 32382–32391. [CrossRef]

12. Cabrita, M.A.; Christofori, G. Sprouty proteins, masterminds of receptor tyrosine kinase signaling. *Angiogenesis* **2008**, *11*, 53–62. [CrossRef]

13. Martinez, N.; Garcia-Dominguez, C.A.; Domingo, B.; Oliva, J.L.; Zarich, N.; Sanchez, A.; Gutierrez-Eisman, S.; Llopis, J.; Rojas, J.M. Sprouty2 binds Grb2 at two different proline-rich regions, and the mechanism of ERK inhibition is independent of this interaction. *Cell Signal.* **2007**, *19*, 2277–2285. [CrossRef]

14. Kovalenko, D.; Yang, X.; Nadeau, R.J.; Harkins, L.K.; Friesel, R. Sef inhibits fibroblast growth factor signaling by inhibiting FGFR1 tyrosine phosphorylation and subsequent ERK activation. *J. Biol. Chem.* **2003**, *278*, 14087–14091. [CrossRef]

15. Tsang, M.; Dawid, I.B. Promotion and attenuation of FGF signaling through the Ras-MAPK pathway. *Sci. STKE* **2004**, *2004*, pe17. [CrossRef]

16. Kostas, M.; Haugsten, E.M.; Zhen, Y.; Sorensen, V.; Szybowska, P.; Fiorito, E.; Lorenz, S.; Jones, N.; de Souza, G.A.; Wiedlocha, A.; et al. Protein Tyrosine Phosphatase Receptor Type G (PTPRG) Controls Fibroblast Growth Factor Receptor (FGFR) 1 Activity and Influences Sensitivity to FGFR Kinase Inhibitors. *Mol. Cell. Proteom.* **2018**, *17*, 850–870. [CrossRef]

17. Lax, I.; Wong, A.; Lamothe, B.; Lee, A.; Frost, A.; Hawes, J.; Schlessinger, J. The docking protein FRS2alpha controls a MAP kinase-mediated negative feedback mechanism for signaling by FGF receptors. *Mol. Cell* **2002**, *10*, 709–719. [CrossRef]

18. Ong, S.H.; Guy, G.R.; Hadari, Y.R.; Laks, S.; Gotoh, N.; Schlessinger, J.; Lax, I. FRS2 proteins recruit intracellular signaling pathways by binding to diverse targets on fibroblast growth factor and nerve growth factor receptors. *Mol. Cell. Biol.* **2000**, *20*, 979–989. [CrossRef]

19. Kouhara, H.; Hadari, Y.R.; Spivak-Kroizman, T.; Schilling, J.; Bar-Sagi, D.; Lax, I.; Schlessinger, J. A lipid-anchored Grb2-binding protein that links FGF-receptor activation to the Ras/MAPK signaling pathway. *Cell* **1997**, *89*, 693–702. [CrossRef]

20. Johnson, G.L.; Lapadat, R. Mitogen-activated protein kinase pathways mediated by ERK, JNK, and p38 protein kinases. *Science* **2002**, *298*, 1911–1912. [CrossRef]

21. Stramucci, L.; Pranteda, A.; Bossi, G. Insights of Crosstalk between p53 Protein and the MKK3/MKK6/p38 MAPK Signaling Pathway in Cancer. *Cancers (Basel)* **2018**, *10*, 131. [CrossRef]

22. Sheridan, D.L.; Kong, Y.; Parker, S.A.; Dalby, K.N.; Turk, B.E. Substrate discrimination among mitogen-activated protein kinases through distinct docking sequence motifs. *J. Biol. Chem.* **2008**, *283*, 19511–19520. [CrossRef]

23. Sorensen, V.; Zhen, Y.; Zakrzewska, M.; Haugsten, E.M.; Walchli, S.; Nilsen, T.; Olsnes, S.; Wiedlocha, A. Phosphorylation of fibroblast growth factor (FGF) receptor 1 at Ser777 by p38 mitogen-activated protein kinase regulates translocation of exogenous FGF1 to the cytosol and nucleus. *Mol. Cell. Biol.* **2008**, *28*, 4129–4141. [CrossRef]

24. Schmidt, O.; Harbauer, A.B.; Rao, S.; Eyrich, B.; Zahedi, R.P.; Stojanovski, D.; Schonfisch, B.; Guiard, B.; Sickmann, A.; Pfanner, N.; et al. Regulation of mitochondrial protein import by cytosolic kinases. *Cell* **2011**, *144*, 227–239. [CrossRef]

25. Harbauer, A.B.; Opalinska, M.; Gerbeth, C.; Herman, J.S.; Rao, S.; Schonfisch, B.; Guiard, B.; Schmidt, O.; Pfanner, N.; Meisinger, C. Mitochondria. Cell cycle-dependent regulation of mitochondrial preprotein translocase. *Science* **2014**, *346*, 1109–1113. [CrossRef]

26. Tanoue, T.; Nishida, E. Molecular recognitions in the MAP kinase cascades. *Cell Signal.* **2003**, *15*, 455–462. [CrossRef]

27. Barsyte-Lovejoy, D.; Galanis, A.; Sharrocks, A.D. Specificity determinants in MAPK signaling to transcription factors. *J. Biol. Chem.* **2002**, *277*, 9896–9903. [CrossRef]

28. Akella, R.; Moon, T.M.; Goldsmith, E.J. Unique MAP Kinase binding sites. *Biochim. Biophys. Acta* **2008**, *1784*, 48–55. [CrossRef]

29. Cowan, K.J.; Storey, K.B. Mitogen-activated protein kinases: New signaling pathways functioning in cellular responses to environmental stress. *J. Exp. Biol.* **2003**, *206*, 1107–1115. [CrossRef]

30. Yordy, J.S.; Muise-Helmericks, R.C. Signal transduction and the Ets family of transcription factors. *Oncogene* **2000**, *19*, 6503–6513. [CrossRef]

31. Kefaloyianni, E.; Gaitanaki, C.; Beis, I. ERK1/2 and p38-MAPK signalling pathways, through MSK1, are involved in NF-kappaB transactivation during oxidative stress in skeletal myoblasts. *Cell Signal.* **2006**, *18*, 2238–2251. [CrossRef]

32. Todd, D.E.; Densham, R.M.; Molton, S.A.; Balmanno, K.; Newson, C.; Weston, C.R.; Garner, A.P.; Scott, L.; Cook, S.J. ERK1/2 and p38 cooperate to induce a p21CIP1-dependent G1 cell cycle arrest. *Oncogene* **2004**, *23*, 3284–3295. [CrossRef]

33. Ng, D.C.; Long, C.S.; Bogoyevitch, M.A. A role for the extracellular signal-regulated kinase and p38 mitogen-activated protein kinases in interleukin-1 beta-stimulated delayed signal tranducer and activator of transcription 3 activation, atrial natriuretic factor expression, and cardiac myocyte morphology. *J. Biol. Chem.* **2001**, *276*, 29490–29498.

34. Sundaramurthy, P.; Gakkhar, S.; Sowdhamini, R. Computational prediction and analysis of impact of the cross-talks between JNK and P38 kinase cascades. *Bioinformation* **2009**, *3*, 250–254. [CrossRef]

35. Burotto, M.; Chiou, V.L.; Lee, J.M.; Kohn, E.C. The MAPK pathway across different malignancies: A new perspective. *Cancer* **2014**, *120*, 3446–3456. [CrossRef]

36. Tolcher, A.W.; Peng, W.; Calvo, E. Rational Approaches for Combination Therapy Strategies Targeting the MAP Kinase Pathway in Solid Tumors. *Mol. Cancer Ther.* **2018**, *17*, 3–16. [CrossRef]

37. Grimaldi, A.M.; Simeone, E.; Festino, L.; Vanella, V.; Palla, M.; Ascierto, P.A. Novel mechanisms and therapeutic approaches in melanoma: Targeting the MAPK pathway. *Discov. Med.* **2015**, *19*, 455–461.

38. Neuzillet, C.; Tijeras-Raballand, A.; de Mestier, L.; Cros, J.; Faivre, S.; Raymond, E. MEK in cancer and cancer therapy. *Pharmacol. Ther.* **2014**, *141*, 160–171. [CrossRef]

39. Haugsten, E.M.; Malecki, J.; Bjorklund, S.M.; Olsnes, S.; Wesche, J. Ubiquitination of fibroblast growth factor receptor 1 is required for its intracellular sorting but not for its endocytosis. *Mol. Biol. Cell* **2008**, *19*, 3390–3403. [CrossRef]

40. Zakrzewska, M.; Krowarsch, D.; Wiedlocha, A.; Otlewski, J. Design of fully active FGF-1 variants with increased stability. *Protein Eng. Des. Sel.* **2004**, *17*, 603–611. [CrossRef]

International Journal of
Molecular Sciences

Article

Scutellariae Radix and Coptidis Rhizoma Improve Glucose and Lipid Metabolism in T2DM Rats via Regulation of the Metabolic Profiling and MAPK/PI3K/Akt Signaling Pathway

Xiang Cui, Da-Wei Qian, Shu Jiang *, Er-Xin Shang, Zhen-Hua Zhu and Jin-Ao Duan *

Jiangsu Collaborative Innovation Center of Chinese Medicinal Resources Industrialization, Nanjing University of Chinese Medicine, 138 Xianlin Road, Nanjing 210023, China; 15951975518@163.com (X.C.); qiandw@njucm.edu.cn (D.-W.Q.); shex@sina.com (E.-X.S.); 18913133908@163.com (Z.-H.Z.)
* Correspondence: jiangshu2020@126.com (S.J.); dja@njucm.edu.cn (J.-A.D.);
  Tel./Fax: +86-25-8581-1516 (S.J.); +86-25-8581-1291 (J.-A.D.)

Received: 11 October 2018; Accepted: 12 November 2018; Published: 18 November 2018

Abstract: *Aim* Scutellariae Radix (SR) and Coptidis Rhizoma (CR) have often been combined to cure type 2 diabetes mellitus (T2DM) in the clinical practice for over thousands of years, but their compatibility mechanism is not clear. Mitogen-activated protein kinase (MAPK) signaling pathway has been suggested to play a critical role during the process of inflammation, insulin resistance, and T2DM. This study was designed to investigate their compatibility effects on T2DM rats and explore the underlying mechanisms by analyzing the metabolic profiling and MAPK/PI3K/Akt signaling pathway. *Methods* The compatibility effects of SR and CR were evaluated with T2DM rats induced by a high-fat diet (HFD) along with a low dose of streptozocin (STZ). Ultra performance liquid chromatography-quadrupole time-of-flight mass spectrometry (UPLC-Q-TOF/MS) was performed to discover potential biomarkers. The levels of pro-inflammatory cytokines; biochemical indexes in serum, and the activities of key enzymes related to glycometabolism in liver were assessed by ELISA kits. qPCR was applied to examine mRNA levels of key targets in MAPK and insulin signaling pathways. Protein expressions of p65; p-p65; phosphatidylinositol-4,5-bisphosphate 3-kinase (PI3K); phosphorylated-PI3K (p-PI3K); protein kinase B (Akt); phosphorylated Akt (p-Akt) and glucose transporter 2 (Glut2) in liver were investigated by Western blot analysis. *Results* Remarkably, hyperglycaemia, dyslipidemia, inflammation, and insulin resistance in T2DM were ameliorated after oral administration of SR and CR, particularly their combined extracts. The effects of SR, CR, low dose of combined extracts (LSC) and high dose of combined extracts (HSC) on pro-inflammatory cytokine transcription in T2DM rats showed that the MAPK pathway might account for the phenomenon with down-regulation of MAPK (P38 mitogen-activated protein kinases (P38), extracellular regulated protein kinases (ERK), and c-Jun N-terminal kinase (JNK)) mRNA, and protein reduction in p-P65. While mRNA levels of key targets such as insulin receptor substrate 1 (IRS1), PI3K, Akt2, and Glut2 in the insulin signaling pathway were notably up-modulated, phosphorylations of PI3K, Akt, and expression of Glut2 were markedly enhanced. Moreover, the increased activities of phosphoenolpyruvate carboxykinase (PEPCK), fructose-1,6-bisphosphatase (FBPase), glucose 6-phosphatase (G6Pase), and glycogen phosphorylase (GP) were highly reduced and the decreased activities of glucokinase (GK), phosphofructokinase (PFK), pyruvate kinase (PK), and glycogen synthase (GS) in liver were notably increased after treatment. Further investigation indicated that the metabolic profiles of plasma and urine were clearly improved in T2DM rats. Fourteen potential biomarkers (nine in plasma and five in urine) were identified. After intervention, these biomarkers returned to normal level to some extent. *Conclusion* The results showed that SR, CR, and combined extract groups were normalized. The effects of combined extracts were more remarkable than single herb treatment. Additionally, this study also showed that the metabonomics method is a promising tool to unravel how traditional Chinese medicines work.

Keywords: SR; CR; Compatibility; T2DM; metabolic profiling; MAPK/PI3K/Akt signaling pathway

---

1. Introduction

Type 2 diabetes mellitus (T2DM), characterized by increased blood glucose level resulting from disturbances of insulin secretion, insulin action or both, is a complex disorder influenced by both lifestyle and genetic factors. T2DM and its serious complications such as nerve damage [1], kidney disease [2], and cardiovascular disease [3,4] have significantly increased society's medical burden over the past few decades and it is estimated that their global numbers in adults will rise to 592 million by 2035 [5]. Thus, its prevalence has become a major public health issue throughout the world, especially in developing countries.

Currently, drugs clinically used to treat T2DM include insulin sensitizers, insulin secreting drugs, insulin, etc. Although current diabetes treatments have exhibited some success in lowering blood glucose levels, their effects are not always sustained and their use may be associated with undesirable side effects such as hypoglycemia [6], gastrointestinal discomfort [7], etc. So it is urgent to discover new and effective drugs with fewer side effects to cure T2DM. Luckily, successful therapies for T2DM have been available from TCM practitioners. Thus, many researchers have focused on developing potent therapies and drugs from Chinese herbs with fewer side effects on T2DM patients.

SR, the dry root of *Scutellaria baicalensis* Georgi, with a number of biological activities such as anti-inflammation [8], anti-cancer [9], and anti-oxidation [10], has been used to treat various types of diseases. The major pharmacologically-active components of SR are flavonoids such as baicalin, wogonoside, baicalein, and wogonin. Accumulating researches have revealed that baicalin can significantly improve insulin resistance [11,12] and suppress gluconeogenesis [13]. CR, the dried rhizome of *Coptis chinensis* Franch, mainly containing alkaloids such as berberine, coptisine, and palmatine, has been used in China for thousands of years to treat diarrhea. Moreover, recent researches have showed that CR has anti-bacterial [14], anti-cancer [15] activities, etc. Especially, berberine has been shown to possess remarkable effects on lowering blood glucose and promoting the secretion of insulin [16,17]. In China, SR and CR are often used together at a ratio of 1:1 to obtain a synergistic effect for treating diabetes and its complications, yet the underlying mechanism on the therapy of T2DM remains unclear.

Due to insulin resistance, the liver excessively releases glucose into the blood as a result of increased glycogenolysis and gluconeogenesis. Besides, decreased glycolysis and glycogenesis lead to reduced consumption of blood glucose, contributing to hyperglycemia eventually [18]. Inflammatory cytokines such as TNF-α and IL-6 promote the development of insulin resistance, and a recent study has reported that anti-inflammation agents showed efficacy in reducing blood glucose level [19]. Moreover, T2DM is often present for years before becoming clinically apparent. Current clinical predictors such as fasting blood glucose are helpful in gauging diabetes risk, but they only reflect extant disease and provide little additional insight regarding pathophysiologic mechanisms. Earlier identification of individuals at risk is necessary and emerging technologies have made it possible through metabolomics. Among the varieties of analytical platforms used for metabonomic analysis, the use of ultra performance liquid chromatography-quadrupole time-of-flight mass spectrometry (UPLC-Q-TOF/MS) is steadily increasing due to its better reproducibility and detection limits, and increased chromatographic resolution, which can assess all metabolites in biological samples and provide insights into the holistic efficacy of TCMs. Besides, the extent of metabolic changes and types of metabolites could be applied as good markers during insulin resistance [20].

Thus, in this study, T2DM rats induced by HFD along with a low dose of STZ were used to assess the efficacy of individual and combined extracts. The levels of biochemical indexes, pro-inflammatory cytokine, key targets in MAPK, and insulin signaling pathways as well as the activities of key enzymes were determined. To elucidate the mechanism, the metabolic changes in plasma and urine from T2DM rats based on UPLC-Q-TOF/MS were investigated. Potential biomarkers and metabolic pathways were also identified. These data would provide sound scientific evidence for the clinical treatment of T2DM.

2. Results

2.1. Ameliorative Effect of SR, CR, and Combined Extracts on Hyperglycemia, Dyslipidemia, and Insulin Resistance in T2DM Rats

To evaluate the compatibility effects of SR and CR, a favorable T2DM rat model with a notable glycolipid metabolism disorder was established by HFD along with low dose of STZ. The significant increase of FBG, FINS, TG, TC, LDL-C, and FFA levels in T2DM rats was observed. By contrast, compared with T2DM rats, the levels of the above biochemical factors were decreased to a range from 51.1% to 85.4% after oral administration of individual and combined extracts. Furthermore, LSC or HSC extracts exerted better effects than the single drug, and the levels of these biochemical indexes declined remarkably after treatment. Metformin demonstrated a similar regulation as HSC in T2DM rats (Figure 1). These results indicated the better effects of combined extracts on the regulation of glucose and lipid homeostasis.

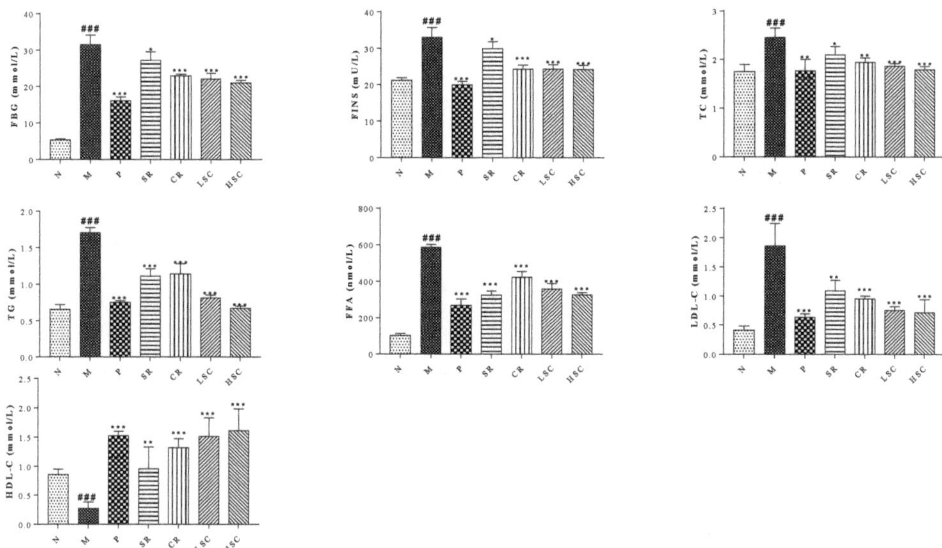

Figure 1. The levels of fasting blood glucose (FBG), fasting insulin (FINS), total cholesterol (TC), triglyceride (TG), free fatty acid (FFA), high-density lipoprotein cholesterol (HDL-C), low-density lipoprotein cholesterol (LDL-C) in serum of the normal group (N), model group (M), and groups gavaged with metformin (P), scutellaria Radix (SR), coptidis Rhizome (CR), low dose of combined extracts group (LSC) and high dose of combined extracts (HSC). The values were shown as mean ± SD. # $p < 0.05$, ## $p < 0.01$, ### $p < 0.001$ vs. normal group; * $p < 0.05$, ** $p < 0.01$, *** $p < 0.001$ vs. model group. Data were analyzed by One-way-ANOVA.

2.2. Pathological Assessment of Some Tissues Related to Insulin Resistance

The liver tissues of normal rats exhibited normal cellular structure with neat liver lobule, liver cords, liver sinusoid, and a clear three pipeline structure of the portal area. While notable changes with severe fatty degeneration were observed in the liver tissues of T2DM rats, this indicated that the model had been successfully established. Compared to model rats, SR and CR group rats showed moderate fatty degeneration, while LSC, HSC, and P group rats showed mild fatty degeneration by

HE staining observation. The islet cells of T2DM rats showed severe atrophy. After treatment, SR and CR group rats showed moderate atrophy, however, LSC, HSC, and P group rats showed mild atrophy. The mean adipocyte diameter of T2DM rats compared with normal rats was seriously increased. After treatment, the mean adipocyte diameter was decreased and HSC showed a normalized effect. The skeletal muscle of T2DM rats was notably infiltrated with inflammatory cells, but P, SR, CR, LSC, and HSC group rats were normalized (Figure 2). The histological examination results were consistent with biochemical data, which further confirmed the synergistic effect of SR and CR.

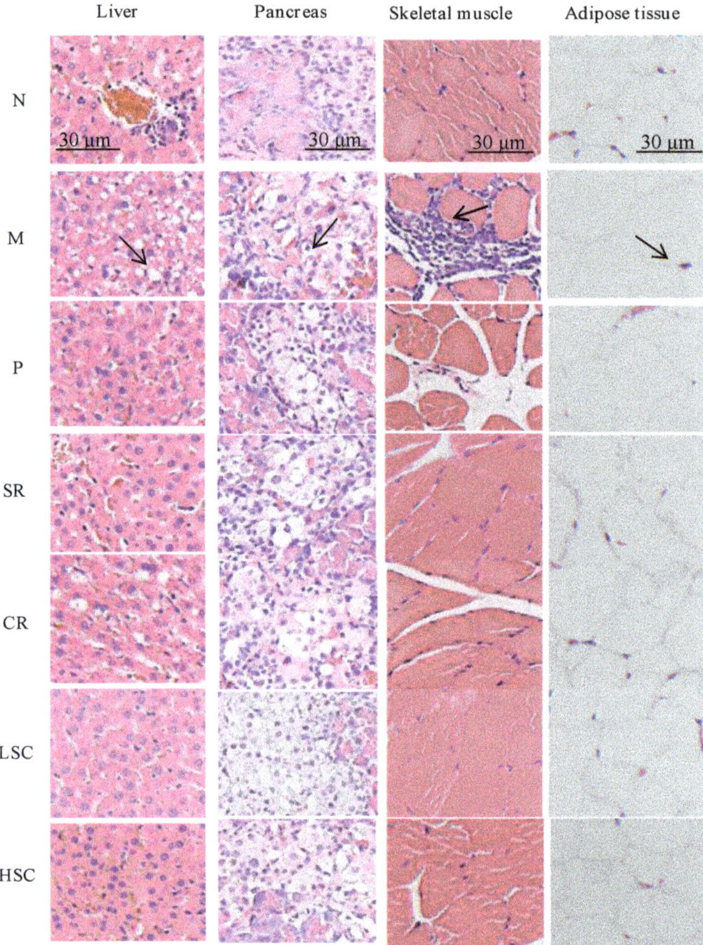

Figure 2. Histopathological observation of liver, pancreas, skeletal muscle and adipose tissue in the normal group (N), model group (M), and groups gavaged with metformin (P), scutellaria Radix (SR), coptidis Rhizome (CR), low dose of combined extracts group (LSC) and high dose of combined extracts (HSC). Liver tissues of M: notable changes with severe fatty degeneration can been seen as shown by the black arrow; Pancreas of M: severe atrophy of pancreatic islet cells is shown by the black arrow; Skeletal muscle of M: notably inflammatory cells infiltration is indicated by the black arrow; Adipose tissues of M: mean adipocyte diameter was seriously increased, as indicated by the black arrow. Samples were stained with H&E and photographed at 400× magnification.

2.3. SR, CR, and Combined Extracts Alleviated Inflammation by Regulation of Pro-Inflammatory Cytokine Expressions through MAPK Signaling Pathway

Recently, many studies have shown that pro-inflammatory macrophage accumulation in liver could trigger chronic low-grade inflammation that promoted the development of insulin resistance [21]. On day 30 after STZ injection, levels of CRP, IFN-γ, TNF-α, IL-6, resistin, SOCS3, IL-1β, and NO in serum were significantly increased in the model group compared to those in the normal group. However, the levels of these pro-inflammatory cytokines were reduced to a range from 96.0% to 58.4% after oral administration of metformin, SR, CR, LSC or HSC (Figure 3). MAPK pathway plays an important role in the inflammatory process [22], so the key targets in this pathway were investigated in this study. As the qPCR results show in Figure 4A–G, mRNA expressions of P38, ERK, c-jun, c-fos, JNK, IKK and P65 were significantly increased in the model group. With the treatment of metformin, SR, CR, LSC, and HSC, mRNA expressions of the above major targets in MAPK pathway were remarkably decreased to a range from 87.9% to 68.8%. Besides, the effects of LSC and HSC were more significant than SR and CR, while metformin showed a similar effect to HSC. NF-κB/p65 is a key transcriptional factor of inflammation regulated gene expression of pro-inflammatory cytokines. Further studies indicated that P65 was activated by phosphorylation in the model group, which could be suppressed by metformin, SR, CR, LSC, and HSC (Figure 4H).

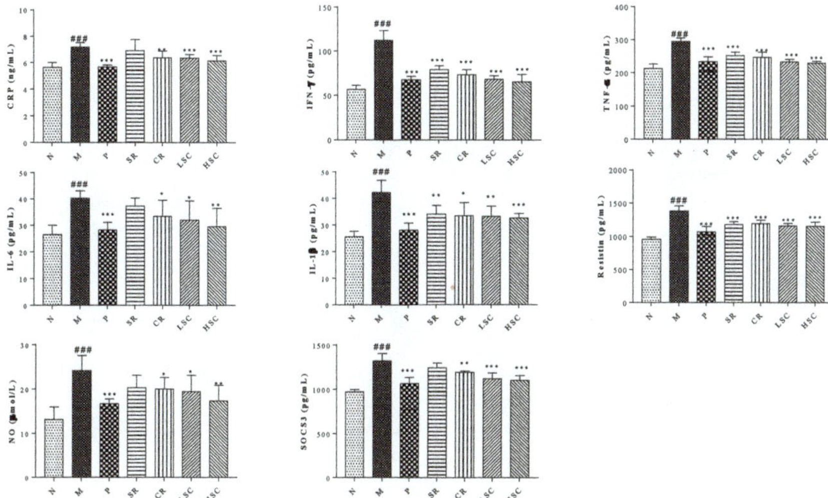

Figure 3. Serum contents of C-reaction protein (CRP), interferon gamma (IFN-γ), tumor necrosis factor-α (TNF-α), interleukin-6 (IL-6), interleukin-1β (IL-1β), resistin, nitric oxide (NO) and suppressor of cytokine signaling 3 (SOCS3) in the normal group (N), model group (M), and groups gavaged with metformin (P), scutellaria Radix (SR), coptidis Rhizome (CR), low dose of combined extracts group (LSC) and high dose of combined extracts (HSC). The values were shown as mean $\pm$ SD. # $p < 0.05$, ## $p < 0.01$, ### $p < 0.001$ vs. normal group; * $p < 0.05$, ** $p < 0.01$, *** $p < 0.001$ vs. model group. Data were analyzed by One-way-ANOVA.

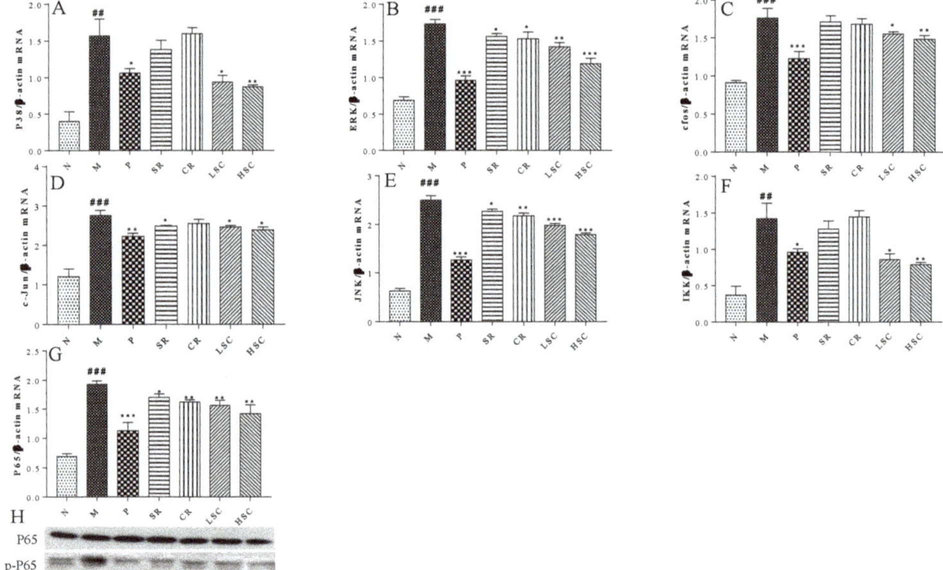

Figure 4. The effect of metformin (P), scutellaria Radix (SR), coptidis Rhizome (CR), low dose of combined extracts group (LSC) and high dose of combined extracts (HSC) on MAPK signaling pathway. (**A–G**) mRNA expression levels of c-jun N-terminal kinase (JNK), p38 mitogen-activated protein kinases (P38), extracellular regulated protein kinases (ERK), c-fos, c-jun, Inhibitor of nuclear factor kappa-B kinase (IKK) and P65 in liver of the N, M, P, SR, CR, LSC, and HSC by qPCR; (**H**) Protein levels of P65, phosphorylated P65 (p-P65) in liver of the N, M, P, SR, CR, LSC, and HSC by Western blot. The values are shown as mean ± SD. # $p < 0.05$, ## $p < 0.01$, ### $p < 0.001$ vs. normal group; * $p < 0.05$, ** $p < 0.01$, *** $p < 0.001$ vs. model group. Data were analyzed by One-way-ANOVA.

2.4. SR, CR and Combined Extracts Activated Insulin Signaling Pathway in Liver

Insulin resistance was derived from systemic inflammation in T2DM [23]. Our researches suggested that LSC and HSC could markedly inhibit inflammation in liver, and subsequent investigation was carried out to determine whether this action contributed to improve insulin signaling. As the qPCR results show in Figure 5A–D, mRNA expressions of IRS1, PI3K, Akt2, and Glut2 were markedly decreased in T2DM rats. Additionally, protein levels of p-PI3K, p-Akt, and Glut2 were notably lower in T2DM rats than those in normal rats (Figure 5E–G). After treatment, mRNA expressions of IRS1, PI3K, Akt2, and Glut2 were remarkably increased to a range from 102% to 171% and protein levels of p-PI3K, p-Akt and Glut2 were notably increased. However, protein levels of PI3K, Akt were not significantly different among groups. Furthermore, LSC and HSC were more effective than individual SR and CR.

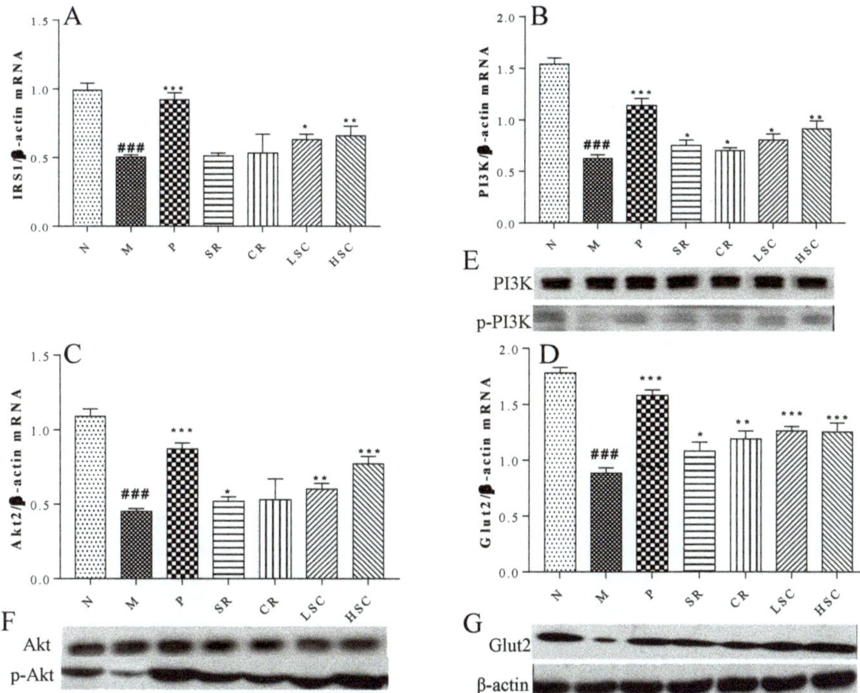

Figure 5. The effect of metformin (P), scutellaria Radix (SR), coptidis Rhizome (CR), low dose of combined extracts group (LSC) and high dose of combined extracts (HSC) on insulin signaling pathway. (**A–D**) mRNA expression levels of Insulin receptor substrate 1 (IRS1), Phosphatidylinositol-4,5-bisphosphate 3-kinase (PI3K), Protein kinase B (Akt2) and Glucose transporter 2 (Glut2) in liver of the N, M, P, SR, CR, LSC, and HSC by qPCR; (**E–G**) Protein expression levels of PI3K, phosphorylated PI3K (p-PI3K), Akt2, phosphorylated Akt2 (p-Akt2), Glut2 in liver of the N, M, P, SR, CR, LSC, and HSC by Western blot. The values are shown as mean ± SD. # $p < 0.05$, ## $p < 0.01$, ### $p < 0.001$ vs. normal group; * $p < 0.05$, ** $p < 0.01$, *** $p < 0.001$ vs. model group. Data were analyzed by One-way-ANOVA.

2.5. SR, CR and Combined Extracts Suppressed Hepatic Glucose Output by Inhibiting Gluconeogenesis and Glycogenolysis as Well as Promoting Glycolysis and Glycogenesis

Gluconeogenesis and glycogenolysis are two major pathways for endogenous glucose production [24,25]. PEPCK, FBPase, G6Pase, and GP are the rate-limiting enzymes controlling gluconeogenesis and glycogenolysis in the liver, respectively. The activities of these enzymes were increased in T2DM rats compared to normal rats but the increased activities were remarkably decreased to a range from 96.4% to 71.1% by SR and CR, especially by their combined extracts (Figure 6).

Glycolysis is regarded as a feeder pathway that prepares glucose for further catabolism and energy production. Besides, glucose storage in the form of glycogen could suppress hepatic glucose output to maintain blood glucose homeostasis. GK, PFK, PK, and GS are the rate-limiting enzymes controlling glycolysis and glycogen synthesis, respectively [26]. However, the activities of GK, PFK, PK, and GS were decreased in T2DM rats. After treatment, the decreased activities were increased to a range from 111.7% to 141.5% by SR and CR, especially their combined extracts (Figure 6).

2.6. Intervention Effects of SR, CR and Combined Extracts on the Metabolic Profiling of T2DM Rats

In this study, UPLC-Q-TOF-MS/MS was applied to generate metabolic profiles of plasma and urine from normal and T2DM rats before and after oral administration of SR, CR, LSC, and HSC in the negative and positive ESI modes. The metabolomics profiles of N and M groups were separated by PCA analysis and orthogonal partial least squares discriminant analysis (OPLS-DA). To investigate the amelioration of SR, CR, LSC, and HSC for curing T2DM, another PCA model was built. The variations of plasma and urine metabolic profiling of SR, CR, LSC, and HSC were restored to the levels of the normal group (Figures 7 and 8). Furthermore, the relative quantities of nine potential biomarkers in plasma and five in urine were significantly regulated by LSC and HSC. There were no obvious effects on the levels of xanthosine and N-acetyl-L-tyrosine for the SR group. The levels of xanthosine and hippuric acid were not regulated by the CR group. It was interesting to note that the levels of xanthosine were neither affected by SR nor CR, but by their combination. The contents of the potential biomarkers in Table S2 were considered as biomarkers for the effects of treatment.

The metabolic pathway was established by importing the potential metabolites into the web-based database MetPA. The pathway impact value calculated from pathway to topology analysis with MetPA above 0.1 was screened out as the potential target pathway. Here in Figure 9, glycerophospholipid metabolism with the impact values of 0.18 were filtered out as the most important metabolic pathways. Moreover, a correlation network of potential metabolites related to T2DM was exhibited in Figure 10. Our results also indicated that administration of LSC and HSC could more effectively modify these biomarkers and their related metabolic pathways including purine metabolism in which SR and CR could not be involved in T2DM rats.

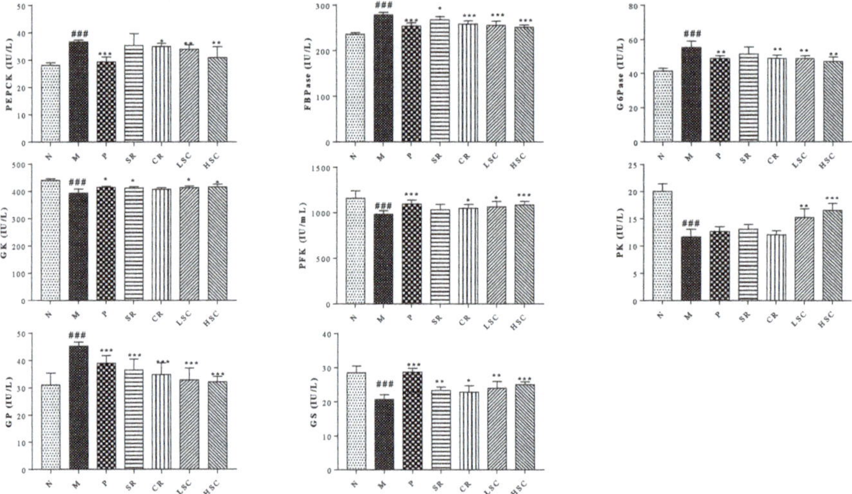

Figure 6. Activities of glucokinase (GK), phosphofructokinase (PFK), pyruvate kinase (PK), phosphoenolpyruvate carboxykinase (PEPCK), fructose-1,6-bisphosphatase (FBPase), glucose 6-phosphatase (G6Pase), glycogen synthase (GS) and glycogen synthase (GP) in livers of the normal group (N), model group (M), and groups gavaged with metformin (P), scutellaria Radix (SR), coptidis Rhizome (CR), low dose of combined extracts group (LSC) and high dose of combined extracts (HSC). The values were shown as mean ± SD. # $p < 0.05$, ## $p < 0.01$, ### $p < 0.001$ vs. normal group; * $p < 0.05$, ** $p < 0.01$, *** $p < 0.001$ vs. model group. Data were analyzed by One-way-ANOVA.

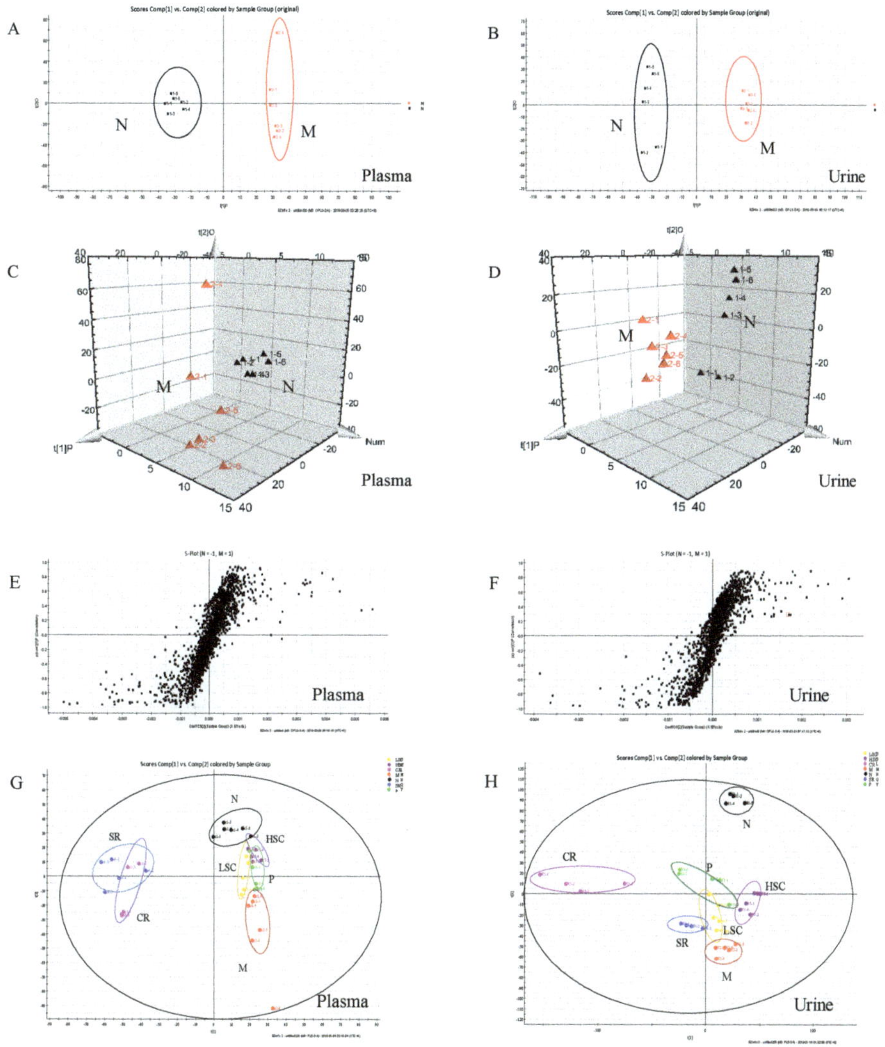

Figure 7. PCA model results between normal (N) and T2DM (M) rats in negative mode ((**A**) 2-D plot of plasma; (**B**) 2-D plot of urine). 3D PLS-DA scores plot of LC–MS spectral data ((**C**) plasma; (**D**) urine). S-plot of OPLS-DA model for M vs. N group ((**E**) plasma; (**F**) urine). PCA analytical results from N and M rats treated with metformin (P), scutellaria Radix (SR), coptidis Rhizome (CR), low dose of combined extracts group (LSC) and high dose of combined extracts (HSC) at negative mode ((**G**) plasma; (**H**) urine).

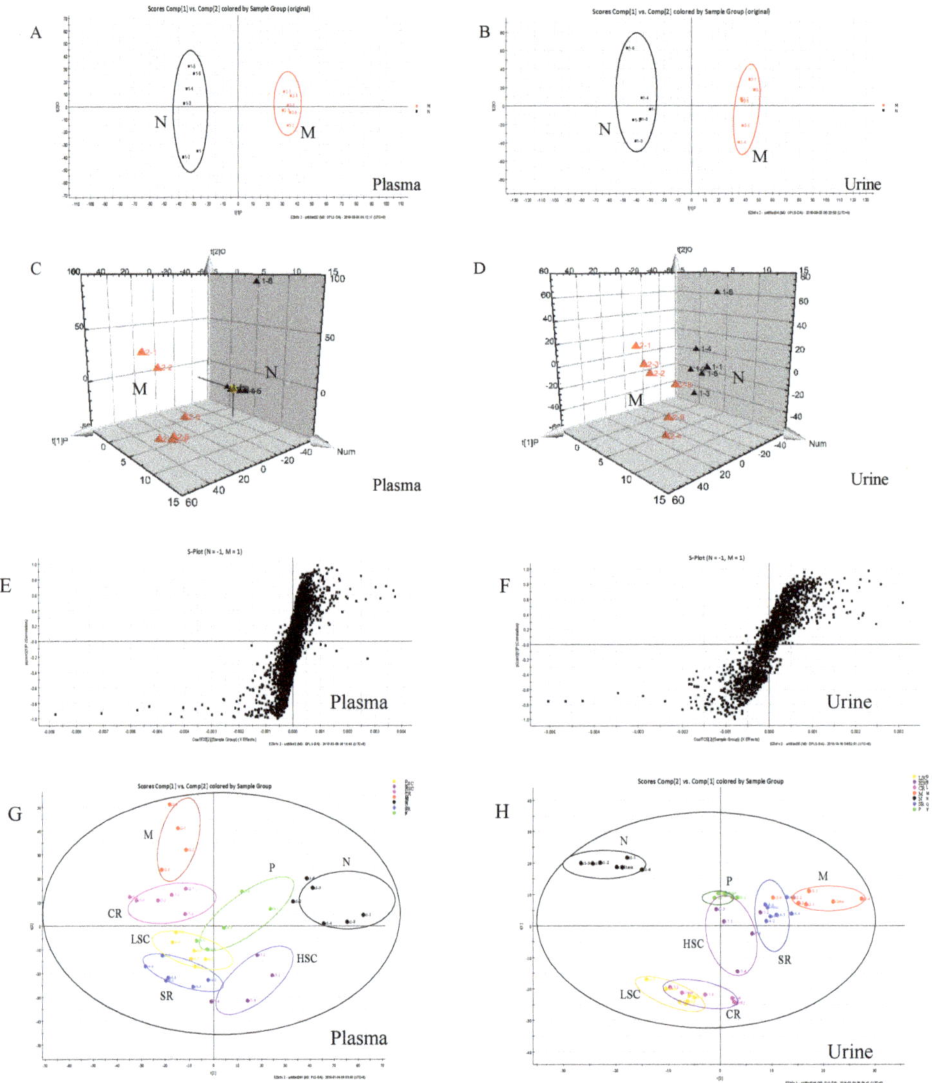

Figure 8. PCA model results between normal (N) and T2DM (M) rats in positive mode ((**A**) 2-D plot of plasma; (**B**) 2-D plot of urine). 3D PLS-DA scores plot of LC-MS spectral data ((**C**) plasma; (**D**) urine). S-plot of OPLS-DA model for M vs. N group ((**E**) plasma; (**F**) urine). PCA analytical results from N and M rats treated with metformin (P), scutellaria Radix (SR), coptidis Rhizome (CR), low dose of combined extracts group (LSC) and high dose of combined extracts (HSC) at positive mode ((**G**) plasma; (**H**) urine).

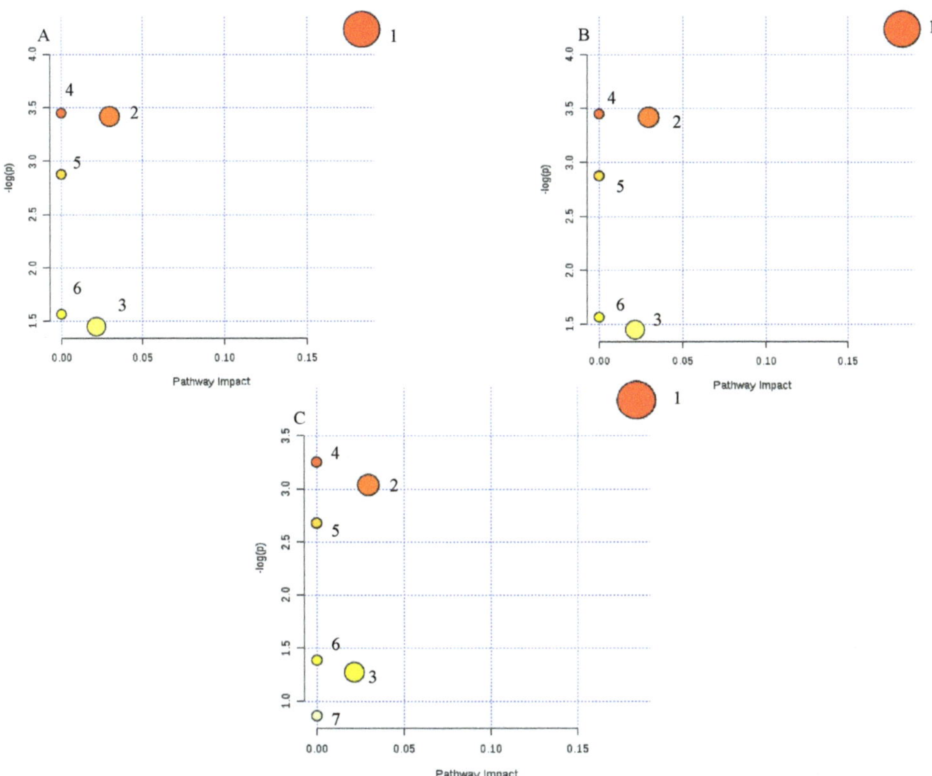

Figure 9. Metabolic pathways involved in potential markers in plasma and urine. (**A**) Scutellaria Radix group (SR); (**B**) Coptidis Rhizome group (CR); (**C**) metformin group (P), Low dose of combined extracts group (LSC) and High dose of combined extracts group (HSC). (**1**) Glycerophospholipid metabolism, (**2**) primary bile acid biosynthesis, (**3**) pyrimidine metabolism, (**4**) linolenic acid metabolism, (**5**) alpha-linolenic acid metabolism, (**6**) arachidonic acid metabolism, (**7**) purine metabolism.

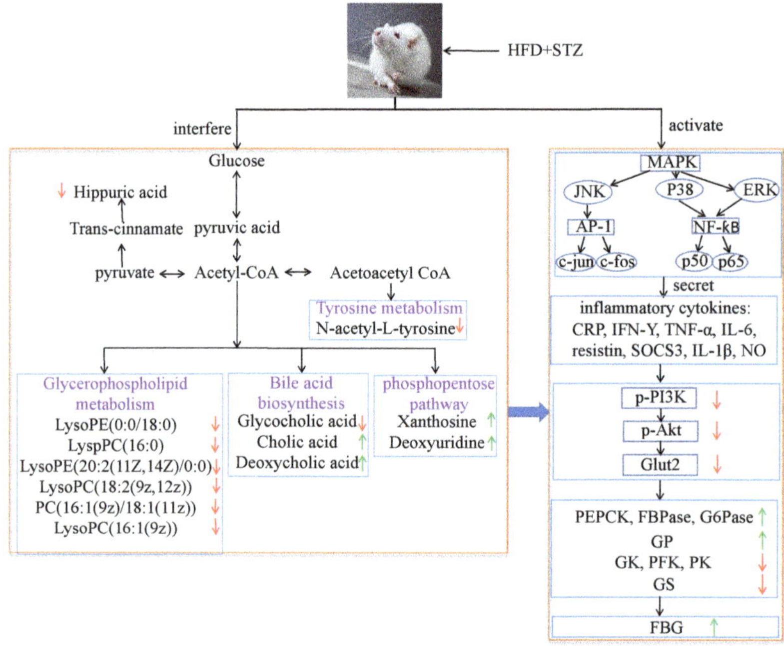

Figure 10. Correlation networks of main potential biomarkers in response to T2DM rats. Compared with the normal group, the red arrows represent the decrease of the contents of endogenous metabolites, while the green arrows represent the increase of the contents of endogenous metabolites.

3. Discussion

SR and CR, especially their combined extract, have been clinically used to treat T2DM for thousands of years; however, their compatibility mechanism remains unknown. Thus, in this study, the amelioration of SR, CR and their combined extracts on T2DM was first investigated, and their compatibility mechanism was further unraveled by the metabolomics and MAPK/PI3K/Akt signaling pathway. Importantly, the induction of the T2DM animal model is widely desired. The genetically modified mice and STZ-induced T2DM rat models have ever been established, however, neither these models nor stable animal models are appropriate because of ignoring dietary factors. Therefore, a suitable and stable T2DM rat model using HFD along with low dose of STZ was successfully established in our study. The levels of FBG, TG, TC, LDL-C, and FFA were remarkably increased, while the level of HDL-C was notably decreased in T2DM rats, which was similar to human T2DM sharing common symptoms (Figure 1).

Additionally, accumulating studies indicated that chronic inflammation was a major contributor to insulin resistance and T2DM [27–30]. Pro-inflammatory cytokines, particularly TNF-α, could enhance the serine phosphorylation of insulin receptor substrate-1, which induced insulin resistance. Some studies have shown glucose lowering effects of specific TNF-α or IL-1β inhibitors [31,32]. As expected, SR and CR could markedly ameliorate inflammation, insulin resistance, hyperglycemia, and hyperlipemia of T2DM rats. Furthermore, their compatibility (LSC and HSC) was more effective than single drug (Figures 1 and 3).

MAPK signaling pathway played a crucial role in inflammation [33]. Especially, NF-κB was a principle factor in regulating inflammatory mediators during the inflammatory process. The stimulation of NF-κB increased the levels of downstream inflammatory factors such as TNF-α, IL-1β, and IL-6, which promoted the insulin resistance in the liver. mRNA expressions of key targets in

MAPK pathway were significantly decreased after treatment. Additionally, protein expression of p-P65 was also decreased. Current data elucidated that LSC and HSC could exert better anti-inflammatory effects via blocking of the MAPK pathway (Figure 4).

The amelioration of inflammation might contribute to activating insulin signaling in liver. Remarkably, LSC and HSC inhibited inflammation in liver, but it was tempting to know whether this action was responsible for improving the insulin signaling pathway. The levels of Glut2, p-PI-3K, and p-Akt are major indexes that reflect the activity of the insulin signaling pathway, play a vital role in insulin-activated glucose uptake in liver, and inhibit glucose release from hepatocytes. As expected, LSC and HSC markedly up-regulated the mRNA expressions of IRS1, PI3K, Akt2, and Glut2, while protein expressions of p-PI3K, p-Akt, and Glut2 were also enhanced. These data suggested that LSC and HSC could overcome the impairment in PI3K/Akt insulin signaling pathway, contributing to the improvement of glucose disposal in T2DM rats (Figure 5).

Subsequently, the influence of activated insulin signaling pathway on glycogenolysis, glycogen synthesis, gluconeogenesis, and glucose metabolism in liver was investigated. The liver is crucial for the maintenance of normal glucose homeostasis—it produces glucose during fasting and stores glucose postprandially. Insulin could inhibit glycogenolysis, stimulate glycogen synthesis, reduce gluconeogenesis, and increase glucose metabolism. However, these hepatic processes were dysregulated in T2DM because the liver became insulin resistant. Gluconeogenesis and glycogenolysis constituted two major pathways for endogenous glucose production. Gluconeogenesis involved the production of glucose from non-carbohydrate precursors such as glycerol, glycogenic amino acid, and lactate during non-availability of dietary glucose, or glycogen stored in the liver was degraded to glucose by glycogenolysis. PEPCK, FBPase, and G6Pase acted irreversibly in the gluconeogenesis pathway. FBPase inhibitors and a decrease in transcription of PEPCK and G6Pase genes could reduce excessive endogenous glucose production in T2DM [34,35]. GP regulated hepatic glycogenolysis and GP inhibitor could lower blood glucose [36]. GS modulated hepatic glycogen synthesis and GSK-3 specific inhibitor could enhance glycogen synthesis by increasing GS activity [37]. GK mainly expressed in hepatocytes had effects on regulating blood glucose level in vivo [38]. Previous study also found that the ability of glucose metabolism in T2DM was reduced when the syntheses of PFK and PK were decreased [39]. In this work, owing to the amelioration of the insulin signaling pathway, PEPCK, FBPase, G6Pase, and GP activities were notably decreased, while prominent increased activities of GK, PFK, PK, and GS were detected after treatment. The above data indicated that the activities of key enzymes related to glycometabolism were modulated to notably lower FBG by SR and CR, especially by their compatibility (Figure 6).

As T2DM was a metabolic disease, the effects of SR, CR, LSC, and HSC on metabolic disorders in T2DM rats were further investigated to depict more insight into their compatibility mechanism. Metabolite profiles are closely related to the development of T2DM [40–43]. Among the most recent investigations on metabolic dysfunctions and plasma metabolites in T2DM, plasma level of bile acids have been confirmed to be closely associated with T2DM [44,45]. Moreover, metabolites of plasma and urine could reflect the changes of systemic physiology. Therefore, based on the comparative analysis of plasma and urine metabolomics in normal and T2DM rats, potential biomarkers and the correlated pathways could be discovered, which would be helpful to clarify the compatibility mechanism of SR and CR in improving metabolic disturbance of T2DM rats. In this study, plasma metabolomics of T2DM rats suggested that cholic acid and deoxycholic acid (secondary bile salts) levels were obviously increased, while the glycocholic acid (conjugated primary bile salts) level was dramatically decreased. Bile acids are major components of bile formed from cholesterol through various enzymatic reactions in hepatocytes. Primary bile acids are synthesized mainly by conjugation with taurine or glycine through the terminal side-chain carboxylic group presenting in the bile acid structure. Bile salt hydrolases, which are produced by many members of gut microbiota, could remove the conjugated amino acid from the primary bile salt to cholate and deoxycholate. Bile acids could exert effects on glucose metabolism via activation of bile acid receptors and G-protein-coupled membrane bound receptors (TGR5) [46–51]. TGR5 was expressed in cells of the hematopoietic system such as monocytes and macrophages and conferred anti-inflammatory properties in vitro and in vivo, decreasing cytokine production in monocytes, macrophages and kupffer cells [52,53]. Recent works revealed that the immune-suppressive actions of TGR5 in macrophages contributed to the prevention of the T2DM process [48]. Thus, we speculated that glycocholic acid might be an endogenous TGR5 ligand, while cholic acid and deoxycholic acid were ligand inhibitors. So glycocholic acid could suppress inflammation by TGR5 activation. Both P38 and ERK were significantly stimulated by deoxycholic acid in a prior study [54]. Supplied with deoxycholic acid, serum glucose and triglyceride levels were increased in mice [55]. After treatment, the levels of cholic acid and deoxycholic acid were markedly decreased and glycocholic acid was notably increased. Furthermore, the combination of SR and CR was more effective than the single drug (Table S2).

Additionally, the glycerophospholipid metabolism was the most important metabolic pathway. The decreased levels of glycerophospholipid metabolites including LysoPE(0:0/18:0), LyspPC(16:0), LysoPE(20:2(11Z,14Z)/0:0), LysoPC(18:2(9z,12z)), PC(16:1(9z)/18:1(11z)), and LysoPC(16:1(9z)) indicated a marked perturbation in the glycerophospholipid metabolism of T2DM rats. PCs, LysoPCs, LysoPEs were largely involved in the pathogenesis of inflammatory and metabolic diseases such as diabetes [56]. PCs and LysoPCs could block ERK, P38 activation and NF-ƙB translocation to the nucleus in peritoneal macrophages, which had an intrinsic anti-inflammatory property [57–59]. Many studies also indicated that levels of lysoPCs are closely associated with oxidative stress and inflammation [60,61]. LysoPCs containing an unsaturated fatty acyl group such as C20:4 and C22:6 were reported to show potent anti-inflammatory activity in in vivo and in vitro models [62,63]. Lower levels of LysoPCs were found to be predictors for T2DM in earlier studies [64] and treatment with LysoPCs could result in a significant increase in the level of GLUT4 at the plasma membranes of 3T3-L1 adipocytes [65]. Thus, the modification of SR, CR, LSC, and HSC on LysoPCs, LysoPEs and PCs levels might contribute to the amelioration of inflammation and insulin resistance. SR and CR combined therapy exhibited better efficacy (Table S2).

By the analysis of urine metabolomics, purine metabolism was notably modified after combined therapy, however, the single drug showed no effect; purine metabolism was terribly dysregulated which was associated with T2DM [66]. Xanthosine, a nucleoside derived from the purine base xanthine and ribose, was elevated in T2DM rats and catabolism of xanthosine could ultimately lead to the production of high uric acid which was linked to diabetic complications [67,68]. Interestingly, xanthosine was affected by neither SR nor CR, but by their combination. Additionally, hippuric acid, which could be converted from dietetic aromatic compounds by gut microbes [69], was significantly decreased in urine of T2DM rats. This might be due to disturbed gut microbiota metabolism under diabetic status [70]. As reported, concentrations of hippuric acid were lower in HFD-induced hyperlipidemic rats than in normal rats [71]. D-glucurono-6,3-lactone, participating in ascorbate and aldarate metabolism, was increased in T2DM rats and was downregulated after treatment of SR, CR and their combined extracts. Pyrimidine metabolism was also disturbed and increased metabolite of deoxyuridine in the urine of T2DM rats was identified. After 30 days therapeutic intervention, these changed trends in T2DM rats were reversed by SR and CR, particularly by their combined extracts (Table S2).

4. Materials and Methods

4.1. Chemicals, Reagents, and Materials

Acetonitrile (HPLC grade) was bought from TEDIA (Fairfield, OH, USA). Formic acid was supplied by Merck KGaA (Darmstadt, Germany). Metformin was purchased from Merck Serono (Shanghai, China). SR and CR were purchased from Nanjing Guo-yao Pharm (Nanjing, Jiangsu, China). Streptozotocin (STZ) was obtained from Sigma chemical (St Louis, MO, USA).

4.2. Extract Preparation

The dry herb pieces of SR (1.0 kg), CR (1.0 kg) and SC (1:1, 3 kg) were extracted with boiling water (1:8) twice, 2 h each time, filtered through gauze and concentrated to 1.0 g/mL, respectively.

4.3. Animals and Induction of T2DM Rats

Pathogen-free male Sprague-Dawley rats (200 ± 20 g) were supplied by the Zhejiang Province Experimental Animal Center. All animals were maintained in cages at 22–26 °C with 55%–65% relative humidity, under a 12 h dark/light cycle, with water and respective diet available ad libitum. The experiments were performed following the principles of the Care and Use of Laboratory Animal and approved by the Animal Ethics Committee of Nanjing University of Chinese Medicine.

After acclimatization for one week, animals were randomly separated into seven groups (6 rats/group): normal group (N), model group (M), metformin group (P), Scutellaria Radix group (SR), Coptidis Rhizome group (CR), Low dose of combined extracts group (LSC), and High dose of combined extracts group (HSC). The normal rats were fed with common pellet diets during the experiment and rats in other groups were fed by a high-fat diet (including 67.5% standard laboratory chow, 20% sucrose, 10% lard oil, and 2.5% egg yolk powder(w/w)). After one month of dietary intervention, a dose of 30 mg/kg STZ (dissolved using 0.1 M citrate buffer, pH 4.2) was given to all the groups except the normal group. The normal rats were given the same dose of sodium buffer. Three days later, all rats were fasted 12 h (free access to water). Fasting blood glucose (FBG) was measured by ONE TOUCH II type blood sugar apparatus. Rats were considered to be T2DM model when their FBG levels exceeded 16.7 mmol/L.

4.4. Drug Administration, Biological Sample Collection, and Preparation

The P group rats were intragastrically given metformin at a dose of 0.09 g/kg (0.09 g metformin per 1 kg rat body weight) for one month. The SR and CR group rats were intragastrically given SR, CR extracts at a dose of 6.3 g/kg (6.3 g crude herbs per 1 kg rat body weight) for one month, respectively. The LSC and HSC group rats were intragastrically given SC extracts at a dose of 6.3 g/kg, 12.6 g/kg (6.3 g crude herbs per 1 kg rat body weight, 12.6 g crude herbs per 1 kg rat body weight) for one month, respectively. All experiments and animal care were approved by the Animal Ethics Committee of Nanjing University of Chinese Medicine.

The rats were fixed in supine position and anesthetized with 10% chloral hydrate by intraperitoneal injection, blood samples were collected in heparinized tubes and no anticoagulant tubes on the 30th day after treatment from abdominal aorta. They were then centrifuged at 3000 r/min for 10 min to obtain the plasma and serum samples stored at −80 °C. The levels of pro-inflammatory cytokines and biochemical indexes in serum were measured using Elisa kits.

Liver tissues were quickly removed, placed on ice, and homogenized in 9 vol. (w/v) of phosphate buffer saline. The homogenate was centrifuged at 4 °C, 13,000 rpm for 10 min and supernatant liquor was then stored at −80 °C until detection of PEPCK, FBPase, G6Pase, GP, GK, PFK, PK, and GS by using Elisa kits according to the instructions of the manufacturer.

One hundred microliters of plasma were added to 300 µL of acetonitrile and two hundred microliters of urine were added to 200 µL of acetonitrile. These mixtures were vortexed for 1 min and centrifuged at 13,000 r/min for 10 min to obtain the supernatant. A 2 µL aliquot of each plasma or urine sample was injected for LC/MS analysis.

4.5. Histological Analysis

Each liver, pancreas, skeletal muscle, adipose tissue was fixed in 4% (w/v) paraformaldehyde over 24 h. With graded ethanol dehydration, specimens were embedded in paraffin. Sequentially, 3 mm sections were rehydrated and stained with hematoxylin and eosin. The program of staining was obtained with the IX51 microscope (Olympus Corporation, Tokyo, Japan).

4.6. Real-Time PCR

Total RNA was isolated from individual liver of each group for analysis of P38, ERK, JNK, IKK, c-jun, c-fos, P65, IRS1, PI3K, Akt2, and Glut2 using Trizol reagent (Invitrogen, Carlsbad, CA, USA) following the protocol provided by the manufacturer. Real-time quantitative PCR was performed by using SYBR Green Master mix and Rox reference dye according to the manufacturer's instructions. The cDNAs were obtained by the reverse transcription of RNA from rat liver. The mRNA levels of individual genes were normalized and calculated using the $\Delta\Delta CT$ method. The primers are listed in Table S1.

4.7. Western Blot

Individual liver of each group for analysis of P65, p-P65, PI3K, p-PI3K, Akt2, p-Akt2, and Glut2 was pulverized, and the proteins were extracted by a total protein extraction kit (Keygen, Shenzhen, China). An equal amount of protein (40 μg) in liver was separated by SDS-PAGE and transferred to polyvinylidene difluoride (PVDF) membranes (Millipore, Billerica, MA, USA). The membranes were blocked with 5% fat-free milk and incubated with different primary antibodies. The bound antibodies were detected using horseradish peroxidase–conjugated anti-rabbit antibodies. Antibody reactivity was detected by ECL Western blotting Detection Systems.

4.8. Metabolic Profiling

4.8.1. Chromatography

Chromatographic experiments were performed on a Thermo Syncronis C_{18} column (2.1 mm i.d. $\times$ 100 mm, 1.7 μm) using an Acquity UPLC^{TM} System (Waters Corp., Milford, MA, USA). The column was maintained at 35 °C, and the mobile phase, at a flow rate of 0.4 mL/min, consisting of solvent A (0.1% formic acid in water) and mobile phase B (acetonitrile). 0–12 min, 5%–20% B; 12–14 min, 20%–25% B; 14–19 min, 25%–55% B; 19–21 min, 55%–70% B; 21–25 min, 70%–95% B.

4.8.2. Mass Spectrometry

MS spectrometry was performed with a Waters Synapt^{TM} Q-TOF/MS (Waters Corp., Milford, CT, USA). The conditions used for the electrospray ion (ESI) source were as follows: capillary voltage of 3.0 kV, sample cone voltage of 30.0 V; extraction cone voltage of 2.0 V, source temperature of 120 °C, desolvation temperature of 350 °C. Nitrogen was used as desolvation and cone gas with a flow rate of 600 and 50 L/h, respectively. Leucine-enkephalin was used as the lock mass generating an $[M + H]^{+}$ ion (m/z 556.2771) and $[M-H]^{-}$ ion (m/z 554.2615) at a concentration of 200 pg/mL and flow rate of 100 μL/min to ensure accuracy during the MS analysis via a syringe pump.

4.8.3. Metabolomic Data Processing and Multivariate Analysis

UPLC/MS data were detected and noise-reduced in both the UPLC and MS domains such that only true analytical peaks were selected for further processing by the MassLynx software (version 4.1, Waters Corp, Milford, MA, USA).

4.8.4. Biomarker Identification and Metabolic Pathway Analysis

The identities of the potential biomarkers were confirmed by comparing their mass spectra and chromatographic retention times with the available reference standards and a full spectral library containing MS/MS data obtained in the positive and/or negative ion modes. The Mass Fragment application manager (Waters MassLynx v4.1, Waters corp., Milford, CT, USA) was used to facilitate the MS/MS fragment ion analysis through the use of chemically intelligent peak-matching algorithms. This information was then used to search multiple databases and to analyze the potential metabolic pathway using MetPA. Potential biological roles were evaluated by an enrichment analysis using MetaboAnalyst (version 3.0, McGill Univeristy, Montreal, Canada).

4.9. Statistical Analysis

Statistical analysis was assessed by SPSS 19.0 (SPSS Inc., Chicago, IL, USA). All data were given as mean $\pm$ SD, and comparison of mean values was evaluated using One-way-ANOVA with Dunnett. In all experiments, p values <0.05 were considered significant difference.

5. Conclusions

Previous studies suggested that macrophage-specific TGR5 signaling in kupffer cells protected liver from inflammation and insulin resistance. An increase of cholic acid and deoxycholic acid was found, while glycocholic acid was decreased in T2DM rats. So combined SR and CR could better modulate inflammation and improve insulin resistance by increasing the levels of glycocholic aid and decreasing the levels of cholic acid and deoxycholic acid. In conclusion, the present studies suggested that combined extracts exerted significant amelioration on T2DM by modulation of proinflammatory cytokines, key target protein expressions in MAPK, and insulin signaling pathways as well as enzymatic activities related to glycometabolism. Thus, administration of combined extract could improve the symptoms of T2DM rats more effectively than the single drug. These results might provide useful hints for T2DM treatment and deserve further clinical investigations.

Supplementary Materials: Supplementary materials can be found at http://www.mdpi.com/1422-0067/19/11/3634/s1.

Author Contributions: X.C. designed the study, performed experiments, analyzed the data, and drafted the manuscript. D.-W.Q., E.-X.S. and Z.-H.Z. contributed significantly to perform experiments and analyze data. S.J. and J.-A.D. designed the study and critically reviewed the manuscript. All authors reviewed and approved the final manuscript. X.C. and S.J. were the guarantors of this work.

Funding: This research received the Nature Science Foundation of China (No. 81673831) and Jiangsu Collaborative Innovation Center of Chinese Medicinal Resources Industrialization (No. ZDXM-1-10).

Acknowledgments: This work was financially supported by the Nature Science Foundation of China (No. 81673831) and Jiangsu Collaborative Innovation Center of Chinese Medicinal Resources Industrialization (No. ZDXM-1-10).

Conflicts of Interest: The authors declare no conflict of interest.

Abbreviations

Akt	Protein kinase B
CR	Coptidis Rhizoma
CRP	C-reactive protein
ERK	Extracellular regulated protein kinases
FBG	Fasting blood glucose
FBPase	Fructose-1,6-bisphosphatase
FFA	Free fatty acid
FINS	Fasting insulin
G6Pase	Glucose 6-phosphatase
GK	Glucokinase
Glut 2	Glucose transporter 2
GP	Glycogen phosphorylase
GS	Glycogen synthase
HDL-C	High-density lipoprotein
HFD	High-fat diet
HSC	How dose of combined extracts group
IFN-γ	Interferon gamma
IKK	Inhibitor of nuclear factor kappa-B kinase

IL-1β	Interleukin 1β
IL-6	Interleukin 6
IRS1	Insulin receptor substrate 1
JNK	C-Jun N-terminal kinase
LDL-C	Low-density lipoprotein
LSC	Low dose of combined extracts group
MAPK	Mitogen-activated protein kinase
NO	Nitric oxide
P	Metformin
P38	P38 mitogen-activated protein kinases
PEPCK	Phosphoenolpyruvate carboxykinase
PFK	Phosphofructokinase
PI3K	Phosphatidylinositol-4,5-bisphosphate 3-kinase
PK	Pyruvate kinase
SOCS3	Suppressor of cytokine signaling 3
SR	Scutellariae Radix
STZ	Streptozocin
T2DM	Type 2 diabetes mellitus
TC	Total chelosterol
TCM	Traditional Chinese medicine
TG	Triglyceride
TNF-α	Tumor necrosis factor alpha
UPLC-Q-TOF/MS	Ultra performance liquid chromatography-quadrupole time-of-flight mass spectrometry

References

1. Kalteniece, A.; Ferdousi, M.; Azmi, S.; Marshall, A.; Soran, H.; Malik, R.A. Keratocyte Density Is Reduced and Related to Corneal Nerve Damage in Diabetic Neuropathy. *Investig. Ophthalmol. Vis. Sci.* **2018**, *59*, 3584–3590. [CrossRef] [PubMed]

2. Cheungpasitporn, W.; Thongprayoon, C.; Vijayvargiya, P.; Anthanont, P.; Erickson, S.B. The risk for new-onset diabetes mellitus after kidney transplantation in patients with autosomal dominant polycystic kidney disease: A systematic review and meta-analysis. *Can. J. Diabetes* **2016**, *6*, 521–528. [CrossRef] [PubMed]

3. Zaoui, P.; Hannedouche, T.; Combe, C. Cardiovascular protection of diabetic patient with chronic renal disease and particular case of end-stage renal disease in elderly patients. *Nephrol. Ther.* **2017**, *13*, 6s16–6s24. [CrossRef]

4. Aldossari, K.K. Cardiovascular outcomes and safety with antidiabetic drugs. *Int. J. Health Sci.* **2018**, *12*, 70–83.

5. Guariguata, L.; Whiting, D.R.; Hambleton, I.; Beagley, J.; Linnenkamp, U.; Shaw, J.E. Global estimates of diabetes prevalence for 2013 and projections for 2035. *Diabetes Res. Clin. Pract.* **2014**, *103*, 137–149. [CrossRef] [PubMed]

6. Meneilly, G.S.; Tessier, D.M. Diabetes, dementia and hypoglycemia. *Can. J. Diabetes* **2016**, *40*, 73–76. [CrossRef] [PubMed]

7. Bonnet, F.; Scheen, A. Understanding and overcoming metformin gastrointestinal intolerance. *Diabetes Obes. Metab.* **2017**, *4*, 473–481. [CrossRef] [PubMed]

8. Yang, X.; Zhang, Q.; Gao, Z.; Yu, C.; Zhang, L. Baicalin alleviates IL-1β-induced inflammatory injury via down-regulating miR-126 in chondrocytes. *Biomed. Pharmacother.* **2018**, *99*, 184–190. [CrossRef] [PubMed]

9. Cheng, C.S.; Chen, J.; Tan, H.Y.; Wang, N.; Chen, Z.; Feng, Y. Scutellaria baicalensis and cancer treatment: Recent progress and perspectives in biomedical and clinical studies. *Am. J. Chin. Med.* **2018**, *46*, 25–54. [CrossRef] [PubMed]

10. Zhang, X.W.; Li, W.F.; Li, W.W.; Ren, K.H.; Fan, C.M.; Chen, Y.Y.; Shen, Y.L. Protective effects of the aqueous extract of Scutellaria baicalensis against acrolein-induced oxidative stress in cultured human umbilical vein endothelial cells. *Pharm. Biol.* **2011**, *49*, 256–261. [CrossRef] [PubMed]

11. Yang, M.D.; Chiang, Y.M.; Higashiyama, R.; Asahina, K.; Mann, J.; Wang, C.C.; Tsukamoto, H. Rosmarinic acid and baicalin epigenetically depress peroxisomal proliferator-activated receptor γ in hepatic stellate cells for their antifibrotic effect. *Hepatology* **2012**, *55*, 1271–1281. [CrossRef] [PubMed]

12. Song, K.H.; Lee, S.H.; Kim, B.Y.; Park, A.Y.; Kim, J.Y. Extracts of Scutellaria baicalensis reduced body weight and blood triglyceride in db/db Mice. *Phytother. Res.* **2013**, *27*, 244–250. [CrossRef] [PubMed]

13. Wang, T.; Jiang, H.; Cao, S.; Chen, Q.; Cui, M.; Wang, Z.; Li, D.; Zhou, J.; Wang, T.; Qiu, F.; et al. Baicalin and its metabolites suppresses gluconeogenesis through activation of AMPK or AKT in insulin resistant HepG-2 cells. *Eur. J. Med. Chem.* **2017**, *141*, 92–100. [CrossRef] [PubMed]

14. Sharma, G.; Nam, J.S.; Sharma, A.R.; Lee, S.S. Antimicrobial Potential of Silver Nanoparticles Synthesized Using Medicinal Herb Coptidis rhizome. *Molecules* **2018**, *23*, 2269. [CrossRef] [PubMed]

15. Chai, F.N.; Ma, W.Y.; Zhang, J.; Xu, H.S.; Li, Y.F.; Zhou, Q.D.; Li, X.G.; Ye, X.L. Coptisine from Rhizoma coptidis exerts an anti-cancer effect on hepatocellular carcinoma by up-regulating miR-122. *Biomed. Pharmacother.* **2018**, *103*, 1002–1011. [CrossRef] [PubMed]

16. Zhang, Z.; Zhang, H.; Li, B.; Meng, X.; Wang, J.; Zhang, Y.; Yao, S.; Ma, Q.; Jin, L.; Yang, J.; et al. Berberine activates thermogenesis in white and brown adipose tissue. *Nat. Commun.* **2014**, *5*, 5493–5508. [CrossRef] [PubMed]

17. Xie, H.; Wang, Q.; Zhang, X.; Wang, T.; Hu, W.; Manicum, T.; Chen, H.; Sun, L. Possible therapeutic potential of berberine in the treatment of STZ plus HFD-induced diabetic osteoporosis. *Biomed. Pharmacother.* **2018**, *108*, 280–287. [CrossRef] [PubMed]

18. Huang, L.; Yue, P.; Wu, X.; Yu, T.; Wang, Y.; Zhou, J.; Kong, D.; Chen, K. Combined intervention of swimming plus metformin ameliorates the insulin resistance and impaired lipid metabolism in murine gestational diabetes mellitus. *PLoS ONE* **2018**, *13*, e0195609. [CrossRef] [PubMed]

19. Wei, X.; Gu, N.; Feng, N.; Guo, X.; Ma, X. Inhibition of p38 mitogen-activated protein kinase exerts a hypoglycemic effect by improving β cell function via inhibition of β cell apoptosis in db/db mice. *J. Enzyme Inhib. Med. Chem.* **2018**, *33*, 1494–1500. [CrossRef] [PubMed]

20. Pace, F.; Carvalho, B.M.; Zanotto, T.M.; Santos, A.; Guadaqnini, D.; Silva, K.L.C.; Mendes, M.C.S.; Rocha, G.Z.; Alegretti, S.M.; Santos, G.A.; et al. Helminth infection in mice improves insulin sensitivity via modulation of gut microbiota and fatty acid metabolism. *Pharmacol. Res.* **2018**, *132*, 33–46. [CrossRef] [PubMed]

21. Jourdan, T.; Nicoloro, S.M.; Zhou, Z.; Shen, Y.; Liu, J.; Coffey, N.J.; Cinar, R.; Godlewski, G.; Gao, B.; Aouadi, M.; et al. Decreasing CB1 receptor signaling in Kupffer cells improves insulin sensitivity in obese mice. *Mol. Metab.* **2017**, *6*, 1517–1528. [CrossRef] [PubMed]

22. Kim, E.A.; Kim, S.Y.; Ye, B.R.; Kim, J.; Ko, S.C.; Lee, W.W.; Kim, K.N.; Choi, I.W.; Jung, W.K.; Heo, S.J. Anti-inflammatory effect of Apo-9′-fucoxanthinone via inhibition of MAPKs and NF-kB signaling pathway in LPS-stimulated RAW 264.7 macrophages and zebrafish model. *Int. Immun. Pharmacol.* **2018**, *59*, 339–346. [CrossRef] [PubMed]

23. Kwon, E.Y.; Choi, M.S. Luteolin Targets the Toll-Like Receptor Signaling Pathway in Prevention of Hepatic and Adipocyte Fibrosis and Insulin Resistance in Diet-Induced Obese Mice. *Nutrients* **2018**, *10*, 1415. [CrossRef] [PubMed]

24. Chung, S.T.; Courville, A.B.; Onuzuruike, A.U.; Galvan-De La Cruz, M.; Mabundo, L.S.; DuBose, C.W.; Kasturi, K.; Cai, H.; Gharib, A.M.; Walter, P.J.; et al. Gluconeogenesis and risk for fasting hyperglycemia in Black and White women. *JCI Insight* **2018**, *3*, 121495. [CrossRef] [PubMed]

25. Godfrey, R.; Quinlivan, R. Skeletal muscle disorders of glycogenolysis and glycolysis. *Nat. Rev. Neurol.* **2016**, *12*, 393–402. [CrossRef] [PubMed]

26. Ros, S.; Zafra, D.; Valles-Ortega, J.; García-Rocha, M.; Forrow, S.; Domínguez, J.; Calb, J.; Guinovart, J.J. Hepatic overexpression of a constitutively active form of liver glycogen synthase improves glucose homeostasis. *J. Biol. Chem.* **2010**, *285*, 37170–37177. [CrossRef] [PubMed]

27. Johnson, A.M.; Olefsky, J.M. The origins and drivers of insulin resistance. *Cell* **2013**, *152*, 673–684. [CrossRef] [PubMed]

28. Ajala-Lawal, R.A.; Aliyu, N.O.; Ajiboye, T.O. Betulinic acid improves insulin sensitivy, hyperglycemia, inflammation and oxidative stress in metabolic syndrome rats via PI3K/Akt pathways. *Arch. Physiol. Biochem.* **2018**, *5*, 1–9. [CrossRef] [PubMed]

29. Solinas, G.; Becattini, B. JNK at the crossroad of obesity, insulin resistance, and cell stress response. *Mol. Metab.* **2016**, *6*, 174–184. [CrossRef] [PubMed]

30. Nandipati, K.C.; Subramanian, S.; Agrawal, D.K. Protein kinases: Mechanisms and downstream targets in inflammation-mediated obesity and insulin resistance. *Mol. Cell. Biochem.* **2017**, *426*, 27–45. [CrossRef] [PubMed]

31. Dou, L.; Wang, S.; Sun, L.; Huang, X.; Zhang, Y.; Shen, T.; Guo, J.; Man, Y.; Tang, W.; Li, J. Mir-338-3p Mediates Tnf-A-Induced Hepatic Insulin Resistance by Targeting PP4r1 to Regulate PP4 Expression. *Cell. Physiol. Biochem.* **2017**, *41*, 2419–2431. [CrossRef] [PubMed]

32. Li, X.; Wang, X.; Wu, D.; Chen, Z.B.; Wang, M.X.; Gao, Y.X.; Gong, C.X.; Qin, M. Interleukin-1β and C-reactive protein level in plasma and gingival crevicular fluid in adolescents with diabetes mellitus. *Beijing Da Xue Xue Bao Yi Xue Ban* **2018**, *50*, 538–542. [PubMed]

33. Zhou, M.M.; Zhang, W.Y.; Li, R.J.; Guo, C.; Wei, S.S.; Tian, X.M.; Luo, J.; Kong, L.Y. Anti-inflammatory activity of Khayandirobilide A from Khaya senegalensis via NF-κB, AP-1 and p38 MAPK/Nrf2/HO-1 signaling pathways in lipopolysaccharide-stimulated RAW 264.7 and BV-2 cells. *Phytomedicine* **2018**, *42*, 152–163. [CrossRef] [PubMed]

34. Van Poelje, P.D.; Potter, S.C.; Erion, M.D. Fructose-1, 6-bisphosphatase inhibitors for reducing excessive endogenous glucose production in type 2 diabetes. *Handb. Exp. Pharmacol.* **2011**, *203*, 279–301.

35. Liu, Q.; Zhang, F.G.; Zhang, W.S.; Pan, A.; Yang, Y.L.; Liu, J.F.; Li, P.; Liu, B.L.; Qi, L.W. Ginsenoside Rg1 inhibits glucagon-induced hepatic gluconeogenesis through Akt-FoxO1 interaction. *Theranostics* **2017**, *16*, 4001–4012. [CrossRef] [PubMed]

36. Pari, L.; Chandramohan, R. Modulatory effects of naringin on hepatic key enzymes of carbohydrate metabolism in high-fat diet/low-dose streptozotocin-induced diabetes in rats. *Gen. Physiol. Biophys.* **2017**, *36*, 343–352. [CrossRef] [PubMed]

37. Hermida, M.A.; Dinesh Kumar, J.; Leslie, N.R. GSK3 and its interactions with the PI3K/AKT/mTOR signalling network. *Adv. Biol. Regul.* **2017**, *65*, 5–15. [CrossRef] [PubMed]

38. Zhang, X.; Jin, Y.; Wu, Y.; Zhang, C.; Jin, D.; Zheng, Q.; Li, Y. Anti-hyperglycemic and anti-hyperlipidemia effects of the alkaloid-rich extract from barks of *Litsea glutinosa* in ob/ob mice. *Sci. Rep.* **2018**, *8*, 12646. [CrossRef] [PubMed]

39. Bockus, L.B.; Matsuzaki, S.; Vadvalkar, S.S.; Young, Z.T.; Giorgione, J.R.; Newhardt, M.F.; Kinter, M.; Humphries, K.M. Cardiac Insulin Signaling Regulates Glycolysis Through Phosphofructokinase 2 Content and Activity. *J. Am. Heart. Assoc.* **2017**, *6*, e007159. [CrossRef] [PubMed]

40. Wang, S.; Yu, X.; Zhang, W.; Ji, F.; Wang, M.; Yang, R.; Li, H.; Chen, W.; Dong, J. Association of serum metabolites with impaired fasting glucose/diabetes and traditional risk factors for metabolic disease in Chinese adults. *Clin. Chim. Acta* **2018**, *487*, 60–65. [CrossRef] [PubMed]

41. Choi, E.; Zhang, X.; Xing, C.; Yu, H. Mitotic checkpoint regulators control insulin signaling and metabolic homeostasis. *Cell* **2016**, *166*, 567–581. [CrossRef] [PubMed]

42. Huypens, P.; Sass, S.; Wu, M.; Dyckhoff, D.; Tschöp, M.; Theis, F.; Marschall, S.; Hrabě de Angelis, M.; Beckers, J. Epigenetic germline inheritance of diet-induced obesity and insulin resistance. *Nat. Genet.* **2016**, *48*, 497–499. [CrossRef] [PubMed]

43. Jang, C.; Oh, S.F.; Wada, S.; Rowe, G.C.; Liu, L.; Chan, M.C.; Rhee, J.; Hoshino, A.; Kim, B.; Ibrahim, A.; et al. A branched-chain amino acid metabolite drives vascular fatty acid transport and causes insulin resistance. *Nat. Med.* **2016**, *22*, 421–426. [CrossRef] [PubMed]

44. Wang, S.; Deng, Y.; Xie, X.; Ma, J.; Xu, M.; Zhao, X.; Gu, W.; Hong, J.; Wang, W.; Xu, G.; et al. Plasma bile acid changes in type 2 diabetes correlated with insulin secretion in two-step hyperglycemic clamp. *J. Diabetes* **2018**, *10*, 874–885. [CrossRef] [PubMed]

45. Nowak, C.; Hetty, S.; Salihovic, S.; Castillejo-Lopez, C.; Ganna, A.; Cook, N.L.; Broeckling, C.D.; Prenni, J.E.; Shen, X.; Giedraitis, V.; et al. Glucose challenge metabolomics implicates medium-chain acylcarnitines in insulin resistance. *Sci. Rep.* **2018**, *8*, 8691. [CrossRef] [PubMed]

46. Zaborska, K.E.; Cummings, B.P. Rethinking Bile Acid Metabolism and Signaling for Type 2 Diabetes Treatment. *Curr. Diabetes Rep.* **2018**, *18*, 109. [CrossRef] [PubMed]

47. Qiao, X.; Ye, M.; Xiang, C. Metabolic regulatory effects of licorice: A bile acid metabonomic study by liquid chromatography coupled with tandem mass spectrometry. *Steroids* **2012**, *77*, 745–755. [CrossRef] [PubMed]

48. Agarwal, S.; Sasane, S.; Kumar, J.; Deshmukh, P.; Bhayani, H.; Giri, P.; Giri, S.; Soman, S.; Kulkarni, N.; Jain, M. Evaluation of novel TGR5 agonist in combination with Sitagliptin for possible treatment of type 2 diabetes. *Bioorg. Med. Chem. Lett.* **2018**, *28*, 1849–1852. [CrossRef] [PubMed]

49. Kumar, D.P.; Rajagopal, S.; Mahavadi, S.; Mirshahi, F.; Grider, J.R.; Murthy, K.S.; Sanyal, A.J. Activation of transmembrane bile acid receptor TGR5 stimulates insulin secretion in pancreatic β cells. *Biochem. Biophys. Res. Commun.* **2012**, *427*, 600–605. [CrossRef] [PubMed]

50. Zhang, X.; Wall, M.; Sui, Z.; Kauffman, J.; Hou, C.; Chen, C.; Du, F.; Kirchner, T.; Liang, Y.; Johnson, D.L.; et al. Discovery of Orally Efficacious Tetrahydrobenzimidazoles as TGR5 Agonists for Type 2 Diabetes. *ACS Med. Chem. Lett.* **2017**, *8*, 560–565. [CrossRef] [PubMed]

51. Lo, S.H.; Li, Y.; Cheng, K.C.; Niu, C.S.; Cheng, J.T.; Niu, H.S. Ursolic acid activates the TGR5 receptor to enhance GLP-1 secretion in type 1-like diabetic rats. *Naunyn Schmiedebergs Arch. Pharmacol.* **2017**, *390*, 1097–1104. [CrossRef] [PubMed]

52. Yoneno, K.; Hisamatsu, T.; Shimamura, K.; Kamada, N.; Ichikawa, R.; Kitazume, M.T.; Mori, M.; Uo, M.; Namikawa, Y.; Matsuoka, K.; et al. TGR5 signaling inhibits the production of pro-inflammatory cytokines by in vitro differentiated inflammatory and intestinal macrophages in Crohn's disease. *Immunology* **2013**, *139*, 19–29. [CrossRef] [PubMed]

53. Pols, T.W.; Nomura, M.; Harach, T.; Lo Sasso, G.; Oosterveer, M.H.; Thomas, C.; Rizzo, G.; Gioiello, A.; Adorini, L.; Pellicciari, R.; et al. TGR5 activation inhibits atherosclerosis by reducing macrophage inflammation and lipid loading. *Cell Metab.* **2011**, *14*, 747–757. [CrossRef] [PubMed]

54. Zeng, H.; Botnen, J.H.; Briske-Anderson, M. Deoxycholic acid and selenium metabolite methylselenol exert common and distinct effects on cell cycle, apoptosis, and MAP kinase pathway in HCT116 human colon cancer cells. *Nutr. Cancer* **2010**, *62*, 85–92. [CrossRef] [PubMed]

55. Kuno, T.; Hirayama-Kurogi, M.; Ito, S.; Ohtsuki, S. Reduction in hepatic secondary bile acids caused by short-term antibiotic-induced dysbiosis decreases mouse serum glucose and triglyceride levels. *Sci. Rep.* **2018**, *8*, 1253. [CrossRef] [PubMed]

56. Zeng, H.; Tong, R.; Tong, W.; Yang, Q.; Qiu, M.; Xiong, A.; Sun, S.; Ding, L.; Zhang, H.; Yang, L.; et al. Metabolic Biomarkers for Prognostic Prediction of Pre-diabetes: Results from a longitudinal cohort study. *Sci. Rep.* **2017**, *7*, 6575. [CrossRef] [PubMed]

57. Carneiro, A.B.; Iaciura, B.M.; Nohara, L.L.; Lopes, C.D.; Veas, E.M.; Mariano, V.S.; Bozza, P.T.; Lopes, U.G.; Atella, G.C.; Almeida, I.C.; et al. Lysophosphatidylcholine triggers TLR2- and TLR4-mediated signaling pathways but counteracts LPS-induced NO synthesis in peritoneal macrophages by inhibiting NF-ƙB translocation and MAPK/ERK phosphorylation. *PLoS ONE* **2013**, *8*, e76233. [CrossRef] [PubMed]

58. Treede, I.; Braun, A.; Jeliaskova, P.; Giese, T.; Füllekrug, J.; Griffiths, G.; Stremmel, W.; Ehehalt, R. TNF-alpha-induced up-regulation of pro-inflammatory cytokines is reduced by phosphatidylcholine in intestinal epithelial cells. *BMC Gastroenterol.* **2009**, *9*, 53. [CrossRef] [PubMed]

59. Treede, I.; Braun, A.; Sparla, R.; Kühnel, M.; Giese, T.; Turner, J.R.; Anes, E.; Kulaksiz, H.; Füllekrug, J.; Stremmel, W.; et al. Anti-inflammatory effects of phosphatidylcholine. *J. Biol. Chem.* **2007**, *282*, 27155–27164. [CrossRef] [PubMed]

60. Suvitaival, T.; Bondia-Pons, I.; Yetukuri, L.; Pöhö, P.; Nolan, J.J.; Hyötyläinen, T.; Kuusisto, J.; Orešič, M. Lipidome as a predictive tool in progression to type 2 diabetes in Finnish men. *Metabolism* **2018**, *78*, 1–12. [CrossRef] [PubMed]

61. Schilling, T.; Eder, C. Importance of lipid rafts for lysophosphatidyl choline-induced caspase-1 activation and reactive oxygen species generation. *Cell. Immunol.* **2010**, *265*, 87–90. [CrossRef] [PubMed]

62. Huang, L.S.; Hung, N.D.; Sok, D.E.; Kim, M.R. Lysophosphatidylcholine containing docosahexaenoic acid at the sn-1 position in anti-inflammatory. *Lipids* **2010**, *45*, 225–236. [CrossRef] [PubMed]

63. Hung, N.D.; Kim, M.R.; Sok, D.E. Mechanisms for anti-inflammatory effects of 1-[15(S)-hydroxyeicosapentaenoyl] lysophosphatidylcholine, administered intraperitoneally, in zymosan A-induced peritonitis. *Br. J. Pharmacol.* **2011**, *162*, 1119–1135. [CrossRef] [PubMed]

64. Wang-Sattler, R.; Yu, Z.; Herder, C. Novel biomarkers for pre-diabetes identified by metabolomics. *Mol. Syst. Biol.* **2012**, *43*, 615–625. [CrossRef] [PubMed]

65. Yea, K.; Kim, J.; Yoon, J.H.; Kwon, T.; Kim, J.H.; Lee, B.D.; Lee, H.J.; Lee, S.J.; Kim, J.I.; Lee, T.G.; et al. Lysophosphatidylcholine activates adipocyte glucose uptake and lowers blood glucose levels in murine models of diabetes. *J. Biol. Chem.* **2009**, *284*, 33833–33840. [CrossRef] [PubMed]

66. Dudzinska, W. Purine nucleotides and their metabolites in patients with type 1 and 2 diabetes mellitus. *J. Biomed. Sci. Eng.* **2014**, *1*, 38–44. [CrossRef]

67. Kushiyama, A.; Tanaka, K.; Hara, S.; Kawazu, S. Linking uric acid metabolism to diabetic complications. *World J. Diabetes* **2014**, *6*, 787–795. [CrossRef] [PubMed]
68. Zoppini, G.; Targher, G.; Chonchol, M.; Ortalda, V.; Abaterusso, C.; Pichiri, I.; Negri, C.; Bonora, E. Serum uric acid levels and incident chronic kidney disease in patients with type 2 diabetes and preserved kidney function. *Diabetes Care* **2012**, *35*, 99–104. [CrossRef] [PubMed]
69. Xiao, Y.; Dong, J.; Yin, Z.; Wu, Q.; Zhou, Y.; Zhou, X. Procyanidin B2 protects against d-galactose-induced mimetic aging in mice: Metabolites and microbiome analysis. *Food. Chem. Toxicol.* **2018**, *119*, 141–149. [CrossRef] [PubMed]
70. Li, L.; Wang, C.; Yang, H.; Liu, S.; Lu, Y.; Fu, P.; Liu, J. Metabolomics reveal mitochondrial and fatty acid metabolism disorders that contribute to the development of DKD in T2DM patients. *Mol. Biosyst.* **2017**, *13*, 2392–2400. [CrossRef] [PubMed]
71. Wu, Q.; Zhang, H.; Dong, X.; Chen, X.F.; Zhu, Z.Y.; Hong, Z.Y.; Chai, Y.F. UPLC-Q-TOF/MS based metabolomic profiling of serum and urine of hyperlipidemic rats induced by high fat diet. *J. Pharm. Anal.* **2014**, *4*, 360–367. [CrossRef] [PubMed]

International Journal of
Molecular Sciences

MDPI

Review

The Regulation of JNK Signaling Pathways in Cell Death through the Interplay with Mitochondrial SAB and Upstream Post-Translational Effects

Sanda Win, Tin Aung Than and Neil Kaplowitz *

Division of Gastrointestinal and Liver Disease, Department of Medicine, Keck School of Medicine,
University of Southern California, Los Angeles, CA 90033, USA; swin@usc.edu (S.W.); tthan@usc.edu (T.A.T.)
* Correspondence: kaplowit@usc.edu

Received: 25 October 2018; Accepted: 17 November 2018; Published: 20 November 2018

Abstract: c-Jun-N-terminal kinase (JNK) activity plays a critical role in modulating cell death, which depends on the level and duration of JNK activation. The kinase cascade from MAPkinase kinase kinase (MAP3K) to MAPkinase kinase (MAP2K) to MAPKinase (MAPK) can be regulated by a number of direct and indirect post-transcriptional modifications, including acetylation, ubiquitination, phosphorylation, and their reversals. Recently, a JNK-mitochondrial SH3-domain binding protein 5 (SH3BP5/SAB)-ROS activation loop has been elucidated, which is required to sustain JNK activity. Importantly, the level of SAB expression in the outer membrane of mitochondria is a major determinant of the set-point for sustained JNK activation. SAB is a docking protein and substrate for JNK, leading to an intramitochondrial signal transduction pathway, which impairs electron transport and promotes reactive oxygen species (ROS) release to sustain the MAPK cascade.

Keywords: reactive oxygen species; PTPN6; SRC; DOK4; p38; MKK4; MKK7; p53; DUSP1; SIRT2

1. Introduction

c-JUN-N-terminal kinase (JNK) is a critical mediator of physiological and pathological responses. An upstream MAP kinase signaling cascade from dual specificity MAP3K (e.g., ASK1) and MAP2K (MKK4/7) to serine/threonine MAPK (e.g., JNK, p38) mediate both the initiation of activation of JNK and its sustained activation [1–3]. There is a critical distinction between transient activation of the signaling pathway lasting minutes versus sustained activation lasting hours or more [4]. In this focused review, we will discuss the recent identification of JNK-mitochondrial SAB (SH3BP5)-ROS activation loop in sustaining the MAP kinase cascade, leading to pathological consequences within the context of our interest in liver disease. In addition, we will discuss a variety of factors that regulate or modulate the MAPK pathway from upstream MAP3K, such as apoptosis signal-regulating kinase 1 (ASK1) and mixed lineage kinase 2/3 (MLK2/3), to downstream JNK, SAB, and mitochondrial ROS. This review highlights the pathophysiological mechanism of sustained activation of JNK through JNK-activation loop and opens possible pharmacological interventions for therapeutic targets in the liver, heart, and brain, where SAB has been shown to have an important role.

2. JNK-SAB-ROS Activation Loop

Interest in the role of JNK signaling in liver injury began nearly 20 years ago with the identification of the protection of cultured mouse hepatocytes from acetaminophen (APAP)-induced necrosis by a JNK inhibitor (SP600125) [5]. This was further supported by the discovery of protective effect in in vivo JNK1 and 2 double knockdown using antisense oligonucleotides [6]. In this work and later in other in vivo and in vitro models of JNK-dependent liver apoptosis (e.g., TNF/D-galactosamine [TNF/GalN], tunicamycin-induced endoplasmic reticulum (ER) stress, and lipoapoptosis), a key important finding, i.e., the association of P-JNK with mitochondria, was discovered. Then, the binding target of P-JNK on mitochondria and the significance of this interaction were uncovered [7–9]. This seemed particularly relevant as earlier works had suggested that mitochondrial ROS played an important role in the sustained activation of JNK and that antioxidants could protect against sustained JNK activation and apoptosis in response to TNF [8,10].

One prior publication by Wiltshire et al. had identified a mitochondrial outer membrane protein—SAB (SH3BP5)—as a target of P-JNK binding and substrate for JNK phosphorylation [11]. These studies were performed in chicken embryotic fibroblast cells, and the functional consequences of the interaction were not further explored [11]. Then, SAB was identified exclusively in the outer membrane of mitochondria in liver [12]. SAB and P-JNK coimmunoprecipitated shortly after toxic stress from APAP prior to overt liver injury [7,13]. The finding was further supported by adenoviral sh-SAB (versus sh-lacZ control)-mediated depletion of SAB in liver. Knockdown of SAB resulted in inhibition of sustained JNK activation and translocation to mitochondria in all the models of JNK-dependent toxicity (APAP, TNF/GalN, ER stress, palmitic acid lipotoxicity) in vivo or in vitro. This was accompanied by striking protection against cell death [7–9]. More recently, these findings have been fully confirmed using hepatocyte-specific inducible knockout of SAB in two-month-old mice either after crossing *SAB*$^{fl/fl}$ mice with transgenic tamoxifen inducible *alb-CRE*$^{+/-}$ mice followed by tamoxifen feeding or by injection of *SAB*$^{fl/fl}$ mice with hepatocyte-targeted CRE viral vectors (adeno-alb-CRE or AAV8-TBG-CRE) [12].

Another important issue to address is the effect of the interaction of P-JNK with SAB on mitochondrial function and ROS production. Using isolated normal liver mitochondria, recombinant P-JNK1 and/or 2 in the presence of ATP was shown to lead to inhibition of oxidative phosphorylation and maximum respiratory capacity [12]. This effect was not observed in the absence of ATP, suggesting that phosphorylation of SAB was required. Furthermore, this effect was accompanied by enhanced O_2 production in MitoSOX-loaded mitochondria [8,14]. The effect of P-JNK + ATP was absent in liver mitochondria from SAB knockout mice and was inhibited by a peptide corresponding to the JNK docking site of SAB, which blocked the interaction of JNK and SAB [8,12].

The topology of SAB can be defined using c- and N-terminal-targeted antisera. The short c-terminus faces the cytoplasm and contains a JNK kinase interaction motif (KIM), which is the docking site. The longer N-terminus faces the intermembrane space [12]. Since there is no evidence that JNK enters the mitochondria, the question relates to how the interaction of P-JNK with SAB and its phosphorylation on the external face lead to impairment of mitochondrial bioenergetics. The mechanism of JNK-SAB-mediated impairment of mitochondrial respiration has been explored [12]. Tyrosine-protein kinase c-SRC, mainly in the P-419-SRC active state, has been shown inside mitochondria of liver and neurons and is required to maintain the function of the electron transport chain. SRC kinase inhibitors reproduce the same effect as P-JNK/ATP on isolated mitochondria. When mitochondria were exposed to P-JNK/ATP, rapid dephosphorylation of P-SRC was observed, but this did not occur in mitochondria from SAB knockout liver and was also inhibited by the KIM blocking peptide [8,9,12]. Furthermore, inactivation (dephosphorylation) of SRC occurred in liver mitochondria after in vivo treatment with APAP or TNF/GalN. Inactivation of SRC was inhibited in isolated mitochondria after treatment with vanadate, which blocked the effect of P-JNK/ATP on mitochondrial respiration. The study indicated that a phosphotyrosine phosphatase (PTP) was responsible for mediating the effect and that intramitochondrial protein tyrosine phosphatase non-receptor type 6 (PTPN6/SHP1) was responsible for inactivation of SRC when JNK interacted with SAB on mitochondrial outer membrane. Mitochondria isolated from PTPN6-depleted mice were resistance to the effects of P-JNK/ATP on P-SRC and mitochondria respiration [12].

Mitochondrial SRC associates with docking protein 4 (DOK4), a kinase and PTP docking protein [15], and DOK4 participates in the JNK activation of intramitochondrial signaling pathway [12]. DOK4 is found exclusively in the mitochondria fraction and is associated with the inner membrane but accessible to the intermembrane space. Knockdown of DOK4 in vivo protected isolated mitochondria from the effect of P-JNK/ATP and protected against liver injury and sustained JNK activation. The effect of DOK4 knockdown was analogous to the knockout of upstream SAB or knockdown of PTPN6. Careful mitochondrial subfractionation and immunoprecipitation studies have revealed that under basal conditions, SAB is in the outer membrane and bound SHP1 (on the intermembrane face), while DOK4 and P-SRC are on the inner membrane. Following toxic stress and JNK activation, PTPN6 coimmunoprecipitation with SAB decreases in the outer membrane fraction and binding to SRC/DOK4 (coimmunoprecipitation) increases on the inner membrane. Thus, when SAB is phosphorylated by P-JNK/ATP on the cytoplasmic face, PTPN6 is released and interacts with P-SRC, dephosphorylating (inactivating) SRC in a DOK4-dependent fashion. Thus, DOK4 appears to serve as a platform, which is required for the interaction of PTPN6 and P-SRC. PTPN6 associated with SAB is inactive (nonphosphorylated); however, when associated with P-SRC, it is phosphoactivated by SRC, which then leads to dephosphorylation of SRC. Inactivation of SRC leads to impaired mitochondrial respiration and increased ROS release from mitochondria. ROS then activates ASK1, and possibly MLK2/3, which sustains activation of MKK4/7, leading to sustained JNK activation (Figure 1). ROS oxidize thioredoxin, relieving ASK1 of inhibition of dimerization and allowing self-activation of ASK1. ROS also activate SRC at or near the plasma membrane, which then activates MLK2/3. These MAP3 kinases then activate MAP2 kinases, which activate JNK.

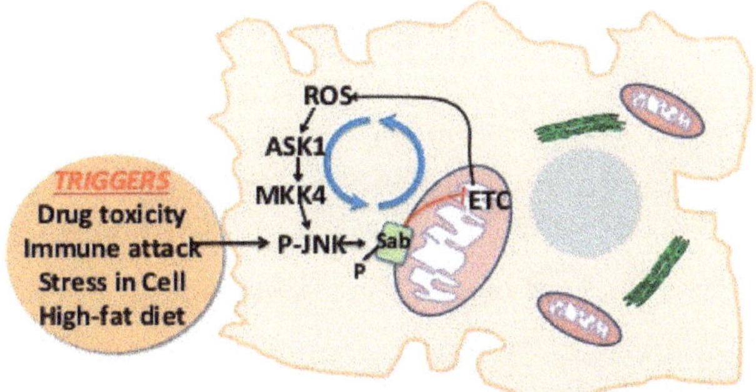

Figure 1. P-JNK-SAB-mitochondria-ROS-mediated JNK activation loop. JNK activation is triggered by physical and chemical stress, including alterations in nutrients, growth factors, cytokines, extracellular matrix, DNA damage, drugs, and toxins. Activated JNK translocates to mitochondria and interacts with SAB, leading to a sequence of events, i.e., inhibition of intramitochondrial c-SRC activity and mitochondrial electron transport chain and thus release of ROS, which further activates ASK1, MKK4/7, and JNK. P-JNK-SAB-ROS activation loop drives sustained JNK activation, and cell death occurs. Black arrows indicate activation. Blue circulating arrows indicate vicious cycle. Red "T" arrow indicates inhibition of electron transport chain (ETC).

The duration and degree of sustained JNK activation mediates many consequences, both through transcriptional regulation by AP-1 targets that modulate expression of many genes involved in proliferation as well as inflammation (cytokines and chemokines), metabolic gene dysregulation e.g., gene repressors, such as nuclear receptor corepressor 1 (NCOR1) action on peroxisome proliferator activated receptor alpha (PPARα) and thioredoxin-disufide reductase (TR), or through direct activation of proapoptotic BH3 family members and inhibition of antiapoptotic Bcl2 family members (see Reference [3] for review).

3. Modulation of JNK Activation Loop

MAP kinase cascade, which senses cellular and extracellular stress, conveys cellular response to regulate cell fate. The timing and duration of JNK activation determine whether cells proliferate or adapt to metabolic or toxic stress or undergo programmed cell death, such as apoptosis, necrosis, and possibly other forms of cell death. Thus, modulators of the JNK-SAB-ROS activation loop (Figure 2) and molecular structure of components in the loop (Figure 3) determine the duration of JNK activation and selectivity and specificity of JNK-mediated cellular responses. As the expression of isoforms of kinases, phosphatases, substrates, inhibitors, and scaffold proteins involved in the JNK activation loop are tissue and cell-type-specific, we will discuss the general principles of modulation of the loop.

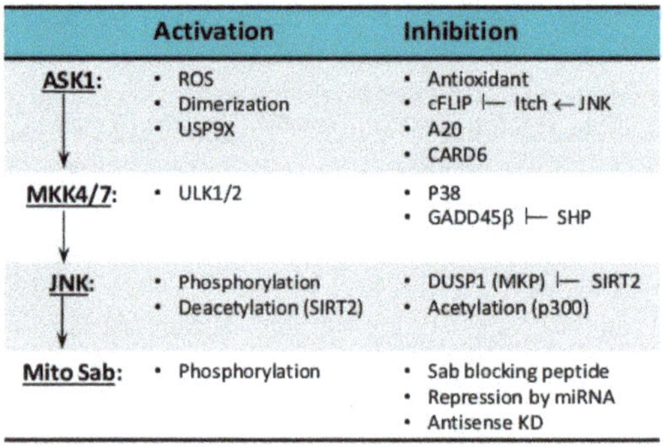

	Activation	Inhibition
ASK1:	• ROS • Dimerization • USP9X	• Antioxidant • cFLIP ⊢— Itch ←JNK • A20 • CARD6
MKK4/7:	• ULK1/2	• P38 • GADD45β ⊢— SHP
JNK:	• Phosphorylation • Deacetylation (SIRT2)	• DUSP1 (MKP) ⊢— SIRT2 • Acetylation (p300)
Mito Sab:	• Phosphorylation	• Sab blocking peptide • Repression by miRNA • Antisense KD

Figure 2. Modulation of JNK activation loop. JNK-SAB-ROS activation loop can be modulated at all level of MAP kinase cascade through phosphorylation by upstream kinase, dephosphorylation by phosphatase, acetylation by sirtuins, deacetylation by p300, protein stabilization by deubiquitinating enzymes, and protein degradation by ubiquitinating enzymes.

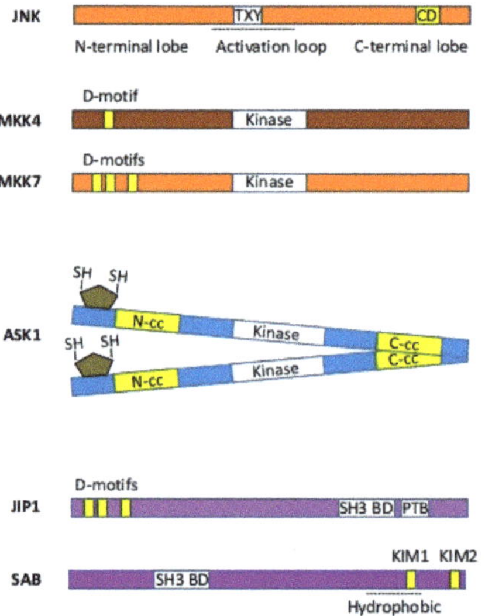

Figure 3. Schematic diagram of molecules involved in JNK activation loop. JNK is activated by dual threonine–tyrosine phosphorylation at (TXY) located within activation loop (indicated as black dotted line) by MKK4/7 through interaction with common docking site (CD) of JNK and docking motif (D-motif; JNK binding site indicated by yellow bars) of MKK4/7. CD of JNK is also shared with JIP and SAB which has one hydrophobic transmembrane spanning domain (indicated as dotted line). Two JNK binding sites on SAB are noted as KIM1 and KIM2 (indicated by yellow bars). Dimerization of ASK1 through interaction of N-terminal coil-coil domains (N-cc) and of C-terminal coil-coil (C-cc) doamins autoactivates ASK1. Removal of thioredoxin (indicated by gray polygon) by oxidation facilitates dimerization of ASK1.

3.1. MAP Kinases—JNK and p38

JNK and p38 are activated by MKK4/7 via dual phosphorylation of a Thr-Pro-Tyr (TPY) motif of the activation loop (A-loop), which connects N-terminal and C-terminal lobes. The ATP binding site lies between the lobes [16]. One critical feature of JNK signaling is the use of its single common docking site (CD) to interact with JNK-binding domain (D-motif) of upstream MKKs, MAP kinase phosphatases, substrates, inhibitors, and scaffold proteins. Thus, JNK activity derives from the ensemble of interactomes that compete for the docking site on JNK. Therefore, reducing the level of a single substrate of MAPK can lead to decreased amounts of active MAPK due to loss of competition and greater access of phosphatases to allow graded response. The specificity of JNK signaling relies on hydrophobic φ-x-φ docking motif (D-motif) on JNK substrates and scaffold proteins, such as c-Jun, ATF2, JIP, and SAB, and their subcellular localization. Preferences for D-motifs on different JNK substrates have been analyzed using 11-mer peptides derived from different substrates [17]. SAB peptide derived from KIM1 D-motif is similar to JIP1 sequence but has 21 folds lower affinity for P-JNK. This appears to be due to adjacent Pro residue. This means that small changes in D-motif and adjacent residues can profoundly impact JNK signaling. Indeed, JIP1 has higher affinity for JNK3 than ATF2 or SAB. Cytosolic JIP is a platform to bring together upstream kinases MKKs, MLKs, and Rac for JNK activation [18–20]. The depletion of JIP in MEF cells prevents fatty-acid-induced JNK activation [20]. JNK phosphorylates c-JUN on Ser63/73 and increases c-JUN-dependent transcription and cell proliferation [1], whereas JNK interaction with less affinity to SAB localized on mitochondria accounts for ROS generation, sustained MAP3K to JNK activation, and cell death [7–9,12,21]. Importantly, JNK phosphorylates p53 on Thr81 and stabilizes and confers its transcriptional activity [22], which could dampen expression of SAB and the JNK activation loop (see below).

A distinct difference of p38 compared to JNK is the activation of downstream MAPK-activated protein kinase (MAPKAPKs), such as MK2/3, MNK1/2 [23]. p38 regulates MK2-mediated TNF-α and IL-6 production by promoting translation and/or stability of their mRNAs [24]. p38 modulation upregulates antioxidant response via NF-κB [25] and interferes with ROS produced by the JNK activation loop. p38-mediated MK2 activation also phosphorylates MDM2 on Ser157 and Ser166, resulting in MDM2 activation and degradation of p53 [26]. p53 expression and activity, contributed to by multiple signaling pathways [27–29], importantly regulates SAB expression and determines susceptibility to promoting the JNK activation loop (abstract, manuscript in preparation). The depletion of p53 or inhibition of p53 by pifithrin leads to higher P-JNK levels and more severe necrosis in acetaminophen-induced acute liver injury [30]. Recently, we uncovered an important regulatory role of p53 in SAB expression and thus the contribution of functional p53 in decreasing JNK-mediated cell death (see below). In addition, human rhabdomyosarcoma cells lacking functional p53 undergo rapid apoptotic cell death in mild cellular stress conditions, such as serum-free condition and inhibition of mTOR, through the ASK1-JNK activation-mediated pathway [31]. The mechanism may be by the upregulation of SAB expression in p53 deficiency, which we will discuss further below. This could be a potential therapeutic pathway to target treatment of functional p53-defective tumor cells.

Dual-specificity protein phosphatases (DUSP), which are also known as mitogen-activated protein kinase phosphatases (MKP), are a family of threonine-tyrosine dual-specificity phosphatases that dephosphorylate and inactivate MAPKs such as extracellular regulated MAP kinase (ERK), p38, and JNK in a context-dependent manner [1,32]. The highest levels of DUSP1 are observed in the heart, lungs, and liver. *DUSP1* or *DUSP5* KO mice were shown to exhibit increased JNK activity due to physiological levels of reactive oxygen species, supporting the role of phosphatase in the JNK activation loop [2,10]. Indeed, DUSP inhibition may be sufficient to induce prolonged activation of JNK following some stimuli. DUSPs localize in cytoplasm and nucleus, but mitochondrial localization or association has not been reported. Indeed, DUSP1-mediated dephosphorylation of JNK is sufficient to inhibit JNK-SAB-ROS activation loop because activation of JNK is required for translocation and interaction with mitochondrial SAB. In breast cancer, an inhibitor of DUSP1 is considered an adjuvant and/or neoadjuvant therapy by enhancing cell apoptosis. The involvement of p38 in modulation of JNK activation loop is more complex because p38/ERK upregulates DUSP1 expression through MAPK-activated protein kinase MSK [23].

In addition to phosphorylation and dephosphorylation, direct regulation of JNK activity also involves participation of acetylation/deacetylation. Cytosolic and nuclear shuttling SIRT2 deacetylates and activates JNK, whereas JNK is inhibited by p300-mediated acetylation. In addition, SIRT2 deacetylates and inhibits DUSP1 [33]. Thus, sirtuin activation occurring in excess nicotinamide adenine dinucleotide (NAD$^+$), such as in nutrient excess or granulocyte-colony stimulating factor (G-CSF) signal transduction, changes the balance of JNK activation. However, JNK activation by sirtuin inducers [34] cannot reach the threshold level to cause sustained JNK activation and therefore does not favor cell death. In addition, sirtuin 1 (*SIRT1*) knockout mice were shown to exhibit less severe liver injury through preconditioned enhanced NF-κB response and dampening sustained JNK activation in the GalN/LPS-induced apoptotic cell death. However, acetaminophen-induced liver injury, which overrides NF-κB-mediated upregulation of antioxidant genes, is not protected [35]. Overall, the contribution of sirtuins in cell death pathways appears to be of minor importance. So far, the role of SIRT3, 4, and 5 in mitochondria have not been explored in the mechanism of JNK-SAB-ROS activation loop [36]. In summary, p38, p53, DUSP1, SIRT2, p300, NF-κB, Gadd45β, and SAB all have the potential to directly or indirectly modulate JNK activity in a complex, context-dependent fashion.

3.2. MKKs—MKK4 and MKK7

MKK4/7 interacts with JNK via D-motif and phosphorylates and activates JNK on JIP platform [37]. In MEF cells, MKK4 also activates p38. MKK7 contains three JNK docking D-motifs within its 100-amino acid regulatory domain. The second docking site of MKK7 binds to JNK via two alternative binding modes [38]. However, the significance and selectivity of the D-motif of MKK7 on JNK needs further exploration. Interestingly, Ser403 of MKK7 in HEK293 cells is phosphoactivated by serine/threonine-protein kinase ULK1/2 (ATG1/2), which is a downstream target of AKT/mTOR signaling pathway. The significance of this crosstalk in the sustained JNK activation loop in disease models requires further exploration as liver-specific deficiency of ULK1/2 in KO mice were shown to delay and partially protect liver injury from acetaminophen-induced hepatoxicity but not GalN/TNF liver injury [39]. Signalosomes, such as receptor complex in TNF-induced JNK activation and TRAFII-mediated JNK activation in ER stress, are important in integration of crosstalk, induction, amplification, and inhibition of the JNK activation loop [40]. The cytosolic JNK interacting protein (JIP) platform is crucial for initial activation of JNK in fatty acid but not in TNF-induced JNK activation [20]. SRC phosphorylation of JIP1 creates phosphotyrosine interaction motifs that bind the SH2 domains of SRC and the guanine nucleotide exchange factor VAV which is required for activation of Rac and downstream activation of MLKs, MKK7 and JNK activation on JIP1 platform in MEF cells, white adipose tissue, and muscle, but other isoforms of JIP could be essential in liver.

The complexity of JNK activation loop is also illustrated by retroinhibition of MAPK cascade by p38a in receptor-mediated cell death [41]. Indeed, p38a is activated by MKK4 and MKK3/6, and p38 then contributes inhibitory crosstalk to the JNK activation loop via inhibitory phosphorylation of TGF-beta activated kinase 1 (TAK1/TAB1), MKK3/6 and/or increased expression of DUSP (MKP) by p38a/MK2 pathway [42–44]. However, this regulatory pathway requires further exploration. In fact, p38 affects several alternative pathways to interfere with the JNK activation loop, such as NF-κB-mediated upregulation of antioxidant genes and 17 kDa Gadd45β (MyD118) [45,46]. Endogenous Gadd45β and MKK7 associate through direct, high-affinity contact. Gadd45β, which is not a phosphatase, inhibits MKK7 by masking the kinase domain. The association is tighter than JIP1 and thus spatially prevents MKK7 activation of JNK. Gadd45β does not inhibit MKK4, MKK3b, or ASK1 activation or phosphorylation of MKK7. Gadd45β expression is suppressed by orphan nuclear receptor small heterodimer partner (SHP), which is a transcriptional corepressor. Depletion of SHP increases Gadd45β expression and prevents sustained JNK activation and liver injury. Additionally, activation of Akt in a parallel survival pathway could activate NF-κB and inhibit MAP3K, such as MLK3 and ASK1 [47–50].

3.3. MAP3K—MLK2/3, ASK1, TAK1

MEKK1-4, DLK, TPL-2, TAO1/2, ASK1, MLK2/3, and TAK1 are common MAP3Ks involved in the activation pathway of JNK and p38, but the unique structure and subcellular distribution of these MAP3Ks reveal important roles in various cells and disease models. ASK1, MLK2/3, and TAK1 are widely studied and discussed in this review. ASK1 triggers cellular responses to redox stress and inflammatory cytokines [51,52] and plays important roles in innate immunity and viral infection [53]. ASK1 has the central kinase domain flanked on either side by coiled-coil domains. The N terminus of the kinase domain contains several regions with regulatory roles that bind to thioredoxin and TNF receptor-associated factors (TRAFs), which regulate the response of ASK1 to ROS and cytokines, respectively [54]. The N-terminal region of ASK1 has also been implicated in binding CIB1 to detect Ca^{2+}-based stress signaling and in binding FBXO21 to trigger innate antiviral signaling [55,56]. The region C terminal to the kinase domain contains a 14-3-3 protein-binding site [57], followed by a region for constitutive oligomerization of ASK1 [58]. Under redox stress, thioredoxin dissociates and TRAF proteins associate with ASK1, which then tightly oligomerizes through its N-terminal coiled-coil (NCC) domain, promoting ASK1 activation and kinase activity via autophosphorylation of Thr845 in its kinase domain [59]. However, oxidative stress induces ubiquitination and subsequent degradation of activated ASK1. On the other hand, the deubiquitinating enzyme USP9X, which has ubiquitin-specific protease activity, interacts and antagonizes ubiquitination and subsequent degradation of activated ASK1 in H2O2-treated cells, resulting in the stabilization of activated ASK1 [60]. However, TNFAIP3 (A20), which has both ubiquitin ligase and deubiquitinase activities, inactivates ASK1 in fatty-acid-induced cells and ameliorates NASH [61]. Overexpression of A20 interacts with ASK1 and reduces stability and promotes the degradation of ASK1 through the ubiquitination process. Thus, overexpression of A20 lowers ASK1 level and preconditions cells to resist stress-induced JNK activation and cell death [62]. A20 is an acute response gene and overall effects of A20 will also depend on context and acute versus chronic disease. Contribution of A20 on JNK-SAB-ROS activation loop appears to be of minor importance. In addition to phosphorylation- and ubiquitination-mediated regulation of ASK1 in basal and response to oxidative stress, cFLIP competes for binding to TRAF binding domain of ASK1 and prevents ASK1 dimerization and activation [63]. However, ITCH promotes Lys48 ubiquitination and degradation of cFLIP. ITCH is activated by JNK [64]. JNK-ITCH-Ask1 signal activation axis does not affect TNF/GalN-induced apoptosis in contexts where SAB is deleted because depletion of SAB or MKK4/7 completely prevents TNFα-induced JNK activation and cell death, indicating requirement of MAP2K, JNK, and SAB to sustain ASK1 activation [7,12,65]. As mitochondria have thioredoxin-2, mitochondria localization of ASK1 and association with thioredoxin-2 has been proposed [66]. However, JNK activation has not been observed in mitochondria and further exploration is required. Another ASK1 regulator that has been recently identified is Caspase Recruitment Domain Protein 6 (CARD6) [67,68]. CARD6 associates with ASK1 and suppresses ASK1 phosphorylation activation and downstream JNK/p38 activation. In high-fat diet-induced fatty liver model, ASK1, MLK3, and TAK1 activation occurs. ASK1 phosphorylation is further increased by CARD6 deficiency but suppressed by CARD6 overexpression. TAK1 phosphorylation is not affected by CARD6, indicating selectivity of regulation. TAK1 is phosphorylated and activated via TLR/IL1 receptor and TRAFs [69]. Phosphorylated TAK1 activates IKK and MAP2K, leading to activation of NF-κB and JNK, respectively. Hepatocyte-specific deletion of TAK1 causes spontaneous hepatocyte death, suggesting hepatoprotective role of TAK1 [70], although further exploration is required.

MLK2/3 are redundant MAP3Ks regulated by small GTPases CDC42 and RAC1 [71]. Recently, JIP1 has been identified as a platform for interaction and signal integration of SRC tyrosine kinase, RAC GTPase, and MLKs for activation of MKK7 and JNK in free fatty acid-induced activation model [19]. As JIP is a JNK-specific scaffold protein, the pathway selectively activates JNK but not p38 in free fatty acid-induced stress in MEF cells and diet-induced mouse models. The role of ASK1 was not examined in this free fatty acid-induced stress in MEF cell model. Indeed, JIP3/4 interacts with ASK1 but cannot mediate JNK activation [72]. The importance of JIP/MAP3K/MAP2K/MAPK signaling pathway in death receptor (TNF)-mediated MAP kinase activation requires further exploration.

3.4. Scaffold Protein—SAB

SAB is a mitochondrial outer membrane protein with N-terminal SH3 domain binding site, one membrane spanning domain, and two D-motif (KIM) on C-terminus [11]. The topology of SAB makes it unique in JNK-mediated signal transduction to mitochondria. The N-terminal of SAB, including SH3-domain-binding site, is in the mitochondria intermembrane space, and C-terminal of SAB with KIM motif is facing the cytoplasm [12]. SAB is the only JNK docking site on mitochondria. The depletion of SAB completely prevents JNK translocation to mitochondria [7,12]. Both JNK and p38 can phosphorylate SAB in cell-free system [73], but in vivo evidence is lacking. The deletion of SAB does not inhibit p38 association with mitochondria [65]. The SH3-domain-binding site of SAB is largely unexplored. The recent identification of the SAB homolog RAB-11-interacting protein-1 (REI-1), a guanine nucleotide exchange factor (GEF) that is homologous to the N-terminal of mammalian SAB, is associated with Rab11 of *C. elegans* [74]. Thus, further explorations are required to examine the role of RAB GTPase in JNK-SAB-ROS activation loop. As PTPN6 dissociates from SAB when JNK interacts with and phosphorylates SAB, there could be possible regulation of the PTPN6 dissociation from SAB. Therefore, RAB like GTPase could be associated with intramitochondrial portion of SAB and might participate in regulation of the JNK-SAB-ROS activation loop. There are other GTPases that have been identified as facing into the intermitochondrial membrane space, such as OPA1, which is regulated by SIRT3 [75], but the association with SAB is not known.

3.5. Regulation of SAB Expression

We have recently begun to address the role of the regulation of SAB expression. We have initially gained insight into this area through overexpression of SAB as well as through exploration of sex differences in susceptibility to acute liver injury in mouse models. We expressed Adeno-SAB in liver-specific SAB knockout mice using increasing doses of adenovirus and found that increasing levels of SAB expression led to increasing susceptibility to injury from a fixed nonlethal dose of APAP. This not only demonstrates that SAB restores susceptibility to liver injury in SAB knockout mice but also that the level of SAB expression determined the severity of liver injury. Furthermore, inducible hepatocyte knockout of JNK1 and 2 (AAV8-TBG-CRE) in *JNK1/2^{fl/fl}* mice markedly protected against APAP injury, which was not increased with concomitant SAB overexpression. This indicates that JNK is required for enhanced susceptibility to APAP injury due to SAB overexpression and that there is no other pathway (other than JNK) for the participation of SAB in the injury process. Furthermore, JNK1/2 deletion did not affect SAB basal expression and vice versa.

It is well known that female mice are very resistant to APAP toxicity in vivo. We confirmed this and found that the resistance applied to TNF/GalN in vivo as well as palmitic acid-induced lipoapoptosis in primary mouse hepatocytes. In all these models, female littermates exhibited markedly decreased levels of sustained JNK activation. This led us to examine SAB expression, which was found to be markedly decreased in females (only 15% of male liver mitochondrial level of SAB). Similar sex difference in SAB expression was observed in normal human liver. We then identified post-transcriptional regulation of SAB expression (repression in females) involving a pathway from estrogen receptor-α to p53 (higher expression in female mouse and human liver) to p53-mediated expression of miR34a-5p, which targets the SAB mRNA coding region, thus repressing SAB expression and decreasing susceptibility to liver injury (abstract, manuscript in preparation). There is currently no information on the transcriptional regulation of SAB expression, and this is an important area we are exploring.

4. Perspectives on the Intervention of the JNK Activation Loop

JNK-SAB-ROS activation loop is an important cell death-promoting pathway in apoptosis and mitochondrial permeability transition pore (MPT)-regulated necrosis (in the context of acetaminophen hepatotoxicity). The pathway is modulated by several parallel survival pathways through crosstalk and negative regulatory feedback. Any adaptation or mechanism changing the balance of survival and death pathways will partially interfere with the JNK-SAB-ROS pathway directly or indirectly and the cell death outcome. Thus, targeting molecules in JNK-SAB-ROS activation loop is a promising strategy to promote cell death, such as in cancer cells [76,77], and to prevent cell death, such as in hepatotoxicity [7,12,65], liver and kidney injury in septic shock, and ischemia/reperfusion injury in heart and brain [78–81]. A selective ASK1 inhibitor, selonsertib (GS-4997), has recently been tested as therapy for NASH in a phase 2 clinical trial (NCT02466516), and patient outcomes were encouraging. Targeting the pivotal role of SAB in JNK activation offers particular promise. Blocking the binding of P-JNK to SAB using KIM1 peptides can be selectively achieved without directly blocking the kinase activity of JNK [8,9,82]. Thus, identification of selective small molecule inhibitors of the binding of P-JNK to SAB seems feasible. Modulating expression of SAB (increase or decrease) may be possible through modulation of the factors that control transcriptional and post-translational regulation (e.g., transcription factors and noncoding RNA that target SAB expression). In addition, antisense oligonucleotides that are cell-type- or organ-specific are being developed to lower SAB expression. These approaches for modification of SAB expression appear to offer the most promise in chronic diseases where sustained JNK activation affects metabolism.

Author Contributions: All authors participated and contributed in discussion, preparation and writing of this review manuscript.

Funding: This work was supported by R01DK067215 (to N.K.), the USC Research Center for Liver Disease's Cell Culture, Cell and Tissue Imaging, Histology and Metabolic/Analytical/Instrumentation Cores (P30DK048522; to N.K.), Baxter Foundation award (S.W.), and Budnick Chair in Liver Disease (N.K.).

Conflicts of Interest: The authors have no relevant conflict of interest to disclose.

Abbreviations

AKT	AKT serine/threonine kinase 1
APAP	acetaminophen
ASK1	apoptosis signal-regulating kinase 1
ATF2	activating transcription factor 2
ATP	adenosine triphosphate
CDC42	cell division cycle 42
cFLIP	CASP8 and FADD like apoptosis regulator
c-JUN	Jun proto-oncogene
DOK4	downstream of tyrosine kinase/docking protein 4
DUSP	dual-specificity phosphatase
ER	endoplasmic reticulum
ERK	extracellular signal-regulated kinase
GalN	*N*-acetyl-galactosamine
ITCH	Itchy E3 ubiquitin protein ligase
JIP	JNK-interacting protein
JNK	c-Jun N-terminal kinases
KIM	kinase interaction motif
KO	knock out
LPS	lipopolysaccharide
MAPK	mitogen-activated protein kinase
MAP2K	mitogen-activated protein kinase kinase
MAP3K	mitogen-activated protein kinase kinase kinase
MKK4	MAPK kinase 4

MKP	mitogen-activated protein kinase phosphatase
MLK	mixed lineage kinase
MPT	mitochondrial permeability transition
NASH	nonalcoholic steatohepatitis
NCOR	nuclear receptor corepressor
NF-κB	nuclear factor kappa-light-chain-enhancer of activated B cells
P-JNK	phosphoactivated JNK
RAC	Rac family small GTPase 1
ROS	reactive oxygen species
SAB (Sh3bp5)	SH3-domain binding protein 5
SH3	SRC Homology 3 Domain
SHP1	SH2 phosphatase 1
SRC	SRC proto-oncogene non-receptor tyrosine kinase
TAK1	TGF-beta activated kinase 1
TNF	tumor necrosis factor

References

1. Bogoyevitch, M.A.; Kobe, B. Uses for JNK: The many and varied substrates of the c-Jun N-terminal kinases. *Microbiol. Mol. Biol. Rev.* **2006**, *70*, 1061–1095. [CrossRef] [PubMed]

2. Weston, C.R.; Davis, R.J. The JNK signal transduction pathway. *Curr. Opin. Cell. Biol.* **2007**, *19*, 142–149. [CrossRef] [PubMed]

3. Win, S.; Than, T.A.; Zhang, J.; Oo, C.; Min, R.W.M.; Kaplowitz, N. New insights into the role and mechanism of c-Jun-N-terminal kinase signaling in the pathobiology of liver diseases. *Hepatology* **2018**, *67*, 2013–2024. [CrossRef] [PubMed]

4. Tobiume, K.; Matsuzawa, A.; Takahashi, T.; Nishitoh, H.; Morita, K.; Takeda, K.; Minowa, O.; Miyazono, K.; Noda, T.; Ichijo, H. ASK1 is required for sustained activations of JNK/p38 MAP kinases and apoptosis. *EMBO Rep.* **2001**, *2*, 222–228. [CrossRef] [PubMed]

5. Gunawan, B.K.; Liu, Z.X.; Han, D.; Hanawa, N.; Gaarde, W.A.; Kaplowitz, N. c-Jun N-terminal kinase plays a major role in murine acetaminophen hepatotoxicity. *Gastroenterology* **2006**, *131*, 165–178. [CrossRef] [PubMed]

6. Hanawa, N.; Shinohara, M.; Saberi, B.; Gaarde, W.A.; Han, D.; Kaplowitz, N. Role of JNK translocation to mitochondria leading to inhibition of mitochondria bioenergetics in acetaminophen-induced liver injury. *J. Biol. Chem.* **2008**, *283*, 13565–13577. [CrossRef] [PubMed]

7. Win, S.; Than, T.A.; Han, D.; Petrovic, L.M.; Kaplowitz, N. c-Jun N-terminal kinase (JNK)-dependent acute liver injury from acetaminophen or tumor necrosis factor (TNF) requires mitochondrial Sab protein expression in mice. *J. Biol. Chem.* **2011**, *286*, 35071–35078. [CrossRef] [PubMed]

8. Win, S.; Than, T.A.; Fernandez-Checa, J.C.; Kaplowitz, N. JNK interaction with Sab mediates ER stress induced inhibition of mitochondrial respiration and cell death. *Cell Death Dis.* **2014**, *5*, e989. [CrossRef] [PubMed]

9. Win, S.; Than, T.A.; Le, B.H.; García-Ruiz, C.; Fernandez-Checa, J.C.; Kaplowitz, N. Sab (Sh3bp5) dependence of JNK mediated inhibition of mitochondrial respiration in palmitic acid induced hepatocyte lipotoxicity. *J. Hepatol.* **2015**, *62*, 1367–1374. [CrossRef] [PubMed]

10. Kamata, H.; Honda, S.; Maeda, S.; Chang, L.; Hirata, H.; Karin, M. Reactive oxygen species promote TNFalpha-induced death and sustained JNK activation by inhibiting MAP kinase phosphatases. *Cell* **2005**, *120*, 649–661. [CrossRef] [PubMed]

11. Wiltshire, C.; Matsushita, M.; Tsukada, S.; Gillespie, D.A.; May, G.H. A new c-Jun N-terminal kinase (JNK)-interacting protein, Sab (SH3BP5), associates with mitochondria. *Biochem. J.* **2002**, *367*, 577–585. [CrossRef] [PubMed]

12. Win, S.; Than, T.A.; Min, R.W.; Aghajan, M.; Kaplowitz, N. c-Jun N-terminal kinase mediates mouse liver injury through a novel Sab (SH3BP5)-dependent pathway leading to inactivation of intramitochondrial Src. *Hepatology* **2016**, *63*, 1987–2003. [CrossRef] [PubMed]

13. Liu, Q.; Rehman, H.; Krishnasamy, Y.; Schnellmann, R.G.; Lemasters, J.J.; Zhong, Z. Improvement of liver injury and survival by JNK2 and iNOS deficiency in liver transplants from cardiac death mice. *J. Hepatol.* **2015**, *63*, 68–74. [CrossRef] [PubMed]

14. Huo, Y.; Win, S.; Than, T.A.; Yin, S.; Ye, M.; Hu, H.; Kaplowitz, N. Antcin H Protects Against Acute Liver Injury Through Disruption of the Interaction of c-Jun-N-Terminal Kinase with Mitochondria. *Antioxid. Redox Signal.* **2017**, *26*, 207–220. [CrossRef] [PubMed]

15. Itoh, S.; Lemay, S.; Osawa, M.; Che, W.; Duan, Y.; Tompkins, A.; Brookes, P.S.; Sheu, S.S.; Abe, J. Mitochondrial Dok-4 recruits Src kinase and regulates NF-kappaB activation in endothelial cells. *J. Biol. Chem.* **2005**, *280*, 26383–26396. [CrossRef] [PubMed]

16. Mishra, P.; Günther, S. New insights into the structural dynamics of the kinase JNK3. *Sci. Rep.* **2018**, *8*, 9435. [CrossRef] [PubMed]

17. Laughlin, J.D.; Nwachukwu, J.C.; Figuera-Losada, M.; Cherry, L.; Nettles, K.W.; LoGrasso, P.V. Structural mechanisms of allostery and autoinhibition in JNK family kinases. *Structure* **2012**, *20*, 2174–2184. [CrossRef] [PubMed]

18. Jaeschke, A.; Czech, M.P.; Davis, R.J. An essential role of the JIP1 scaffold protein for JNK activation in adipose tissue. *Genes Dev.* **2004**, *18*, 1976–1980. [CrossRef] [PubMed]

19. Morel, C.; Standen, C.L.; Jung, D.Y.; Gray, S.; Ong, H.; Flavell, R.A.; Kim, J.K.; Davis, R.J. Requirement of JIP1-mediated c-Jun N-terminal kinase activation for obesity-induced insulin resistance. *Mol. Cell. Biol.* **2010**, *30*, 4616–4625. [CrossRef] [PubMed]

20. Kant, S.; Standen, C.L.; Morel, C.; Jung, D.Y.; Kim, J.K.; Swat, W.; Flavell, R.A.; Davis, R.J. A Protein Scaffold Coordinates SRC-Mediated JNK Activation in Response to Metabolic Stress. *Cell Rep.* **2017**, *20*, 2775–2783. [CrossRef] [PubMed]

21. Chambers, J.W.; LoGrasso, P.V. Mitochondrial c-Jun N-terminal kinase (JNK) signaling initiates physiological changes resulting in amplification of reactive oxygen species generation. *J. Biol. Chem.* **2011**, *286*, 16052–16062. [CrossRef] [PubMed]

22. Fuchs, S.Y.; Adler, V.; Pincus, M.R.; Ronai, Z. MEKK1/JNK signaling stabilizes and activates p53. *Proc. Natl. Acad. Sci. USA* **1998**, *95*, 10541–10546. [CrossRef] [PubMed]

23. Cargnello, M.; Roux, P.P. Activation and function of the MAPKs and their substrates, the MAPK-activated protein kinases. *Microbiol. Mol. Biol. Rev.* **2011**, *75*, 50–83. [CrossRef] [PubMed]

24. Sabio, G.; Davis, R.J. TNF and MAP kinase signaling pathways. *Semin. Immunol.* **2014**, *26*, 237–245. [CrossRef] [PubMed]

25. Pérez, S.; Rius-Pérez, S.; Tormos, A.M.; Finamor, I.; Nebreda, Á.R.; Taléns-Visconti, R.; Sastre, J. Age-dependent regulation of antioxidant genes by p38α MAPK in the liver. *Redox Biol.* **2018**, *16*, 276–284. [CrossRef] [PubMed]

26. Weber, H.O.; Ludwig, R.L.; Morrison, D.; Kotlyarov, A.; Gaestel, M.; Vousden, K.H. HDM2 phosphorylation by MAPKAP kinase 2. *Oncogene* **2005**, *24*, 1965–1972. [CrossRef] [PubMed]

27. Malmlöf, M.; Roudier, E.; Högberg, J.; Stenius, U. MEK-ERK-mediated phosphorylation of Mdm2 at Ser-166 in hepatocytes. Mdm2 is activated in response to inhibited Akt signaling. *J. Biol. Chem.* **2007**, *282*, 2288–2296. [CrossRef] [PubMed]

28. Alarcon-Vargas, D.; Ronai, Z. p53-Mdm2–the affair that never ends. *Carcinogenesis* **2002**, *23*, 541–547. [CrossRef] [PubMed]

29. Carr, M.I.; Jones, S.N. Regulation of the Mdm2-p53 signaling axis in the DNA damage response and tumorigenesis. *Transl. Cancer Res.* **2016**, *5*, 707–724. [CrossRef] [PubMed]

30. Huo, Y.; Yin, S.; Yan, M.; Win, S.; Aung Than, T.; Aghajan, M.; Hu, H.; Kaplowitz, N. Protective role of p53 in acetaminophen hepatotoxicity. *Free Radic Biol. Med.* **2017**, *106*, 111–117. [CrossRef] [PubMed]

31. Huang, S.; Shu, L.; Dilling, M.B.; Easton, J.; Harwood, F.C.; Ichijo, H.; Houghton, P.J. Sustained activation of the JNK cascade and rapamycin-induced apoptosis are suppressed by p53/p21(Cip1). *Mol. Cell* **2003**, *11*, 1491–1501. [CrossRef]

32. Huang, C.Y.; Tan, T.H. DUSPs, to MAP kinases and beyond. *Cell Biosci.* **2012**, *2*, 24. [CrossRef] [PubMed]

33. Sarikhani, M.; Mishra, S.; Desingu, P.A.; Kotyada, C.; Wolfgeher, D.; Gupta, M.P.; Singh, M.; Sundaresan, N.R. SIRT2 regulates oxidative stress-induced cell death through deacetylation of c-Jun NH2-terminal kinase. *Cell Death Differ.* **2018**, *25*, 1638–1656. [CrossRef] [PubMed]

34. Villalba, J.M.; Alcaín, F.J. Sirtuin activators and inhibitors. *Biofactors* **2012**, *38*, 349–359. [CrossRef] [PubMed]

35. Cui, X.; Chen, Q.; Dong, Z.; Xu, L.; Lu, T.; Li, D.; Zhang, J.; Zhang, M.; Xia, Q. Inactivation of Sirt1 in mouse livers protects against endotoxemic liver injury by acetylating and activating NF-κB. *Cell Death Dis.* **2016**, *7*, e2403. [CrossRef] [PubMed]

36. Singh, C.K.; Chhabra, G.; Ndiaye, M.A.; Garcia-Peterson, L.M.; Mack, N.J.; Ahmad, N. The Role of Sirtuins in Antioxidant and Redox Signaling. *Antioxid. Redox Signal.* **2018**, *28*, 643–661. [CrossRef] [PubMed]

37. Ho, D.T.; Bardwell, A.J.; Abdollahi, M.; Bardwell, L. A docking site in MKK4 mediates high affinity binding to JNK MAPKs and competes with similar docking sites in JNK substrates. *J. Biol. Chem.* **2003**, *278*, 32662–32672. [CrossRef] [PubMed]

38. Kragelj, J.; Palencia, A.; Nanao, M.H.; Maurin, D.; Bouvignies, G.; Blackledge, M.; Jensen, M.R. Structure and dynamics of the MKK7-JNK signaling complex. *Proc. Natl. Acad. Sci. USA* **2015**, *112*, 3409–3414. [CrossRef] [PubMed]

39. Sun, Y.; Li, T.Y.; Song, L.; Zhang, C.; Li, J.; Lin, Z.Z.; Lin, S.C.; Lin, S.Y. Liver-specific deficiency of unc-51 like kinase 1 and 2 protects mice from acetaminophen-induced liver injury. *Hepatology* **2018**, *67*, 2397–2413. [CrossRef] [PubMed]

40. Zeke, A.; Misheva, M.; Reményi, A.; Bogoyevitch, M.A. JNK Signaling: Regulation and Functions Based on Complex Protein-Protein Partnerships. *Microbiol. Mol. Biol. Rev.* **2016**, *80*, 793–835. [CrossRef] [PubMed]

41. Heinrichsdorff, J.; Luedde, T.; Perdiguero, E.; Nebreda, A.R.; Pasparakis, M. p38 alpha MAPK inhibits JNK activation and collaborates with IkappaB kinase 2 to prevent endotoxin-induced liver failure. *EMBO Rep.* **2008**, *9*, 1048–1054. [CrossRef] [PubMed]

42. Zarubin, T.; Han, J. Activation and signaling of the p38 MAP kinase pathway. *Cell Res.* **2005**, *15*, 11–18. [CrossRef] [PubMed]

43. Girnius, N.; Davis, R.J. TNFα-Mediated Cytotoxic Responses to IAP Inhibition Are Limited by the p38α MAPK Pathway. *Cancer Cell* **2016**, *29*, 131–133. [CrossRef] [PubMed]

44. Lalaoui, N.; Hänggi, K.; Brumatti, G.; Chau, D.; Nguyen, N.N.; Vasilikos, L.; Spilgies, L.M.; Heckmann, D.A.; Ma, C.; Ghisi, M.; et al. Targeting p38 or MK2 Enhances the Anti-Leukemic Activity of Smac-Mimetics. *Cancer Cell* **2016**, *30*, 499–500. [CrossRef] [PubMed]

45. De Smaele, E.; Zazzeroni, F.; Papa, S.; Nguyen, D.U.; Jin, R.; Jones, J.; Cong, R.; Franzoso, G. Induction of gadd45beta by NF-kappaB downregulates pro-apoptotic JNK signaling. *Nature* **2001**, *414*, 308–313. [CrossRef] [PubMed]

46. Papa, S.; Zazzeroni, F.; Bubici, C.; Jayawardena, S.; Alvarez, K.; Matsuda, S.; Nguyen, D.U.; Pham, C.G.; Nelsbach, A.H.; Melis, T.; et al. Gadd45 beta mediates the NF-kappa B suppression of JNK signaling by targeting MKK7/JNKK2. *Nat. Cell Biol.* **2004**, *6*, 146–153. [CrossRef] [PubMed]

47. Bai, D.; Ueno, L.; Vogt, P.K. Akt-mediated regulation of NFkappaB and the essentialness of NFkappaB for the oncogenicity of PI3K and Akt. *Int. J. Cancer* **2009**, *125*, 2863–2870. [CrossRef] [PubMed]

48. Barthwal, M.K.; Sathyanarayana, P.; Kundu, C.N.; Rana, B.; Pradeep, A.; Sharma, C.; Woodgett, J.R.; Rana, A. Negative regulation of mixed lineage kinase 3 by protein kinase B/AKT leads to cell survival. *J. Biol. Chem.* **2003**, *278*, 3897–3902. [CrossRef] [PubMed]

49. Lin, Z.; Liu, T.; Kamp, D.W.; Wang, Y.; He, H.; Zhou, X.; Li, D.; Yang, L.; Zhao, B.; Liu, G. AKT/mTOR and c-Jun N-terminal kinase signaling pathways are required for chrysotile asbestos-induced autophagy. *Free Radic. Biol. Med.* **2014**, *72*, 296–307. [CrossRef] [PubMed]

50. Song, J.J.; Lee, Y.J. Dissociation of Akt1 from its negative regulator JIP1 is mediated through the ASK1-MEK-JNK signal transduction pathway during metabolic oxidative stress: A negative feedback loop. *J. Cell Biol.* **2005**, *170*, 61–72. [CrossRef] [PubMed]

51. Soga, M.; Matsuzawa, A.; Ichijo, H. Oxidative Stress-Induced Diseases via the ASK1 Signaling Pathway. *Int. J. Cell Biol.* **2012**, *2012*, 439587. [CrossRef] [PubMed]

52. Ichijo, H.; Nishida, E.; Irie, K.; ten Dijke, P.; Saitoh, M.; Moriguchi, T.; Takagi, M.; Matsumoto, K.; Miyazono, K.; Gotoh, Y. Induction of apoptosis by ASK1, a mammalian MAPKKK that activates SAPK/JNK and p38 signaling pathways. *Science* **1997**, *275*, 90–94. [CrossRef] [PubMed]

53. Matsuzawa, A.; Saegusa, K.; Noguchi, T.; Sadamitsu, C.; Nishitoh, H.; Nagai, S.; Koyasu, S.; Matsumoto, K.; Takeda, K.; Ichijo, H. ROS-dependent activation of the TRAF6-ASK1-p38 pathway is selectively required for TLR4-mediated innate immunity. *Nat. Immunol.* **2005**, *6*, 587–592. [CrossRef] [PubMed]

54. Weijman, J.F.; Kumar, A.; Jamieson, S.A.; King, C.M.; Caradoc-Davies, T.T.; Ledgerwood, E.C.; Murphy, J.M.; Mace, P.D. Structural basis of autoregulatory scaffolding by apoptosis signal-regulating kinase 1. *Proc. Natl. Acad. Sci. USA* **2017**, *114*, E2096–E2105. [CrossRef] [PubMed]

55. Yoon, K.W.; Cho, J.H.; Lee, J.K.; Kang, Y.H.; Chae, J.S.; Kim, Y.M.; Kim, J.; Kim, E.K.; Kim, S.E.; Baik, J.H.; et al. CIB1 functions as a Ca(2+)-sensitive modulator of stress-induced signaling by targeting ASK1. *Proc. Natl. Acad. Sci. USA* **2009**, *106*, 17389–17394. [CrossRef] [PubMed]

56. Yu, Z.; Chen, T.; Li, X.; Yang, M.; Tang, S.; Zhu, X.; Gu, Y.; Su, X.; Xia, M.; Li, W.; et al. Lys29-linkage of ASK1 by Skp1-Cullin 1-Fbxo21 ubiquitin ligase complex is required for antiviral innate response. *Elife* **2016**, *5*, e14087. [CrossRef] [PubMed]

57. Cockrell, L.M.; Puckett, M.C.; Goldman, E.H.; Khuri, F.R.; Fu, H. Dual engagement of 14-3-3 proteins controls signal relay from ASK2 to the ASK1 signalosome. *Oncogene* **2010**, *29*, 822–830. [CrossRef] [PubMed]

58. Kawarazaki, Y.; Ichijo, H.; Naguro, I. Apoptosis signal-regulating kinase 1 as a therapeutic target. *Expert Opin. Ther. Targets* **2014**, *18*, 651–664. [CrossRef] [PubMed]

59. Cebula, M.; Schmidt, E.E.; Arnér, E.S. TrxR1 as a potent regulator of the Nrf2-Keap1 response system. *Antioxid. Redox Signal.* **2015**, *23*, 823–853. [CrossRef] [PubMed]

60. Nagai, H.; Noguchi, T.; Homma, K.; Katagiri, K.; Takeda, K.; Matsuzawa, A.; Ichijo, H. Ubiquitin-like sequence in ASK1 plays critical roles in the recognition and stabilization by USP9X and oxidative stress-induced cell death. *Mol. Cell* **2009**, *36*, 805–818. [CrossRef] [PubMed]

61. Zhang, P.; Wang, P.X.; Zhao, L.P.; Zhang, X.; Ji, Y.X.; Zhang, X.J.; Fang, C.; Lu, Y.X.; Yang, X.; Gao, M.M.; et al. The deubiquitinating enzyme TNFAIP3 mediates inactivation of hepatic ASK1 and ameliorates nonalcoholic steatohepatitis. *Nat. Med.* **2018**, *24*, 84–94. [CrossRef] [PubMed]

62. Won, M.; Park, K.A.; Byun, H.S.; Sohn, K.C.; Kim, Y.R.; Jeon, J.; Hong, J.H.; Park, J.; Seok, J.H.; Kim, J.M.; et al. Novel anti-apoptotic mechanism of A20 through targeting ASK1 to suppress TNF-induced JNK activation. *Cell Death Differ.* **2010**, *17*, 1830–1841. [CrossRef] [PubMed]

63. Wang, P.X.; Ji, Y.X.; Zhang, X.J.; Zhao, L.P.; Yan, Z.Z.; Zhang, P.; Shen, L.J.; Yang, X.; Fang, J.; Tian, S.; et al. Targeting CASP8 and FADD-like apoptosis regulator ameliorates nonalcoholic steatohepatitis in mice and nonhuman primates. *Nat. Med.* **2017**, *23*, 439–449. [CrossRef] [PubMed]

64. Chang, L.; Kamata, H.; Solinas, G.; Luo, J.L.; Maeda, S.; Venuprasad, K.; Liu, Y.C.; Karin, M. The E3 ubiquitin ligase itch couples JNK activation to TNFalpha-induced cell death by inducing c-FLIP(L.) turnover. *Cell* **2006**, *124*, 601–613. [CrossRef] [PubMed]

65. Zhang, J.; Min, R.W.M.; Le, K.; Zhou, S.; Aghajan, M.; Than, T.A.; Win, S.; Kaplowitz, N. The role of MAP2 kinases and p38 kinase in acute murine liver injury models. *Cell Death Dis.* **2017**, *8*, e2903. [CrossRef] [PubMed]

66. Zhang, R.; Al-Lamki, R.; Bai, L.; Streb, J.W.; Miano, J.M.; Bradley, J.; Min, W. Thioredoxin-2 inhibits mitochondria-located ASK1-mediated apoptosis in a JNK-independent manner. *Circ. Res.* **2004**, *94*, 1483–1491. [CrossRef] [PubMed]

67. Qin, J.J.; Mao, W.; Wang, X.; Sun, P.; Cheng, D.; Tian, S.; Zhu, X.Y.; Yang, L.; Huang, Z.; Li, H. Caspase recruitment domain 6 protects against hepatic ischemia/reperfusion injury by suppressing ASK1. *J. Hepatol.* **2018**, *69*, 1110–1122. [CrossRef] [PubMed]

68. Sun, P.; Zeng, Q.; Cheng, D.; Zhang, K.; Zheng, J.; Liu, Y.; Yuan, Y.F.; Tang, Y.D. Caspase Recruitment Domain Protein 6 protects against hepatic steatosis and insulin resistance by suppressing Ask1. *Hepatology* **2018**. [CrossRef]

69. Takaesu, G.; Ninomiya-Tsuji, J.; Kishida, S.; Li, X.; Stark, G.R.; Matsumoto, K. Interleukin-1 (IL-1) receptor-associated kinase leads to activation of TAK1 by inducing TAB2 translocation in the IL-1 signaling pathway. *Mol. Cell. Biol.* **2001**, *21*, 2475–2484. [CrossRef] [PubMed]

70. Malato, Y.; Willenbring, H. The MAP3K TAK1: A chock block to liver cancer formation. *Hepatology* **2010**, *52*, 1506–1509. [CrossRef] [PubMed]

71. Sharma, M.; Urano, F.; Jaeschke, A. Cdc42 and Rac1 are major contributors to the saturated fatty acid-stimulated JNK pathway in hepatocytes. *J. Hepatol.* **2012**, *56*, 192–198. [CrossRef] [PubMed]

72. Kelkar, N.; Standen, C.L.; Davis, R.J. Role of the JIP4 scaffold protein in the regulation of mitogen-activated protein kinase signaling pathways. *Mol. Cell. Biol.* **2005**, *25*, 2733–2743. [CrossRef] [PubMed]

73. Court, N.W.; Kuo, I.; Quigley, O.; Bogoyevitch, M.A. Phosphorylation of the mitochondrial protein Sab by stress-activated protein kinase 3. *Biochem. Biophys. Res. Commun.* **2004**, *319*, 130–137. [CrossRef] [PubMed]

74. Sakaguchi, A.; Sato, M.; Sato, K.; Gengyo-Ando, K.; Yorimitsu, T.; Nakai, J.; Hara, T.; Sato, K.; Sato, K. REI-1 Is a Guanine Nucleotide Exchange Factor Regulating RAB-11 Localization and Function in C. elegans Embryos. *Dev. Cell* **2015**, *35*, 211–221. [CrossRef] [PubMed]

75. Samant, S.A.; Zhang, H.J.; Hong, Z.; Pillai, V.B.; Sundaresan, N.R.; Wolfgeher, D.; Archer, S.L.; Chan, D.C.; Gupta, M.P. SIRT3 deacetylates and activates OPA1 to regulate mitochondrial dynamics during stress. *Mol. Cell. Biol.* **2014**, *34*, 807–819. [CrossRef] [PubMed]

76. Paudel, I.; Hernandez, S.M.; Portalatin, G.M.; Chambers, T.P.; Chambers, J.W. Sab Concentrations Indicate Chemotherapeutic Susceptibility in Ovarian Cancer Cell Lines. *Biochem. J.* **2018**, *475*, 3471–3492. [CrossRef] [PubMed]

77. Chambers, T.P.; Portalatin, G.M.; Paudel, I.; Robbins, C.J.; Chambers, J.W. Sub-chronic administration of LY294002 sensitizes cervical cancer cells to chemotherapy by enhancing mitochondrial JNK signaling. *Biochem. Biophys. Res. Commun.* **2015**, *463*, 538–544. [CrossRef] [PubMed]

78. Chambers, J.W.; Pachori, A.; Howard, S.; Iqbal, S.; LoGrasso, P.V. Inhibition of JNK mitochondrial localization and signaling is protective against ischemia/reperfusion injury in rats. *J. Biol. Chem.* **2013**, *288*, 4000–4011. [CrossRef] [PubMed]

79. Chambers, T.P.; Santiesteban, L.; Gomez, D.; Chambers, J.W. Sab mediates mitochondrial dysfunction involved in imatinib mesylate-induced cardiotoxicity. *Toxicology* **2017**, *382*, 24–35. [CrossRef] [PubMed]

80. Chambers, J.W.; Howard, S.; LoGrasso, P.V. Blocking c-Jun N-terminal kinase (JNK) translocation to the mitochondria prevents 6-hydroxydopamine-induced toxicity in vitro and in vivo. *J. Biol. Chem.* **2013**, *288*, 1079–1087. [CrossRef] [PubMed]

81. Chambers, J.W.; Pachori, A.; Howard, S.; Ganno, M.; Hansen, D.J.; Kamenecka, T.; Song, X.; Duckett, D.; Chen, W.; Ling, Y.Y.; et al. Small Molecule c-jun-N-terminal Kinase (JNK) Inhibitors Protect Dopaminergic Neurons in a Model of Parkinson's Disease. *ACS Chem. Neurosci.* **2011**, *2*, 198–206. [CrossRef] [PubMed]

82. Chambers, J.W.; Cherry, L.; Laughlin, J.D.; Figuera-Losada, M.; Lograsso, P.V. Selective inhibition of mitochondrial JNK signaling achieved using peptide mimicry of the Sab kinase interacting motif-1 (KIM1). *ACS Chem. Biol.* **2011**, *6*, 808–818. [CrossRef] [PubMed]

International Journal of
Molecular Sciences

Article

Novel Functions of Death-Associated Protein Kinases through Mitogen-Activated Protein Kinase-Related Signals

Mohamed Elbadawy [1,2,†], Tatsuya Usui [1,*,†], Hideyuki Yamawaki [3] and Kazuaki Sasaki [1]

[1] Laboratory of Veterinary Pharmacology, Department of Veterinary Medicine, Faculty of Agriculture, Tokyo University of Agriculture and Technology, 3-5-8 Saiwai-cho, Fuchu, Tokyo 183-8509, Japan; mohamed.elbadawy@fvtm.bu.edu.eg (M.E.); skazuaki@cc.tuat.ac.jp (K.S.)

[2] Department of Pharmacology, Faculty of Veterinary Medicine, Benha University, Moshtohor, Elqaliobiya, Toukh 13736, Egypt

[3] Laboratory of Veterinary Pharmacology, School of Veterinary Medicine, Kitasato University, Towada, Aomori 034-8628, Japan; yamawaki@vmas.kitasato-u.ac.jp

* Correspondence: fu7085@go.tuat.ac.jp; Tel./Fax: +81-42-367-5769

† These authors contributed equally to this work.

Received: 13 September 2018; Accepted: 1 October 2018; Published: 4 October 2018

Abstract: Death associated protein kinase (DAPK) is a calcium/calmodulin-regulated serine/threonine kinase; its main function is to regulate cell death. DAPK family proteins consist of DAPK1, DAPK2, DAPK3, DAPK-related apoptosis-inducing protein kinases (DRAK)-1 and DRAK-2. In this review, we discuss the roles and regulatory mechanisms of DAPK family members and their relevance to diseases. Furthermore, a special focus is given to several reports describing cross-talks between DAPKs and mitogen-activated protein kinases (MAPK) family members in various pathologies. We also discuss small molecule inhibitors of DAPKs and their potential as therapeutic targets against human diseases.

Keywords: MAPK; DAPK; ERK; p38; JNK

---

1. Introduction: DAPKs, MAPKs

Death-associated protein kinase (DAPK) family proteins are closely related, Ca^{2+}/calmodulin (CaM)-regulated serine/threonine kinases, whose members not only possess significant homology in their catalytic domains but also share cell death-associated functions [1,2]. DAPK family proteins include DAPK1, DAPK2, DAPK3, and DAPK-related apoptosis-inducing protein kinases (DRAK-1 and DRAK-2) [3–7] (Figure 1). DAPK1 has multiple complex domains including an N-terminal kinase domain, a Ca^{2+}/CaM-binding domain, a series of ankyrin repeats, a cytoskeleton binding domain, and a carboxyl-terminal death domain. DAPK2 contains an N-terminal kinase domain with high homology to DAPK1 catalytic domain [5], a conserved CaM-binding autoregulatory domain, and a C-terminal tail with no homology to any known proteins [6]. DAPK3 has an N-terminal kinase domain, a leucine zipper domain, and two putative nuclear localization sequences (NLS) [8]. The kinase domain and death domain are both critical for its pro-apoptotic activity [1,2,9]. All of these kinases are closely related to each other, sharing about 80% identity in their kinase domains [3,6], except for DRAK-1 and DRAK-2, whose kinase domains are only 50% identical to DAPK1 [10].

Mitogen-activated protein kinases (MAPKs) are an important sub-family of non-receptor serine-threonine kinases. MAPKs mediate signal transduction pathways that are involved in cellular responses to a diverse range of stimuli, such as mitogens, hormones, osmotic stress, heat shock, proinflammatory cytokines, and significant developmental changes in organisms. They mediate cellular functions including proliferation, differentiation, mitosis, gene expression, and apoptosis [11]. Extracellular stimuli such as growth factors result in a sequential phosphorylation cascade that ultimately leads to activation of MAPKs. Once activated, MAPKs activate downstream signals and transcription factors. MAPKs mainly consist of the extracellular signal-regulated kinases (ERK1–8), p38 MAPKs (p38α–δ), and c-Jun N-terminal kinases (JNK1–3) [12]. While the ERKs are mainly activated in response to proliferative signals, p38s and JNKs are activated in response to various stresses. Although there are several MAPK isoforms, the best-investigated ones are ERK1/2, JNK1/2, and p38α.

The mutual regulation between MAPK and DAPK family proteins plays a role in apoptosis regulation and several diseases. In this review, we introduce the reports showing the regulatory mechanisms and various functions of DAPK family proteins. Furthermore, we refer to reports indicating the relationship between DAPK and MAPK family proteins in multiple diseases.

2. Cellular Functions of DAPK Family Proteins

Increased activity of DAPK family proteins results in pronounced death-associated cellular changes, which include cell rounding, membrane blebbing, detachment from extracellular matrix, and formation of autophagic vesicles [1,2,4–6,9,13–19].

Among DAPK family proteins, DAPK1 controls cell cycle, apoptosis, autophagy, tumor metastasis, and oxidative stress. Several reports demonstrate that complex regulation of DAPK1 activity by various signaling pathways modulates the balance between pro-apoptotic and pro-survival pathways [20]. Furthermore, DAPK1 has been implicated in autophagy induction upon endoplasmic reticulum (ER) stress [21].

DAPK2 is known to be involved in pro-inflammatory responses mediated by granulocytes, which might be linked to the mechanism of myosin light chain (MLC) phosphorylation by DAPK2 [22]. Besides, DAPK2 has also been associated with differentiation processes in the erythropoietic lineage. DAPK2 knock-in mice showed a decreased response to erythropoietin treatment, suggesting that DAPK2 might exert fundamental regulatory effects on pro-erythroblast development [23].

The biological role of DAPK3 has been gradually investigated [24]. DAPK3 is pro-apoptotic [3] and executes this function either by inducing apoptosis or activating autophagy with or without the involvement of caspase proteins [25,26]. DAPK3 was also shown to mediate inflammatory signals including L13a (ribosome protein), ERK, and interferon (IFN)-γ-activated inhibition of translation [27].

3. Regulation of DAPK Family Proteins

DAPK1, DAPK2, and DAPK3 are all ubiquitously expressed in various tissues, such as heart, lung, spleen, and brain. In particular, DAPK1 is highly expressed in the hippocampus [1,3–5].

DAPK1 acts as a positive mediator of apoptosis induced by several death stimuli, such as interferon (IFN)-γ, Fas, transforming growth factor (TGF)-β, tumor necrosis factor (TNF)-α, ceramide, oncogene expression, and DNA damaging agents [2,9,13,14,16,28–30]. In normal conditions, DAPK1 is auto-phosphorylated at Ser308, which blocks its CaM binding site and keeps it inactivated [31] (Figure 1). Once stimulated, DAPK1 is dephosphorylated and promotes CaM binding to the site, which induces apoptotic responses [32–34]. In these papers, it was also suggested that DAPK1 dephosphorylation is caused by a class III phosphoinositide (PI)3-kinase-dependent phosphatase.

DAPK2 is regarded as a tumor suppressor in non-solid tumors and is implicated in apoptotic cell death [22]. After stimulation, DAPK2 at Ser318 is dephosphorylated and promotes CaM binding to the autoregulatory domain. These processes release the inhibitory binding domain to DAPK2, which leads to the access of the substrate of DAPK2. Dephosphorylation of DAPK2 also enhances homodimerization of DAPK2, which promotes membrane blebbing [35]. Recently, DAPK2 was shown to be phosphorylated at Ser299 by cyclic guanosine monophosphate (cGMP)-dependent protein kinase 1 [36]. Phosphorylation of DAPK2 at Ser299 increases DAPK2 activity independently of CaM binding (Figure 1).

On the contrary, DAPK3 lacks a Ca^{2+}/CaM regulatory domain, and its activity is regulated independently of intracellular Ca^{2+} levels. There are reports showing that DAPK3 is activated via interacting with DAPK1, forming a death-associated multi-protein complex [1,21]. DAPK3 was also shown to regulate DAPK1-induced apoptosis in HEK293T cells [37], and to promote starvation-induced autophagy through the regulation of Atg9-mediated autophagosome formation [38]. DAPK3 is activated in response to stress signals, and the cytosolic localization of DAPK3 may be a critical determinant in its pro-apoptotic activity [3,4,37]. The cytoplasmic distribution of DAPK3 may be regulated through its phosphorylation by DAPK1 [37]. This regulation of DAPK3 by DAPK1 suggests a mechanism by which a death signal can be transferred from one kinase to another in a catalytic amplification loop. Moreover, DAPK2 is capable of phosphorylating DAPK1 and DAPK3 [37], which is associated with cytoskeletal remodeling [1].

These studies suggest that the activity of DAPK proteins is regulated not only by the upstream signals but also the interactions by themselves.

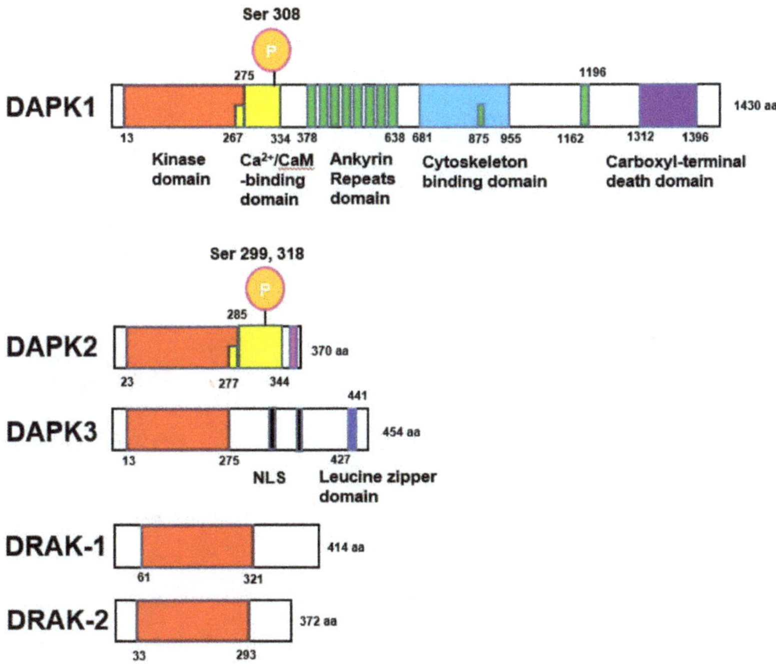

Figure 1. Structures of death associated protein kinase (DAPK) family proteins. DAPK1 possesses an N-terminal kinase domain, a Ca²⁺/Calmodulin (CaM)-binding domain, an ankyrin repeats domain, a cytoskeleton binding domain, and a carboxyl-terminal death domain. DAPK1 is auto-phosphorylated at Ser308, which blocks its CaM binding site and keeps the site inactivated. DAPK2 contains an N-terminal kinase domain, a Ca²⁺/CaM-binding domain, and a C-terminal tail with no homology to any known proteins. DAPK2 at Ser318 is dephosphorylated and promotes CaM binding to the autoregulatory domain. Phosphorylation of DAPK2 at Ser299 increases DAPK2 activity independently of CaM binding. DAPK3 has an N-terminal kinase domain, a leucine zipper domain, and two nuclear localization sequences (NLS). DRAK-1 and DRAK-2 have an N-terminal kinase domain. The location and total number of amino acids (aa) are shown. The recent domain information on human DAPK family proteins was collected from http://www.uniprot.org.

4. Cellular Functions of DAPK Family Proteins

Increased activity of DAPK family proteins results in pronounced death-associated cellular changes, which include cell rounding, membrane blebbing, detachment from the extracellular matrix, and formation of autophagic vesicles [1,2,4–6,9,13–19].

Among DAPK family proteins, DAPK1 controls cell cycle, apoptosis, autophagy, tumor metastasis, and oxidative stress. Several reports demonstrate that complex regulation of DAPK1 activity by various signaling pathways modulates the balance between pro-apoptotic and pro-survival pathways [20]. Furthermore, DAPK1 has been implicated in autophagy induction upon endoplasmic reticulum (ER) stress [21].

DAPK2 is known to be involved in pro-inflammatory responses mediated by granulocytes, which might be linked to the mechanism of myosin light chain (MLC) phosphorylation by DAPK2 [22]. Besides, DAPK2 has also been associated with differentiation processes in the erythropoietic lineage. DAPK2 knock-in mice showed a decreased response to erythropoietin treatment, suggesting that DAPK2 might exert fundamental regulatory effects on pro-erythroblast development [23].

The biological role of DAPK3 has been gradually investigated [24]. DAPK3 is pro-apoptotic [3] and executes this function either by inducing apoptosis or activating autophagy with or without the involvement of caspase proteins [25,26]. DAPK3 was also shown to mediate inflammatory signals including L13a (ribosome protein), ERK, and IFN-γ-activated inhibition of translation [27].

5. DAPKs in Disease

Among DAPK proteins, the relationship between DAPK1 and diseases has been often reported (Figure 2A). In the developing and adult brain, DAPK1 is widely expressed [39,40]. In addition, elevated DAPK1 activity is detected following brain injury due to ischemia [41,42], seizure [43,44], epilepsy [45], and Alzheimer's disease (AD) [46,47] as well as in response to ceramide and glutamate toxicity [28,48], indicating a close relationship between DAPK1 and neuronal cell death [49].

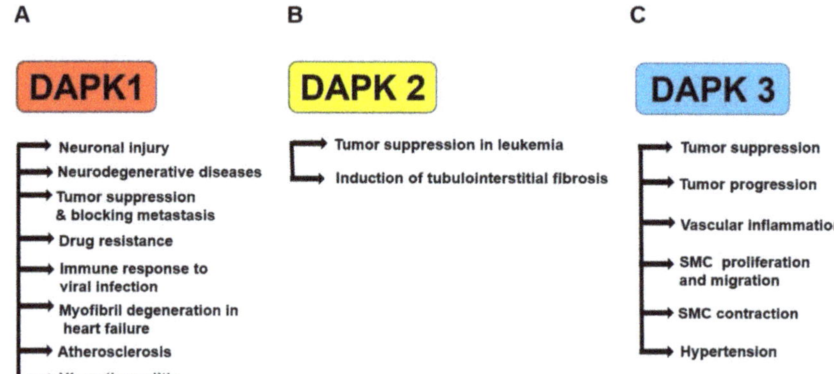

Figure 2. Roles of DAPK family proteins in various diseases. DAPK1 regulates neuronal injury, neurodegenerative diseases, tumor suppression, metastasis blocking, resistance to anti-cancer drugs, immune response during antiviral infection, myofibril degeneration in heart failure, atherosclerosis, and ulcerative colitis (**A**). DAPK2 regulates tumor suppression in leukemia and induction of tubulointerstitial fibrosis (**B**). DAPK3 regulates tumor suppression or progression, vascular inflammation, smooth muscle cell (SMC) proliferation and migration, SMC contraction, and hypertension (**C**).

Two main characteristics of AD are amyloid-beta (Aβ) senile plaques and tau neurofibrillary tangles. The metabolism of amyloid precursor protein (APP) results in extracellular deposition of Aβ protein that leads to the formation of Aβ senile plaques. In AD, tau is commonly hyperphosphorylated prior to tangle formation and neurodegeneration. [50,51]. However, how tau accumulation and phosphorylation are deregulated in AD is not fully understood.

Several studies showed that DAPK1 contributes to the pathogenesis of AD through excessive processing of APP and [46] triggering hyperphosphorylation of tau. It was also shown that DAPK1 inhibits microtubule assembly and stability through activation of microtubule-affinity regulating kinase (MARK)1 and MARK2, which leads to phosphorylation of tau at Ser262 [52]. It was also demonstrated that DAPK1 expression is highly upregulated in human AD hippocampus tissues. In the report, DAPK1 knock-out mice exhibited a decrease in expression and stability of tau protein [53].

In addition to regulating APP and tau proteins, DAPK1 mediates the neuronal cell death in AD model animals. DAPK1 binds N-myc downstream regulated gene 2 (NDRG2), which triggers its phosphorylation at Ser350 and induces neuronal cell death in AD model mice. In contrast, DAPK1 inhibition prevented NDRG2-mediated neuronal cell death [54]. These data suggest that DAPK1 might be a novel therapeutic target for the treatment of human AD [53].

In a wide variety of tumors, DAPK1 expression is frequently suppressed and the tumor-suppressive function of DAPK1 is linked to its role in cell death via apoptosis and autophagy [20]. It was reported that DAPK1 expression was lost in tumors due to hypermethylation of the DAPK1 gene [55]. Moreover, it was shown that DAPK1 was capable of suppressing oncogenic transformation caused by c-Myc and E2F, which blocks the tumor metastasis [13,14]. DAPK1 is also mediated in anti-cancer drug resistance to 5-fluorouracil in endometrial adenocarcinoma cells [56], anti-epidermal growth factor receptor antibodies in lung cancer cells [57], gemcitabine in pancreatic cancer cells [58], and cisplatin in cervical squamous cancer cells [59].

DAPK1 also has a role in cellular antiviral immune response. Once the viral infections occur, the viruses inhibit INF-mediated signals including INF-α/β for their proliferation [60,61]. After viral infection, DAPK1 enhances the activation of IFN-mediated signals through the interaction with IFN regulatory factor 3 and 7. Moreover, IFN-β increases the expression and activation of DAPK1 through the regulation of phosphorylation levels at Ser308 [62].

Besides, DAPK1 is involved in the regulation of myofibril degeneration and myocyte apoptosis induced by chronic stimulation with β1-adrenergic receptors. The result implies that DAPK1 activation might contribute to the pathogenesis of β-adrenergic receptor-related signaling during the development of heart failure [63].

DAPK1 seems to have both pro- and anti-inflammatory functions. It positively contributes to production and secretion of interleukin (IL)-1β in macrophages [64], while it negatively regulates inflammation in purified human T cells [65], monocytes [27], and mouse lung tissues [66]. In human diseases, ulcerative colitis (UC) was found to be closely related to DAPK1 function through inhibition of inflammation. It was also shown that DAPK1 promoter methylation led to the decrease in DAPK1 protein expression and enhanced the severity of inflammation in UC, suggesting an anti-inflammatory role of DAPK1 in UC [67].

Another disease that DAPK1 may be involved in is atherosclerosis. Although DAPK1 expression was increased in atherosclerotic plaques, the detailed mechanisms are not known [68]. In addition, shear stress has been reported to regulate DAPK1 expression and activity, which promotes TNF-α-induced apoptosis in cultured bovine aortic endothelial cells (ECs) [69,70]. These reports might suggest a potential role of DAPK1 in the regulation of diseases through vascular ECs.

In contrast to DAPK1 and DAPK3, DAPK2 has not been identified as a tumor suppressor in solid tumors. Interestingly, DAPK2 was predominantly found in the hematopoietic compartment [23] and emerged as a tumor suppressor in several types of leukemia [71]. Most recently, DAPK2 has also been linked to the induction of tubulointerstitial fibrosis in mice kidneys upon chronic cisplatin exposure [72] (Figure 2B).

DAPK3 is proposed to be a tumor suppressor, suggesting that mutations in DAPK3 could result in the loss of function. It was reported that the DAPK3 gene is frequently methylated or mutated in various types of cancer [73], resulting in loss of tumor suppression via DAPK3 in cancer. In view of this evidence, DAPK3 has been regarded as a tumor suppressor.

On the other hand, a recent study demonstrated that DAPK3 promotes cancer cell proliferation rather than the promotion of apoptosis in many types of cancer cells. In prostate cancer cells, DAPK3 promoted proliferation [74]. Furthermore, knockdown of the DAPK3 gene prevented proliferation in colon cancer cells through the inhibition of Wnt/β-catenin signals [75]. In our previous study, it was demonstrated that knockdown of the DAPK3 gene also blocked non-small cell lung cancer (NSCLC) progression via cellular signaling [76]. It was found that DAPK3 regulates proliferation, migration, and invasion through ERK/c-Myc signaling in A549 cells. These data suggest the possibility of DAPK3 as a novel therapeutic target for many types of cancer.

In our previous study, it was found that the expression level of DAPK3 protein was increased in the mesenteric artery from spontaneously hypertensive rats (SHR) [77]. Moreover, it was found that DAPK3 promoted reactive oxygen species (ROS)-dependent vascular inflammation and thereby mediated the development of hypertension in SHR [78]. Cho et al. demonstrated in mesenteric arteries of SHR that DAPK3 modulated calyculin A-induced contraction via increasing Ca^{2+}-independent Myosin light chain kinase (MLCK) activity [79]. In our previous study, it was also demonstrated that DAPK3 mediated vascular structural remodeling via stimulating smooth muscle cell (SMC) proliferation and migration [80]. Furthermore, it was demonstrated that phosphorylation of MLC2 by DAPK3 promoted smooth muscle contraction and motility [81]. These data suggest that DAPK3 might become a pharmaceutical target for prevention of hypertensive cardiovascular diseases (Figure 2C).

Nevertheless, the data of DAPK proteins in diseases are preliminary. Further studies are needed to clarify the expression and functional correlation of DAPK proteins with these diseases.

6. Cross-Talk between DAPKs and ERK Signaling

ERKs are a group of MAPKs (ERK1–8) and ERK1/2 are the first discovered members of the MAPK family. Once activated (phosphorylated), ERK1/2 moves from the cytoplasm to the nucleus, which is critical for many cellular functions, such as gene transcription, cell proliferation, and differentiation [82]. Since ERK1/2 was found to be upregulated in human tumors, inhibitors of this pathway have been used for cancer therapeutics [83].

ERK1/2 was identified as a DAPK1-interacting protein [84] (Figure 3). DAPK1 interacts with ERK1/2 through a docking sequence within its death domain, which promotes apoptotic cell death [9,55]. On the other hand, a recent study reported that phosphorylation of DAPK1 at Ser735 by ERK1/2 leads to apoptosis in human fibroblasts [84]. This mutual regulation between DAPK1 and ERK1/2 constitutes a positive feedback circuit that ultimately enhances the apoptotic activity of DAPK1.

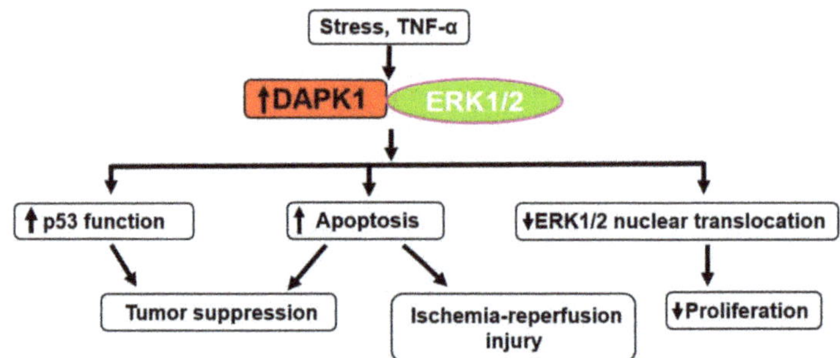

Figure 3. Interaction between DAPK1 and extracellular signal-regulated kinase (ERK)1/2. DAPK1-ERK1/2 interaction promotes tumor suppression and ischemia-reperfusion injury through upregulation of p53 function or propagating apoptosis. It also inhibits cell proliferation through prevention of ERK1/2 nuclear translocation. TNF = tumor necrosis factor. Up arrow indicates the increased responses, expression of protein and activity. Signal arrow indicates the promotional effects.

Mechanistically, it was shown that DAPK1 prevented activation of ribosomal S6 kinase through inhibition of ERK1/2 nuclear translocation, indicating that DAPK1 may play an inhibitory role in the survival function of ERK1/2 signaling [85]. It was recently shown that a germline mutation in the death domain of DAPK1 (N1347S) prevented the binding of DAPK1 to ERK1/2, and attenuated TNF-α-induced apoptosis [86].

ERK1/2 regulates activation of the p53 pathway [87], and DAPK1 activates p53 in an oncogenic signaling pathway [28]. Therefore, ERK1/2 might mediate p53 and DAPK1 pathways to maintain p53 function as a tumor suppressor.

DAPK1-ERK1/2 signals also regulate neuronal apoptosis following ischemia-reperfusion [88]. In the report, ischemia-reperfusion led to activation of DAPK1 and ERK1/2. DAPK1 was also proved to bind ERK1/2 during reperfusion following oxygen-glucose deprivation. Prevention of DAPK1- ERK1/2 binding by knockdown of the DAPK1 gene reduced neuronal apoptosis through the promotion of nuclear translocation of ERK1/2. These results reveal the potential mechanism of the DAPK1-ERK1/2 signal in the contribution to neuronal apoptosis in response to ischemia-reperfusion. Prevention of this signal pathway might become a promising therapeutic target against stroke [88].

7. DAPKs as Upstream Activators of Stress-Activated Protein Kinases p38 and JNK

p38 plays roles in inflammation, cell proliferation, differentiation, survival, and cell death [89–92]. While several reports showed a tumor suppressive role for p38 [93], p38 has also been involved in tumor progression by promoting cell migration, angiogenesis, and inflammatory responses [93].

Several studies provided direct evidence for the interaction of DAPK1 with p38 signaling in inflammation-associated colorectal cancer cells [94]. That study identified for the first time that p38 co-localized and interacted with DAPK1 and triggered DAPK1-mediated apoptosis in HCT116 cells. In human colorectal cancer tissues, the co-expression of DAPK1 and p38 was associated with apoptotic cell death. These results imply that a DAPK1-p38 interaction has a role in tumor suppression in colorectal cancer (Figure 4A).

In our previous study, it was clarified that DAPK3 mediates platelet-derived growth factor (PDGF)-BB-induced proliferation and migration of SMC through activation of p38/heat shock protein (HSP)27 signals, which leads to vascular structural remodeling including neointimal hyperplasia [80] (Figure 4B). Since SMC migration and vascular remodeling are important processes for the development of hypertension, these data suggest that the DAPK3/p38/HSP27 axis might be a potential pharmaceutical target for the prevention of hypertensive cardiovascular diseases [80]. These data seem to be opposite to the previous reports demonstrating that DAPK3 promoted tumor cell death [73]. This might be because DAPK3 has different functions depending on the types of cells.

JNK pathways are activated by various stress stimuli, such as heat shock, osmotic shock, ultraviolet irradiation, and cytokines [90–92], which mediate various functions by acting on downstream targets that include transcription factors, such as Elk1, p53, AP1, and ATF2, as well as the anti-apoptotic protein Bcl-2 [93].

JNK is a key node of the cell death network activated under oxidative stress [95–97], and DAPK1 plays a central role in oxidative stress-induced JNK signaling. It was also shown that DAPK1 mediates oxidative stress-induced JNK phosphorylation through the activation of protein kinase D, which mediates cell death [98] (Figure 5A).

In contrast, DAPK2 inhibits JNK activation through the prevention of oxidative stress in osteosarcoma and lung cancer cells [99] (Figure 5B). In the report, knockdown of the DAPK2 gene promoted mitochondrial membrane depolarization through increasing levels of mitochondrial reactive oxygen species (ROS) and phosphorylation of JNK. These data imply that DAPK2 is vital to maintaining mitochondrial integrity and cellular metabolism in cancer cells.

In our previous study, it was demonstrated that DAPK3 mediated TNF-α-induced activation of JNK, p38, and Akt through a generation of ROS [78] (Figure 4B,5C). In the report, the role of DAPK3 in vascular inflammatory responses and development of hypertension was investigated. As a result, it was found that the inhibition of DAPK3 prevented TNF-α-induced vascular cell adhesion molecule (VCAM)-1 expression and activation of JNK, p38, and Akt, as well as ROS production in vascular SMCs. It was also demonstrated that the inhibition of DAPK3 prevented TNF-α-induced expression of

VCAM-1, e-selectin, and cyclooxygenase-2, as well as ROS production, in vascular endothelial cells. Since DAPK inhibitor blocked the development of hypertension in SHR and vascular inflammation, it was suggested that DAPK3 regulates hypertensive disease through vascular inflammation.

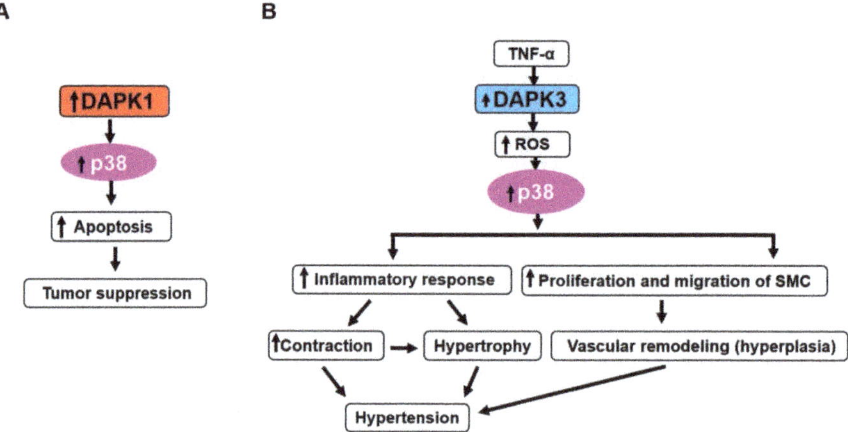

Figure 4. Interaction between DAPK proteins and p38. DAPK1-p38 plays a role in tumor suppression via promotion of apoptosis (**A**). DAPK3-p38 might mediate hypertension through increasing vascular contraction and hypertrophy via reactive oxygen species (ROS) generation and activation of p38-dependent inflammatory response (**B**). Besides, DAPK3 mediates vascular remodeling including hyperplasia through ROS generation, p38 activation, and promotion of proliferation and migration of SMC, which might lead to hypertension. Arrow indicates promotional effects.

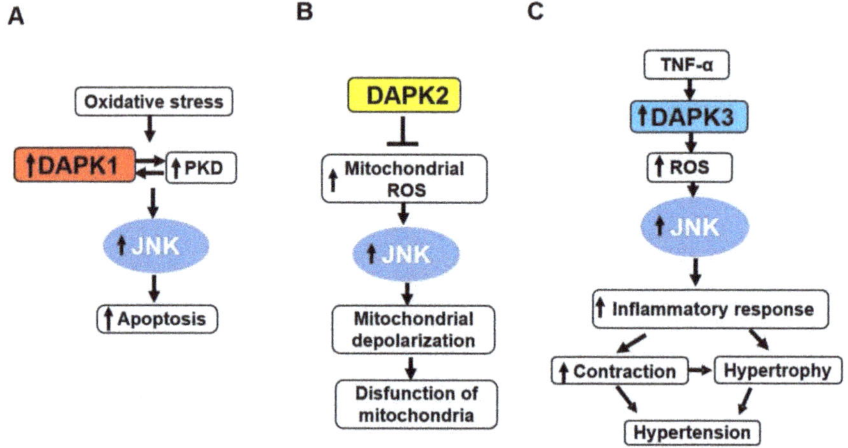

Figure 5. Interaction between DAPK proteins and c-Jun N-terminal kinase (JNK). DAPK1-protein kinase D (PKD) interaction promotes oxidative stress-mediated apoptosis (**A**). DAPK2 regulates the level of mitochondrial ROS through activation of JNK, which might maintain the mitochondrial functions (**B**). DAPK3 might mediate hypertension through propagating vascular contraction and hypertrophy via ROS generation and activation of JNK-dependent inflammatory response (**C**). Arrow indicates promotional effects. T bar indicates the suppressive effects.

8. DAPKs are Therapeutic Targets of Small Molecule Inhibitors

Discovery of DAPK inhibitors has been performed using previously described experimental approaches [100,101]. Several small molecules and plant-derived compounds have been developed as inhibitors of the DAPK family kinases (Table 1). Some are selective in their action, others are not, and only a few of them have been tested in cells. It is generally considered that inhibition of DAPK proteins intercepts cell death and prevents further damage of ischemic regions in cerebral infarction and other ischemic diseases.

Alkylated 3-amino-6-phenylpyridazine inhibited DAPK1 catalytic activity in cultured neuron cells and mouse models [41]. Furthermore, the single intraperitoneal injection of this compound reduced brain injury in an animal model when administered 6 h after the insult (hypoxia-ischemia induced injury) [41].

Another compound, (4Z)-4-(3-pyridylmethylene)-2-styryl-oxazol-5-one, was known to inhibit DAPK1 activity effectively and selectively (IC_{50} = 69 nM) [26,53,102]. This drug regulated DAPK1 activity-dependent neurite outgrowth and microtubule polymerization by affecting tau function [53].

The role of DAPK activity in the histone deacetylase inhibitor (LBH589)-induced apoptosis in HCT116 wildtype colon cancer cells was clarified by using a selective DAPK1 inhibitor, 2-phenyl-4-(pyridin-3-ylmethylidene)-4,5-dihydro-1,3-oxazol-5-one, known as TC-DAPK6, for inhibition of DAPK1 catalytic activity [26,103]. It was found that pre-treatment with TC-DAPK6 did not influence the LBH589-induced caspase 3-mediated cell death [103]. TC-DAPK6 acted as a novel DAPK1 and DAPK3 inhibitor (IC_{50} = 69 and 225 nM, respectively), which makes it the most potent DAPK1 inhibitor discovered to date [104].

Eighteen analogs exhibiting anti-proliferative and pro-apoptotic properties on an acute T cell leukemia cell line were obtained [105]. These analogs have a benzyloxy group at the C6 position and 9-tert-butyl-6-(benzyloxy)-8-phenyl-9*H*-purine (6d).

Several other selective inhibitors of DAPK have also been investigated [26]. The non-selective inhibitors targeting multiple protein kinases make it challenging to dissect precisely the role of DAPK1 on certain biological processes. Among them, growing evidence of the role in DAPK1 in cancer processes are appearing [106]. Therefore, the use of selective inhibitors might shed light on the exact participation of DAPK1 in tumor development.

Besides, natural DAPK inhibitors have been developed. The chloroform fraction of *Laurus nobilis*, a DAPK1 non-selective inhibitor of plant origin, significantly improved the ischemic neuronal death through keeping DAPK1 in an inactivate form after oxygen and glucose deprivation in human SH-SY5Y neuroblastoma cell lines and organotypic hippocampal slice tissue. Moreover, it also reduced the infarct size and neurological deficit of middle cerebral artery occlusion in vivo [107].

Recently, a dual Pim/DAPK3 inhibitor (HS56) was synthesized using crystal structure-guided medicinal chemistry techniques [108]. HS56 reduced Pim kinase-induced myosin phosphorylation and the contractility of vascular smooth muscle in spontaneously hypertensive RenTG mice, suggesting a novel multi-target strategy for hypertensive diseases [108]. In the same report, selective DAPK3 inhibitors (HS94 and HS148) were also developed [108].

While specific inhibitors for DAPK2 remain to be developed, DAPK1 and DAPK3 inhibitors have been tested for DAPK2 activity. Since the catalytic domain of the three kinases shares high-sequence homology, the most potent derivative showed comparable IC_{50} value to that of DAPK1 and DAPK3 [109].

The potency of most DAPK inhibitors in in vitro kinase assays toward DAPK proteins is not very high. Improvement of the structures of these compounds is needed to gain more potent inhibitors and to perform clinical trials. Although there are only a few detailed studies focusing on DAPK proteins as a drug target, it would be helpful to use selective inhibitors against unmet medical needs.

Table 1. Information on DAPK inhibitors.

Name	Structure	Types of Inhibitors	IC$_{50}$
Alkylated 3-amino-6-phenylpyridazine		ATP-competitive	13, 22 mM for DAPK1
(4Z)-4-(3-pyridylmethylene)-2-styryl-oxazol-5-one		ATP-competitive	69 and 225 nM against DAPK1 and DAPK3, respectively
(4Z)-2-[(E)-2-Phenylethenyl)-4-(3-pyridinylmethylene)-5(4H)-oxazolone (TC-DAPK6)		ATP-competitive	69 and 225 nM against DAPK1 and DAPK3, respectively
6-benzyloxy-9-tert-butyl-8-phenyl-9H-purine (6d)		ATP-competitive	2.5 mM for DAPK1
HS56		ATP-competitive	1 mM for DAPK3

Adenosine triphosphate (ATP).

9. Conclusions

Several recent studies have unraveled pathologically relevant mechanisms involving DAPK-MAPK cross-talks in neuronal diseases, cancer, and cardiovascular diseases. With the availability of more specific and effective DAPK inhibitors, the interface between DAPK and MAPKs might become promising targets in the treatment of various diseases.

Funding: This study was supported in part by a Grant for Scientific Research from the Japan Society for the Promotion of Science (17K15370 TU).

Conflicts of Interest: The authors declare no conflict of interest.

References

1. Bialik, S.; Kimchi, A. The death-associated protein kinases: Structure, function, and beyond. *Annu. Rev. Biochem.* **2006**, *75*, 189–210. [CrossRef] [PubMed]
2. Cohen, O.; Feinstein, E.; Kimchi, A. Dap-kinase is a ca2+/calmodulin-dependent, cytoskeletal-associated protein kinase, with cell death-inducing functions that depend on its catalytic activity. *Embo. J.* **1997**, *16*, 998–1008. [CrossRef] [PubMed]
3. Kawai, T.; Matsumoto, M.; Takeda, K.; Sanjo, H.; Akira, S. Zip kinase, a novel serine/threonine kinase which mediates apoptosis. *Mol. Cell. Biol.* **1998**, *18*, 1642–1651. [CrossRef] [PubMed]
4. Kogel, D.; Plottner, O.; Landsberg, G.; Christian, S.; Scheidtmann, K.H. Cloning and characterization of dlk, a novel serine/threonine kinase that is tightly associated with chromatin and phosphorylates core histones. *Oncogene* **1998**, *17*, 2645–2654. [CrossRef] [PubMed]
5. Kawai, T.; Nomura, F.; Hoshino, K.; Copeland, N.G.; Gilbert, D.J.; Jenkins, N.A.; Akira, S. Death-associated protein kinase 2 is a new calcium/calmodulin-dependent protein kinase that signals apoptosis through its catalytic activity. *Oncogene* **1999**, *18*, 3471–3480. [CrossRef] [PubMed]

6. Inbal, B.; Shani, G.; Cohen, O.; Kissil, J.L.; Kimchi, A. Death-associated protein kinase-related protein 1, a novel serine/threonine kinase involved in apoptosis. *Mol. Cell. Biol.* **2000**, *20*, 1044–1054. [CrossRef] [PubMed]

7. Koen, T.; Bertrand, S.; Matthias, W. Structural and functional diversity in the activity and regulation of dapk-related protein kinases. *FEBS J.* **2013**, *280*, 5533–5550.

8. Detjen, K.M.; Brembeck, F.H.; Welzel, M.; Kaiser, A.; Haller, H.; Wiedenmann, B.; Rosewicz, S. Activation of protein kinase calpha inhibits growth of pancreatic cancer cells via p21(cip)-mediated g(1) arrest. *J. Cell. Sci.* **2000**, *113*, 3025–3035. [PubMed]

9. Cohen, O.; Inbal, B.; Kissil, J.L.; Raveh, T.; Berissi, H.; Spivak-Kroizaman, T.; Feinstein, E.; Kimchi, A. Dap-kinase participates in tnf-alpha- and fas-induced apoptosis and its function requires the death domain. *J. Cell. Biol.* **1999**, *146*, 141–148. [CrossRef] [PubMed]

10. Sanjo, H.; Kawai, T.; Akira, S. Draks, novel serine/threonine kinases related to death-associated protein kinase that trigger apoptosis. *J. Biol. Chem.* **1998**, *273*, 29066–29071. [CrossRef] [PubMed]

11. Pearson, G.; Robinson, F.; Beers Gibson, T.; Xu, B.-e.; Karandikar, M.; Berman, K.; Cobb, M.H. Mitogen-activated protein (map) kinase pathways: Regulation and physiological functions*. *Endocr. Rev.* **2001**, *22*, 153–183. [CrossRef] [PubMed]

12. Hazzalin, C.A.; Mahadevan, L.C. Mapk-regulated transcription: A continuously variable gene switch? *Nat. Rev. Mol. Cell Biol.* **2002**, *3*, 30–40. [CrossRef] [PubMed]

13. Inbal, B.; Cohen, O.; Polak-Charcon, S.; Kopolovic, J.; Vadai, E.; Eisenbach, L.; Kimchi, A. Dap kinase links the control of apoptosis to metastasis. *Nature* **1997**, *390*, 180–184. [CrossRef] [PubMed]

14. Raveh, T.; Droguett, G.; Horwitz, M.S.; DePinho, R.A.; Kimchi, A. Dap kinase activates a p19arf/p53-mediated apoptotic checkpoint to suppress oncogenic transformation. *Nat. Cell Biol* **2001**, *3*, 1–7. [CrossRef] [PubMed]

15. Yamamoto, M.; Hioki, T.; Ishii, T.; Nakajima-Iijima, S.; Uchino, S. Dap kinase activity is critical for c(2)-ceramide-induced apoptosis in pc12 cells. *Eur J. Biochem* **2002**, *269*, 139–147. [CrossRef] [PubMed]

16. Jang, C.W.; Chen, C.H.; Chen, C.C.; Chen, J.Y.; Su, Y.H.; Chen, R.H. Tgf-beta induces apoptosis through smad-mediated expression of dap-kinase. *Nat. Cell. Biol.* **2002**, *4*, 51–58. [CrossRef] [PubMed]

17. Wang, W.J.; Kuo, J.C.; Yao, C.C.; Chen, R.H. Dap-kinase induces apoptosis by suppressing integrin activity and disrupting matrix survival signals. *J. Cell. Biol.* **2002**, *159*, 169–179. [CrossRef] [PubMed]

18. Bialik, S.; Kimchi, A. Lethal weapons: Dap-kinase, autophagy and cell death: Dap-kinase regulates autophagy. *Curr. Opin. Cell. Biol.* **2010**, *22*, 199–205. [CrossRef] [PubMed]

19. Levin-Salomon, V.; Bialik, S.; Kimchi, A. Dap-kinase and autophagy. *Apoptosis* **2014**, *19*, 346–356. [CrossRef] [PubMed]

20. Gade, P.; Roy, S.K.; Li, H.; Nallar, S.C.; Kalvakolanu, D.V. Critical role for transcription factor c/ebp-beta in regulating the expression of death-associated protein kinase 1. *Mol. Cell. Biol.* **2008**, *28*, 2528–2548. [CrossRef] [PubMed]

21. Gozuacik, D.; Bialik, S.; Raveh, T.; Mitou, G.; Shohat, G.; Sabanay, H.; Mizushima, N.; Yoshimori, T.; Kimchi, A. Dap-kinase is a mediator of endoplasmic reticulum stress-induced caspase activation and autophagic cell death. *Cell Death Differ.* **2008**, *15*, 1875–1886. [CrossRef] [PubMed]

22. Geering, B.; Stoeckle, C.; Rozman, S.; Oberson, K.; Benarafa, C.; Simon, H.U. Dapk2 positively regulates motility of neutrophils and eosinophils in response to intermediary chemoattractants. *J. Leukoc Biol.* **2014**, *95*, 293–303. [CrossRef] [PubMed]

23. Fang, J.; Menon, M.; Zhang, D.; Torbett, B.; Oxburgh, L.; Tschan, M.; Houde, E.; Wojchowski, D.M. Attenuation of epo-dependent erythroblast formation by death-associated protein kinase-2. *Blood* **2008**, *112*, 886–890. [CrossRef] [PubMed]

24. Usui, T.; Okada, M.; Yamawaki, H. Zipper interacting protein kinase (zipk): Function and signaling. *Apoptosis* **2014**, *19*, 387–391. [CrossRef] [PubMed]

25. Van Eldik, L.J. Structure and enzymology of a death-associated protein kinase. *Trends Pharmacol. Sci.* **2002**, *23*, 302–304. [CrossRef]

26. Okamoto, M.; Takayama, K.; Shimizu, T.; Muroya, A.; Furuya, T. Structure-activity relationship of novel dapk inhibitors identified by structure-based virtual screening. *Bioorg. Med. Chem.* **2010**, *18*, 2728–2734. [CrossRef] [PubMed]

27. Mukhopadhyay, R.; Ray, P.S.; Arif, A.; Brady, A.K.; Kinter, M.; Fox, P.L. Dapk-zipk-l13a axis constitutes a negative-feedback module regulating inflammatory gene expression. *Mol. Cell.* **2008**, *32*, 371–382. [CrossRef] [PubMed]

28. Pelled, D.; Raveh, T.; Riebeling, C.; Fridkin, M.; Berissi, H.; Futerman, A.H.; Kimchi, A. Death-associated protein (dap) kinase plays a central role in ceramide-induced apoptosis in cultured hippocampal neurons. *J. Biol. Chem.* **2002**, *277*, 1957–1961. [CrossRef] [PubMed]

29. Martoriati, A.; Doumont, G.; Alcalay, M.; Bellefroid, E.; Pelicci, P.G.; Marine, J.C. Dapk1, encoding an activator of a p19arf-p53-mediated apoptotic checkpoint, is a transcription target of p53. *Oncogene* **2005**, *24*, 1461–1466. [CrossRef] [PubMed]

30. Chawla-Sarkar, M.; Lindner, D.J.; Liu, Y.F.; Williams, B.R.; Sen, G.C.; Silverman, R.H.; Borden, E.C. Apoptosis and interferons: Role of interferon-stimulated genes as mediators of apoptosis. *Apoptosis* **2003**, *8*, 237–249. [CrossRef] [PubMed]

31. De Diego, I.; Kuper, J.; Bakalova, N.; Kursula, P.; Wilmanns, M. Molecular basis of the death-associated protein kinase-calcium/calmodulin regulator complex. *Sci. Signal.* **2010**, *3*, ra6. [CrossRef] [PubMed]

32. Shani, G.; Henis-Korenblit, S.; Jona, G.; Gileadi, O.; Eisenstein, M.; Ziv, T.; Admon, A.; Kimchi, A. Autophosphorylation restrains the apoptotic activity of drp-1 kinase by controlling dimerization and calmodulin binding. *Embo. J.* **2001**, *20*, 1099–1113. [CrossRef] [PubMed]

33. Shohat, G.; Spivak-Kroizman, T.; Cohen, O.; Bialik, S.; Shani, G.; Berrisi, H.; Eisenstein, M.; Kimchi, A. The pro-apoptotic function of death-associated protein kinase is controlled by a unique inhibitory autophosphorylation-based mechanism. *J. Biol. Chem.* **2001**, *276*, 47460–47467. [CrossRef] [PubMed]

34. Shang, T.; Joseph, J.; Hillard, C.J.; Kalyanaraman, B. Death-associated protein kinase as a sensor of mitochondrial membrane potential: Role of lysosome in mitochondrial toxin-induced cell death. *J. Biol. Chem.* **2005**, *280*, 34644–34653. [CrossRef] [PubMed]

35. Inbal, B.; Bialik, S.; Sabanay, I.; Shani, G.; Kimchi, A. Dap kinase and drp-1 mediate membrane blebbing and the formation of autophagic vesicles during programmed cell death. *J. Cell. Biol.* **2002**, *157*, 455–468. [CrossRef] [PubMed]

36. Isshiki, K.; Matsuda, S.; Tsuji, A.; Yuasa, K. Cgmp-dependent protein kinase i promotes cell apoptosis through hyperactivation of death-associated protein kinase 2. *Biochem. Biophys. Res. Commun.* **2012**, *422*, 280–284. [CrossRef] [PubMed]

37. Shani, G.; Marash, L.; Gozuacik, D.; Bialik, S.; Teitelbaum, L.; Shohat, G.; Kimchi, A. Death-associated protein kinase phosphorylates zip kinase, forming a unique kinase hierarchy to activate its cell death functions. *Mol. Cell. Biol.* **2004**, *24*, 8611–8626. [CrossRef] [PubMed]

38. Liu, C.; Kelnar, K.; Liu, B.; Chen, X.; Calhoun-Davis, T.; Li, H.; Patrawala, L.; Yan, H.; Jeter, C.; Honorio, S.; et al. The microrna mir-34a inhibits prostate cancer stem cells and metastasis by directly repressing cd44. *Nat. Med.* **2011**, *17*, 211–215. [CrossRef] [PubMed]

39. Sakagami, H.; Kondo, H. Molecular cloning and developmental expression of a rat homologue of death-associated protein kinase in the nervous system. *Mol. Brain Res.* **1997**, *52*, 249–256. [CrossRef]

40. Yamamoto, M.; Takahashi, H.; Nakamura, T.; Hioki, T.; Nagayama, S.; Ooashi, N.; Sun, X.; Ishii, T.; Kudo, Y.; Nakajima-Iijima, S.; et al. Developmental changes in distribution of death-associated protein kinase mrnas. *J. Neurosci. Res.* **1999**, *58*, 674–683. [CrossRef]

41. Velentza, A.V.; Wainwright, M.S.; Zasadzki, M.; Mirzoeva, S.; Schumacher, A.M.; Haiech, J.; Focia, P.J.; Egli, M.; Watterson, D.M. An aminopyridazine-based inhibitor of a pro-apoptotic protein kinase attenuates hypoxia-ischemia induced acute brain injury. *Bioorg. Med. Chem. Lett.* **2003**, *13*, 3465–3470. [CrossRef]

42. Schumacher, A.M.; Velentza, A.V.; Watterson, D.M.; Wainwright, M.S. Dapk catalytic activity in the hippocampus increases during the recovery phase in an animal model of brain hypoxic-ischemic injury. *Biochim. Biophys. Acta* **2002**, *1600*, 128–137. [CrossRef]

43. Henshall, D.C.; Araki, T.; Schindler, C.K.; Shinoda, S.; Lan, J.Q.; Simon, R.P. Expression of death-associated protein kinase and recruitment to the tumor necrosis factor signaling pathway following brief seizures. *J. Neurochem.* **2003**, *86*, 1260–1270. [CrossRef] [PubMed]

44. Araki, T.; Shinoda, S.; Schindler, C.K.; Quan-Lan, J.; Meller, R.; Taki, W.; Simon, R.P.; Henshall, D.C. Expression, interaction, and proteolysis of death-associated protein kinase and p53 within vulnerable and resistant hippocampal subfields following seizures. *Hippocampus* **2004**, *14*, 326–336. [CrossRef] [PubMed]

45. Mor, I.; Carlessi, R.; Ast, T.; Feinstein, E.; Kimchi, A. Death-associated protein kinase increases glycolytic rate through binding and activation of pyruvate kinase. *Oncogene* **2012**, *31*, 683–693. [CrossRef] [PubMed]
46. Li, Y.; Grupe, A.; Rowland, C.; Nowotny, P.; Kauwe, J.S.; Smemo, S.; Hinrichs, A.; Tacey, K.; Toombs, T.A.; Kwok, S.; et al. Dapk1 variants are associated with Alzheimer's disease and allele-specific expression. *Hum. Mol. Genet.* **2006**, *15*, 2560–2568. [CrossRef] [PubMed]
47. Li, Y.; Grupe, A.; Rowland, C.; Nowotny, P.; Kauwe, J.S.K.; Smemo, S.; Hinrichs, A.; Tacey, K.; Toombs, T.A.; Kwok, S.; et al. Dapk1 variants are associated with Alzheimer's disease and allele-specific expression. *Hum. Mol. Genet.* **2006**, *15*, 2560–2568. [CrossRef] [PubMed]
48. Tu, W.; Xu, X.; Peng, L.; Zhong, X.; Zhang, W.; Soundarapandian, M.M.; Balel, C.; Wang, M.; Jia, N.; Zhang, W.; et al. Dapk1 interaction with nmda receptor nr2b subunits mediates brain damage in stroke. *Cell* **2010**, *140*, 222–234. [CrossRef] [PubMed]
49. Shamloo, M.; Soriano, L.; Wieloch, T.; Nikolich, K.; Urfer, R.; Oksenberg, D. Death-associated protein kinase is activated by dephosphorylation in response to cerebral ischemia. *J. Biol. Chem.* **2005**, *280*, 42290–42299. [CrossRef] [PubMed]
50. Lee, V.M.; Balin, B.J.; Otvos, L., Jr.; Trojanowski, J.Q. A68: A major subunit of paired helical filaments and derivatized forms of normal tau. *Science* **1991**, *251*, 675–678. [CrossRef] [PubMed]
51. Goedert, M.; Spillantini, M.G.; Cairns, N.J.; Crowther, R.A. Tau proteins of alzheimer paired helical filaments: Abnormal phosphorylation of all six brain isoforms. *Neuron* **1992**, *8*, 159–168. [CrossRef]
52. Wu, P.R.; Tsai, P.I.; Chen, G.C.; Chou, H.J.; Huang, Y.P.; Chen, Y.H.; Lin, M.Y.; Kimchi, A.; Chien, C.T.; Chen, R.H. Dapk activates mark1/2 to regulate microtubule assembly, neuronal differentiation, and tau toxicity. *Cell Death Differ.* **2011**, *18*, 1507–1520. [CrossRef] [PubMed]
53. Kim, B.M.; You, M.H.; Chen, C.H.; Lee, S.; Hong, Y.; Hong, Y.; Kimchi, A.; Zhou, X.Z.; Lee, T.H. Death-associated protein kinase 1 has a critical role in aberrant tau protein regulation and function. *Cell Death Dis.* **2014**, *5*, e1237. [CrossRef] [PubMed]
54. Martin, H.G.; Wang, Y.T. Blocking the deadly effects of the nmda receptor in stroke. *Cell* **2010**, *140*, 174–176. [CrossRef] [PubMed]
55. Cohen, O.; Kimchi, A. Dap-kinase: From functional gene cloning to establishment of its role in apoptosis and cancer. *Cell Death Differ.* **2001**, *8*, 6–15. [CrossRef] [PubMed]
56. Tanaka, T.; Bai, T.; Toujima, S.; Utsunomiya, T.; Utsunomiya, H.; Yukawa, K.; Tanaka, J. Impaired death-associated protein kinase-mediated survival signals in 5-fluorouracil-resistant human endometrial adenocarcinoma cells. *Oncol. Rep.* **2012**, *28*, 330–336. [CrossRef] [PubMed]
57. Ogawa, T.; Liggett, T.E.; Melnikov, A.A.; Monitto, C.L.; Kusuke, D.; Shiga, K.; Kobayashi, T.; Horii, A.; Chatterjee, A.; Levenson, V.V.; et al. Methylation of death-associated protein kinase is associated with cetuximab and erlotinib resistance. *Cell Cycle* **2012**, *11*, 1656–1663. [CrossRef] [PubMed]
58. Guo, Q.; Chen, Y.; Wu, Y. Enhancing apoptosis and overcoming resistance of gemcitabine in pancreatic cancer with bortezomib: A role of death-associated protein kinase-related apoptosis-inducing protein kinase 1. *Tumori J.* **2009**, *95*, 796–803. [CrossRef]
59. Bai, T.; Tanaka, T.; Yukawa, K.; Umesaki, N. A novel mechanism for acquired cisplatin-resistance: Suppressed translation of death-associated protein kinase mrna is insensitive to 5-aza-2′-deoxycitidine and trichostatin in cisplatin-resistant cervical squamous cancer cells. *Int. J. Oncol.* **2006**, *28*, 497–508. [CrossRef] [PubMed]
60. Parkin, J.; Cohen, B. An overview of the immune system. *Lancet* **2001**, *357*, 1777–1789. [CrossRef]
61. Zhang, J.; Hu, M.-M.; Shu, H.-B.; Li, S. Death-associated protein kinase 1 is an irf3/7-interacting protein that is involved in the cellular antiviral immune response. *Cell. Mol. Immunol.* **2014**, *11*, 245. [CrossRef] [PubMed]
62. McGargill, M.A.; Wen, B.G.; Walsh, C.M.; Hedrick, S.M. A deficiency in drak2 results in a t cell hypersensitivity and an unexpected resistance to autoimmunity. *Immunity* **2004**, *21*, 781–791. [CrossRef] [PubMed]
63. Guo, X.; Hooshdaran, B.; Rafiq, K.; Xi, H.; Kolpakov, M.; Liu, S.; Zhang, X.; Chen, X.; Houser, S.R.; Koch, W.J.; et al. Abstract 16144: Death associated protein kinase mediates myofibril degeneration and myocyte apoptosis induced by beta-adrenergic receptors. *Circulation* **2016**, *134*, A16144–A16144.
64. Chuang, Y.-T.; Lin, Y.-C.; Lin, K.-H.; Chou, T.-F.; Kuo, W.-C.; Yang, K.-T.; Wu, P.-R.; Chen, R.-H.; Kimchi, A.; Lai, M.-Z. Tumor suppressor death-associated protein kinase is required for full IL-1beta production. *Blood* **2011**, *117*, 960–970. [CrossRef] [PubMed]

65. Chuang, Y.T.; Fang, L.W.; Lin-Feng, M.H.; Chen, R.H.; Lai, M.Z. The tumor suppressor death-associated protein kinase targets to tcr-stimulated nf-kappa b activation. *J. Immunol.* **2008**, *180*, 3238–3249. [CrossRef] [PubMed]

66. Nakav, S.; Cohen, S.; Feigelson, S.W.; Bialik, S.; Shoseyov, D.; Kimchi, A.; Alon, R. Tumor suppressor death-associated protein kinase attenuates inflammatory responses in the lung. *Am. J. Respir. Cell. Mol. Biol.* **2012**, *46*, 313–322. [CrossRef] [PubMed]

67. Kuester, D.; Guenther, T.; Biesold, S.; Hartmann, A.; Bataille, F.; Ruemmele, P.; Peters, B.; Meyer, F.; Schubert, D.; Bohr, U.R.; et al. Aberrant methylation of dapk in long-standing ulcerative colitis and ulcerative colitis-associated carcinoma. *Pathol. Res. Pract.* **2010**, *206*, 616–624. [CrossRef] [PubMed]

68. Martinet, W.; Schrijvers, D.M.; De Meyer, G.R.; Thielemans, J.; Knaapen, M.W.; Herman, A.G.; Kockx, M.M. Gene expression profiling of apoptosis-related genes in human atherosclerosis: Upregulation of death-associated protein kinase. *Arterioscler. Thromb. Vasc. Biol.* **2002**, *22*, 2023–2029. [CrossRef] [PubMed]

69. Rennier, K.; Ji, J.Y. Shear stress regulates expression of death-associated protein kinase in suppressing tnfalpha-induced endothelial apoptosis. *J. Cell. Physiol* **2012**, *227*, 2398–2411. [CrossRef] [PubMed]

70. Rennier, K.; Ji, J.Y. Effect of shear stress and substrate on endothelial dapk expression, caspase activity, and apoptosis. *BMC Res. Notes* **2013**, *6*, 10. [CrossRef] [PubMed]

71. Rizzi, M.; Tschan, M.P.; Britschgi, C.; Britschgi, A.; Hugli, B.; Grob, T.J.; Leupin, N.; Mueller, B.U.; Simon, H.U.; Ziemiecki, A.; et al. The death-associated protein kinase 2 is up-regulated during normal myeloid differentiation and enhances neutrophil maturation in myeloid leukemic cells. *J. Leukoc. Biol.* **2007**, *81*, 1599–1608. [CrossRef] [PubMed]

72. Guay, J.A.; Wojchowski, D.M.; Fang, J.; Oxburgh, L. Death associated protein kinase 2 is expressed in cortical interstitial cells of the mouse kidney. *BMC Res. Notes* **2014**, *7*, 345. [CrossRef] [PubMed]

73. Brognard, J.; Zhang, Y.W.; Puto, L.A.; Hunter, T. Cancer-associated loss-of-function mutations implicate dapk3 as a tumor-suppressing kinase. *Cancer Res.* **2011**, *71*, 3152–3161. [CrossRef] [PubMed]

74. Leister, P.; Felten, A.; Chasan, A.I.; Scheidtmann, K.H. Zip kinase plays a crucial role in androgen receptor-mediated transcription. *Oncogene* **2008**, *27*, 3292–3300. [CrossRef] [PubMed]

75. Togi, S.; Ikeda, O.; Kamitani, S.; Nakasuji, M.; Sekine, Y.; Muromoto, R.; Nanbo, A.; Oritani, K.; Kawai, T.; Akira, S.; et al. Zipper-interacting protein kinase (zipk) modulates canonical wnt/beta-catenin signaling through interaction with nemo-like kinase and t-cell factor 4 (nlk/tcf4). *J. Biol. Chem.* **2011**, *286*, 19170–19177. [CrossRef] [PubMed]

76. Kake, S.; Usui, T.; Ohama, T.; Yamawaki, H.; Sato, K. Death-associated protein kinase 3 controls the tumor progression of a549 cells through erk mapk/c-myc signaling. *Oncol. Rep.* **2017**, *37*, 1100–1106. [CrossRef] [PubMed]

77. Usui, T.; Okada, M.; Hara, Y.; Yamawaki, H. Exploring calmodulin-related proteins, which mediate development of hypertension, in vascular tissues of spontaneous hypertensive rats. *Biochemical. Biophysical. Res. Commun.* **2011**, *405*, 47–51. [CrossRef] [PubMed]

78. Usui, T.; Okada, M.; Hara, Y.; Yamawaki, H. Death-associated protein kinase 3 mediates vascular inflammation and development of hypertension in spontaneously hypertensive rats. *Hypertension* **2012**, *60*, 1031–1039. [CrossRef] [PubMed]

79. Cho, Y.E.; Ahn, D.S.; Morgan, K.G.; Lee, Y.H. Enhanced contractility and myosin phosphorylation induced by ca(2+)-independent mlck activity in hypertensive rats. *Cardiovasc. Res.* **2011**, *91*, 162–170. [CrossRef] [PubMed]

80. Usui, T.; Sakatsume, T.; Nijima, R.; Otani, K.; Kazama, K.; Morita, T.; Kameshima, S.; Okada, M.; Yamawaki, H. Death-associated protein kinase 3 mediates vascular structural remodelling via stimulating smooth muscle cell proliferation and migration. *Clin. Sci.* **2014**, *127*, 539–548. [CrossRef] [PubMed]

81. Ihara, E.; MacDonald, J.A. The regulation of smooth muscle contractility by zipper-interacting protein kinase. *Can. J. Physiol. Pharmacol.* **2007**, *85*, 79–87. [CrossRef] [PubMed]

82. Gottesman, M.M.; Fojo, T.; Bates, S.E. Multidrug resistance in cancer: Role of atp-dependent transporters. *Nat. Rev. Cancer* **2002**, *2*, 48–58. [CrossRef] [PubMed]

83. Sebolt-Leopold, J.S.; Herrera, R. Targeting the mitogen-activated protein kinase cascade to treat cancer. *Nat. Rev. Cancer* **2004**, *4*, 937–947. [CrossRef] [PubMed]

84. Adler, S.S.; Afanasiev, S.; Aidala, C.; Ajitanand, N.N.; Akiba, Y.; Alexander, J.; Amirikas, R.; Aphecetche, L.; Aronson, S.H.; Averbeck, R.; et al. Measurement of transverse single-spin asymmetries for midrapidity production of neutral pions and charged hadrons in polarized p + p collisions at square root(s) = 200 gev. *Phys. Rev. Lett.* **2005**, *95*, 202001. [CrossRef] [PubMed]

85. Ballif, B.A.; Blenis, J. Molecular mechanisms mediating mammalian mitogen-activated protein kinase (mapk) kinase (mek)-mapk cell survival signals. *Cell Growth Differ.* **2001**, *12*, 397–408. [PubMed]

86. Stevens, C.; Lin, Y.; Sanchez, M.; Amin, E.; Copson, E.; White, H.; Durston, V.; Eccles, D.M.; Hupp, T. A germ line mutation in the death domain of dapk-1 inactivates erk-induced apoptosis. *J. Biol. Chem.* **2007**, *282*, 13791–13803. [CrossRef] [PubMed]

87. Agarwal, M.L.; Ramana, C.V.; Hamilton, M.; Taylor, W.R.; DePrimo, S.E.; Bean, L.J.; Agarwal, A.; Agarwal, M.K.; Wolfman, A.; Stark, G.R. Regulation of p53 expression by the ras-map kinase pathway. *Oncogene* **2001**, *20*, 2527–2536. [CrossRef] [PubMed]

88. Xiong, W.; Wu, Y.; Xian, W.; Song, L.; Hu, L.; Pan, S.; Liu, M.; Yao, S.; Pei, L.; Shang, Y. Dapk1-erk signal mediates oxygen glucose deprivation reperfusion induced apoptosis in mouse n2a cells. *J. Neurol. Sc.i* **2018**, *387*, 210–219. [CrossRef] [PubMed]

89. Gardner, A.M.; Johnson, G.L. Fibroblast growth factor-2 suppression of tumor necrosis factor alpha-mediated apoptosis requires ras and the activation of mitogen-activated protein kinase. *J. Biol. Chem.* **1996**, *271*, 14560–14566. [CrossRef] [PubMed]

90. Kayali, A.G.; Austin, D.A.; Webster, N.J. Stimulation of mapk cascades by insulin and osmotic shock: Lack of an involvement of p38 mitogen-activated protein kinase in glucose transport in 3t3-l1 adipocytes. *Diabetes* **2000**, *49*, 1783–1793. [CrossRef] [PubMed]

91. Puls, A.; Eliopoulos, A.G.; Nobes, C.D.; Bridges, T.; Young, L.S.; Hall, A. Activation of the small gtpase cdc42 by the inflammatory cytokines tnf(alpha) and il-1, and by the epstein-barr virus transforming protein lmp1. *J. Cell Sci.* **1999**, *112*, 2983–2992. [PubMed]

92. Fritz, G.; Kaina, B. Activation of c-jun n-terminal kinase 1 by uv irradiation is inhibited by wortmannin without affecting c-iun expression. *Mol. Cell. Biol.* **1999**, *19*, 1768–1774. [CrossRef] [PubMed]

93. Wagner, E.F.; Nebreda, A.R. Signal integration by jnk and p38 mapk pathways in cancer development. *Nat. Rev. Cancer* **2009**, *9*, 537–549. [CrossRef] [PubMed]

94. Bajbouj, K.; Poehlmann, A.; Kuester, D.; Drewes, T.; Haase, K.; Hartig, R.; Teller, A.; Kliche, S.; Walluscheck, D.; Ivanovska, J.; et al. Retracted: Identification of phosphorylated p38 as a novel dapk-interacting partner during tnfalpha-induced apoptosis in colorectal tumor cells. *Am. J. Pathol.* **2009**, *175*, 557–570. [CrossRef] [PubMed]

95. Finkel, T.; Holbrook, N.J. Oxidants, oxidative stress and the biology of ageing. *Nature* **2000**, *408*, 239–247. [CrossRef] [PubMed]

96. Sakon, S.; Xue, X.; Takekawa, M.; Sasazuki, T.; Okazaki, T.; Kojima, Y.; Piao, J.H.; Yagita, H.; Okumura, K.; Doi, T.; et al. Nf-kappab inhibits tnf-induced accumulation of ros that mediate prolonged mapk activation and necrotic cell death. *Embo J.* **2003**, *22*, 3898–3909. [CrossRef] [PubMed]

97. Kamata, H.; Honda, S.; Maeda, S.; Chang, L.; Hirata, H.; Karin, M. Reactive oxygen species promote tnfalpha-induced death and sustained jnk activation by inhibiting map kinase phosphatases. *Cell* **2005**, *120*, 649–661. [CrossRef] [PubMed]

98. Eisenberg-Lerner, A.; Kimchi, A. Dap kinase regulates jnk signaling by binding and activating protein kinase d under oxidative stress. *Cell Death Differ.* **2007**, *14*, 1908–1915. [CrossRef] [PubMed]

99. Schlegel, C.R.; Georgiou, M.L.; Misterek, M.B.; Stocker, S.; Chater, E.R.; Munro, C.E.; Pardo, O.E.; Seckl, M.J.; Costa-Pereira, A.P. Dapk2 regulates oxidative stress in cancer cells by preserving mitochondrial function. *Cell Death Dis* **2015**, *6*, e1671. [CrossRef] [PubMed]

100. Watterson, D.M.; Mirzoeva, S.; Guo, L.; Whyte, A.; Bourguignon, J.J.; Hilbert, M.; Haiech, J.; Van Eldik, L.J. Ligand modulation of glial activation: Cell permeable, small molecule inhibitors of serine-threonine protein kinases can block induction of interleukin 1 beta and nitric oxide synthase ii. *Neurochem. Int.* **2001**, *39*, 459–468. [CrossRef]

101. Mirzoeva, S.; Sawkar, A.; Zasadzki, M.; Guo, L.; Velentza, A.V.; Dunlap, V.; Bourguignon, J.-J.; Ramstrom, H.; Haiech, J.; Van Eldik, L.J.; et al. Discovery of a 3-amino-6-phenyl-pyridazine derivative as a new synthetic antineuroinflammatory compound. *J. Med. Chem.* **2002**, *45*, 563–566. [CrossRef] [PubMed]

102. Okamoto, M.; Takayama, K.; Shimizu, T.; Ishida, K.; Takahashi, O.; Furuya, T. Identification of death-associated protein kinases inhibitors using structure-based virtual screening. *J. Mèd. Chem.* **2009**, *52*, 7323–7327. [CrossRef] [PubMed]

103. Gandesiri, M.; Chakilam, S.; Ivanovska, J.; Benderska, N.; Ocker, M.; Di Fazio, P.; Feoktistova, M.; Gali-Muhtasib, H.; Rave-Frank, M.; Prante, O.; et al. Dapk plays an important role in panobinostat-induced autophagy and commits cells to apoptosis under autophagy deficient conditions. *Apoptosis* **2012**, *17*, 1300–1315. [CrossRef] [PubMed]

104. Li, S.X.; Han, Y.; Xu, L.Z.; Yuan, K.; Zhang, R.X.; Sun, C.Y.; Xu, D.F.; Yuan, M.; Deng, J.H.; Meng, S.Q.; et al. Uncoupling dapk1 from nmda receptor glun2b subunit exerts rapid antidepressant-like effects. *Mol. Psychiatry* **2018**, *23*, 597–608. [CrossRef] [PubMed]

105. de las Infantas, M.J.P.; Torres-Rusillo, S.; Unciti-Broceta, J.D.; Fernandez-Rubio, P.; Luque-Gonzalez, M.A.; Gallo, M.A.; Unciti-Broceta, A.; Molina, I.J.; Diaz-Mochon, J.J. Synthesis of 6,8,9 poly-substituted purine analogue libraries as pro-apoptotic inducers of human leukemic lymphocytes and dapk-1 inhibitors. *Org. Biomol. Chem.* **2015**, *13*, 5224–5234. [CrossRef] [PubMed]

106. Huang, Y.D.; Chen, L.; Guo, L.B.; Hupp, T.R.; Lin, Y. Evaluating dapk as a therapeutic target. *Apoptosis* **2014**, *19*, 371–386. [CrossRef] [PubMed]

107. Cho, E.Y.; Lee, S.J.; Nam, K.W.; Shin, J.; Oh, K.B.; Kim, K.H.; Mar, W. Amelioration of oxygen and glucose deprivation-induced neuronal death by chloroform fraction of bay leaves (*Laurus nobilis*). *Biosci. Biotechnol. Biochem.* **2010**, *74*, 2029–2035. [CrossRef] [PubMed]

108. Carlson, D.A.; Singer, M.R.; Sutherland, C.; Redondo, C.; Alexander, L.T.; Hughes, P.F.; Knapp, S.; Gurley, S.B.; Sparks, M.A.; MacDonald, J.A.; et al. Targeting pim kinases and dapk3 to control hypertension. *Cell Chem. Biol.* **2018**. [CrossRef] [PubMed]

109. MacDonald, J.A.; Sutherland, C.; Carlson, D.A.; Bhaidani, S.; Al-Ghabkari, A.; Sward, K.; Haystead, T.A.; Walsh, M.P. A small molecule pyrazolo [3,4-d]pyrimidinone inhibitor of zipper-interacting protein kinase suppresses calcium sensitization of vascular smooth muscle. *Mol. Pharmacol.* **2016**, *89*, 105–117. [CrossRef] [PubMed]

International Journal of
Molecular Sciences

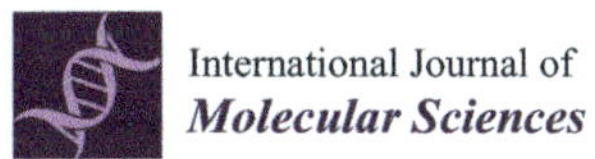

Article

Effects of Sunitinib and Other Kinase Inhibitors on Cells Harboring a *PDGFRB* Mutation Associated with Infantile Myofibromatosis

Martin Sramek [1,2,3], **Jakub Neradil** [1,2,3], **Petra Macigova** [1,2], **Peter Mudry** [2], **Kristyna Polaskova** [2,3], **Ondrej Slaby** [4], **Hana Noskova** [4], **Jaroslav Sterba** [2,3] and **Renata Veselska** [1,2,3,*]

[1] Laboratory of Tumor Biology, Department of Experimental Biology, Faculty of Science, Masaryk University, 61137 Brno, Czech Republic; martin.sramek@mail.muni.cz (M.S.); jneradil@sci.muni.cz (J.N.); macigova@med.muni.cz (P.M.)

[2] Department of Pediatric Oncology, University Hospital Brno and Faculty of Medicine, Masaryk University, 66263 Brno, Czech Republic; mudry.peter@fnbrno.cz (P.M.); polaskova.kristyna@fnbrno.cz (K.P.); sterba.jaroslav@fnbrno.cz (J.S.)

[3] International Clinical Research Center, St. Anne's University Hospital, 65691 Brno, Czech Republic

[4] Central European Institute of Technology, Masaryk University, 62500 Brno, Czech Republic; ondrej.slaby@ceitec.muni.cz (O.S.); hana.noskova@ceitec.muni.cz (H.N.)

*   Correspondence: veselska@sci.muni.cz; Tel.: +420-549-49-7905

Received: 1 August 2018; Accepted: 29 August 2018; Published: 1 September 2018

Abstract: Infantile myofibromatosis represents one of the most common proliferative fibrous tumors of infancy and childhood. More effective treatment is needed for drug-resistant patients, and targeted therapy using specific protein kinase inhibitors could be a promising strategy. To date, several studies have confirmed a connection between the p.R561C mutation in gene encoding platelet-derived growth factor receptor beta (PDGFR-beta) and the development of infantile myofibromatosis. This study aimed to analyze the phosphorylation of important kinases in the NSTS-47 cell line derived from a tumor of a boy with infantile myofibromatosis who harbored the p.R561C mutation in PDGFR-beta. The second aim of this study was to investigate the effects of selected protein kinase inhibitors on cell signaling and the proliferative activity of NSTS-47 cells. We confirmed that this tumor cell line showed very high phosphorylation levels of PDGFR-beta, extracellular signal-regulated kinases (ERK) 1/2 and several other protein kinases. We also observed that PDGFR-beta phosphorylation in tumor cells is reduced by the receptor tyrosine kinase inhibitor sunitinib. In contrast, MAPK/ERK kinases (MEK) 1/2 and ERK1/2 kinases remained constitutively phosphorylated after treatment with sunitinib and other relevant protein kinase inhibitors. Our study showed that sunitinib is a very promising agent that affects the proliferation of tumor cells with a p.R561C mutation in PDGFR-beta.

Keywords: infantile myofibromatosis; receptor tyrosine kinases; platelet-derived growth factor receptor; protein kinase inhibitors; sunitinib; erlotinib; FR180204; U0126; targeted therapy

---

1. Introduction

Infantile myofibromatosis (IM; [MIM#228550]) is a disorder of mesenchymal proliferation characterized by the development of nonmetastatic tumors [1] that present as firm, flesh-colored to purple nodules usually located in the skin, subcutaneous tissues, bone, muscle or visceral organs [2,3]. This disease was described under different names, the name "infantile myofibromatosis" was first used in 1981 [4]. Although rare, with an incidence of 1 in 400,000 children, IM represents the most common proliferative fibrous tumor of infancy [5,6]. Myofibromas are usually present at birth or develop shortly thereafter, and almost 90% of the tumors are diagnosed before the age of two years, with a median age of three months [6–8]. A male predominance has been reported, and the ratio of male to female patients varies from 1.5:1 to 1.8:1 [5].

IM clinically presents in three main forms: (1) Solitary, (2) multicentric without visceral involvement, and (3) multicentric with visceral involvement [6]. The prognosis is excellent in solitary or multicentric nonvisceral forms with a possibility of spontaneous regression of the lesions but is poor when detected in the viscera [9]. Surgical excision of a single lesion is the standard of care [8]. Multiple lesions or surgically unresectable lesions are treated using various therapeutics, such as anti-inflammatory drugs, interferon-alpha, vinblastine, vincristine, dactinomycin, cyclophosphamide and methotrexate [6,8].

The molecular pathogenesis of IM is not completely understood. Familial forms exhibiting autosomal dominant and recessive transmission have been reported over the past two decades [10]. In 2013, several point mutations in the platelet-derived growth factor receptor beta (PDGFR-beta) gene (*PDGFRB*) were identified to be associated with familial IM. A study of nine unrelated families diagnosed with IM revealed two disease-causing mutations in *PDGFRB*: c.1978C>A (p.P660T) and c.1681C>T (p.R561C) [1]. Interestingly, one family did not have either of these *PDGFRB* mutations, but all affected individuals had a c.4556T>C (p.L1519P) mutation in *NOTCH3*. The germline mutation c.1681C>T (p.R561C) in *PDGFRB* was also detected in 11 individuals with familial IM [7]. In addition, one individual harbored a c.1998C>A (p.N666K) somatic mutation. Very recently, a novel *PDGFRB* mutation (c.1679C>T; p.P560L) was identified in a 3-generation family with multicentric IM [11].

Platelet-derived growth factors (PDGFs) and PDGF receptors (PDGFRs) have important functions in the regulation of cell growth and survival [12]. The PDGF family consists of four structurally related single polypeptide units that constitute five functional homo- or heterodimers: PDGF-AA, PDGF-BB, PDGF-AB, PDGF-CC, and PDGF-DD [13]. PDGFs act via two receptor tyrosine kinases (RTKs), PDGFR-alpha and PDGFR-beta [14]. Both receptors can activate many major signal transduction pathways, including the Ras/MAPK, PI3K/Akt and phospholipase C-gamma pathways [15].

Moreover, other genes were associated with IM etiology, which demonstrates the possible genetic heterogeneity of this disease. As mentioned above, a connection between a c.4556T>C (p.L1519P) mutation in *NOTCH3* and IM was described in one study [1]. Human cells express four different Notch receptors, Notch 1–4, each encoded by a different gene [16]. The expression of *PDGFRB* can be regulated by Notch activity, as PDGFR-beta expression can be robustly upregulated by Notch 1 and Notch 3 signaling [17]. Another example is a c.511G>C (p.V171L) mutation in the potential tumor suppressor *NDRG4* that was associated with IM in one case [18]. In the same year, it was demonstrated that the c.1276G>A (p.V426M) mutation in *PTPRG* (protein tyrosine phosphatase, receptor type G) was able to substantially influence the penetrance of a c.1681C>T (p.R561C) mutation in *PDGFRB* [19]. *PTPRG* encodes an enzyme that could dephosphorylate PDGFR-beta and thus reduce PDGFR-beta activity [19,20].

A recent work revealed that two IM-associated mutations in *PDGFRB*, c.1681C>T (p.R561C) and c.1998C>A (p.N666K), constitutively activate PDGFR-beta and can induce cancer development in vivo [21]. The same study showed that cells harboring p.R561C and p.N666K mutations are sensitive to specific tyrosine kinase inhibitors, which were able to decrease PDGFR-beta phosphorylation and downstream signaling. These results suggested that blocking PDGFR-beta activity would offer a therapeutic option for IM treatment. Indeed, in a recently published study, targeted treatment with sunitinib and low-dose vinblastine led to a robust response in a child with refractory multiple IM and a c.1681C>T (p.R561C) mutation in *PDGFRB* [8].

In this work, we demonstrate for the first time the efficacy of sunitinib, erlotinib, U0126 and FR180204 on the cell line harboring a c.1681C>T (p.R561C) *PDGFRB* mutation found in patients with IM. Sunitinib is known as an inhibitor of several kinases, including PDGFR-beta [22], erlotinib is an inhibitor of epidermal growth factor receptor (EGFR) [23], U0126 inhibits MEK1/2 phosphorylation [24], and FR180204 inhibits ERK1/2 phosphorylation. These inhibitors were chosen on the basis of our previous findings [8] as well as on the results of subsequent phosphoprotein profiling of the NSTS-47 cell line.

2. Results

2.1. Germline Mutations in PDGFRB Were Identified in Both Children, and the Same Mutation in PDGFRB Was Confirmed in NSTS-47 Cells

Genetic analyses revealed that both siblings harbor a heterozygous germline c.1681C>T (p.R561C) mutation in the *PDGFRB* gene (Table 1). It was also confirmed that NSTS-47 cell line harbors the same heterozygous germline mutation c.1681C>T (p.R561C) in *PDGFRB*.

Table 1. Germline mutations identified in patients.

Gender	Age	*PDGFRB* Mutation
Male	3.5 months	c.1681C>T (p.R561C)
Female	8 years	c.1681C>T (p.R561C)

2.2. PDGFR-Beta, EGFR and ERK1/2 Kinases Are Highly Phosphorylated in Cells Harboring c.1681C>T (p.R561C) Mutation in PDGFRB

Given that both siblings and NSTS-47 cells harbor the c.1681C>T (p.R561C) mutation in *PDGFRB* and that PDGFR-beta c.1681C>T (p.R561C) mutants are constitutively phosphorylated and can activate various signaling pathways [21], we assessed the phosphorylation level of 49 RTKs and 26 other signaling proteins in tumor samples as well as in NSTS-47 cells. NSTS-47 cells were harvested, and phosphorylation levels were analyzed after cultivation for 24 h in Dulbecco's modified Eagle's medium (DMEM) without fetal calf serum (FCS) to eliminate the effects of various serum growth factors on the phosphorylation of the studied proteins. The screening of all 75 proteins showed that PDGFR-beta, EGFR (Figure 1) and ERK1/2 (Figure 2) kinases exhibited very high levels of phosphorylation in all samples. High levels of phosphorylation were also observed for ROR2, AXL (Figure 1), HSP27 and p38-gamma (Figure 2). These results confirmed that some kinases (namely, PDGFR-beta, EGFR and ERK1/2) were constitutively activated, as the high phosphorylation levels of these proteins were easily detectable in both tumor samples and in NSTS-47 cells after cultivation under serum-free conditions for 24 h.

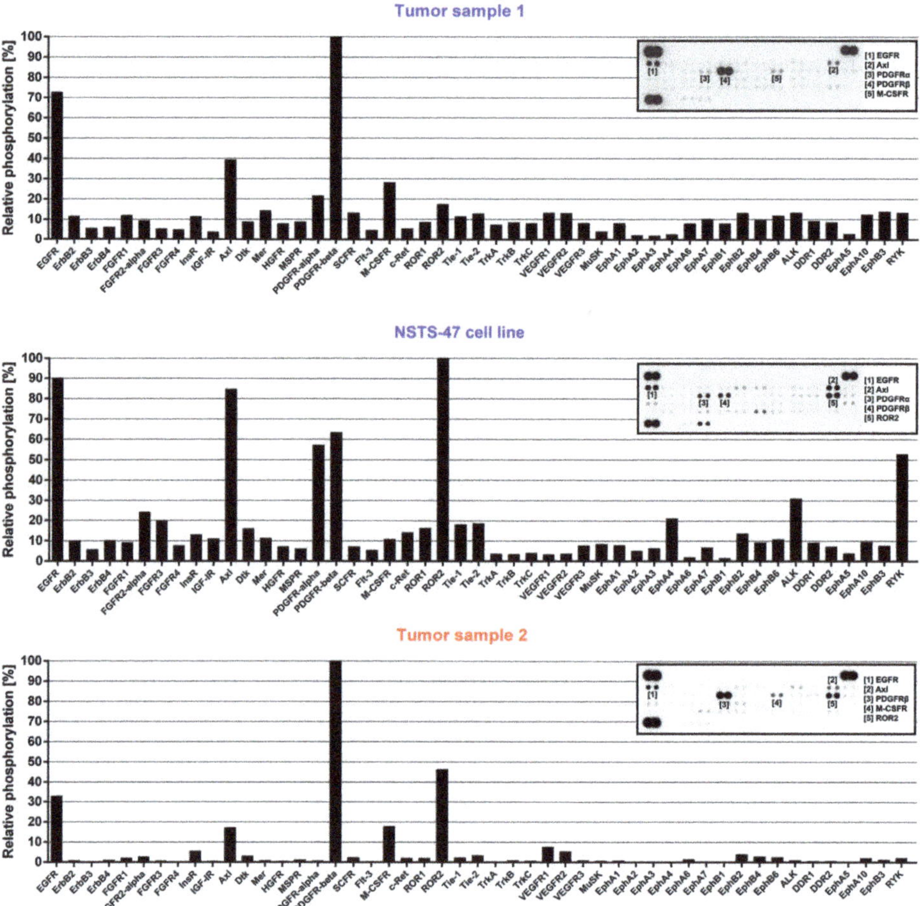

Figure 1. Phospho-receptor tyrosine kinases (RTK) array analysis. The relative phosphorylation of 49 RTKs was analyzed in tumor tissue obtained from the boy when he was 3.5 months old (Tumor sample 1), in the NSTS-47 cell line (derived from a tumor tissue of the boy obtained when he was 1 year and 7 months old) and in the tumor tissue of his 8-year-old sister (Tumor sample 2). platelet-derived growth factor receptor beta (PDGFR-beta) and epidermal growth factor receptor (EGFR) exhibited high levels of phosphorylation in all cases. Phosphorylation in NSTS-47 cells was measured after 24 h of serum-free cultivation. The array images captured using X-ray film are shown for each sample, and the five most phosphorylated receptor tyrosine kinases (RTKs) are marked. The upper part of the figure (Tumor sample 1) was already published in our previous case report [8] under the Creative Commons Attribution 4.0 International License.

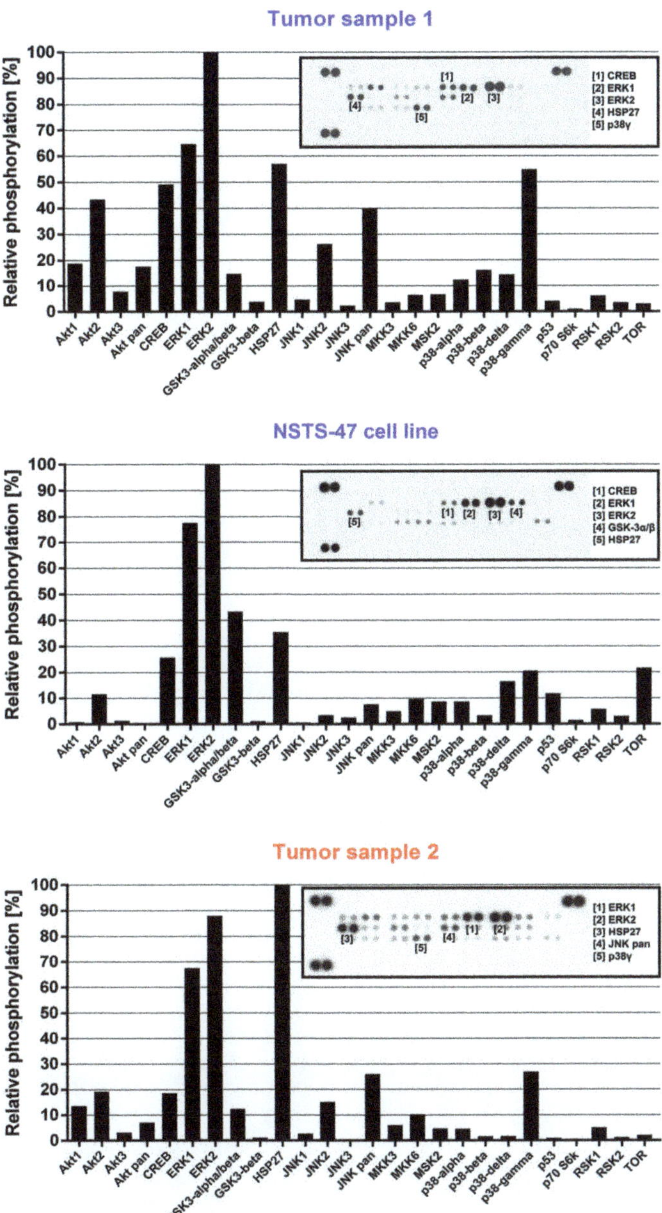

Figure 2. Phospho-mitogen-activated protein kinase (MAPK) array analysis. The relative phosphorylation of 26 signaling proteins, including 9 MAPKs, was detected in tumor tissue obtained from the boy when he was 3.5 months old (Tumor sample 1), in the NSTS-47 cell line (derived from a tumor tissue of the boy obtained when he was 1 year and 7 months old) and in the tumor tissue of his 8-year-old sister (Tumor sample 2). ERK1/2 exhibited high levels of phosphorylation in all cases. Phosphorylation levels in NSTS-47 cells was measured after 24 h of serum-free cultivation. The array images captured using X-ray film are shown for each sample, and the five most phosphorylated proteins are marked.

2.3. NSTS-47 Cells Are Sensitive to Sunitinib and Erlotinib

It was confirmed that cells with the mutation c.1681C>T (p.R561C) in *PDGFRB* are sensitive to the tyrosine kinase inhibitors imatinib, nilotinib and ponatinib [21]. Given the phosphorylation profile in the NSTS-47 cell line, whether specific tyrosine kinase inhibitors could affect the proliferation of this cell line was assessed. NSTS-47 cells were first treated with sunitinib. Sunitinib was chosen for several reasons: (1) The NSTS-47 cell line harbors a c.1681C>T (p.R561C) mutation in *PDGFRB*, and PDGFR-beta was substantially phosphorylated in these cells; (2) sunitinib treatment inhibits PDGFR-beta phosphorylation [25]; and (3) sunitinib was successfully used to treat the boy with IM whose tumor tissue was used to generate the NSTS-47 cell line [8].

Cells were treated for six days with various concentrations of sunitinib, and after incubation, the proliferative activity was determined using the MTT assay. At sunitinib concentrations of 50 and 100 nM, which can be achieved in the plasma of children treated with sunitinib [26], the proliferative activity of NSTS-47 cells was significantly decreased (Figure 3A). In addition, 50 nM and 100 nM sunitinib decreased the proliferative activity of NSTS-47 cells to 75% and 73%, respectively, after six days.

To verify whether the observed effect of sunitinib is robust, NSTS-47 cells were cultivated with sunitinib in medium supplemented with PDGF-BB. A significant decrease in proliferative activity was observed after sunitinib treatment even when the cells grew in medium supplemented with PDGF-BB at a high concentration of 10 ng/mL (Figure 3B). In some experiments, the cultivation medium was changed every 24 h, and new medium with fresh inhibitor and fresh PDGF-BB was added (at medium changes) to prevent the potential degradation of sunitinib and PDGF-BB (Figure 3C).

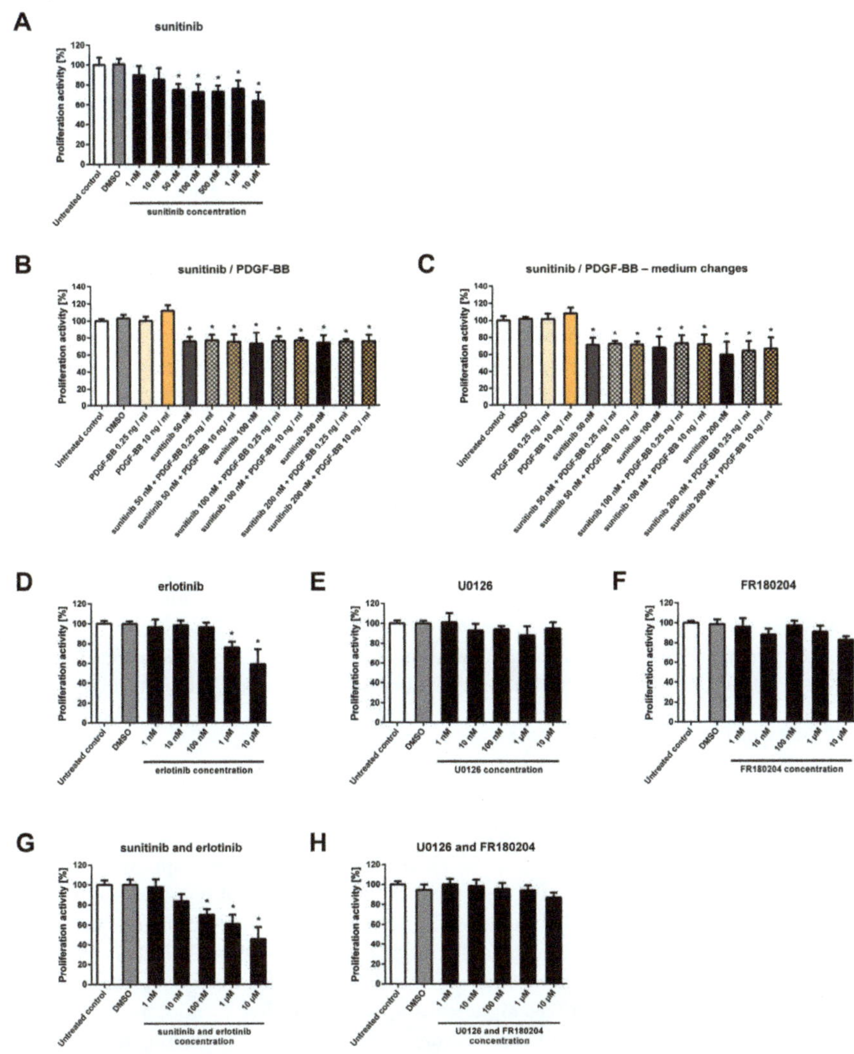

Figure 3. Proliferative activity of NSTS-47 cells after various experimental treatments. Proliferative activity was measured using an MTT assay after 6 days of incubation. The data represent the mean ± SD. Experiments were repeated three times in hexaplicate (**A,D–H**) or in triplicate (**B,C**). * $p < 0.05$ indicates a significant difference compared to control cells. (**A**) Sunitinib significantly decreased the proliferative activity of NSTS-47 cells. (**B**) NSTS-47 cells were sensitive to sunitinib, and this effect was not influenced by the presence of PDGF-BB at a high concentration (10 ng/mL). (**C**) Medium containing inhibitor and PDGF-BB was changed every 24 h during cultivation, which had no significant effect on the efficacy of the inhibitor. (**D**) NSTS-47 cells were also sensitive to erlotinib, as this inhibitor significantly affected cell proliferation. (**E**) No significant effect was observed after U0126 treatment. (**F**) FR180204 also did not significantly affect proliferative activity. (**G**) The combination of erlotinib and sunitinib significantly decreased the proliferative activity of NSTS-47 cells. (**H**) The combination of U0126 and FR180204 did not have a significant effect on NSTS-47 cell proliferation.

Next, NSTS-47 cells were treated with erlotinib, U0126 and FR180204. These three inhibitors were chosen based on EGFR and ERK1/2 phosphorylation in NSTS-47 cells (Figures 1 and 2). The ability of the combination of sunitinib and erlotinib to block both highly phosphorylated RTKs was tested, and a combination of U0126 and FR180204 was used to block the MEK/ERK signaling pathway.

At an erlotinib concentration of 1 μM, which can be achieved in the plasma of children treated with erlotinib [27], the proliferative activity of the NSTS-47 cell line was significantly decreased to 75% after 6 days of cultivation (Figure 3D). In contrast, NSTS-47 cells were not sensitive to U0126 and FR180204 because treatment of the NSTS-47 cell line with these inhibitors did not induce a significant decrease in proliferative activity (Figure 3E,F).

The combination of erlotinib and sunitinib also significantly decreased the proliferative activity of NSTS-47 cells (Figure 3G), but the effect of this combined treatment was similar to the effects of sunitinib or erlotinib alone. For instance, 100 nM sunitinib and 100 nM erlotinib decreased the proliferative activity to 70% (Figure 3G), but 100 nM sunitinib alone decreased the proliferative activity to 73% (Figure 3A). Another example is the combination of 1 μM erlotinib and 1 μM sunitinib; this treatment decreased the proliferative activity to 61% (Figure 3G), but 1 μM erlotinib alone decreased the proliferative activity to 75% (Figure 3D), and 1 μM sunitinib decreased the proliferative activity to 76% after six days (Figure 3A). Therefore, the combination of sunitinib and erlotinib did not have a significant additional effect on the reduction of NSTS-47 cell proliferation. In addition, the combination of U0126 and FR180204 did not show any significant effect on proliferative activity (Figure 3H).

Taken together, our results demonstrate that sunitinib and erlotinib can significantly decrease the proliferative activity of NSTS-47 cells, which harbor a c.1681C>T (p.R561C) mutation in *PDGFRB*, at concentrations that are achievable for these inhibitors in children plasma. However, the combination of sunitinib and erlotinib did not show an additional significant effect on cell proliferation. The inhibitors FR180204 and U0126 also did not have a significant effect on NSTS-47 cell proliferation.

2.4. PDGFR-Beta and EGFR Exhibited Ligand-Dependent Tyrosine Phosphorylation

Considering that only some kinase inhibitors significantly decreased the proliferative activity of the NSTS-47 cell line, detailed analyses of target kinases that should be affected by previously used inhibitors were performed using Western blotting. First, it was observed that the constitutively phosphorylated receptors PDGFR-beta and EGFR in NSTS-47 cells can respond to their ligands: Our results show that phosphorylation of both receptors was considerably increased in response to PDGF-BB or EGF (Figure 4A,B). Cell populations were serum starved for 24 h and then stimulated for 15, 30 or 60 min using two different concentrations of PDGF-BB or EGF. The cells that were serum starved for only 24 h and cells that were cultivated with FCS were used as negative controls. Receptor phosphorylation was significantly increased after 15 min, and then decreased in a time-dependent manner. Surprisingly, serum-starved cells that were not stimulated with PDGF-BB or EGF also exhibited an increase in receptor phosphorylation, in comparison to serum-cultivated cells. These experiments demonstrated that both receptors were functional and were able to activate downstream signaling molecules.

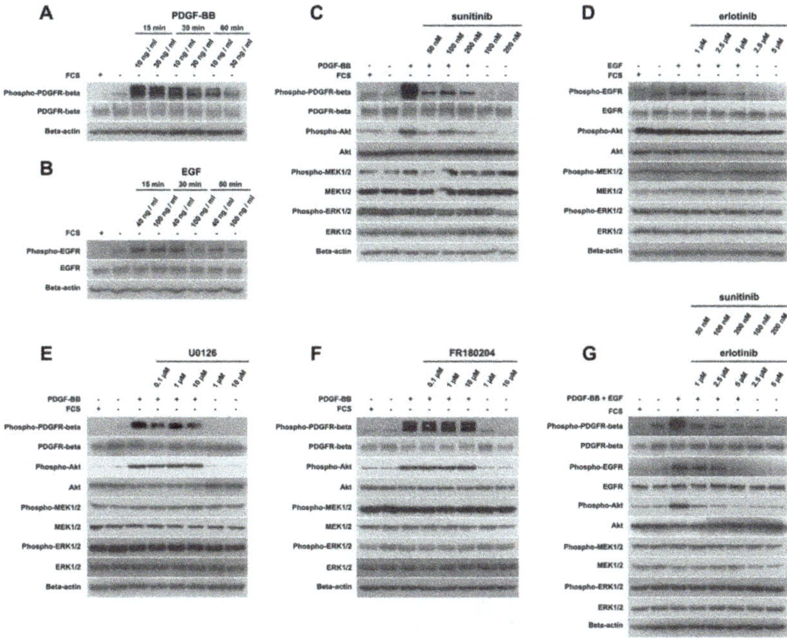

Figure 4. Analysis of protein phosphorylation. (**A**) PDGFR-beta phosphorylation is increased in response to PDGF-BB. Cells were stimulated for 15, 30 or 60 min using two different concentrations (10 ng/mL and 30 ng/mL) of PDGF-BB. (**B**) EGFR phosphorylation is increased in response to epidermal growth factor (EGF). Cells were stimulated for 15, 30 or 60 min using two different concentrations (40 ng/mL and 100 ng/mL) of EGF. (**C**) Sunitinib was able to decrease PDGFR-beta and Akt phosphorylation but not MEK1/2 and ERK1/2 phosphorylation. (**D**) Erlotinib decreased EGFR and Akt phosphorylation but had no effect on MEK1/2 and ERK1/2 phosphorylation. (**E**) U0126 treatment did not decrease MEK1/2 phosphorylation. (**F**) FR180204 treatment did not cause any changes in ERK1/2 phosphorylation. (**G**) The combination of sunitinib and erlotinib decreased PDGFR-beta, EGFR and Akt phosphorylation, but MEK1/2 and ERK1/2 phosphorylation was not affected.

2.5. Detailed Analysis of Signaling Pathways Revealed Constitutive Phosphorylation of MEK1/2 and ERK1/2 Proteins

In the next step, we analyzed the phosphorylation of PDGFR-beta, EGFR and downstream kinases, which can be activated by these RTKs after treatment with kinase inhibitors. In all experiments, cells were cultivated for 24 h in medium containing an inhibitor but not FCS. After 24 h, some cells were stimulated with PDGF-BB or/and EGF for 15 min to observe the effects of inhibitors on ligand-stimulated cells. Cells that were serum starved for only 24 h, and cells that were cultivated with FCS were used as negative controls.

Sunitinib alone decreased the phosphorylation of PDGFR-beta (Figure 4C). Akt phosphorylation was also decreased after sunitinib treatment, but a substantial decrease in MEK1/2 and ERK1/2 phosphorylation was not observed. Erlotinib decreased the phosphorylation of EGFR, but only at higher concentrations, and Akt phosphorylation was also slightly decreased (Figure 4D). No effect of erlotinib on MEK1/2 and ERK1/2 phosphorylation was observed. Surprisingly, U0126 did not decrease the phosphorylation of MEK1/2 (Figure 4E). Similarly, FR180204 treatment had no effect on ERK1/2 phosphorylation (Figure 4F). As expected, the combination of sunitinib and erlotinib markedly decreased the phosphorylation of PDGFR-beta, EGFR and Akt, but no effect was observed on MEK1/2 and ERK1/2 phosphorylation (Figure 4G).

Altogether, sunitinib and erlotinib showed inhibitory effects on RTKs and Akt. Interestingly, no substantial changes in MEK1/2 and ERK1/2 phosphorylation were observed after treatment with any inhibitor.

2.6. Serum Starvation of NSTS-47 Cells Induces an Increase in PDGFA Expression

In some cases, our data indicated higher phosphorylation of PDGFR-beta and EGFR in serum-starved cells than in cells cultivated in DMEM supplemented with FCS (Figure 4A,B). Therefore, the expression of selected EGFR and PDGFR-beta ligands was measured to investigate whether there is a possible autocrine PDGF/PDGFR or EGF (TGF-alpha)/EGFR signaling loop that could contribute to the higher phosphorylation of RTKs. Expression of *EGF*, *PDGFA*, *PDGFB* and *TGFA* was analyzed under normal serum conditions (DMEM supplemented with 20% FCS) and under serum starvation conditions using qPCR. Substantial differences were observed in the transcriptional response of serum-starved cells (Figure 5). qPCR analyses also showed increased levels of *PDGFA* expression, while *EGF* and *PDGFB* mRNA levels were not significantly influenced by serum starvation, and *TGFA* expression was considerably decreased.

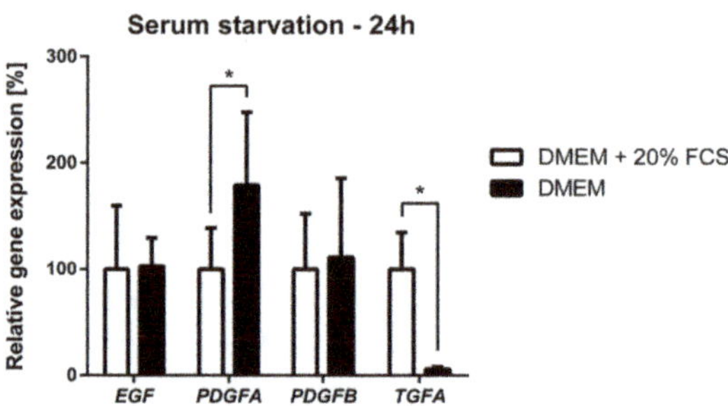

Figure 5. Effect of serum starvation on *EGF*, *PDGFA*, *PDGFB* and *TGFA* expression in the NSTS-47 cell line. Cells were cultivated in Dulbecco's modified Eagle's medium (DMEM) supplemented with 20% FCS or in DMEM without FCS. After 24 h, cells were harvested, and the expression of selected genes was analyzed using qPCR. The results represent the mean ± SD of nine (six in case of *PDGFB*) independent experiments. * $p < 0.05$ indicates statistically significant differences.

3. Discussion

IM is a rare disorder of mesenchymal proliferation that is characterized by the development of nonmetastatic tumors [1]. Several studies have confirmed that specific point mutations in the *PDGFRB* gene are involved in the pathogenesis of IM [1,7,11]. However, mutations in the *PDGFRB* gene presumably show incomplete penetrance and variable expressivity, and other genes may be involved in the pathogenesis of IM [1,18,19].

The main goal of this study was to analyze the effects of various protein kinase inhibitors (PKIs) on the NSTS-47 cell line, which harbors the IM-associated c.1681C>T (p.R561C) mutation in *PDGFRB*. The results showed that sunitinib, a potent inhibitor of PDGFR-beta phosphorylation, can significantly decrease the proliferation of NSTS-47 cells.

Previously published results [21] show that PDGFR-beta p.R561C mutant cells have constitutively phosphorylated PDGFR-beta and are able to induce the phosphorylation of ERK1/2, PLC-gamma, STAT3, STAT5 and Akt in the absence of PDGF. These results are in accordance with our observations. We found that PDGFR-beta and ERK1/2 kinases were highly phosphorylated in both s and even in NSTS-47 cells that were serum starved for 24 h. We also detected increased phosphorylation of Akt2 in Tumor Sample 1.

The same study that revealed a role for the p.R561C mutation in PDGFR-beta [21] showed that imatinib, nilotinib, and ponatinib can decrease PDGFR-beta phosphorylation and inhibit cell proliferation. We studied the effects of sunitinib, a multi-tyrosine kinase inhibitor that is able to target PDGFR-beta. Sunitinib was chosen because siblings from whom tumor tissue samples were obtained responded very well to treatment with this inhibitor [8]. Sunitinib alone significantly decreased the proliferative activity of the NSTS-47 cell line, and this finding could explain the response of the siblings to the targeted therapy.

Western blot analyses showed that sunitinib is able to decrease the phosphorylation of mutant PDGFR-beta even in the presence of high PDGF-BB levels and can also decrease the phosphorylation of Akt. Because activated Akt is a well-established survival factor [28], these effects of sunitinib on PDGFR-beta and Akt phosphorylation can explain why sunitinib reduced the proliferative activity of NSTS-47 cells.

A similar inhibitory effect was observed for EGFR and erlotinib (the inhibitor of EGFR phosphorylation). Erlotinib also decreased NSTS-47 cell proliferation, and Western blot analysis showed that it was able to decrease EGFR and Akt phosphorylation. However, neither sunitinib nor erlotinib inhibited the phosphorylation of the corresponding receptor completely, and some receptor molecules remained phosphorylated even when high doses of those inhibitors were used.

Surprisingly, phosphorylation of MEK1/2 and ERK1/2 proteins was not significantly influenced by any inhibitor. This observation could explain why sunitinib and erlotinib incompletely decreased proliferative activity and why U0126 and FR180204 did not influence proliferative activity. MEK1/2 and ERK1/2 belong to the Ras/MAPK signaling cascade, which transmits signals from receptors and participate in regulating the cell cycle, apoptosis and differentiation [29]. All tyrosine kinase inhibitors have been previously shown to be able to simultaneously decrease PDGFR-beta and ERK1/2 phosphorylation, which resulted in the inhibition of proliferative activity [21]. In NSTS-47 cells, sunitinib and erlotinib decreased the phosphorylation of PDGFR-beta, EGFR and Akt, but for yet unknown reasons, MEK1/2 and ERK1/2 kinases remained phosphorylated at levels that were comparable with those detected in untreated cells.

Interestingly, incomplete penetrance of the c.1681C>T (p.R561C) mutation was found in a family with two children suffering from IM [19]. Genetic analyses revealed a c.1681C>T (p.R561C) mutation in *PDGFRB* in both siblings and, surprisingly, also in their healthy mother. However, both siblings had inherited a heterozygous c.1276G>A (p.V426M) mutation in *PTPRG* from their healthy father. The *PTPRG* gene encodes a protein called receptor-type tyrosine-protein phosphatase gamma that can dephosphorylate PDGFR-beta [19,20]. Therefore, the mutation in *PTPRG* could probably decrease the efficiency of the phosphatase to dephosphorylate its substrates and thus positively influence the phosphorylation of PDGFR-beta and the penetrance of mutant *PDGFRB* [19].

Finally, our analyses of gene expression showed that the phosphorylation status of PDGFRs in NSTS-47 cells was not influenced by only mutations in PDGFR-beta. We analyzed the gene expression levels of *EGF*, *PDGFA*, *PDGFB* and *TGFA* in NSTS-47 cells that were serum starved for 24 h. The expression of *TGFA* decreased, but no difference was observed in the expression of *EGF* and *PDGFB*; however, *PDGFA* gene expression was significantly increased. The increase in *PDGFA* expression was unexpected and could result in the stimulation of cells via an autocrine mechanism, an increase in PDGFR-alpha phosphorylation and improved survival of NSTS-47 cells in the absence of serum.

4. Materials and Methods

4.1. Tumor Samples

Two tumor samples and one tumor-derived cell line were used in this study. Tumor Sample 1 was obtained from a 3.5-month-old infant boy suffering from inborn generalized IM, and Tumor Sample 2 was obtained from his 8-year-old sister who was suffering from a skull base tumor and had a history of spontaneous regression of subcutaneous lesions. The Research Ethics Committee of the School of Medicine (Masaryk University, Brno, Czech Republic) approved the study protocol, and written informed consent was obtained from legal guardians of the siblings. A case report concerning these siblings was published recently [8].

4.2. Cell Line and Cell Culture

The NSTS-47 cell line was established in our laboratory with the procedure previously described [30]. A tumor sample was obtained from the same boy mentioned in the previous paragraph during curative surgical procedure when he was 1 year and 7 months old. Cells were grown in Dulbecco's modified Eagle's medium (DMEM), supplemented with 20% fetal calf serum (FCS), 2 mM glutamine, 100 IU/mL penicillin and 100 µg/mL streptomycin (all purchased from GE Healthcare Europe GmbH, Freiburg, Germany). The cell line was maintained under standard conditions at 37 °C in a humidified atmosphere containing 5% CO_2 and subcultured one or two times per week. Cells from passage number 8 to 19 were used for experiments.

4.3. Genetic Analyses

The mutation in *PDGFRB* was identified by Sanger sequencing using an ABI 3130 Genetic Analyzer (Applied Biosystems, Foster City, CA, USA) and confirmed by whole exome sequencing (WES). In all cases, WES was performed using the TruSeq Exome Kit, NextSeq® 500/550 Mid Output Kit v2 and NextSeq 500 (all Illumina, San Diego, CA, USA).

4.4. Chemicals

Sunitinib, erlotinib, U0126 (all purchased from Cell Signaling Technology, Danvers, MA, USA) and FR180204 (Sigma-Aldrich, St. Louis, MO, USA) were prepared as a 20 mM stock solution in dimethyl sulfoxide (DMSO) and stored at −20 °C. PDGF-BB (Cell Signaling Technology) was prepared at a concentration of 100 µg/mL in 20 mM citric acid (pH 3.0) supplemented with 0.8% BSA (bovine serum albumin) and stored at 4 °C. EGF (Sigma-Aldrich) was prepared at a concentration of 100 µg/mL in 10 mM HCl and stored at 4 °C. For the determination of proliferative activity, concentrations of protein kinase inhibitors (PKIs) ranging from 0.001 to 10 µM and PDGF-BB concentrations of 0.25 and 10 ng/mL were tested. For Western blot analyses, PKI concentrations ranging from 0.05 to 10 µM, PDGF-BB concentrations of 10 and 30 ng/mL and EGF concentrations of 40 and 100 ng/mL were used.

4.5. Phospho-RTK and Phospho-MAPK Array Analysis

The relative phosphorylation levels of 49 RTKs were analyzed using the Human Phospho-RTK Array kit (R&D Systems, Minneapolis, MN, USA), and the relative phosphorylation levels of 26 proteins, including 9 MAPKs, were determined using the Human Phospho-MAPK Array kit (R&D Systems) according to the manufacturer's protocol. The levels of phosphorylation were quantified using ImageJ software [31] and normalized to control spots and the background. The analysis was performed as described in previous studies [8,32].

4.6. MTT Assay

The MTT assay was used to determine the proliferative activity of the NSTS-47 cell line. A total of 10^3 cells were seeded in 200 µL of culture medium into each well of 96-well microplates, and cells were allowed to adhere overnight. The next day, the medium was carefully removed, and fresh medium containing various concentrations of chemicals described above or control medium was added. The microplates were incubated under standard conditions. To evaluate changes in cell proliferation, the medium was removed and replaced with 200 µL of fresh DMEM containing 3-(4-dimethylthiazol-2-yl)-2,5-diphenyltetrazolium bromide (MTT) at a concentration of 0.5 mg per mL. The microplates were then incubated at 37 °C for 3.5 h. The medium was carefully removed, and the formazan crystals were dissolved in 200 µL of DMSO. The absorbance was measured at 570 nm using a Sunrise Absorbance Reader (Tecan, Männedorf, Switzerland), with a reference absorbance at 620 nm.

4.7. Western Blotting and Immunodetection

Whole-cell extracts were loaded onto 10% polyacrylamide gels, electrophoresed, and blotted on polyvinylidene difluoride membranes (Bio-Rad Laboratories, Munich, Germany). The membranes were blocked with 5% nonfat dry milk in phosphate buffered saline (PBS) containing 0.1% Tween-20 and incubated overnight with the corresponding primary antibody. The primary and secondary antibodies used in this study are shown in Table 2. Membranes were incubated with corresponding secondary antibodies for 1 h. ECL-Plus detection was performed according to the manufacturer's instructions (GE Healthcare, Little Chalfont, UK).

Table 2. Primary and secondary antibodies.

Primary Antibodies			
Antigen	Manufacturer	Catalog No.	Dilution
Beta-actin	Sigma-Aldrich	A5441	1:20,000
Akt (pan)	Cell Signaling Technology	4691	1:1000
Phospho-Akt (Ser473)	Cell Signaling Technology	4060	1:2000
ERK1/2	Cell Signaling Technology	4695	1:1000
Phospho-ERK1/2 (Thr202/Tyr204)	Cell Signaling Technology	4370	1:2000
MEK1/2	Cell Signaling Technology	9122	1:1000
Phospho-MEK1/2 (Ser217/221)	Cell Signaling Technology	9121	1:1000
EGFR	Cell Signaling Technology	2646	1:1000
Phospho-EGFR (Tyr1068)	Cell Signaling Technology	2236	1:1000
PDGFR-beta	Cell Signaling Technology	3169	1:1000
Phospho-PDGFR-beta (Tyr751)	Cell Signaling Technology	4549	1:1000
Secondary antibodies			

Specificity	Conjugate	Manufacturer	Catalog No.	Dilution
Anti-Mouse IgG	horseradish peroxidase	Cell Signaling Technology	7076	1:2000–1:20,000
Anti-Rabbit IgG	horseradish peroxidase	Cell Signaling Technology	7074	1:2000

4.8. RT-qPCR

The relative expression levels of selected genes were studied using RT-qPCR. Total RNA was extracted using the GenElute™ Mammalian Total RNA Miniprep kit (Sigma-Aldrich), and RNA concentration and purity were determined spectrophotometrically. For all samples, equal amounts of RNA were reverse transcribed into cDNA using M-MLV reverse transcriptase (Top-Bio, Prague, Czech Republic). RT-qPCR was carried out in 10 µL reaction volumes using the KAPA SYBR® FAST qPCR Kit (Kapa Biosystems, Wilmington, MA, USA) and analyzed using the 7500 Fast Real-Time PCR System and 7500 Software v. 2.0.6 (both Life Technologies, Carlsbad, CA, USA). Changes in the transcript levels were determined using the $2^{-\Delta\Delta CT}$ method [33]. The housekeeping gene *HSP90AB1* was used as an endogenous reference control. The primers used in this study are listed in Table 3.

Table 3. Primers.

Gene	Gene Symbol	Primer Sequence
Epidermal growth factor	*EGF*	F: 5′-AGGATTGACACAGAAGGAACCAA-3′ R: 5′-ACATACTCTCTCTTGCCTTGACC-3′
Heat shock protein 90 alpha family class B member 1	*HSP90AB1*	F: 5′-CGCATGAAGGAGACACAGAA-3′ R: 5′-TCCCATCAAATTCCTTGAGC-3′
Platelet derived growth factor subunit A	*PDGFA*	F: 5′-TCCGTAGGGAGTGAGGATTCTTT-3′ R: 5′-GGCTTCTTCCTGACGTATTCCA-3′
Platelet derived growth factor subunit B	*PDGFB*	F: 5′-GATCCGCTCCTTTGATGATCTCC-3′ R: 5′-ATCTCGATCTTTCTCACCTGGAC-3′
Transforming growth factor alpha	*TGFA*	F: 5′-TGCCACTCAGAAACAGTGGTC-3′ R: 5′-AGTCCGTCTCTTTGCAGTTCTT-3′

F, forward primer; R, reverse primer.

4.9. Statistical Analysis

Quantitative data are shown as the mean ± standard deviation (SD). Data from MTT assays were analyzed using one-way ANOVA followed by Dunnett's test; $p < 0.05$ was considered statistically significant. The qPCR data were analyzed using the Mann-Whitney test (two-tailed); $p < 0.05$ was considered statistically significant.

5. Conclusions

To conclude, our work demonstrated that tumor cells with the c.1681C>T (p.R561C) mutation in *PDGFRB* show high levels of PDGFR-beta and ERK1/2 phosphorylation. Furthermore, our data support the use of specific tyrosine kinase inhibitors targeting PDGFR-beta phosphorylation as a treatment suitable for IM. This is the first study to show that sunitinib is able to reduce the proliferative activity of IM cells with a c.1681C>T (p.R561C) mutation in vitro.

Author Contributions: J.N., R.V. and J.S. designed the study. J.S., P.M. (Peter Mudry) and K.P. provided tumor samples and relevant clinical data. H.N. and O.S. performed genetic analyses. M.S., P.M. (Petra Macigova) and J.N. designed and performed experiments with NSTS-47 cell line. M.S. and R.V. composed the manuscript. All authors reviewed and approved the final version of the manuscript.

Funding: This study was supported by projects No. 16-34083A and No. 16-33209A from the Ministry of Healthcare of the Czech Republic and by project No. LQ1605 from the National Program of Sustainability II (MEYS CR).

Acknowledgments: The authors thank Johana Maresova for her technical assistance and dr. Petr Chlapek for the derivation of the NSTS-47 cell line.

Conflicts of Interest: The authors declare no conflict of interest.

Abbreviations

DMEM	Dulbecco's modified Eagle's medium
DMSO	dimethyl sulfoxide
EGFR	epidermal growth factor receptor
ERK	extracellular signal-regulated kinase
FCS	fetal calf serum
IM	infantile myofibromatosis
MAPK	mitogen-activated protein kinase
MEK	MAPK/ERK kinase
MTT	3-(4-dimethylthiazol-2-yl)-2,5-diphenyltetrazolium bromide
PDGFR	platelet-derived growth factor receptor
PKIs	protein kinase inhibitors
RTKs	receptor tyrosine kinases
TGFA	transforming growth factor alpha
WES	whole exome sequencing

References

1. Martignetti, J.A.; Tian, L.; Li, D.; Ramirez, M.C.; Camacho-Vanegas, O.; Camacho, S.C.; Guo, Y.; Zand, D.J.; Bernstein, A.M.; Masur, S.K.; et al. Mutations in PDGFRB cause autosomal-dominant infantile myofibromatosis. *Am. J. Hum. Genet.* **2013**, *92*, 1001–1007. [CrossRef] [PubMed]
2. Levine, E.; Fréneaux, P.; Schleiermacher, G.; Brisse, H.; Pannier, S.; Teissier, N.; Mesples, B.; Orbach, D. Risk-adapted therapy for infantile myofibromatosis in children. *Pediatr. Blood Cancer* **2012**, *59*, 115–120. [CrossRef] [PubMed]
3. Kim, E.J.; Wang, K.C.; Lee, J.Y.; Phi, J.H.; Park, S.H.; Cheon, J.E.; Jang, Y.E.; Kim, S.K. Congenital solitary infantile myofibromatosis involving the spinal cord. *J. Neurosurg. Pediatr.* **2013**, *11*, 82–86. [CrossRef] [PubMed]
4. Venkatesh, V.; Kumar, B.P.; Kumar, K.A.; Mohan, A.P. Myofibroma-a rare entity with unique clinical presentation. *J. Maxillofac. Oral Surg.* **2015**, *14*, 64–68. [CrossRef] [PubMed]

5.   Hausbrandt, P.A.; Leithner, A.; Beham, A.; Bodo, K.; Raith, J.; Windhager, R. A rare case of infantile myofibromatosis and review of literature. *J. Pediatr. Orthop. B* **2010**, *19*, 122–126. [CrossRef] [PubMed]

6.   Weaver, M.S.; Navid, F.; Huppmann, A.; Meany, H.; Angiolillo, A. Vincristine and Dactinomycin in Infantile Myofibromatosis with a Review of Treatment Options. *J. Pediatr. Hematol. Oncol.* **2015**, *37*, 237–241. [CrossRef] [PubMed]

7.   Cheung, Y.H.; Gayden, T.; Campeau, P.M.; LeDuc, C.A.; Russo, D.; Nguyen, V.H.; Guo, J.; Qi, M.; Guan, Y.; Albrecht, S.; et al. A recurrent PDGFRB mutation causes familial infantile myofibromatosis. *Am. J. Hum. Genet.* **2013**, *92*, 996–1000. [CrossRef] [PubMed]

8.   Mudry, P.; Slaby, O.; Neradil, J.; Soukalova, J.; Melicharkova, K.; Rohleder, O.; Jezova, M.; Seehofnerova, A.; Michu, E.; Veselska, R.; et al. Case report: Rapid and durable response to PDGFR targeted therapy in a child with refractory multiple infantile myofibromatosis and a heterozygous germline mutation of the PDGFRB gene. *BMC Cancer* **2017**, *17*, 119. [CrossRef] [PubMed]

9.   Gatibelza, M.E.; Vazquez, B.R.; Bereni, N.; Denis, D.; Bardot, J.; Degardin, N. Isolated infantile myofibromatosis of the upper eyelid: Uncommon localization and long-term results after surgical management. *J. Pediatr. Surg.* **2012**, *47*, 1457–1459. [CrossRef] [PubMed]

10. Murray, N.; Hanna, B.; Graf, N.; Fu, H.; Mylene, V.; Campeau, P.M.; Ronan, A. The spectrum of infantile myofibromatosis includes both non-penetrance and adult recurrence. *Eur. J. Med. Genet.* **2017**, *60*, 353–358. [CrossRef] [PubMed]

11. Lepelletier, C.; Al-Sarraj, Y.; Bodemer, C.; Shaath, H.; Fraitag, S.; Kambouris, M.; Hamel-Teillac, D.; Shanti, H.E.; Hadj-Rabia, S. Heterozygous PDGFRB Mutation in a Three-generation Family with Autosomal Dominant Infantile Myofibromatosis. *Acta Derm. Venereol.* **2017**, *97*, 858–859. [CrossRef] [PubMed]

12. Heldin, C.H. Targeting the PDGF signaling pathway in tumor treatment. *Cell. Commun. Signal.* **2013**, *11*, 97. [CrossRef] [PubMed]

13. Cao, Y. Multifarious functions of PDGFs and PDGFRs in tumor growth and metastasis. *Trends Mol. Med.* **2013**, *19*, 460–473. [CrossRef] [PubMed]

14. Andrae, J.; Gallini, R.; Betsholtz, C. Role of platelet-derived growth factors in physiology and medicine. *Genes Dev.* **2008**, *22*, 1276–1312. [CrossRef] [PubMed]

15. Hoch, R.V.; Soriano, P. Roles of PDGF in animal development. *Development* **2003**, *130*, 4769–4784. [CrossRef] [PubMed]

16. Aster, J.C.; Pear, W.S.; Blacklow, S.C. The Varied Roles of Notch in Cancer. *Annu. Rev. Pathol* **2017**, *12*, 245–275. [CrossRef] [PubMed]

17. Jin, S.; Hansson, E.M.; Tikka, S.; Lanner, F.; Sahlgren, C.; Farnebo, F.; Baumann, M.; Kalimo, H.; Lendahl, U. Notch signaling regulates platelet-derived growth factor receptor-beta expression in vascular smooth muscle cells. *Circ. Res.* **2008**, *102*, 1483–1491. [CrossRef] [PubMed]

18. Linhares, N.D.; Freire, M.C.; Cardenas, R.G.; Pena, H.B.; Bahia, M.; Pena, S.D. Exome sequencing identifies a novel homozygous variant in NDRG4 in a family with infantile myofibromatosis. *Eur. J. Med. Genet.* **2014**, *57*, 643–648. [CrossRef] [PubMed]

19. Linhares, N.D.; Freire, M.C.; Cardenas, R.G.; Bahia, M.; Puzenat, E.; Aubin, F.; Pena, S.D. Modulation of expressivity in PDGFRB-related infantile myofibromatosis: A role for PTPRG? *Genet. Mol. Res.* **2014**, *13*, 6287–6292. [CrossRef] [PubMed]

20. Mirenda, M.; Toffali, L.; Montresor, A.; Scardoni, G.; Sorio, C.; Laudanna, C. Protein tyrosine phosphatase receptor type γ is a JAK phosphatase and negatively regulates leukocyte integrin activation. *J. Immunol.* **2015**, *194*, 2168–2179. [CrossRef] [PubMed]

21. Arts, F.A.; Chand, D.; Pecquet, C.; Velghe, A.; Constantinescu, S.; Hallberg, B.; Demoulin, J.B. PDGFRB mutants found in patients with familial infantile myofibromatosis or overgrowth syndrome are oncogenic and sensitive to imatinib. *Oncogene* **2016**, *35*, 3239–3248. [CrossRef] [PubMed]

22. Faivre, S.; Demetri, G.; Sargent, W.; Raymond, E. Molecular basis for sunitinib efficacy and future clinical development. *Nat. Rev. Drug Discov.* **2007**, *6*, 734–745. [CrossRef] [PubMed]

23. Sos, M.L.; Koker, M.; Weir, B.A.; Heynck, S.; Rabinovsky, R.; Zander, T.; Seeger, J.M.; Weiss, J.; Fischer, F.; Frommolt, P.; et al. PTEN loss contributes to erlotinib resistance in EGFR-mutant lung cancer by activation of Akt and EGFR. *Cancer Res.* **2009**, *69*, 3256–3261. [CrossRef] [PubMed]

24. Yap, J.L.; Worlikar, S.; MacKerell, A.D.; Shapiro, P.; Fletcher, S. Small-molecule inhibitors of the ERK signaling pathway: Towards novel anticancer therapeutics. *ChemMedChem* **2011**, *6*, 38–48. [CrossRef] [PubMed]

Int. J. Mol. Sci. **2018**, *19*, 2599

25. Abouantoun, T.J.; Castellino, R.C.; MacDonald, T.J. Sunitinib induces PTEN expression and inhibits PDGFR signaling and migration of medulloblastoma cells. *J. Neurooncol.* **2011**, *101*, 215–226. [CrossRef] [PubMed]

26. Wetmore, C.; Daryani, V.M.; Billups, C.A.; Boyett, J.M.; Leary, S.; Tanos, R.; Goldsmith, K.C.; Stewart, C.F.; Blaney, S.M.; Gajjar, A. Phase II evaluation of sunitinib in the treatment of recurrent or refractory high-grade glioma or ependymoma in children: A children's Oncology Group Study ACNS1021. *Cancer Med.* **2016**, *5*, 1416–1424. [CrossRef] [PubMed]

27. Jakacki, R.I.; Hamilton, M.; Gilbertson, R.J.; Blaney, S.M.; Tersak, J.; Krailo, M.D.; Ingle, A.M.; Voss, S.D.; Dancey, J.E.; Adamson, P.C. Pediatric phase I and pharmacokinetic study of erlotinib followed by the combination of erlotinib and temozolomide: A Children's Oncology Group Phase I Consortium Study. *J. Clin. Oncol.* **2008**, *26*, 4921–4927. [CrossRef] [PubMed]

28. Altomare, D.A.; Testa, J.R. Perturbations of the AKT signaling pathway in human cancer. *Oncogene* **2005**, *24*, 7455–7464. [CrossRef] [PubMed]

29. McCubrey, J.A.; Steelman, L.S.; Chappell, W.H.; Abrams, S.L.; Wong, E.W.; Chang, F.; Lehmann, B.; Terrian, D.M.; Milella, M.; Tafuri, A.; et al. Roles of the Raf/MEK/ERK pathway in cell growth, malignant transformation and drug resistance. *Biochim. Biophys. Acta* **2007**, *1773*, 1263–1284. [CrossRef] [PubMed]

30. Veselska, R.; Kuglik, P.; Cejpek, P.; Svachova, H.; Neradil, J.; Loja, T.; Relichova, J. Nestin expression in the cell lines derived from glioblastoma multiforme. *BMC Cancer* **2006**, *6*, 32. [CrossRef] [PubMed]

31. Schneider, C.A.; Rasband, W.S.; Eliceiri, K.W. NIH Image to ImageJ: 25 Years of image analysis. *Nat. Methods* **2012**, *9*, 671–675. [CrossRef] [PubMed]

32. Skoda, J.; Neradil, J.; Zitterbart, K.; Sterba, J.; Veselska, R. EGFR signaling in the HGG-02 glioblastoma cell line with an unusual loss of EGFR gene copy. *Oncol. Rep.* **2014**, *31*, 480–487. [CrossRef] [PubMed]

33. Schmittgen, T.D.; Livak, K.J. Analyzing real-time PCR data by the comparative C(T) method. *Nat. Protoc.* **2008**, *3*, 1101–1108. [CrossRef] [PubMed]

International Journal of
Molecular Sciences

Article

Sirt1 Protects against Oxidative Stress-Induced Apoptosis in Fibroblasts from Psoriatic Patients: A New Insight into the Pathogenetic Mechanisms of Psoriasis

Matteo Becatti [1,†], Victoria Barygina [1,†], Amanda Mannucci [1], Giacomo Emmi [2], Domenico Prisco [2], Torello Lotti [3], Claudia Fiorillo [1,*,‡] and Niccolò Taddei [1,‡]

[1] Department of Experimental and Clinical Biomedical Sciences "Mario Serio", University of Florence, 50134 Florence, Italy; matteo.becatti@unifi.it (M.B.); v.barygina@gmail.com (V.B.); amanda.mannucci@unifi.it (A.M.); niccolo.taddei@unifi.it (N.T.)
[2] Department of Experimental and Clinical Medicine, University of Florence, 50134 Florence, Italy; giacomo.emmi@unifi.it (G.E.); domenico.prisco@unifi.it (D.P.)
[3] Department of Dermatology, University of Rome "G. Marconi", 00146 Rome, Italy; professor@torellolotti.it
* Correspondence: claudia.fiorillo@unifi.it; Tel.: +39-05-5275-1221
† These authors contributed equally to this work.
‡ These authors contributed equally to this work.

Received: 20 April 2018; Accepted: 23 May 2018; Published: 25 May 2018

Abstract: Psoriasis, a multisystem chronic disease characterized by abnormal keratinocyte proliferation, has an unclear pathogenesis where systemic inflammation and oxidative stress play mutual roles. Dermal fibroblasts, which are known to provide a crucial microenvironment for epidermal keratinocyte function, represented the selected experimental model in our study which aimed to clarify the potential role of SIRT1 in the pathogenetic mechanisms of the disease. We firstly detected the presence of oxidative stress (lipid peroxidation and total antioxidant capacity), significantly reduced SIRT1 expression level and activity, mitochondrial damage and apoptosis (caspase-3, -8 and -9 activities) in psoriatic fibroblasts. Upon SIRT1 activation, redox balance was re-established, mitochondrial function was restored and apoptosis was no longer evident. Furthermore, we examined p38, ERK and JNK activation, which was strongly altered in psoriatic fibroblasts, in response to SIRT1 activation and we measured caspase-3 activity in the presence of specific MAPK inhibitors demonstrating the key role of the SIRT1 pathway against apoptotic cell death via MAPK modulation. Our results clearly demonstrate the involvement of SIRT1 in the protective mechanisms related to fibroblast injury in psoriasis. SIRT1 activation exerts an active role in restoring both mitochondrial function and redox balance via modulation of MAPK signaling. Hence, SIRT1 can be proposed as a specific tool for the treatment of psoriasis.

Keywords: SIRT1; MAPK; oxidative stress; psoriasis

---

1. Introduction

Psoriasis, a chronic, inflammatory multisystemic disease affecting 2–3% of the world population, is characterized by abnormal keratinocyte proliferation resulting in the formation of raised, itchy and well-demarcated erythematous lesions on the skin. The pathogenesis of psoriasis is not well understood, but recent data suggest that both innate and adaptive immunity have a key role in the disease [1]. Aberrant activation and metabolism of epidermal keratinocytes leading to strongly enhanced keratinocyte proliferation and anomalous terminal differentiation are the hallmark of psoriatic skin. The critical role of dermal fibroblasts in providing a crucial microenvironment for epidermal keratinocyte function [2] and in regulating epidermal morphogenesis was already suggested [3]. Moreover, recent reports demonstrated that, in vivo, improvement of aged dermal fibroblast function significantly increases epidermal keratinocyte proliferation with consequent enlargement of epidermal thickness [4]. However, the molecular features and the factors responsible for the complex interactions between dermal stromal cells and epidermal keratinocytes in both physiologic and pathologic conditions are still to be elucidated. On these bases, we focused on dermal fibroblasts which exert specific functions in the microenvironment of epidermal keratinocytes and, together with infiltrating PMNs, induce a marked redox imbalance in psoriatic derma by extensively producing reactive oxygen species (ROS), such as superoxide and H_2O_2, and thus displaying a significant role in the anomalous keratinocyte growth which characterizes psoriasis [5].

The key role of ROS in the pathogenesis of psoriasis has been already reported [6,7]. Indeed, several studies have revealed increased levels of oxidative stress markers and decreased activity of the main antioxidant enzymes in the plasma of psoriatic patients [8–12]. In addition, it has been demonstrated that mitogen-activated protein kinases (MAPK) pathways, such as extracellular signal-regulated kinase (ERK), c-Jun N-terminal kinase (JNK) and p38 MAPK, are involved in the pathogenesis of psoriasis [7,13–17]. However, the molecular regulatory mechanisms of these cellular signal transduction pathways and their possible interplay are not fully elucidated. A variety of stress stimuli, including oxidative stress, induces functional changes of these kinases similar to those found to be associated with apoptosis [18]. Similarly, NAD^+-dependent protein deacetylase SIRT1 has been shown to be strongly upregulated by oxidative stress [19]. SIRT1, the most extensively studied member of the sirtuin family, plays a key role in metabolism, stress responses, and many other cellular processes [20]. Additionally, our previous data demonstrated that SIRT1 upregulation affects the MAPK pathway and inhibits pro-apoptotic molecules, reducing oxidative stress and apoptosis in several cellular models [21,22].

Here, for the first time, we investigated in primary fibroblasts from lesional psoriatic skin, the possible involvement of SIRT1 in MAPK pathways.

2. Results

2.1. SIRT1 Expression and Activity in Fibroblasts from Healthy and Psoriatic Subjects

We evaluated SIRT1 expression and activity in fibroblasts from healthy and psoriatic patients. As previously demonstrated [23], SIRT1 expression (Figure 1A) was markedly reduced in fibroblasts from psoriatic patients compared to healthy subjects (-51% vs. control fibroblasts, $p < 0.01$). Similarly, SIRT1 activity (Figure 1B) in psoriatic fibroblasts exhibited a significant decrease in comparison with healthy cells (306 ± 99 vs. 855 ± 188, $p < 0.01$).

2.2. Dose-Dependent Effects of SIRT1720 on SIRT1 Activity in Psoriatic Fibroblasts

A dose-dependent test was performed in psoriatic fibroblasts treated with SRT1720 concentrations from 1 to 50 µM (Figure 1C) to evaluate the effect of SRT1720 on SIRT1 activity. Treatment (24 h) with 10 µM SRT1720 induced a dramatic increase in SIRT1 activity (3.85 ± 0.29 fold increase). Hence, 10 µM SRT1720 was used for the programmed experiments. Interestingly, the addition of SIRT1 siRNA to SRT1720-treated cells induced a complete abolishment of the observed increase ($p < 0.01$) (Figure 1C).

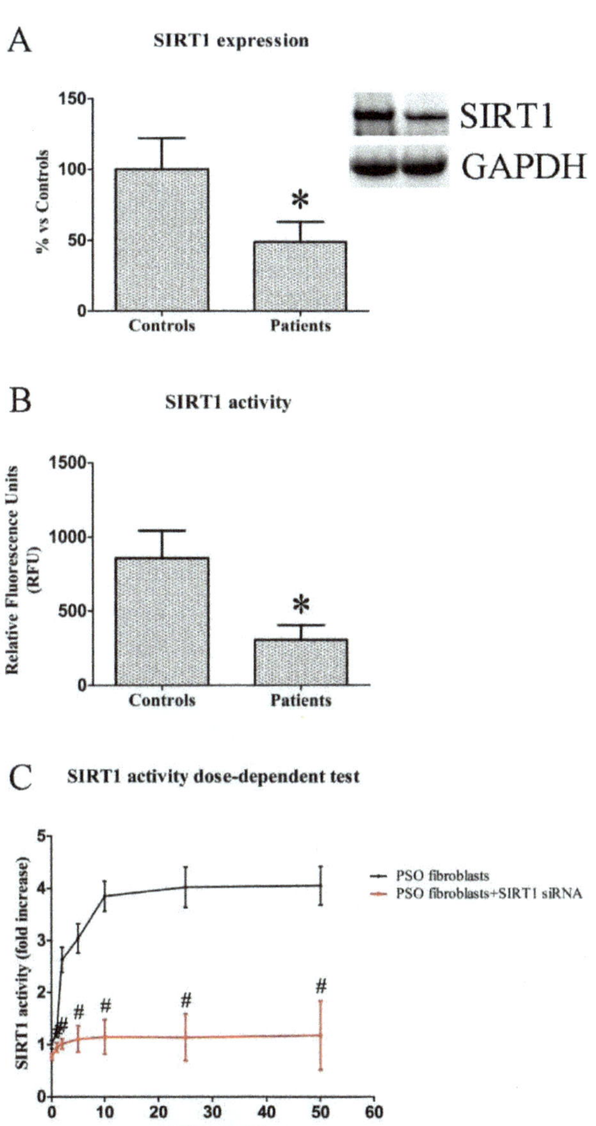

Figure 1. (**A**) Representative Western blot analysis of SIRT1 expression in fibroblasts from controls and psoriatic patients. Histogram represents data from controls (n = 4 biopsies) and patients (n = 4 biopsies); (**B**) SIRT1 activity in fibroblasts from controls (n = 4 biopsies) and psoriatic patients (n = 4 biopsies); (**C**) SIRT1 activity in fibroblasts from lesional psoriatic skin (n = 4 biopsies) after 24 h of incubation with different concentrations of SRT1720. Each experiment was performed in triplicate. * Significant difference ($p \leq 0.05$) vs. fibroblasts from healthy patients. # Significant difference ($p \leq 0.01$) vs. PSO fibroblasts.

2.3. SIRT1 Activation Decreases Oxidative Stress in Fibroblasts from Psoriatic Patients

Figure 2A shows a significant total antioxidant capacity (TAC) decrease in psoriatic fibroblasts with respect to controls (-45%, $p < 0.01$).

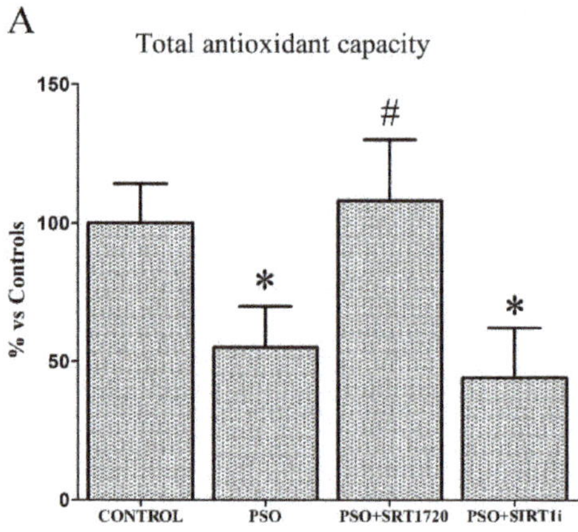

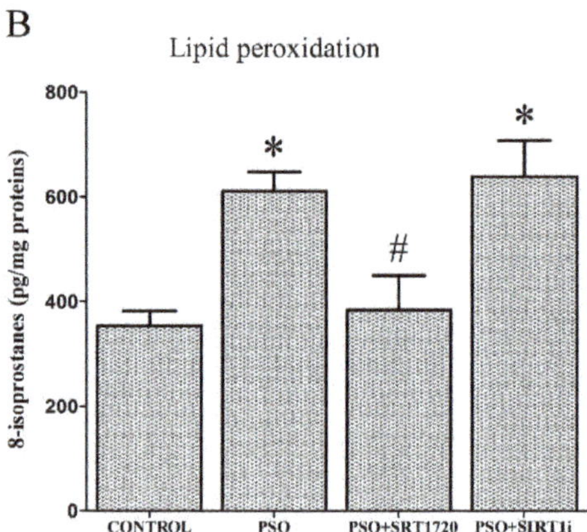

Figure 2. (**A**) Evaluation of total antioxidant capacity (TAC) and (**B**) 8-isoprostanes in fibroblasts from controls (n = 4 biopsies) and psoriatic patients (n = 4 biopsies) in the presence of SRT1720 or the SIRT1 inhibitor (SIRT1i). Each experiment was performed in triplicate. * Significant difference ($p \leq 0.05$) vs. fibroblasts from healthy patients. # Significant difference ($p \leq 0.05$) vs. fibroblasts from psoriatic patients.

SIRT1 activation effectively restored intracellular TAC levels (+53% vs. untreated PSO cells, $p < 0.01$); interestingly, this effect was abrogated by the SIRT1 inhibitor, demonstrating the key role of SIRT1 in improving antioxidant defense systems. Increased levels of 8-isoprostanes (lipid peroxidation markers) were also found in psoriatic fibroblasts with respect to control fibroblasts (+73%, $p < 0.01$). SRT1720-treated psoriatic fibroblasts showed significantly lower 8-isoprostanes levels (-47% vs. untreated PSO cells, $p < 0.01$), confirming the pivotal role of SIRT1 pathways in cell redox balance (Figure 2B). Similar results were found when lipid peroxidation was measured using BODIPY by flow cytometry and confocal microscopy (Figure 3). Similarly, the fluorescent probe H_2DCFDA was used for determining intracellular ROS production (Figure 3). SRT1720-treated fibroblasts displayed less marked ROS production, thus indicating a strong protective effect exerted by SIRT1 pathways against ROS. Analogous results were found when we evaluated NO production (Figure 3).

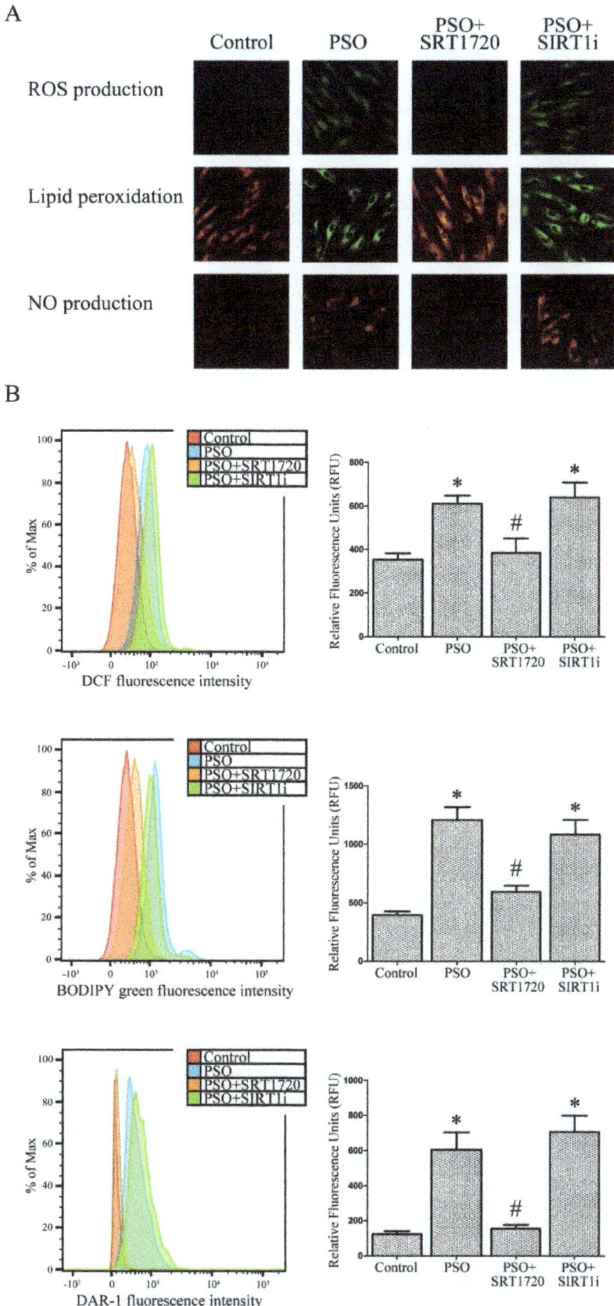

Figure 3. (**A**) Confocal microscope analysis (63× magnification) and (**B**) FACS analysis of ROS production, lipoperoxidation and NO production in fibroblasts from controls (n = 4 biopsies) and psoriatic patients (n = 4 biopsies) in the presence of SRT1720 or the SIRT1 inhibitor (SIRT1i). Each experiment was performed in triplicate. * Significant difference ($p \leq 0.05$) vs. fibroblasts from healthy patients. # Significant difference ($p \leq 0.05$) vs. fibroblasts from psoriatic patients.

2.4. SIRT1 Activation Protects Psoriatic Fibroblasts from Mitochondrial Damage

In order to ascertain whether SIRT1 activation can protect against mitochondrial damage, we analyzed the mitochondrial permeability transition pore opening, mitochondrial membrane polarization and mithocondrial superoxide production. Confocal microscope analysis (Figure 4A) indicated marked alterations in mitochondrial permeability transition pore (mPTP) opening and mitochondrial membrane depolarization (TMRM probe) in fibroblasts from psoriatic patients. These changes were not evident in control cells. SIRT1 activation by SRT1720 efficiently restored mitochondrial function.

In line with these results, psoriatic fibroblast treatment with the SIRT1 inhibitor did not cause any significant change compared to psoriatic untreated cells. Moreover, we evaluated the mitochondrial superoxide production by confocal microscope analysis with the mitochondrial superoxide-specific fluorescent probe MitoSOX. MitoSOX is a selective indicator of mitochondrial superoxide, is selectively targeted to mitochondria and is able to efficiently compete with superoxide dismutase (SOD) for superoxide. Following mitochondrial oxidation, oxidized MitoSOX quickly migrates to the nucleus and becomes highly fluorescent upon binding to DNA. High levels of mitochondrial superoxide production in psoriatic fibroblasts compared to control cells are evident whereas SRT1720 treatment efficiently reverts this effect (Figure 4). In the presence of the SIRT1 inhibitor, high fluorescence, indicating high levels of mitochondrial superoxide, is evident. To further confirm and quantify these results, mitochondrial permeability transition pore opening, mitochondrial membrane polarization and mithocondrial superoxide production were also analyzed by flow cytometry (Figure 4B).

Mitochondrial number can influence experiments where mitochondrial function is investigated. We therefore counted mitochondria number in the SRT1720-treated or untreated fibroblasts from psoriatic patients (by confocal microscopy and FACS analysis) to ensure that altered mitochondrial function was not due to a mere numeric change of mitochondria. In our psoriatic fibroblasts, no significant variation in mitochondrial number upon treatment was evident (data not shown). Therefore, all of the observed mitochondrial function and cell viability alterations are not ascribable to changes in mitochondrial number.

2.5. SIRT1 Activation Protects Psoriatic Fibroblasts from Apoptosis

Caspase-3, -8 and -9 activities, which play a central role in apoptosis, were measured to verify the occurrence of apoptosis under our experimental conditions. As shown in Figure 5A, confocal microscope analysis clearly demonstrated marked caspase-3, -8 and -9 activities in fibroblasts from psoriatic patients and this increase was abrogated by SRT1720 treatment.

These important findings were also quantified by FACS analysis (Figure 5B). In particular, caspase-3 activity (Figure 5B) was strongly increased in psoriatic fibroblasts (+540% vs. fibroblasts from healthy subjects, $p < 0.01$) whereas SIRT1 activation strongly inhibited this effect (-376% vs. untreated PSO fibroblasts, $p < 0.01$). This effect was completely neutralized when using the SIRT1 inhibitor. Caspase-8 activity (Figure 5B), which plays a pivotal role in the extrinsic apoptotic signaling pathway via death receptors, was significantly increased in psoriatic fibroblasts (+233% vs. control fibroblasts, $p < 0.01$). SIRT1 activation clearly abrogated this effect (-71% vs. untreated PSO cells, $p < 0.01$). Once again, SIRT1 inhibitor treatment reverted this effect. Finally, we evaluated caspase-9 activity, a key player in the intrinsic or mitochondrial apoptotic pathway. As expected, fibroblasts from psoriatic patients showed increased levels of caspase-9 activity (+348% vs. control fibroblasts, $p < 0.01$) and SRT1720 treatment strongly reduced this effect (-188% vs. untreated PSO fibroblasts). Moreover, when SIRT1 signaling was inhibited, caspase-9 activity was even more marked than that observed in untreated PSO cells (+62% vs. untreated PSO fibroblasts, $p < 0.01$). All these findings point to a protective role of SIRT1 activation against apoptotic cell death in fibroblasts from psoriatic patients.

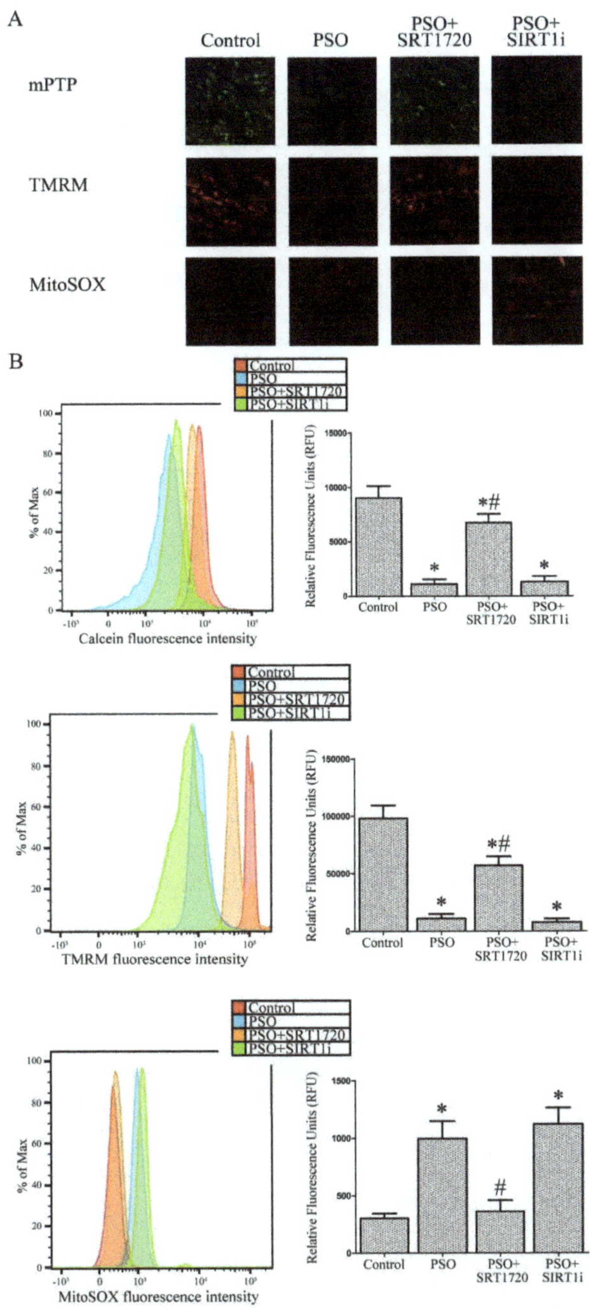

Figure 4. (**A**) Confocal microscope analysis (63× magnification) and (**B**) FACS analysis of mitochondrial permeability transition pore opening (mPTP), mitochondrial depolarisation (TMRM) and mitochondrial superoxide production (MitoSOX) in fibroblasts from controls ($n = 4$ biopsies) and psoriatic patients ($n = 4$ biopsies) in the presence of SRT1720 or the SIRT1 inhibitor (SIRT1i). Each experiment was performed in triplicate. * Significant difference ($p \leq 0.05$) vs. fibroblasts from healthy patients. # Significant difference ($p \leq 0.05$) vs. fibroblasts from psoriatic patients.

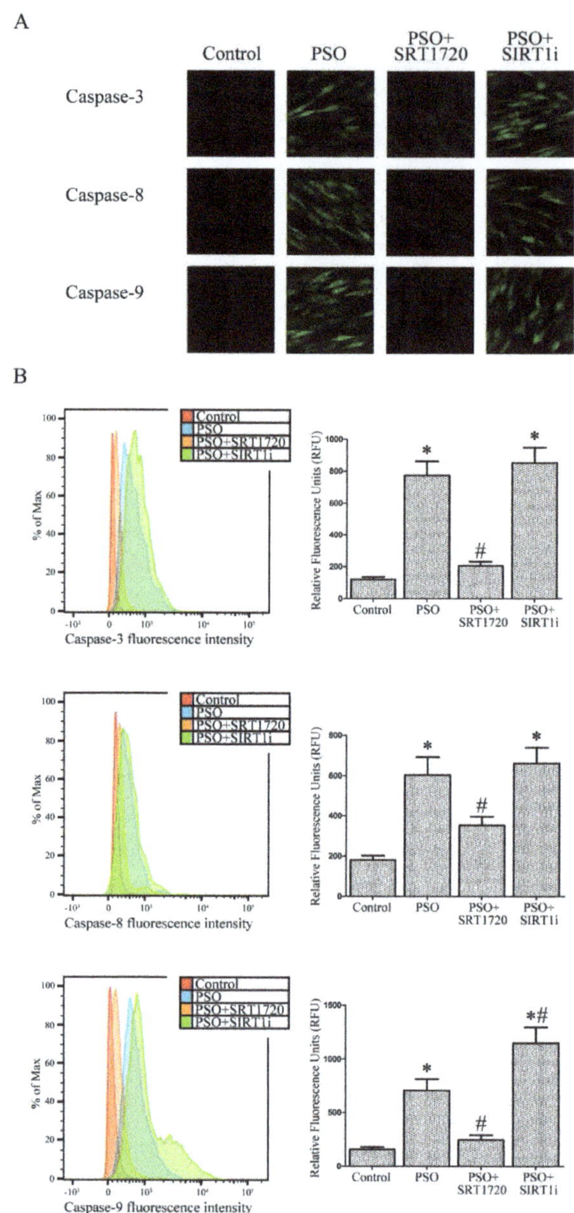

Figure 5. (**A**) Confocal microscope analysis (63× magnification) and (**B**) flow cytometry analysis of caspases-3, 8 and 9 activation in fibroblasts from controls ($n = 4$ biopsies) and psoriatic patients ($n = 4$ biopsies) in the presence of SRT1720 or the SIRT1 inhibitor (SIRT1i). Each experiment was performed in triplicate. * Significant difference ($p \leq 0.05$) vs. fibroblasts from healthy patients. # Significant difference ($p \leq 0.05$) vs. fibroblasts from psoriatic patients.

2.6. SIRT1 siRNA-Treatment of Fibroblasts from Healthy Subjects

ROS production, lipid peroxidation, caspase-3 activity and mitochondrial membrane polarization were measured in untreated and in SIRT1 siRNA-treated control fibroblasts, challenged or not with

H$_2$O$_2$ (500 µM for 3 h) to elucidate the role of SIRT1 in oxidative-mediated cell injury. Without H$_2$O$_2$, the biochemical modifications displayed by SIRT1 siRNA-treated fibroblasts did not differ from untreated fibroblasts (Figure 6) in agreement with our previous data [21,22].

In the presence of H$_2$O$_2$ treatment, the observed alterations of the SIRT1 siRNA-treated fibroblasts paralleled those observed in fibroblasts from psoriatic skin (Figure 6). These findings suggest that psoriatic fibroblast damage can be mediated by SIRT1.

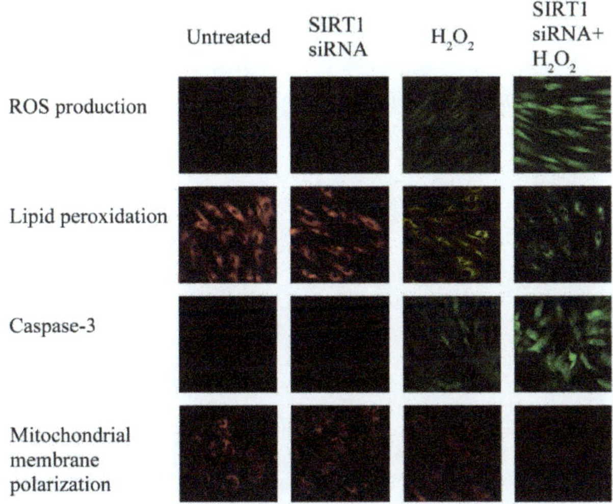

Figure 6. SIRT1 effects in oxidative stress protection. ROS production, lipid peroxidation, caspase-3 activity and mitochondrial membrane polarization were measured by confocal microscopy (63× magnification) in untreated and in SIRT1 siRNA-treated control fibroblasts, challenged or not with H$_2$O$_2$ (500 µM for 3 h) to elucidate the role of SIRT1 in oxidative-mediated cell injury. Fibroblasts were obtained from healthy skin (n = 4 biopsies). These parameters were measured in untreated and in SIRT1 siRNA-treated control fibroblasts, challenged or not with H$_2$O$_2$ (500 µM for 3 h).

2.7. SIRT1 Activation Modulates MAPK Pathways in Fibroblasts from Psoriatic Patients

In our previous studies, we demonstrated the central role of MAPK pathways in inducing cell damage and apoptosis [21,22,24]. Here, we examined p38, ERK and JNK activation in response to SIRT1 activation.

In Figure 7, the levels of ERK phosphorylation, whose anti-apoptotic effect is well documented, are shown. In psoriatic fibroblasts, low levels of ERK phosphorylation were observed with respect to control fibroblasts (−437% vs. control, $p < 0.01$). SIRT1 activation induced a strong and significant increase in ERK phosphorylation (+217% vs. untreated PSO fibroblasts, $p < 0.01$) whereas SIRT1 inhibitor treatment significantly decreased ERK phosphorylation (−54% vs. untreated PSO fibroblasts, $p < 0.01$). At the same time, high levels of JNK phosphorylation were observed in psoriatic cells (about twenty-fold increase vs. the control, $p < 0.01$). SIRT1 activation completely abrogated JNK phosphorylation (−1340% vs. untreated PSO cells, $p < 0.01$) whereas the SIRT1 inhibitor led to an increase in JNK phosphorylation (+51% vs. untreated PSO cells, $p < 0.01$), demonstrating a role of SIRT1 in the JNK pathway.

p38 phosphorylation was also markedly enhanced in psoriatic fibroblasts (3.7-fold increase vs. control fibroblasts, $p < 0.01$). SIRT1 activation significantly reduced p38 phosphorylation (−142% vs. untreated PSO fibroblasts, $p < 0.01$) whilst the SIRT1 inhibitor completely abolished this effect (no significant difference vs. untreated PSO cells was evident).

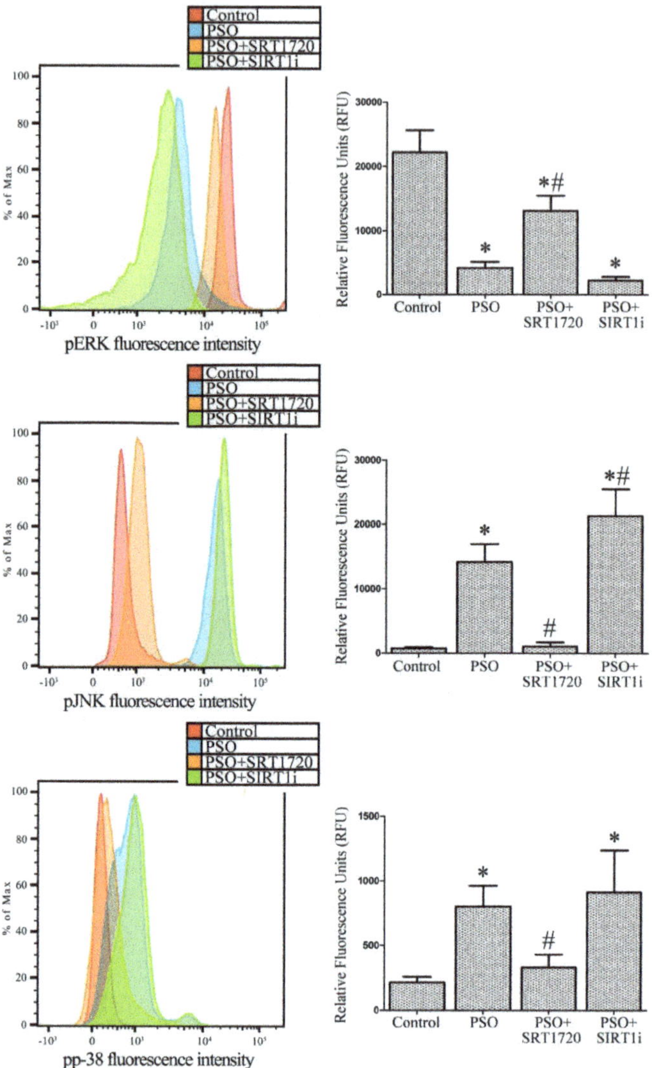

Figure 7. MAPK phosphorylation in fibroblasts from controls ($n = 4$ biopsies) and psoriatic patients ($n = 4$ biopsies) in the presence of SRT1720 or the SIRT1 inhibitor (SIRT1i). Each experiment was performed in triplicate. * Significant difference ($p \leq 0.05$) vs. fibroblasts from healthy patients. # Significant difference ($p \leq 0.05$) vs. fibroblasts from psoriatic patients.

To further investigate the role of SIRT1 in the molecular pathways of psoriatic fibroblasts, we measured caspase-3 activity in the presence of specific MAPK inhibitors. In Figure 8, we show that in the presence of p38 or the JNK inhibitor, caspase-3 activity significantly decreased ($p < 0.01$ vs. untreated PSO fibroblasts), indicating the involvement of these pathways in the induction of apoptosis in psoriatic fibroblasts. Remarkably, in the presence of ERK inhibitor, untreated fibroblasts displayed the highest caspase-3 activity, demonstrating the key role of ERK in the protection against apoptotic cell death.

Upon SIRT1 activation with the above inhibitors, a further decrease in caspase-3 activity was observed ($p < 0.01$ vs. untreated psoriatic fibroblasts). On the contrary, the SIRT1 inhibitor together with p38 or JNK inhibitors induced an increase in caspase-3 activity. Simultaneous treatment with the three inhibitors did not protect against cell death (data not shown). All these data demonstrated the key role of the SIRT1 pathway against apoptotic cell death via MAPK modulation.

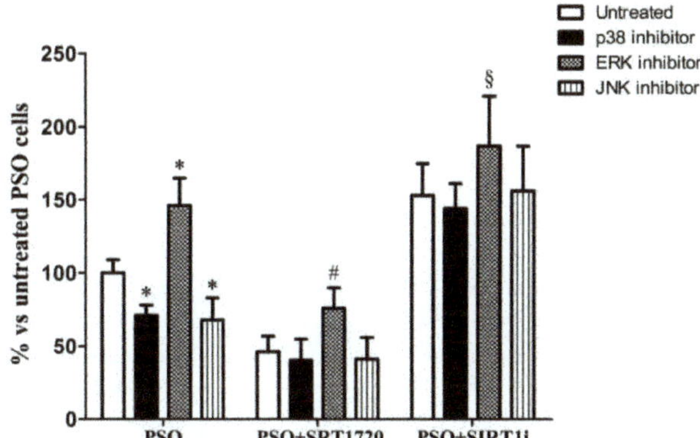

Figure 8. The caspase-3 activity measured using flow cytometry analysis in fibroblasts from psoriatic patients (n = 3 biopsies) in the presence of specific MAPK inhibitors. Each experiment was performed in triplicate. * Significant difference ($p \leq 0.05$) vs. untreated fibroblasts from psoriatic patients. # Significant difference ($p \leq 0.05$) vs. MAPK inhibitors untreated PSO + SRT1720 fibroblasts. § Significant difference ($p \leq 0.05$) vs. MAPK inhibitors untreated PSO + SIRT1i fibroblasts.

3. Discussion

The present study was undertaken to explore the potential contribution of SIRT-1 signaling to the pathogenetic mechanisms of psoriasis. Both innate and adaptive immune systems play a central role in the disease which can be defined as an inflammatory skin disorder characterized by abnormal keratinocyte proliferation and differentiation [1,25]. The key role of dermal fibroblasts in regulating epidermal microenviroment, immune cell behaviour [26,27] and keratinocyte function has been already suggested [1,5,26,27]. Indeed, impaired fibroblasts can extensively produce superoxide and H_2O_2 (modifying the redox balance of psoriatic derma), promote inflammatory mechanisms [28] and may contribute to the epidermal overgrowth by inducing keratinocyte proliferation [29–32]. Based on this background, fibroblast cultures derived from psoriatic lesions represent our selected experimental model.

Mammalian skin is equipped with efficient antioxidant defence mechanisms that prevent oxidative injury to lipids and proteins, contributing to barrier integrity, which is essential for healthy skin condition. Thus, cellular redox balance is essential for skin homeostasis and an imbalance between pro-oxidant and antioxidant mechanisms can result in skin diseases, including psoriasis [33].

In the blood of psoriatic patients, redox imbalance has been witnessed by the presence of lipid peroxidation (increased levels of malondialdehyde) associated with decreased erythrocyte-superoxide dismutase activity, catalase activity and total antioxidant status [12,34]. In line with these observations, we recently identified NADPH oxidase as one of the possible upregulated ROS sources in lesional fibroblasts from psoriatic patients [35]. However, the molecular mechanisms underlying this alteration in psoriatic fibroblasts are still poorly understood.

Oxidative stress conditions and other metabolic and genotoxic stress are sensed and counteracted by sirtuins, a family of NAD$^+$-dependent deacetylases which contribute to react to these damaging conditions through diverse pathways. Sirtuin 1 (SIRT1), in particular, is involved in the regulation of cellular survival, cellular senescence/aging, inflammation-immune function, endothelial functions, and circadian rhythms [21,22,36,37]. Moreover, SIRT1 is able to promote the differentiation of human keratinocytes [38] and its activation has been shown to inhibit keratinocyte proliferation [39]. Recently, some authors reported significantly decreased SIRT1 levels in lesional skin from psoriatic samples compared to controls, suggesting its potential involvement in the pathogenesis of psoriasis [40]. Furthermore, in a recent randomized, double-blind, placebo-controlled Phase IIa study, the effects of a selective small molecule SIRT1 activator in 40 patients with moderate-to-severe psoriasis were investigated [41]. On the basis of these observations and of our recent findings [23], here we demonstrate that, in fibroblasts from lesional psoriatic skin, SIRT1 activity is strongly reduced compared to healthy fibroblasts, suggesting a possible involvement of SIRT1 in the pathogenesis of psoriasis.

Another important finding, consistent with our previously obtained results [42], is the presence of oxidative stress in psoriatic fibroblasts, as suggested by enhanced lipoperoxidation and TAC reduction. Interestingly, psoriatic fibroblasts treated with the SIRT1 activator showed reduced oxidative stress levels, an effect that was reversed by the SIRT1 inhibitor, thus demonstrating a prominent role of SIRT1 in the maintenance of redox homeostasis. These findings are in line with a study reporting SIRT1-induced resistance to oxidative stress through FoxO in fibroblasts [19], where, upon SIRT1 overexpression, catalase expression was stimulated. In the presence of FoxO1a dominant negative, SIRT1 upregulation was inhibited [19]. The key role of SIRT1 in the protection against oxidative stress-induced cellular damage has been also confirmed by our experiments performed in SIRT1 siRNA-treated healthy skin fibroblasts.

In psoriatic fibroblasts, mitochondrial dysfunction, enhanced ROS production and signs of oxidative stress were present: indeed, mPTP opening and aberrant mitochondrial depolarisation were associated with significantly higher mitochondrial superoxide generation. Several studies indicate that SIRT1 activation reduces oxidative stress and maintains mitochondrial function. It has been reported that renal injury after ischemia-reperfusion is reduced by SIRT1, which is able to decrease nitrosative stress and inflammation and enhance energy metabolism by stimulating mitochondrial biogenesis [43]. An animal experimental model of hemorrhagic shock and reperfusion clearly demonstrated that SIRT1 activation induces p53 deacetylation, inhibits mPTP opening, suppresses a mitochondria-mediated apoptotic pathway and attenuates renal injury [44]. Other studies have also demonstrated that SIRT1 can stimulate the expression of antioxidants and, via an auto-feedback loop, can also potentiate SIRT1 expression [45–49]. Finally, it has been shown that SIRT1 overexpression protects human skin fibroblasts from UVB-induced cellular damage, increasing the resistance to oxidative stress [50]. Our results clearly demonstrate a protective effect of activated SIRT1 in the mitochondria of psoriatic fibroblasts where superoxide production, mitochondrial depolarization and impaired mPTP opening were inhibited by SRT1720 treatment. These findings are in line with our previous data demonstrating a protective role of SIRT1 in perilesional vitiligo keratinocytes [22], where oxidative stress and mitochondrial dysfunction were removed by treatment with resveratrol (a SIRT1 activator compound).

Our data clearly show that in psoriatic fibroblasts, the mitochondrial alterations (associated with an increase in ROS production) are responsible for the strong activation of the apoptotic process. In particular, caspase-3, -8 and -9, which play a key role in apoptotic cell death, displayed increased activities in psoriatic fibroblasts. Treatment with SRT1720 induced a significant downregulation of the apoptotic pathway and the presence of SIRT1 inhibitor counteracted this reduction suggesting a pivotal role for SIRT1 in the modulation of apoptosis in psoriatic fibroblasts.

To examine in depth the role of SIRT1 signaling in psoriasis, we performed experiments aimed at highlighting the relationship between SIRT1 and the redox-sensitive MAPK pathway, which has been shown to be altered in psoriasis [13,14,51,52]. MAPKs (mitogen-activated protein kinases) are signaling proteins activated by substantially diverse extracellular stimuli (cytokines, hormones, growth factors, UV irradiation, oxidative stress) and which regulate fundamental cellular processes (gene expression, cell proliferation, growth, survival, migration, and apoptosis) [53,54]. MAPKs are activated by phosphorylation of both threonine and tyrosine residues, and in turn they phosphorylate other downstream intracellular kinases and transcription factors. Extracellular signal-regulated kinase 1/2 (ERK1/2), p38 MAPK, and c-Jun N-terminal kinase (JNK) are the members of MAPKs whose involvement in oxidative stress signalling has been extensively investigated. JNK, ERK and p38 kinases exhibit distinct effects on apoptotic cell death. In the present study, using inhibitors for each kinase, we observed that JNK and p38 kinases promote apoptosis, while ERK displays anti-apoptotic effects. In particular, in psoriatic fibroblasts, we observed a decrease in p-ERK, which was reversed by SRT1720 treatment. In contrast, previous studies reported an increased ERK activation in human psoriatic lesion as compared with that in normal human epidermis [14] and in cell extracts from lesional psoriatic skin [13]. To further investigate the molecular pathways underlying the protective effects induced by SIRT1 activation, caspase-3 activity was measured in psoriatic fibroblasts in the presence of the ERK inhibitor, demonstrating the key role of ERK activation in protection against apoptotic cell death. In particular, the effect of the ERK inhibitor was suppressed by SRT1720 treatment, indicating a pivotal role of SIRT1 in protection against apoptosis. Our data are in agreement with previous studies reporting that SIRT1 activation promotes ERK phosphorylation in human fibroblasts [55] and that the histone deacetylase inhibitor suppresses the Ras-MAP kinase signaling pathway [56], suggesting that SIRT1 may stimulate the ERK pathway.

It has been demonstrated that JNK is not active in healthy human epidermis, but its activity is increased in psoriasis [17]; moreover, experimental data highlighted that TNF-α and UV light exert their pro-inflammatory effects in part via JNK activation [57,58]. In an elegant study, Gazel and co-workers demonstrated that JNK inhibition in epidermal keratinocytes is sufficient to initiate their differentiation, suggesting that attenuating JNK activity could be a differentiation therapy-based approach for treating psoriasis [59]. Here, we show a significant increase in JNK activity in psoriatic fibroblasts and interestingly, we report that SIRT1 activation prevents JNK activation. Since the p38 inhibitor prevents fibroblast apoptosis, a pivotal role for JNK MAPK in this pathway can be proposed.

The role of p38 MAPK in the development of psoriasis is well documented [13,14,52]. In particular, it has been shown that p38 MAPK activity is increased in lesional compared to non-lesional psoriatic skin [13]; moreover, the antimicrobial peptide S100A8, known to be upregulated in lesional psoriatic skin and believed to play a role in the pathogenesis of psoriasis, was found to be regulated by a p38 MAPK-dependent mechanism in cultured human keratinocytes [60]. p38 MAPK has been found to be activated in keratinocytes treated with H_2O_2, tumor necrosis factor (TNF)-α and interleukin (IL)-1β, suggesting a key role of p38 MAPK in mediating keratinocyte responses to cellular stress [61,62]. Because lesional psoriatic skin is characterized by increased expression of inflammatory cytokines [63], it is possible that p38 MAPK has a role in the inflammatory aspect of psoriasis. Interestingly, we found that p38 MAPK activity was upregulated in psoriatic fibroblasts compared to healthy fibroblasts and, upon SIRT1 activation, p38 activation was dramatically reduced. Moreover, in the presence of a p38 inhibitor we observed a significant reduction in apoptosis, confirming the key role of p38 MAPK in this pathway. Our data are in agreement with a previous study of Soegaard-Madsen and co-workers [52], demonstrating that p38 inhibition may be a mechanism by which the anti-TNF-α agent Adalimumab mediates its anti-psoriatic effect.

This study is not without limitations. First, the study population is small and it requires verification in a larger cohort (twelve patients and seven controls were enrolled in this study but each experiment is related to fibroblasts derived from three or four patients/controls only). In addition, the psoriatic subjects included in the study were characterized by moderate psoriasis (PASI = 12.4 $\pm$ 0.5). Therefore, the results need confirmation using a similar study design involving patients affected by severe disease. Another important consideration must be taken into account: even if fibroblasts and endothelial cells contribute to the pathogenesis of psoriasis, further studies on SIRT1 signaling are also needed on keratinocytes, which display a crucial role in the development of psoriatic lesions.

The involvement of SIRT1 in the protective mechanisms related to stress responses via its interactions with different substrates has been already studied [64] but data on skin biology are almost unknown. In particular, this is the first study exploring SIRT1 signaling in psoriasis. The present data demonstrate that in fibroblasts from psoriatic patients, SIRT1 activation displays an active role in restoring both mitochondrial function and redox balance via modulation of MAPK signaling. Therefore, SIRT1 can be proposed as a novel molecular target for the treatment of psoriasis.

4. Materials and Methods

4.1. Patients

The Local Ethics Committee approved the present study. All experiments were performed on twelve patients (six males and six females) affected by moderate psoriasis (PASI = 12.4 ± 0.5) with a mean age of 41.4 ± 7.6 years and with a mean disease duration of 14.1 years (from 8 to 25 years). The demographic and clinical data for each patient are summarized in Table 1. Seven healthy controls (four males and three females), matched for age and body mass index, were also enrolled in the study. No subject was subjected to any systemic therapy before or during the study, or had a history of any disease, e.g., diabetes mellitus and atherosclerosis, which might affect blood redox status. All subjects provided signed informed consent. The study was carried out according to the Helsinki Declaration.

Table 1. Demographic and clinical data of psoriatic patients involved in the study.

Patient	Age	Body Mass Index (BMI)	Duration of Disease (Years)	PASI
M1	33	25	8	13
M2	41	24	12	13
M3	29	25	10	12
M4	52	23	21	12
M5	48	26	11	12
M6	45	25	10	13
F1	51	21	25	13
F2	45	23	20	13
F3	38	26	18	12
F4	32	24	12	12
F5	37	25	9	12
F6	46	26	13	12
Mean ± SD	41.4 ± 7.6	24.4 ± 1.5	14.1 ± 5.5	12.4 ± 0.5

4.2. Fibroblast Isolation and Setting Up of Cell Cultures

Lesional skin punch biopsies from twelve patients affected by plaque psoriasis and from seven healthy controls were used to obtain primary fibroblasts cell cultures. Briefly, skin biopsies were incubated with dispase II (2 U/mL, Sigma-Aldrich, Milan, Italy) and the derma was digested with collagenase (3 mg/mL, Sigma-Aldrich) for 45 min at 37 °C. Then, cells were filtered through a 70 μm filter to remove debris. Cells were cultured in DMEM medium (Sigma-Aldrich) with 10% FBS (Sigma-Aldrich) and 1% penicillin/streptomycin (Sigma-Aldrich). Fibroblasts were characterized by a high Vimentin (Sigma-Aldrich) expression by FACS and confocal microscopy analyses. Cells up to the fourth passage were used for experiments.

4.3. Preparation of Cell Homogenates

After trypsinization, fibroblasts (1×10^6) were resuspended in 100 μL of lysis buffer (20 mM Tris-HCl pH8, 1% Triton X-100, 10% (v/v) glycerol, 137 mM NaCl, 2 mM EDTA and 6 M urea supplemented with 0.2 mM PMSF, 10 mg/mL leupeptin + aprotinin). Samples were then twice sonicated in ice for 5 s, centrifuged at 14,000× g for 10 min at 4 °C, and the supernatant collected. Protein concentration was determined according to the Bradford method [65].

4.4. Western Blot Analysis of SIRT1

SIRT1 expression levels were assessed by Western blot analysis. Briefly, equal amounts of homogenates (50 µg) were separated on 4–12% SDS-PAGE gels (Criterion XT, Bio-Rad Laboratories, Milan, Italy). After blotting into PVDF Hybond membranes and incubation overnight at 4 °C with (rabbit) anti-SIRT1 antibody (Santa Cruz Biotechnology Inc., Santa Cruz, CA, USA), membranes were washed and incubated for 1 h with peroxidase-conjugated secondary antibody. Then, bands detected with a SuperSignal West Dura (Pierce, Rockford, IL, USA) were quantified using Quantity-One software (Bio-Rad, Milan, Italy) [66]. SIRT1 expression levels were calculated as ratios between the densitometry of the corresponding band and the loading control (GAPDH).

4.5. Determination of Cellular SIRT1 Activity

SIRT1 activity was determined according to the method described by Fulco et al. [67] with some modifications. SIRT1 activity was determined by the SIRT1 Direct Fluorescent Screening Assay Kit (Cayman, Ann Arbor, MI, USA) as previously reported [22]. The fluorescence was detected using a Perkin-Elmer (Milan, Italy), LS 55 luminescence spectrometer (ex: 360 nm; em: 460 nm).

4.6. Cell Treatments

Fibroblasts from psoriatic patients or controls were grown for 24 h in the presence of 10 µM SRT1720 (a selective small molecule activator of SIRT1).

In another set of experiments, cells were treated with 10 µM p38 kinase inhibitor (SB203580), 10 µM JNK inhibitor (SP600125), 10 µM MEK inhibitor (PD98059), and 1 µM specific SIRT1 inhibitor (6-Chloro-2,3,4,9-tetrahydro-1*H*-Carbazole-1-carboxamide) for 3 h. All reagents were purchased from Sigma at the highest purity available.

4.7. Determination of Total Antioxidant Capacity (TAC)

As previously reported, intracellular TAC was measured in cell lysates by a chemiluminescent assay using an Abel Antioxidant Test Kit (Knight Scientific Limited, Plymouth, UK) [68]. The protein content was measured by the Bradford method [65].

4.8. Determination of Lipid Peroxidation

8-Isoprostane levels (lipid peroxidation index) were measured in cell lysates using the 8-isoprostane EIA kit (Cayman Chemical Co., Ann Arbor, MI, USA), following the manufacturer's instructions. Moreover, lipid peroxidation was investigated by confocal microscopy using BODIPY 581/591 C11 (Life Technologies, Carlsbad, CA, USA) [69]. The fluorescent probe BODIPY 581/591 C11 shifts its fluorescence from red to green in the presence of oxidizing agents. Fluorescence was detected using a confocal Leica TCS SP5 scanning microscope (Mannheim, Germany) using a Leica Plan Apo 63X oil immersion objective and then projected as a single composite image by superimposition. Lipid peroxidation was also quantified by FACS analysis. Cells were incubated in DMEM with BODIPY 581/591 C11 (2 µM) for 30 min at 37 °C [70] and analyzed using a FACSCanto flow cytometer (Becton-Dickinson, San Jose, CA, USA).

4.9. Assessment of Intracellular ROS, NO Production and Mitochondrial Superoxide

After seeding on glass cover slips, fibroblasts were loaded for 15 min at 37 °C with the following fluorescent probes: MitoSOX (mitochondrial superoxide-specific probe, 3 µM), DAR-1 (NO probe, 1 µM) and H_2DCF-DA (ROS production probe, 1 µM) all purchased from Life Technologies, Carlsbad, CA, USA. Cells were fixed in 2.0% buffered paraformaldehyde for 10 min at RT, and the fluorescence was detected by a Leica TCS SP5 confocal scanning microscope (Mannheim, Germany). Mitochondrial superoxide, NO and ROS generation were also monitored using the same fluorescent probes (MitoSOX: 0.5 µM; H_2DCF-DA: 1 µM; DAR-1: 1 µM) by FACSCanto flow cytometer (Becton-Dickinson) [70].

4.10. Mitochondrial Number

MitoTracker Deep Red 633 (Life Technologies, Carlsbad, CA, USA) was used to stain mitochondria by confocal microscopy as previously described [21]. Mitochondrial number was also monitored by flow cytometry. Cells were incubated with MitoTracker Deep Red 633 (200 nM) for 20 min at 37 °C and immediately analysed by a FACSCanto flow cytometer (Becton-Dickinson).

4.11. Mitochondrial Permeability Transition Pore Opening

The fluorescent calcein-AM probe was used to assess mitochondrial permeability, an indicator of mitochondrial dysfunction and early apoptosis, as described by Petronilli et al. [24], albeit with minor modifications [71]. Calcein-AM, after entering the cell and following deesterification, emits fluorescence. Cell cobalt chloride co-loading quenches cell fluorescence except for mitochondria where cobalt cannot enter (living cells). In contrast, in the case of mitochondrial permeability transition pore opening (mPTP), cobalt enters mitochondria and quenches calcein fluorescence (apoptotic cells). Thus, decreased mitochondrial calcein fluorescence represents a measure of the extent of mPTP induction. Mitochondrial permeability transition pore opening was monitored by confocal microscopy and FACS analysis [71].

4.12. Mitochondrial Membrane Potential

Tetramethylrhodamine methyl ester perchlorate (TMRM) was used to assess the mitochondrial membrane potential. For confocal microscope analysis, cells were cultured on glass cover slips and loaded for 20 min at 37 °C with 100 nM TMRM (Life Technologies, Carlsbad, CA, USA). After fixing, cells were analyzed using a confocal Leica TCS SP5 scanning microscope (Mannheim, Germany). Mitochondrial membrane potential was also quantified by flow cytometry. Cells were incubated for 20 min at 37 °C with TMRM (100 nM) in DMEM and analyzed using a FACSCanto flow cytometer (Becton-Dickinson).

4.13. Determination of Caspase Activity

Caspase-3 and caspase-9 activity were analysed by confocal microscopy and FACS analysis. Fibroblasts were loaded with FAM-FLICA™ Caspases solution (Caspase FLICA kit FAM-DEVD-FMK) for 1 h at 37 °C, washed twice with PBS and analysed by a Leica TCS SP5 confocal laser scanning microscope and FACSCanto flow cytometer (Becton-Dickinson) [72]. In another set of experiments, aimed to assess the role of ERK, p38 and JNK signalling pathways, cells were treated with 10 μM SP600125 (specific JNK inhibitor), 10 μM PD98059 (MEK inhibitor) or 1 μM SIRT1 inhibitor for 3 h prior (or not) to SRT1720 treatment.

4.14. SIRT1 RNA Interference (RNAi) Experiments

Fibroblasts from healthy subjects were cultured in complete medium without antibiotics for 2 days. Cells were seeded into a six-well plate. Then, 8 μl of LipofectamineTM LTX and 3 μL PLUSTM Reagent (Life Technologies, Carlsbad, CA, USA) were diluted in 90 μL of culture medium. Subsequently, 12 μL SIRT1 siRNA (siRNA for SIRT1-sc-40986- from Santa Cruz Biotechnology) was mixed with the medium containing Lipofectamine together with PLUS reagent and incubated for 30 min at RT for complex formation. Finally, cells were incubated with a final SIRT1 siRNA concentration of 100 nM. After 48 h, SIRT1 protein expression was determined by Western blot. To study the possible role of SIRT1 in the oxidative-mediated cell injury, untreated and SIRT1 RNAi-treated fibroblasts obtained from healthy subjects were challenged for 3 h with 500 μM H_2O_2. ROS production, lipid peroxidation, caspase-3 activity and mitochondrial membrane polarization were evaluated by confocal microscope analysis.

4.15. Assessment of MAPK Activity by FACS Analysis

Fibroblasts were fixed and permeabilized using the BD Cytofix/Cytoperm buffer (Becton-Dickinson) following the manufacturer's instructions. Anti-Phospho-p38 MAPK (Thr180/Tyr182)

(28B10) Mouse mAb (Alexa Fluor® 488 Conjugate), anti-Phospho-SAPK/JNK (Thr183/Tyr185) (G9) Mouse mAb (PE Conjugate), and anti-Phospho-p44/42 MAPK (Erk1/2) (T hr202/Tyr204) (D13.14.4E) XP® Rabbit mAb (Alexa Fluor® 488 Conjugate) were used at 1:50 dilution for 1 h at RT according to manufacturer's instructions.

4.16. Statistical Analysis

All data are expressed as the mean $\pm$ SD. Comparisons between groups were analyzed using one-Way Analysis of variance (ANOVA) followed by the Bonferroni *t*-test. A *p* value <0.05 was accepted as statistically significant.

Author Contributions: M.B. and C.F. conceived and designed the experiments; M.B., V.B., A.M. and G.E. performed the experiments; M.B., C.F., V.B., A.M., G.E., D.P. and N.T. analyzed the results; T.L. enrolled patients; M.B. and C.F. drafted the paper.

Acknowledgments: This study was supported by "Fondi di Ateneo" research funding from the University of Florence awarded to Claudia Fiorillo.

Conflicts of Interest: The authors declare no conflict of interest.

References

1. Boehncke, W.H.; Schön, M.P. Psoriasis. *Lancet* **2015**, *386*, 983–994. [CrossRef]
2. Blanpain, C.; Fuchs, E. Epidermal homeostasis: A balancing act of stem cells in the skin. *Nat. Rev. Mol. Cell Biol.* **2009**, *10*, 207–217. [CrossRef] [PubMed]
3. El Ghalbzouri, A.; Lamme, E.; Ponec, M. Crucial role of fibroblasts in regulating epidermal morphogenesis. *Cell Tissue Res.* **2002**, *310*, 189–199. [CrossRef] [PubMed]
4. Quan, T.; Wang, F.; Shao, Y.; Rittié, L.; Xia, W.; Orringer, J.S.; Voorhees, J.J.; Fisher, G.J. Enhancing structural support of the dermal microenvironment activates fibroblasts, endothelial cells, and keratinocytes in aged human skin in vivo. *J. Investig. Dermatol.* **2013**, *133*, 658–667. [CrossRef] [PubMed]
5. Trouba, K.J.; Hamadeh, H.K.; Amin, R.P.; Germolec, D.R. Oxidative stress and its role in skin disease. *Antioxid. Redox Signal.* **2002**, *4*, 665–673. [CrossRef] [PubMed]
6. Barygina, V.; Becatti, M.; Lotti, T.; Moretti, S.; Taddei, N.; Fiorillo, C. Treatment with low-dose cytokines reduces oxidative-mediated injury in perilesional keratinocytes from vitiligo skin. *J. Dermatol. Sci.* **2015**, *79*, 163–170. [CrossRef] [PubMed]
7. Zhou, Q.; Mrowietz, U.; Rostami-Yazdi, M. Oxidative stress in the pathogenesis of psoriasis. *Free Radic. Biol. Med.* **2009**, *47*, 891–905. [CrossRef] [PubMed]
8. Gabr, S.A.; Al-Ghadir, A.H. Role of cellular oxidative stress and cytochrome c in the pathogenesis of psoriasis. *Arch. Dermatol. Res.* **2012**, *304*, 451–457. [CrossRef] [PubMed]
9. Pujari, V.M.; Ireddy, S.; Itagi, I.; Kumar, H.S. The serum levels of malondialdehyde, vitamin e and erythrocyte catalase activity in psoriasis patients. *J. Clin. Diagn. Res.* **2014**, *8*, CC14–CC16. [CrossRef] [PubMed]
10. Kadam, D.P.; Suryakar, A.N.; Ankush, R.D.; Kadam, C.Y.; Deshpande, K.H. Role of oxidative stress in various stages of psoriasis. *Indian J. Clin. Biochem.* **2010**, *25*, 388–392. [CrossRef] [PubMed]
11. Basavaraj, K.H.; Vasu Devaraju, P.; Rao, K.S. Studies on serum 8-hydroxy guanosine (8-OHdG) as reliable biomarker for psoriasis. *J. Eur. Acad. Dermatol. Venereol.* **2013**, *27*, 655–657. [CrossRef] [PubMed]
12. Rocha-Pereira, P.; Santos-Silva, A.; Rebelo, I.; Figueiredo, A.; Quintanilha, A.; Teixeira, F. Dislipidemia and oxidative stress in mild and in severe psoriasis as a risk for cardiovascular disease. *Clin. Chim. Acta* **2001**, *303*, 33–39. [CrossRef]
13. Johansen, C.; Kragballe, K.; Westergaard, M.; Henningsen, J.; Kristiansen, K.; Iversen, L. The mitogen-activated protein kinases p38 and ERK1/2 are increased in lesional psoriatic skin. *Br. J. Dermatol.* **2005**, *152*, 37–42. [CrossRef] [PubMed]
14. Yu, X.J.; Li, C.Y.; Dai, H.Y.; Cai, D.X.; Wang, K.Y.; Xu, Y.H.; Chen, L.M.; Zhou, C.L. Expression and localization of the activated mitogen-activated protein kinase in lesional psoriatic skin. *Exp. Mol. Pathol.* **2007**, *83*, 413–418. [CrossRef] [PubMed]
15. Wang, S.; Uchi, H.; Hayashida, S.; Urabe, K.; Moroi, Y.; Furue, M. Differential expression of phosphorylated extracellular signal-regulated kinase 1/2, phosphorylated p38 mitogen-activated protein kinase and nuclear

factor-kappaB p105/p50 in chronic inflammatory skin diseases. *J. Dermatol.* **2009**, *36*, 534–540. [CrossRef] [PubMed]

16. Haase, I.; Hobbs, R.M.; Romero, M.R.; Broad, S.; Watt, F.M. A role for mitogen-activated protein kinase activation by integrins in the pathogenesis of psoriasis. *J. Clin. Investig.* **2001**, *108*, 527–536. [CrossRef] [PubMed]

17. Takahashi, H.; Ibe, M.; Nakamura, S.; Ishida-Yamamoto, A.; Hashimoto, Y.; Iizuka, H. Extracellular regulated kinase and c-Jun N-terminal kinase are activated in psoriatic involved epidermis. *J. Dermatol. Sci.* **2002**, *30*, 94–99. [CrossRef]

18. Kyriakis, J.M.; Avruch, J. Mammalian mitogen-activated protein kinase signal transduction pathways activated by stress and inflammation. *Physiol. Rev.* **2001**, *81*, 807–869. [CrossRef] [PubMed]

19. Alcendor, R.R.; Gao, S.; Zhai, P.; Zablocki, D.; Holle, E.; Yu, X.; Tian, B.; Wagner, T.; Vatner, S.F.; Sadoshima, J. Sirt1 regulates aging and resistance to oxidative stress in the heart. *Circ. Res.* **2007**, *100*, 1512–1521. [CrossRef] [PubMed]

20. Rahman, S.; Islam, R. Mammalian Sirt1: Insights on its biological functions. *Cell. Commun. Signal* **2011**, *9*, 11. [CrossRef] [PubMed]

21. Becatti, M.; Taddei, N.; Cecchi, C.; Nassi, N.; Nassi, P.A.; Fiorillo, C. SIRT1 modulates MAPK pathways in ischemic-reperfused cardiomyocytes. *Cell. Mol. Life Sci.* **2012**, *69*, 2245–2260. [CrossRef] [PubMed]

22. Becatti, M.; Fiorillo, C.; Barygina, V.; Cecchi, C.; Lotti, T.; Prignano, F.; Silvestro, A.; Nassi, P.; Taddei, N. SIRT1 regulates MAPK pathways in vitiligo skin: Insight into the molecular pathways of cell survival. *J. Cell. Mol. Med.* **2014**, *18*, 514–529. [CrossRef] [PubMed]

23. Becatti, M.; Barygina, V.; Emmi, G.; Silvestri, E.; Taddei, N.; Lotti, T.; Fiorillo, C. SIRT1 activity is decreased in lesional psoriatic skin. *Intern. Emerg. Med.* **2016**, *11*, 891–893. [CrossRef] [PubMed]

24. Petronilli, V.; Miotto, G.; Canton, M.; Brini, M.; Colonna, R.; Bernardi, P.; Di Lisa, F. Transient and long-lasting openings of the mitochondrial permeability transition pore can be monitored directly in intact cells by changes in mitochondrial calcein fluorescence. *Biophys. J.* **1999**, *765*, 725–754. [CrossRef]

25. Sabat, R.; Philipp, S.; Höflich, C.; Kreutzer, S.; Wallace, E.; Asadullah, K.; Volk, H.D.; Sterry, W.; Wolk, K. Immunopathogenesis of psoriasis. *Exp. Dermatol.* **2007**, *16*, 779–798. [CrossRef] [PubMed]

26. Boyman, O.; Conrad, C.; Tonel, G.; Gilliet, M.; Nestle, F.O. The pathogenic role of tissue-resident immune cells in psoriasis. *Trends Immunol.* **2007**, *28*, 51–57. [CrossRef] [PubMed]

27. Dimon-Gadal, S.; Gerbaud, P.; Thérond, P.; Guibourdenche, J.; Anderson, W.B.; Evain-Brion, D.; Raynaud, F. Increased oxidative damage to fibroblasts in skin with and without lesions in psoriasis. *J. Investig. Dermatol.* **2000**, *114*, 984–989. [CrossRef] [PubMed]

28. Er-raki, A.; Charveron, M.; Bonafé, J.L. Increased superoxide anion production in dermal fibroblasts of psoriatic patients. *Skin Pharmacol.* **1993**, *6*, 253–258. [CrossRef] [PubMed]

29. Debets, R.; Hegmans, J.P.; Buurman, W.A.; Benner, R.; Prens, E.P. Expression of cytokines and their receptors by psoriatic fibroblasts. II. Decreased TNF receptor expression. *Cytokine* **1996**, *8*, 80–88. [CrossRef] [PubMed]

30. Miura, H.; Sanso, S.; Higashiyama, M.; Yoshikawa, K.; Itami, S. Involvement of insulin-like growth factor-I in psoriasis as a paracrine growth factor: Dermal fibroblasts play a regulatory role in developing psoriatic lesions. *Arch. Dermatol. Res.* **2000**, *292*, 590–597. [CrossRef] [PubMed]

31. Ishikawa, H.; Mitsuhashi, Y.; Takagi, Y.; Hashimoto, I. Psoriatic fibroblasts enhance cornified envelope formation in normal keratinocytes in vitro. *Arch. Dermatol. Res.* **1997**, *289*, 551–553. [CrossRef] [PubMed]

32. Cunliffe, I.A.; Richardson, P.S.; Rees, R.C.; Rennie, I.G. Effect of TNF, IL-1, and IL-6 on proliferation of human Tenon's capsule fibroblasts in tissue culture. *Br. J. Ophthalmol.* **1995**, *79*, 590–595. [CrossRef] [PubMed]

33. Briganti, S.; Picardo, M. Antioxidant activity, lipid peroxidation and skin diseases: What's new. *J. Eur. Acad. Dermatol. Venereol.* **2003**, *17*, 663–669. [CrossRef] [PubMed]

34. Kökçam, I.; Naziroğlu, M. Antioxidants and lipid peroxidation status in the blood of patients with psoriasis. *Clin. Chim. Acta* **1999**, *289*, 23–31. [CrossRef]

35. Barygina, V.; Becatti, M.; Lotti, T.; Taddei, N.; Fiorillo, C. Low dose cytokines reduce oxidative stress in primary lesional fibroblasts obtained from psoriatic patients. *J. Dermatol. Sci.* **2016**, *83*, 242–244. [CrossRef] [PubMed]

36. Alcendor, R.R.; Kirshenbaum, L.A.; Imai, S.; Vatner, S.F.; Sadoshima, J. Silent information regulator 2alpha, a longevity factor and class III histone deacetylase, is an essential endogenous apoptosis inhibitor in cardiac myocytes. *Circ. Res.* **2004**, *95*, 971–980. [CrossRef] [PubMed]

37. Vachharajani, V.T.; Liu, T.; Wang, X.; Hoth, J.J.; Yoza, B.K.; McCall, C.E. Sirtuins Link Inflammation and Metabolism. *J. Immunol. Res.* **2016**, *2016*, 8167273. [CrossRef] [PubMed]

38. Blander, G.; Bhimavarapu, A.; Mammone, T.; Maes, D.; Elliston, K.; Reich, C.; Matsui, M.S.; Guarente, L.; Loureiro, J.J. SIRT1 promotes differentiation of normal human keratinocytes. *J. Investig. Dermatol.* **2009**, *129*, 41–49. [CrossRef] [PubMed]

39. Wu, Z.; Uchi, H.; Morino-Koga, S.; Shi, W.; Furue, M. Resveratrol inhibition of human keratinocyte proliferation via SIRT1/ARNT/ERK dependent downregulation of aquaporin 3. *J. Dermatol. Sci.* **2014**, *75*, 16–23. [CrossRef] [PubMed]

40. Rasheed, H.; El-Komy, M.; Hegazy, R.A.; Gawdat, H.I.; AlOrbani, A.M.; Shaker, O.G. Expression of sirtuins 1, 6, tumor necrosis factor, and interferon-γ in psoriatic patients. *Int. J. Immunopathol. Pharmacol.* **2016**, *29*, 764–768. [CrossRef] [PubMed]

41. Krueger, J.G.; Suárez-Fariñas, M.; Cueto, I.; Khacherian, A.; Matheson, R.; Parish, L.C.; Leonardi, C.; Shortino, D.; Gupta, A.; Haddad, J.; et al. A Randomized, Placebo-Controlled Study of SRT2104, a SIRT1 Activator, in Patients with Moderate to Severe Psoriasis. *PLoS ONE* **2015**, *10*, e0142081. [CrossRef] [PubMed]

42. Barygina, V.V.; Becatti, M.; Soldi, G.; Prignano, F.; Lotti, T.; Nassi, P.; Wright, D.; Taddei, N.; Fiorillo, C. Altered redox status in the blood of psoriatic patients: Involvement of NADPH oxidase and role of anti-TNF-α therapy. *Redox Rep.* **2013**, *18*, 100–106. [CrossRef] [PubMed]

43. Khader, A.; Yang, W.L.; Kuncewitch, M.; Jacob, A.; Prince, J.M.; Asirvatham, J.R.; Nicastro, J.; Coppa, G.F.; Wang, P. Sirtuin 1 activation stimulates mitochondrial biogenesis and attenuates renal injury after ischemia-reperfusion. *Transplantation* **2014**, *98*, 148–156. [CrossRef] [PubMed]

44. Zeng, Z.; Chen, Z.; Xu, S.; Zhang, Q.; Wang, X.; Gao, Y.; Zhao, K.S. Polydatin Protecting Kidneys against Hemorrhagic Shock-Induced Mitochondrial Dysfunction via SIRT1 Activation and p53 Deacetylation. *Oxid. Med. Cell. Longev.* **2016**, *2016*, 1737185. [CrossRef] [PubMed]

45. Brunet, A.; Sweeney, L.B.; Sturgill, J.F.; Chua, K.F.; Greer, P.L.; Lin, Y.; Tran, H.; Ross, S.E.; Mostoslavsky, R.; Cohen, H.Y.; et al. Stress-dependent regulation of FOXO transcription factors by the SIRT1 deacetylase. *Science* **2004**, *303*, 2011–2015. [CrossRef] [PubMed]

46. Van der Horst, A.; Tertoolen, L.G.; de Vries-Smits, L.M.; Frye, R.A.; Medema, R.H.; Burgering, B.M. FOXO4 is acetylated upon peroxide stress and deacetylated by the longevity protein hSir2SIRT1. *J. Biol. Chem.* **2004**, *279*, 28873–28879. [CrossRef] [PubMed]

47. Sengupta, A.; Molkentin, J.D.; Paik, J.H.; DePinho, R.A.; Yutzey, K.E. FoxO transcription factors promote cardiomyce survival upon induction of oxidative stress. *J. Biol. Chem.* **2011**, *286*, 7468–7478. [CrossRef] [PubMed]

48. Xiong, S.; Salazar, G.; Patrushev, N.; Alexander, R.W. FoxO1 mediates an autofeedback loop regulating SIRT1 expression. *J. Biol. Chem.* **2011**, *286*, 5289–5299. [CrossRef] [PubMed]

49. Yamamoto, T.; Sadoshima, J. Protection of the heart against ischemia/reperfusion by silent information regulator 1. *Trends Cardiovasc. Med.* **2011**, *21*, 27–32. [CrossRef] [PubMed]

50. Chung, K.W.; Choi, Y.J.; Park, M.H.; Jang, E.J.; Kim, D.H.; Park, B.H.; Yu, B.P.; Chung, H.Y. Molecular Insights into SIRT1 Protection Against UVB-Induced Skin Fibroblast Senescence by Suppression of Oxidative Stress and p53 Acetylation. *J. Gerontol. A Biol. Sci Med. Sci.* **2015**, *70*, 959–968. [CrossRef] [PubMed]

51. Arasa, J.; Terencio, M.C.; Andrés, R.M.; Valcuende-Cavero, F.; Montesinos, M.C. Decreased SAPK/JNK signalling affects cytokine release and STAT3 activation in psoriatic fibroblasts. *Exp. Dermatol.* **2015**, *24*, 800–802. [CrossRef] [PubMed]

52. Soegaard-Madsen, L.; Johansen, C.; Iversen, L.; Kragballe, K. Adalimumab therapy rapidly inhibits p38 mitogen-activated protein kinase activity in lesional psoriatic skin preceding clinical improvement. *Br. J. Dermatol.* **2010**, *162*, 1216–1223. [CrossRef] [PubMed]

53. Kyriakis, J.M.; Avruch, J. Mammalian MAPK signal transduction pathways activated by stress and inflammation: A 10-year update. *Physiol. Rev.* **2012**, *92*, 689–737. [CrossRef] [PubMed]

54. Qi, M.; Elion, E.A. MAP kinase pathways. *J. Cell Sci.* **2005**, *118*, 3569–3572. [CrossRef] [PubMed]

55. Huang, J.; Gan, Q.; Han, L.; Li, J.; Zhang, H.; Sun, Y.; Zhang, Z.; Tong, T. SIRT1 Overexpression Antagonizes Cellular senescence with Activated ERK/S6k1 Signaling in Human Diploid Fibroblasts. *PLoS ONE* **2008**, *3*, e1710. [CrossRef] [PubMed]

56. Kobayashi, Y.; Ohtsuki, M.; Murakami, T.; Kobayashi, T.; Sutheesophon, K.; Kitayama, H.; Kano, Y.; Kusano, E.; Nakagawa, H.; Furukawa, Y. Histone deacetylase inhibitor FK228 suppresses the Ras-MAP

kinase signaling pathway by upregulating Rap1 and induces apoptosis in malignant melanoma. *Oncogene* **2006**, *25*, 512–524. [CrossRef] [PubMed]

57. Sluss, H.K.; Barrett, T.; Dérijard, B.; Davis, R.J. Signal transduction by tumor necrosis factor mediated by JNK protein kinases. *Mol. Cell. Biol.* **1994**, *14*, 8376–8384. [CrossRef] [PubMed]

58. Fisher, G.J.; Talwar, H.S.; Lin, J.; Lin, P.; McPhillips, F.; Wang, Z.; Li, X.; Wan, Y.; Kang, S.; Voorhees, J.J. Retinoic acid inhibits induction of c-Jun protein by ultraviolet radiation that occurs subsequent to activation of mitogen-activated protein kinase pathways in human skin in vivo. *J. Clin. Investig.* **1998**, *101*, 1432–1440. [CrossRef] [PubMed]

59. Gazel, A.; Banno, T.; Walsh, R.; Blumenberg, M. Inhibition of JNK promotes differentiation of epidermal keratinocytes. *J. Biol. Chem.* **2006**, *281*, 20530–20541. [CrossRef] [PubMed]

60. Mose, M.; Kang, Z.; Raaby, L.; Iversen, L.; Johansen, C. TNFα- and IL-17A-mediated S100A8 expression is regulated by p38 MAPK. *Exp. Dermatol.* **2013**, *22*, 476–481. [CrossRef] [PubMed]

61. Zhang, Q.S.; Maddock, D.A.; Chen, J.P.; Heo, S.; Chiu, C.; Lai, D.; Souza, K.; Mehta, S.; Wan, Y.S. Cytokine-induced p38 activation feedback regulates the prolonged activation of AKT cell survival pathway initiated by reactive oxygen species in response to UV irradiation in human keratinocytes. *Int. J. Oncol.* **2001**, *19*, 1057–1061. [CrossRef] [PubMed]

62. Garmyn, M.; Mammone, T.; Pupe, A.; Gan, D.; Declercq, L.; Maes, D. Human keratinocytes respond to osmotic stress by p38 map kinase regulated induction of HSP70 and HSP27. *J. Investig. Dermatol.* **2001**, *117*, 1290–1295. [CrossRef] [PubMed]

63. Baliwag, J.; Barnes, D.H.; Johnston, A. Cytokines in psoriasis. *Cytokine* **2015**, *73*, 342–350. [CrossRef] [PubMed]

64. Santos, L.; Escande, C.; Denicola, A. Potential Modulation of Sirtuins by Oxidative Stress. *Oxid. Med. Cell. Longev.* **2016**, *2016*, 9831825. [CrossRef] [PubMed]

65. Bradford, M.M. A rapid and sensitive method for the quantitation of microgram quantities of protein utilizing the principle of protein-dye binding. *Anal. Biochem.* **1976**, *72*, 248–254. [CrossRef]

66. Fiorillo, C.; Becatti, M.; Pensalfini, A.; Cecchi, C.; Lanzilao, L.; Donzelli, G.; Nassi, N.; Giannini, L.; Borchi, E.; Nassi, P. Curcumin protects cardiac cells against ischemia-reperfusion injury: Effects on oxidative stress, NF-kappaB, and JNK pathways. *Free Radic. Biol. Med.* **2008**, *45*, 839–846. [CrossRef] [PubMed]

67. Fulco, M.; Cen, Y.; Zhao, P.; Hoffman, E.P.; McBurney, M.W.; Sauve, A.A.; Sartorelli, V. Glucose restriction inhibits skeletal myoblast differentiation by activating SIRT1 through AMPK-mediated regulation of Nampt. *Dev. Cell* **2008**, *14*, 661–673. [CrossRef] [PubMed]

68. Fiorillo, C.; Becatti, M.; Attanasio, M.; Lucarini, L.; Nassi, N.; Evangelisti, L.; Porciani, M.C.; Nassi, P.; Gensini, G.F.; Abbate, R.; et al. Evidence for oxidative stress in plasma of patients with Marfan syndrome. *Int. J. Cardiol.* **2010**, *145*, 544–546. [CrossRef] [PubMed]

69. Drummen, G.P.; Gadella, B.M.; Post, J.A.; Brouwers, J.F. Mass spectrometric characterization of the oxidation of the fluorescent lipid peroxidation reporter molecule C11-BODIPY(581/591). *Free Radic. Biol. Med.* **2004**, *36*, 1635–1644. [CrossRef] [PubMed]

70. Pini, A.; Boccalini, G.; Baccari, M.C.; Becatti, M.; Garella, R.; Fiorillo, C.; Calosi, L.; Bani, D.; Nistri, S. Protection from cigarette smoke-induced vascular injury by recombinant human relaxin-2 (serelaxin). *J. Cell. Mol. Med.* **2016**, *20*, 891–902. [CrossRef] [PubMed]

71. Becatti, M.; Boccalini, G.; Pini, A.; Fiorillo, C.; Bencini, A.; Bani, D.; Nistri, S. Protection of coronary endothelial cells from cigarette smoke-induced oxidative stress by a new Mn(II)-containing polyamine-polycarboxilate scavenger of superoxide anion. *Vascul. Pharmacol.* **2015**, *75*, 19–28. [CrossRef] [PubMed]

72. Becatti, M.; Prignano, F.; Fiorillo, C.; Pescitelli, L.; Nassi, P.; Lotti, T.; Taddei, N. The involvement of Smac/DIABLO, p53, NF-kB, and MAPK pathways in apoptosis of keratinocytes from perilesional vitiligo skin: Protective effects of curcumin and capsaicin. *Antioxid. Redox Signal.* **2010**, *13*, 1309–1321. [CrossRef] [PubMed]

International Journal of
Molecular Sciences

Article

Cecropin A Modulates Tight Junction-Related Protein Expression and Enhances the Barrier Function of Porcine Intestinal Epithelial Cells by Suppressing the MEK/ERK Pathway

Zhenya Zhai [1], Xiaojun Ni [1], Chenglong Jin [1], Wenkai Ren [1], Jie Li [1], Jinping Deng [1,*], Baichuan Deng [1,*] and Yulong Yin [1,2,*]

[1]   Guangdong Provincial Key Laboratory of Animal Nutrition Control, Subtropical Institute of Animal Nutrition and Feed, College of Animal Science, South China Agricultural University, Guangzhou 510642, Guangdong, China; zhai@stu.scau.edu.cn (Z.Z.); nnnxjun@stu.scau.edu.cn (X.N.); jin@stu.scau.edu.cn (C.J.); renwenkai19@126.com (W.R.); lijiedk@stu.scau.edu.cn (J.L.)
[2]   National Engineering Laboratory for Pollution Control and Waste Utilization in Livestock and Poultry Production, Institute of Subtropical Agriculture, The Chinese Academy of Sciences, Changsha 410125, Hunan, China
*   Correspondence: dengjinping@scau.edu.cn (J.D.); dengbaichuan@scau.edu.cn (B.D.); yinyulong@isa.ac.cn (Y.Y.); Tel.: +86-20-8528-0547 (B.D.)

Received: 4 June 2018; Accepted: 29 June 2018; Published: 2 July 2018

Abstract: Inflammatory bowel disease (IBD) in humans and animals is associated with bacterial infection and intestinal barrier dysfunction. Cecropin A, an antimicrobial peptide, has antibacterial activity against pathogenic bacteria. However, the effect of cecropin A on intestinal barrier function and its related mechanisms is still unclear. Here, we used porcine jejunum epithelial cells (IPEC-J2) as a model to investigate the effect and mechanism of cecropin A on intestinal barrier function. We found that cecropin A reduced *Escherichia coli* (*E. coli*) adherence to IPEC-J2 cells and downregulated mRNA expression of tumor necrosis factor α (*TNF-α*), interleukin-6 (*IL-6*), and interleukin-8 (*IL-8*). Furthermore, cecropin A elevated the transepithelial electrical resistance (TER) value while reducing the paracellular permeability of the IPEC-J2 cell monolayer barrier. Finally, by using Western blotting, immunofluorescence and pathway-specific antagonists, we demonstrated that cecropin A increased ZO-1, claudin-1 and occludin protein expression and regulated membrane distribution and F-actin polymerization by increasing CDX2 expression. We conclude that cecropin A enhances porcine intestinal epithelial cell barrier function by downregulating the mitogen-activated protein kinase (MEK)/extracellular signal-regulated kinase (ERK) pathway. We suggest that cecropin A has the potential to replace antibiotics in the treatment of IBD due to its antibacterial activity on gram-negative bacteria and its enhancement effect on intestinal barrier function.

Keywords: antimicrobial peptide; cecropin A; tight junction protein; MEK/ERK signaling; porcine intestinal epithelial cell

---

1. Introduction

Anti-infective drugs play important roles in the prevention and treatment of inflammatory bowel disease (IBD) in humans and animals. IBD is a complex gastrointestinal disease, mainly induced by infection with gram-negative bacteria such as *Escherichia coli* (*E. coli*) and *Salmonella* [1]. In recent decades, an increasing prevalence of antibiotic resistance has threatened human health [2]. Finding effective alternatives to antibiotics has become an increasingly urgent task. Among the potential alternatives, antimicrobial peptides (AMPs) are particularly important, due to their broad spectrum antibacterial activity and decreased likelihood of inducing antibiotic resistance. Currently, more than 2800 AMPs have been found in animals, plants or microorganisms [3]. In animals, AMPs play important roles in host defense and are crucial in the immune system [4]. The activities of AMPs vary greatly due to their different sequences and structures. In addition to antimicrobial activity, some AMPs also have wound healing abilities through promoting cell proliferation, reducing inflammation or enhancing intestinal barrier function [5,6].

Cecropins are a group of peptides with an α-helical structure and were initially found in insects. Currently, there are more than 30 records of cecropins in the Antimicrobial Peptide Database (APD), including naturally discovered and artificially synthesized cecropins [3]. Cecropin A is one of the earliest discovered cecropins by Steiner et al. from *Hyalophora cecropia* [7]. Over the past decades, the antibacterial mechanisms of cecropin A have been extensively researched [8,9]. In addition, cecropin A is also a commonly used template for peptide molecular hybrids to enhance the antibacterial activity of AMPs [10]. Although the antibacterial activity of cecropin A has been demonstrated for decades, to our best knowledge, the effect of cecropin A on intestinal barrier function is still unknown.

IBD is caused by pathogenic bacterial infection and intestinal mucosal barrier disruption. Intestinal mucosal surfaces consist of epithelial cells, such as absorptive cells, endocrine cells and Paneth cells [11]. The epithelial cells form a selectively leaky barrier, which is crucial for nutrient substance exchange and host defense [12–15]. These functions depend on intact intestinal epithelial cell layers, which are composed of cell–cell attachments at the cell lateral membrane by tight junctions (TJs) and subjacent adherens junctions [13]. The TJs consist of transmembrane proteins such as claudins, occludin and junctional adhesion molecules (JAMs). These proteins are clustered and stabilized by cytoplasmic scaffolding proteins called zonula occludens (ZOs) and cytoskeletons such as F-actin. Different tight junction proteins play various roles in barrier function. Claudins and occludin are located at apical and basal positions of the lateral membrane, respectively [16]. ZOs, such as ZO-1, can interact with cytoskeleton, claudin-1 and occludin [17]. To summarize, the TJ-cytoskeleton structure is essential for the intestinal barrier.

The regulation of TJ expression and membrane distribution is complex. The mitogen-activated protein kinase (MAPK) pathways, which contain three downstream pathways including extracellular signal-regulated kinase (ERK), p38 and c-jun, are responsible for cell proliferation, proliferation and immune reaction in the gastrointestinal tract [18]. ERK, which may be activated by mitogen-activated protein kinase (MEK), is one of the most important pathways for maintaining gastrointestinal tract homeostasis and regulating the intestinal barrier. However, according to previous studies, the effect of MEK/ERK on intestinal barrier function is controversial. Piegholdt et al. [19] showed that biochanin A and prunetin may improve epithelial function through downregulation of ERK, while Wang et al. [20] showed an improvement of the intestinal epithelial barrier through upregulation of the ERK pathway by polyphenol-rich propolis extracts. The regulatory effect of the MEK/ERK signaling pathway on the intestinal barrier is unclear.

In this study, we evaluated the effects of cecropin A on intestinal barrier function in an IPEC-J2 cell monolayer model. We also detected the TJ protein level and membrane distribution by using Western blotting and cell immunofluorescence, respectively. Finally, the changes in the MEK/ERK signaling pathway were detected to reveal the regulatory mechanism of cecropin A on the barrier function.

2. Results

2.1. The Antibacterial Activity of Different AMPs

Seven AMPs were selected from the APD database, antibacterial activities including minimum inhibitory concentration (MIC) and minimum bactericidal concentration (MBC) were tested by using 11 specific bacterial strains (Table S1) and the information of origin source, peptide length, net charge were also described (Table S2). The results showed that cecropin A possessed the best antibacterial activity (MIC and MBC between 1.5 and 6.25 µg/mL) to gram-negative bacterial strains, such as *E. coli*, *Salmonella* and *Pseudomonas aeruginosa*.

2.2. Cytotoxicity to IPEC-J2 Cells

The cytotoxicity of AMPs to the pig intestinal epithelium cell IPEC-J2 was evaluated by using an MTT (3-(4,5-dimethylthiazol-2-yl)-2,5-diphenyltetrazolium bromide) assay (Figure S1A–G). The concentration was 1.5–100 µg/mL. The results showed that IPEC-J2 cell viability was not reduced after treatment with cecropin A (1.5–12.5 µg/mL, Figure S1C) for 8 h compared to that of the control group. Six other AMPs reduced cell viability in a dose-dependent manner.

2.3. Cecropin A Inhibits E. coli Adherence and Ameliorates Inflammation

IPEC-J2 cells pretreated with 3.125, 6.25, and 12.5 µg/mL cecropin A displayed reduced *E. coli* adherence in a dose-dependent manner (Figure 1A). In addition, after coculture with *E. coli*, the TNF-α, IL-6, and IL-8 mRNA expression in IPEC-J2 cells was also downregulated after cecropin A treatment for 48 h, compared to that in the non-treated cells (Figure 1B). The results suggested that the defense capability of IPEC-J2 cells against bacteria was increased.

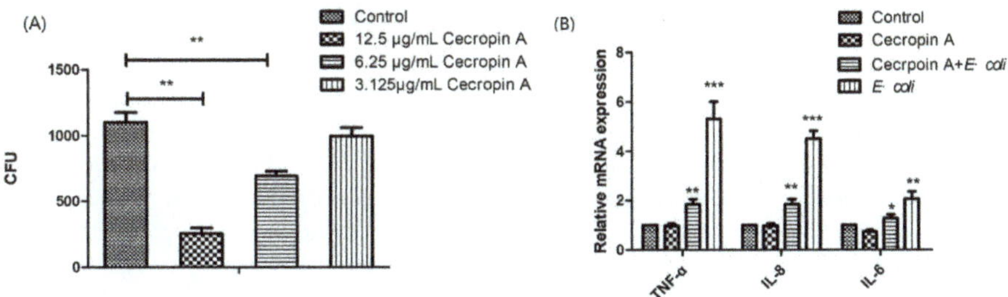

Figure 1. Cecropin A inhibited bacterial adherence and reduced the expression of inflammatory factors. The porcine jejunum epithelial cells (IPEC-J2) cells were pretreated with cecropin A (3.125, 6.25, and 12.5 µg/mL) for 48 h, and the CFU of adherent bacteria were counted ((**A**), $n = 3$). The relative mRNA expression of *TNF-α*, *IL-8*, and *IL-6* were tested by using qPCR ((**B**), $n = 6$). Control: control group; cecropin A: cells were pretreated with cecropin A; cecropin A + *E. coli*: cells were pretreated with cecropin A and then cocultured with *E. coli*; *E. coli*: cells cocultured with *E. coli*. The results were confirmed by three independent experiments per treatment. Representative results of the three independent experiments are shown. Data (mean ± SEM) were analyzed with one-way ANOVA. * $p < 0.05$, ** $p < 0.01$, *** $p < 0.001$.

2.4. Cecropin A Increases the TER and Decreases the Paracellular Diffusion of FITC-Dextran through the IPEC-J2 Monolayer

The integrity of the intestinal barrier may increase the defense capability of the host and reduce the adherence of pathogenic bacteria. We examined whether cecropin A could enhance the intestinal monolayer barrier function. Transepithelial electrical resistance (TER) values were assessed at 24, 48 and 72 h after cecropin A treatment. Our data showed that, compared to those of the control cells, the TER values of cecropin A-treated cells were significantly increased at 48 h and 72 h (Figure 2A; $p < 0.01$). In addition, we measured the permeability of large solutes by using 4 kDa FITC-dextran as a tracer. The data showed that, after cell incubation with FITC-dextran, the concentrations of FITC-dextran in the basal compartment were significantly higher than those in cecropin A-treated cells at 48 h and 72 h (Figure 2B, $p < 0.05$).

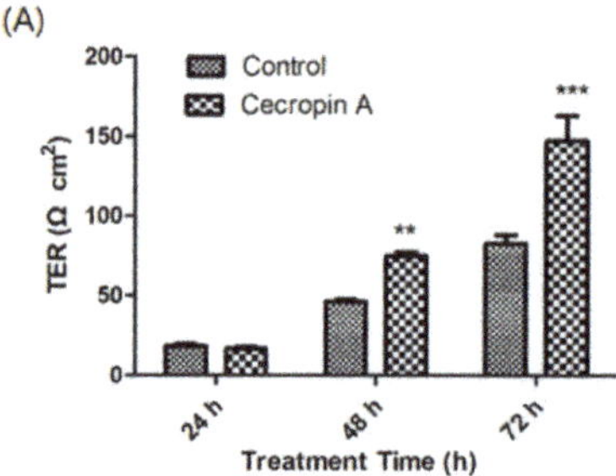
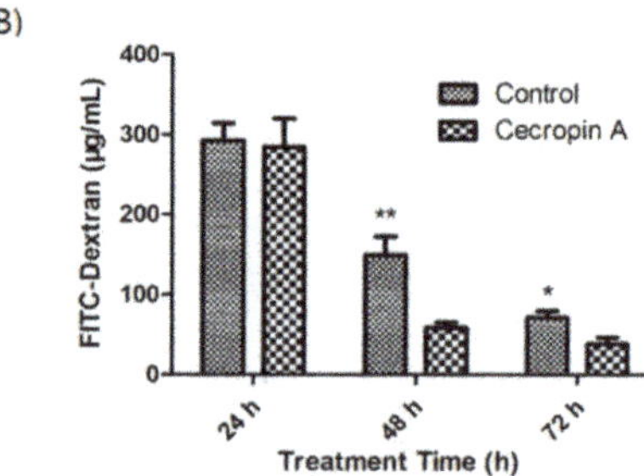

Figure 2. Cecropin A enhances the transepithelial electrical resistance (TER) and reduces the permeability of the IPEC-J2 cell monolayer. The TER ((**A**), $n = 6$) and permeability ((**B**), $n = 6$) of the cell monolayer were tested after treatment with cecropin A for 24 h, 48 h, and 72 h. The results were confirmed by three independent experiments per treatment. Representative results of the three independent experiments are shown. Data (mean ± SEM) were analyzed with the Student's *t*-test. * $p < 0.05$, ** $p < 0.01$, *** $p < 0.001$.

2.5. Cecropin A Regulates TJ Protein Expression Levels, Membrane Distribution and F-Actin Polymerization

To elucidate how cecropin A increases the TER value and decreases paracellular permeability, we measured the protein levels of ZO-1, claudin-1 and occludin. The results showed that the protein levels of TJs were significantly upregulated (Figure 3A,B). We also detected TJ membrane distribution by using cell immunofluorescence. The results showed that ZO-1, claudin-1 and occludin were much more polymerized at the cell–cell boundary in the cecropin A group than in the control group. (Figure 3C). The integrity of the intestinal monolayer barrier is coordinated by the connection between TJs and the cytoskeleton. Representative F-actin staining indicated that the cecropin A group had more extensive F-actin than the control group (Figure 3C). In addition, F-actin was much more polymerized at the cell–cell boundary in the cecropin A group than in the control group. In accordance with the TJ distribution, the results suggested that the better organized F-actin-TJ structure may help increase the monolayer barrier function and the defense capability against the adherence of *E. coli*.

2.6. Cecropin A Regulates the Intestinal Barrier by Downregulating the MEK/ERK Pathways

To elucidate how cecropin A regulates the TER and TJs, the MEK/ERK signaling pathway was detected using Western blotting. The data showed that phosphorylation of MEK and ERK in control group cells was significantly higher than that in the cecropin A group cells ($p < 0.01$, Figure 4). Caudal type homeobox 2 (CDX2), a transcriptional factor that regulates the differentiation of intestinal cells, was also detected ($p < 0.001$). The results showed that the CDX2 protein level was upregulated in the cecropin A group compared with that in the control group. In addition, the results suggested that cecropin A may regulate TJ expression through regulating the phosphorylation of MEK and ERK and the expression of CDX2.

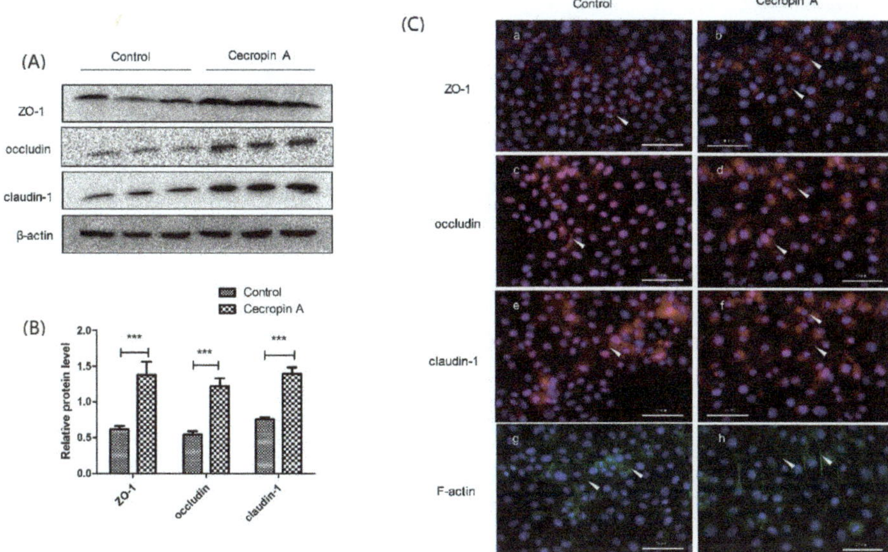

Figure 3. The effect of cecropin A on tight junction (TJ) protein expression, membrane distribution and F-actin polymerization. Western blotting analysis of zonula occludens-1 (ZO-1), occludin and claudin-1 expression were upregulated by cecropin A treatment ((**A**), *n* = 3); quantification of ZO-1, occludin, and claudin-1 protein expression was shown ((**B**), *n* = 3); cell immunofluorescence (**C**) (400×) showed that the cecropin A induced the TJs polymerized at the cell–cell boundary (**a–f**) and indicated F-actin polymerization (**g,h**) in cells ((**c**), *n* = 3). The results were confirmed by three independent experiments per treatment. Representative results of the three independent experiments are shown. Data (mean ± SEM) were analyzed with the Student's *t*-test. Cell nuclei were stained by 4′,6-Diamidino-2-phenylindole dihydrochloride (DAPI) and are shown in blue. Claudin-1, ZO-1 and occludin are shown in red and pointed out by white arrow heads. F-actin is shown in green. Scale bar is 50 μm. *** $p < 0.001$.

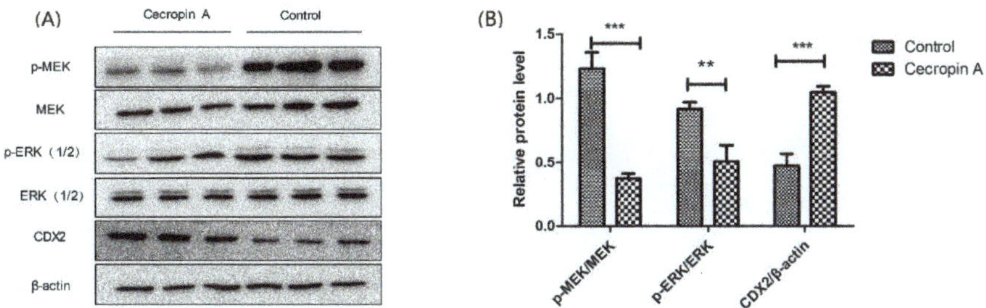

Figure 4. Cecropin A downregulates mitogen-activated protein kinase (MEK)/extracellular signal-regulated kinase (ERK) phosphorylation and upregulates CDX2 expression, *n* = 3. Western blotting analysis of p-MEK, MEK, p-ERK(1/2), ERK (1/2) and CDX2 showed that cecropin A downregulated the MEK/ERK pathway and increased CDX2 protein level (**A,B**). The results were confirmed by three independent experiments per treatment. Representative results of the three independent experiments are shown. Data (mean ± SEM) were analyzed with the Student's *t*-test. Control: control group; Cecropin A: cells treated with cecropin A group. ** $p < 0.01$, *** $p < 0.001$.

2.7. Inhibition of the MEK/ERK Pathway Increases TER, TJ Expression, Membrane Distribution and F-Actin Polymerization

To confirm the regulatory effect of the MEK/ERK pathway on the intestinal barrier and TJs, a specific inhibitor (PD184352) of ERK 1/2 was used (Figure 5A). After treatment with the inhibitor for 48 h, the adherence of *E. coli* was reduced compared to that in the cecropin A group (Figure 5B). The TER values of the cecropin A-treated group were significantly higher than those of the control group, while the TER values of the PD184352 + cecropin A group were higher than those of the cecropin A group (Figure 5C, $p < 0.01$). In addition, protein levels of CDX2, ZO-1, claudin-1, and occludin in the PD184352 + cecropin A group were significantly higher than those in the cecropin A group ($p < 0.001$, Figure 6A,B). Moreover, the TJs and F-actin in the PD184352 + cecropin A group were distributed at the cell-cell adjacent position (Figure 6C). The results showed that the inhibition of MEK/ERK may increase intestinal barrier function by increasing the TJ expression level and membrane distribution.

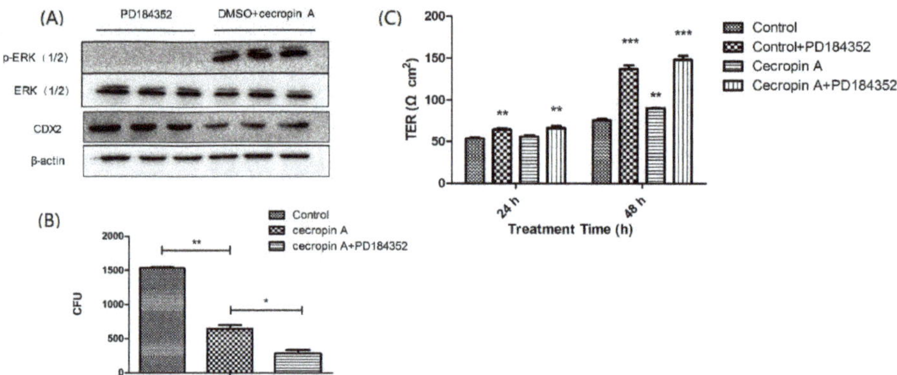

Figure 5. The effect of inhibiting ERK phosphorylation on TER and *E. coli* adherence. TER was upregulated ((**B**), $n = 3$) through inhibition of ERK phosphorylation ((**A**), $n = 3$). IPEC-J2 cells were treated with the ERK-specific inhibitor PD184352, cecropin A and cecropin A + PD184352 for 24 and 48 h, and the CFUs of adherent *E. coli* were decreased after the PD184352 treatment for 48 h ($n = 3$). Data (mean ± SEM) were analyzed with the Student's *t*-test (**A**) and one-way ANOVA (**B,C**). The results were confirmed by three independent experiments per treatment. Representative results of the three independent experiments are shown. * $p < 0.05$, ** $p < 0.01$, *** $p < 0.001$.

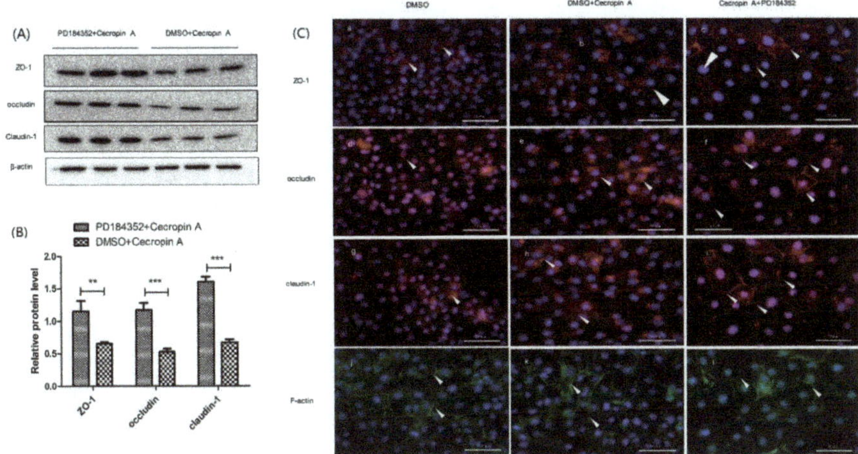

Figure 6. The inhibitory effects of MEK/ERK on TJ protein expression, membrane distribution and F-actin polymerization. Western blotting analysis of ZO-1, occludin and claudin-1 expression ((**A**,**B**), $n = 3$); cell immunofluorescence ((**C**), $n = 3$, 400×) showed the membrane distribution (**a–i**) and F-actin polymerization (**j–l**). The results were confirmed by three independent experiments per treatment. Representative results of the three independent experiments are shown. Data (mean ± SEM) were analyzed with the Student's *t*-test. Cell nuclei were stained by DAPI and are shown in blue. Claudin-1, ZO-1 and occludin are shown in red and pointed out by white arrow heads. F-actin is shown in green. Scale bar is 50 μm. ** $p < 0.01$, *** $p < 0.001$.

3. Discussion

In recent decades, thousands of AMPs have been found in organisms. The antibacterial activities of AMPs differ greatly from each other because of their different sequences and structures [21,22]. Although several antibacterial theories for AMPs have been put forward, their potential mechanisms are still unclear. According to previous studies, most of the AMPs have hemolytic activity or cytotoxicity when used at high concentrations [23]. In this study, we have shown that cecropin A had better antibacterial activity and lower cytotoxicity compared to those of other AMPs (PG-1, LL-37, etc.) selected from the APD3 database. Interestingly, we also found that the antibacterial activity of cecropin A is much higher against *E. coli*, *Salmonella typhimurium* and *Pseudomonas aeruginosa* than *Staphylococcus aureus*, which may suggest that cecropin A was more effective against gram-negative bacteria.

In addition to the antibacterial activities, we also showed that cecropin A may enhance the barrier function of the IPEC-J2 monolayer and increase the defense capability against pathogenic bacteria. In this study, we showed that after treatment with cecropin A for 48 h, adherent *E. coli* were significantly reduced and the mRNA levels of TNF-α, IL-6 and IL-8 were downregulated, which indicates that inflammation may be alleviated. The integrity of intestinal barrier function and morphology may affect the colonization of harmful bacteria and play an important role in protecting hosts from microorganism infection and inflammatory diseases induced by pathogenesis [24,25]. To evaluate the intestinal barrier function, TER and cell monolayer permeability experiments were performed. The data showed that cecropin A could significantly increase TER and reduce permeability. The TJs (ZOs, occludin, claudins, etc.) are the most important parts of the epithelial barrier and are essential for cell–cell adhesion [13]. Our results showed that protein expression levels of ZO-1, occludin and claudin-1 were increased. In addition, immunofluorescence showed that the TJs we detected were distributed on the cell membranes after cecropin A treatment. ZOs (mainly ZO-1) are connected to the cytoskeleton (F-actin). On the other hand, ZO-1 also connected to the intracellular loops of claudins and occludin. Claudins and occludin are responsible for adjacent cell connections. The cytoskeleton-TJ structure may prevent pathogenic bacterial adherence or invasion into cells. The function of TJs depends on the protein expression level and membrane distribution [26]. In this study, we evaluated the expression and location of claudin-1, ZO-1, and occludin by using cell immunofluorescence. The results showed that the membrane distribution of claudin-1, ZO-1 and occludin was significantly increased by cecropin A stimulation, which suggests that cecropin A may regulate the barrier function through regulation of TJ expression and membrane expression.

To elucidate the mechanism of cecropin A regulating the intestinal monolayer barrier, the MEK/ERK signaling pathway was detected. The results showed that cecropin A may enhance the barrier function, regulate TJ expression and membrane distribution by downregulating the MEK/ERK pathway. The MEK/ERK pathway is conserved among eukaryotes, and one of the most important roles of MEK/ERK is to regulate cell proliferation and inhibit differentiation in epithelial cells and tumor cells [18,21]. In the MEK/ERK signaling pathway, ERK 1/2 may be activated by MEK, and then ERK 1/2 may regulate downstream transcriptional factors and widely regulate cell physiological processes, such as proliferation, differentiation, migration and apoptosis [18]. In addition, previous studies have also showed that MEK/ERK inhibition may induce upregulation of TJs in undifferentiated cells, such as embryonic stem cells, intestinal epithelial cells (IEC-6, caco-2) or tumor cells [27,28]. Similarly, ERK activation may induce blood–brain barrier injury [29]. The related physiological processes are regulated by the transcriptional factors downstream of ERK. CDX2, a caudal-related homeobox gene, is an essential regulator of gene transcription and tumor suppression in gastrointestinal tract development and homeostasis [30,31]. Previous studies have shown that CDX2 may play important roles in cell differentiation and proliferation and that it is regulated by the MEK/ERK signaling pathway [27]. In addition, CDX2 also plays an important role in TJ regulation in the intestinal epithelium. Previously, studies showed that in rat intestinal epithelium-derived line, IEC-6, caco-2 and colorectal carcinoma cells, downregulation of CDX2 by the MEK/ERK signaling pathway may decrease the protein expression levels of claudin-1, occludin and ZO-1 [27,32,33]. Consistent with this, our results showed that cecropin A may downregulate MEK and ERK phosphorylation, upregulate CDX2 expression, and upregulate protein levels of ZO-1, claudin-1 and occludin. Interestingly, previously studies showed that LL-37 and human beta defensin-3 (hBD-3) may activate phosphatidylinositide 3 kinases (PI3K)-Protein kinase B (Akt), PKC (protein kinase C) and Glycogen synthase kinase 3β (GSK-3β) and upregulate TJ expression and membrane distribution in human skin cells [26,34], which may be involved in the cell adherence and junction remodeling pathway, suggesting that more than one pathway exists to regulate TJ expression. Overall, in this study, we found that cecropin A enhanced the barrier function in the IPEC-J2 cell monolayer model by upregulating the TJ protein level (ZO-1, claudin-1 and occludin) and membrane polymerization, which was negatively regulated by the MEK/ERK signaling pathway. Our results suggested that cecropin A has the potential to replace antibiotics in the treatment of IBD due to its antibacterial activity on gram-negative bacteria and enhancement effect on intestinal barrier function.

4. Materials and Methods

4.1. Bacterial Strains

The *Escherichia coli* ATCC 35401, ATCC 35150, ATCC 25922, and SSI 82000, *Pseudomonas aeruginosa* ATCC 9027 and ATCC 27853, *Salmonella* ATCC13312 and ATCC9120, *Salmonella typhimurium* ATCC 14028, and *Staphylococcus aureus* ATCC 29213 strains were purchased from Guangdong Culture Collection Center. The *Escherichia coli* W25K strain was isolated from a piglet with diarrhea as described previously [35]. The strains were cultured in LB medium to logarithmic growth period (OD = 0.5) and then were transferred to Mueller-Hinton Broth (MHB) medium for the minimum inhibitory concentration (MIC) or minimum bactericidal concentration (MBC) test.

4.2. Peptide Synthesis

The 7 AMPs were synthesized and purified by a Chinese peptide company (DgPeptides Co., Ltd., Hangzhou, China), and the sequences were confirmed via matrix-assisted laser desorption/ionization time-of-flight mass spectrometry (MALDI-TOF MS). The purity of the AMPs was higher than 95%, which was measured by reversed-phase high-performance liquid chromatography.

4.3. Cell Culture

IPEC-J2 cells were cultured at 37 °C in 100% humidity and 5% CO_2 conditions with DMEM/F12 (Thermo, Waltham, MA, USA) supplemented with 5% fetal bovine serum (Gibco, Waltham, MA, USA) and 1% penicillin/streptomycin. The medium was changed every other day.

4.4. MIC/MBC Test

The MIC test was assessed according to the process of the Clinical and Laboratory Standards Institute. The bacterial strains were diluted with MHB medium, and the concentration was 5×10^5 CFU/mL. Then, 180 μL bacterial suspension was transferred to 96-well cell plates. The AMPs were diluted with phosphate-buffered solution (PBS), then 20 μL AMPs was added to the bacterial suspension, and the final concentrations were 200, 100, 50, 25, 12.5, 6.25, 3.125, and 1.56 μg/mL. Then, the bacteria were cultured at 37 °C for 8 h.

4.5. Cell Vitality Assay

Thiazolyl blue tetrazolium bromide (MTT) was purchased from Sigma-Aldrich (St. Louis, MI, USA). The MTT was dissolved in PBS (5 mg/mL). The cytotoxicity of AMPs to IPEC-J2 cells was tested by using an MTT assay. A density of 1×10^5 cells in 180 μL medium per well was seeded in 96-well plates (Corning, New York, NY, USA), and then 20 μL AMPs was added. The final concentrations were 200, 100, 50, 25, 12.5, 6.25, 3.125, were 1.56 μg/mL. After 8 h culture, 20 μL MTT was added, and then the cells were continued to culture for 4 h. The supernatant was discarded, and 150 μL DMSO was added. After 30 min shaking at room temperature, the absorbance was measured at a wavelength of 490 nm. The cell viability was calculated by using the following equation:

$$\text{Cell viability} = (OD_{\text{Control}} - OD_{\text{AMP}})/OD_{\text{Control}} \times 100\%$$

4.6. Quantifying Adhe7rent Bacteria

To calculate the number of adherent *E. coli*, *E. coli* were cocultured with IPEC-J2 cells in DMEM/high-glucose medium (no penicillin/streptomycin, no FBS). Then, the cells were washed with PBS six times and lysed with 1% Triton X-100 for 20 min at room temperature. Next, 5 μL lysates were plated on MacConkey agar plates overnight. The total number of bacteria was quantified as CFUs.

4.7. TER and Permeability Measurement

To evaluate the barrier integrity of IPEC-J2 cells, the transepithelial resistance (TER) and was measured, and a permeability assay was performed. The initial TER was tested before the cells were seeded. IPEC-J2 cells were then seeded in a Transwell membrane insert (12 mm diameter, 0.4 μm pore size, Corning) at a density of 7×10^5 cells/well. Then, 200 μL and 500 μL medium was added to the apical and basal compartments, respectively. Cecropin A (12.5 μg/mL) was added to the apical and basal compartments. The TER values were measured every day by using an ohm-meter fitted with chopstick electrodes (Millipore ESR-2; Burlington, MA, USA). Before each test, the plates were placed at room temperature for 30 min. The TER was calculated by using the following equation:

$$\text{TER } (\Omega \cdot cm^2) = (\text{TER} - \text{TER}_{\text{initial}}) \times 0.3$$

To evaluate the permeability of the monolayer intestinal cells, FITC-dextran was used. FITC-dextran (Sigma-Aldrich, St. Louis, MO, USA) was dissolved in PBS at 5 mg/mL. After the medium was discarded, 200 μL FITC-dextran was added to the apical compartment, and 500 μL PBS was added to the basal compartment and cultured for 2 h. Then, 100 μL liquid from each well was transferred to 96-well plates, and the absorbance was read at 480 nm excitation and 520 nm emission wavelengths. Then, the content of the FITC-dextran in the basal compartment was calculated using a standard curve.

4.8. qPCR

Total RNA was extracted using TRIzol reagent according to the manufacturer's instructions. Concentration and purity of RNA was checked by using a NanoDrop 2000. cDNA was generated from 1 μg total RNA using a First Strand cDNA Synthesis Kit (Thermo Scientific, Waltham, MA, USA). Quantitative PCR (qPCR) was performed to quantify mRNA expression levels of IL-6, IL-8 and TNF-α relative to that of a housekeeping gene, glyceraldehyde-3-phosphate dehydrogenase (GAPDH) by using SYBR Green mix (ABI) according to the manufacturer's instructions. The forward and reverse primers are shown in Table S3.

4.9. Western Blotting

The total protein was extracted with lysis buffer. The concentration of protein was tested by using a BCA protein assay kit (Thermo Scientific, Waltham, MA, USA) and mixed with 5× loading buffer. 20 μg protein sample was loaded in each well. The supernatant was then separated by 10% SDS-PAGE and transferred onto a PVDF membrane (Millipore, Burlington, MA, USA). After blocking with 5% skimmed milk powder, the membrane was incubated with the appropriate primary antibodies overnight at 4 °C, followed by incubation with a horseradish peroxidase (HRP)-conjugated secondary antibody for 2 h. Bands were detected using an ECL Western Blotting Substrate (Thermo Scientific, Waltham, MA, USA). Band intensity was quantified using ImageJ software. The primary antibodies including β-actin, p-MEk, MEK, p-ERK, ERK, CDX2 were purchased from Cell Signaling Technology (CST, Danvers, MA, USA). To specifically inhibit the phosphorylation of ERK, PD184352 (CST) was dissolved in DMSO and then used at a concentration of 10 μM for 48 h.

4.10. Cell Immunofluorescence Aassay

To test the expression and location of tight junction proteins (ZO-1, occludin and claudin-1) [31] and the cytoskeleton (F-actin), cell immunofluorescence was used. The IPEC-J2 cells were cultured and became confluent on the slide. After treatment, the cells were washed with PBS three times, fixed with 4% polyoxymethylene for 30 min, and then washed by PBS again three times. For F-actin staining, 0.5% Triton X-100 was added for 20 min. After that, 0.5% bovine serum albumin was added for 1 h, and then the primary antibodies were added and incubated with the cells at 4 °C overnight. The slides were then washed with PBST three times, and the FITC-labeled secondary antibody was added for 2 h. At last, the cell nuclei were stained by using DAPI (Santa Cruz Biotechnology, Dallas, TX, USA). The slides were then observed by using a fluorescence microscope. Primary antibodies including occludin (Abcam, Cambridge, UK), claudin-1 (CST, Danvers, MA, USA) and ZO-1 (Thermo Fisher Scientific, Waltham, MA, USA) were used. For staining F-actin, FITC-phalloidin (Sigma-Aldrich, St. Louis, MI, USA) was used.

4.11. Statistics

Data are expressed as the mean ± SEM. The Student's *t*-test was conducted to determine the differences between 2 groups using SAS (version 9.2, SAS Institute Inc., Cary, NC, USA), and a one-way ANOVA was used to determine differences among groups. Differences were considered statistically significant when $p < 0.05$.

Int. J. Mol. Sci. **2018**, *19*, 1941

Supplementary Materials: Supplementary materials can be found at http://www.mdpi.com/1422-0067/19/7/1941/s1.

Author Contributions: Study concept and design: Z.Z.; data acquisition: Z.Z., X.N., J.L. and C.J.; W.R. provided the bacteria strain for this study; Z.Z., J.D., B.D. and Y.Y. wrote and modify the manuscript.

Funding: The authors gratefully thank the Nation K&D Program of China (2018YFD0500603), the National Natural Science Foundation of China (Grant Nos. 31790411) and the Natural Science Foundation of Guangdong Province (Grant Nos. 2017A030310410 and 2017A030310398) for supporting of the projects. The studies meet with the approval of the university's review board.

Conflicts of Interest: The authors declare no conflicts of interests.

Abbreviations

IBD	Inflammatory bowel disease
AMP	Antimicrobial peptides
TJ	Tight junction
E. coli	*Escherichia coli*
IPEC-J2	Porcine jejunum epithelial cells
TER	Trans-epithelial electrical resistance
MEK	Mitogen-activated protein kinase kinase
ERK	Extracellular signal-regulated kinase
AA	Amino acid
JAMs	Junctional adhesion molecules
ZO	Zonula occludens
MAPK	Mitogen-activated protein kinase

References

1. Sheehan, D.; Shanahan, F. The Gut Microbiota in Inflammatory Bowel Disease. *Gastroenterol. Clin.* **2017**, *46*, 143. [CrossRef] [PubMed]

2. Perez, F.; Adachi, J.; Bonomo, R.A. Antibiotic-Resistant Gram-Negative Bacterial Infections in Patients with Cancer. *Clin. Infect. Dis.* **2014**, *59*, S335–S339. [CrossRef] [PubMed]

3. Wang, G.S.; Li, X.; Wang, Z. APD3: The antimicrobial peptide database as a tool for research and education. *Nucleic Acids Res.* **2016**, *44*, D1087–D1093. [CrossRef] [PubMed]

4. Choi, H.; Rangarajan, N.; Weisshaar, J.C. Lights, Camera, Action! Antimicrobial Peptide Mechanisms Imaged in Space and Time. *Trends Microbiol.* **2016**, *24*, 111–122. [CrossRef] [PubMed]

5. Liu, H.; Duan, Z.L.; Tang, J.; Lv, Q.M.; Rong, M.Q.; Lai, R. A short peptide from frog skin accelerates diabetic wound healing. *FEBS J.* **2014**, *281*, 4633–4643. [CrossRef] [PubMed]

6. Zhang, H.W.; Xia, X.; Han, F.F.; Jiang, Q.; Rong, Y.L.; Song, D.G.; Wang, Y.Z. Cathelicidin-BF, a Novel Antimicrobial Peptide from *Bungarus fasciatus*, Attenuates Disease in a Dextran Sulfate Sodium Model of Colitis. *Mol. Pharm.* **2015**, *12*, 1648–1661. [CrossRef] [PubMed]

7. Steiner, H.; Hultmark, D.; Engström, Å.; Bennich, H.; Boman, H.G. Sequence and specificity of two antibacterial proteins involved in insect immunity. *Nature* **1981**, *292*, 246–248. [CrossRef] [PubMed]

8. Rangarajan, N.; Bakshi, S.; Weisshaar, J.C. Localized Permeabilization of *E. coli* Membranes by the Antimicrobial Peptide Cecropin A. *Biochemistry* **2013**, *52*, 6584–6594. [CrossRef] [PubMed]

9. Fu, H.M.; Bjorstad, A.; Dahlgren, C.; Bylund, J. A bactericidal cecropin-A peptide with a stabilized alpha-helical structure possess an increased killing capacity but no proinflammatory activity. *Inflammation* **2004**, *28*, 337–343. [CrossRef] [PubMed]

10. Ryu, S.; Acharya, S.; Gurley, C.; Park, Y.; Armstrong, C.A.; Song, P.I. Antimicrobial and anti-inflammatory effects of newly designed synthetic Cecropin A (1–8)-Magainin 2 (1–12) hybrid peptide CA-MA analogue P5 against Malassezia furfur. *J. Investig. Dermatol.* **2010**, *130*, S123–S123.

11. Van der Flier, L.G.; Clevers, H. Stem Cells, Self-Renewal, and Differentiation in the Intestinal Epithelium. *Annu. Rev. Physiol.* **2009**, *71*, 241–260. [CrossRef] [PubMed]

12. Miner-Williams, W.M.; Moughan, P.J. Intestinal barrier dysfunction: Implications for chronic inflammatory conditions of the bowel. *Nutr. Res. Rev.* **2016**, *29*, 40–59. [CrossRef] [PubMed]

13. Turner, J.R. Intestinal mucosal barrier function in health and disease. *Nat. Rev. Immunol.* **2009**, *9*, 799–809. [CrossRef] [PubMed]

14. Shen, L.; Weber, C.R.; Raleigh, D.R.; Yu, D.; Tumer, J.R. Tight, Junction Pore and Leak Pathways: A Dynamic Duo. *Annu. Rev. Physiol.* **2011**, *73*, 283–309. [CrossRef] [PubMed]

15. Yang, F.J.; Wang, A.N.; Zeng, X.F.; Hou, C.L.; Liu, H.; Qiao, S.Y. Lactobacillus reuteri I5007 modulates tight junction protein expression in IPEC-J2 cells with LPS stimulation and in newborn piglets under normal conditions. *BMC Microbiol.* **2015**, *15*, 32. [CrossRef] [PubMed]

16. Gunzel, D.; Yu, A.S.L. Claudins and the Modulation of Tight Junction Permeability. *Physiol. Rev.* **2013**, *93*, 525–569. [CrossRef] [PubMed]

17. Odenwald, M.A.; Choi, W.; Buckley, A.; Shashikanth, N.; Joseph, N.E.; Wang, Y.; Warren, M.H.; Buschmann, M.M.; Pavlyuk, R.; Hildebrand, J.; et al. ZO-1 interactions with F-actin and occludin direct epithelial polarization and single lumen specification in 3D culture. *J. Cell Sci.* **2017**, *130*, 243–259. [CrossRef] [PubMed]

18. Osaki, L.H.; Gama, P. MAPKs and Signal Transduction in the Control of Gastrointestinal Epithelial Cell Proliferation and Differentiation. *Int. J. Mol. Sci.* **2013**, *14*, 10143–10161. [CrossRef] [PubMed]

19. Piegholdt, S.; Pallauf, K.; Esatbeyoglu, T.; Speck, N.; Reiss, K.; Ruddigkeit, L.; Stocker, A.; Huebbe, P.; Rimbach, G. Biochanin A and prunetin improve epithelial barrier function in intestinal CaCo-2 cells via downregulation of ERK, NF-kappa B, and tyrosine phosphorylation. *Free Radic. Biol. Med.* **2014**, *70*, 255–264. [CrossRef] [PubMed]

20. Wang, K.; Jin, X.L.; Chen, Y.F.; Song, Z.H.; Jiang, X.S.; Hu, F.L.; Conlon, M.A.; Topping, D.L. Polyphenol-Rich Propolis Extracts Strengthen Intestinal Barrier Function by Activating AMPK and ERK Signaling. *Nutrients* **2016**, *8*, 272. [CrossRef] [PubMed]

21. Kubicek-Sutherland, J.Z.; Lofton, H.; Vestergaard, M.; Hjort, K.; Ingmer, H.; Andersson, D.I. Antimicrobial peptide exposure selects for *Staphylococcus aureus* resistance to human defence peptides. *J. Antimicrob. Chemother.* **2017**, *72*, 115–127. [CrossRef] [PubMed]

22. Joo, H.S.; Fu, C.I.; Otto, M. Bacterial strategies of resistance to antimicrobial peptides. *Philos. Trans. R. Soc. B* **2016**, *371*, 20150292. [CrossRef] [PubMed]

23. Rai, A.; Pinto, S.; Evangelista, M.B.; Gil, H.; Kallip, S.; Ferreira, M.G.S.; Ferreira, L. High-density antimicrobial peptide coating with broad activity and low cytotoxicity against human cells. *Acta Biomater.* **2016**, *33*, 64–77. [CrossRef] [PubMed]

24. Kirschner, N.; Brandner, J.M. Barriers and more: Functions of tight junction proteins in the skin. *Ann. N. Y. Acad. Sci.* **2012**, *1257*, 158–166. [CrossRef] [PubMed]

25. Liu, S.F.; Yang, W.; Shen, L.; Turner, J.R.; Coyne, C.B.; Wang, T.Y. Tight Junction Proteins Claudin-1 and Occludin Control Hepatitis C Virus Entry and Are Downregulated during Infection to Prevent Superinfection. *J. Virol.* **2009**, *83*, 2011–2014. [CrossRef] [PubMed]

26. Akiyama, T.; Niyonsaba, F.; Kiatsurayanon, C.; Nguyen, T.T.; Ushio, H.; Fujimura, T.; Ueno, T.; Okumura, K.; Ogawa, H.; Ikeda, S. The Human Cathelicidin LL-37 Host Defense Peptide Upregulates Tight Junction-Related Proteins and Increases Human Epidermal Keratinocyte Barrier Function. *J. Innate Immun.* **2014**, *6*, 739–753. [CrossRef] [PubMed]

27. Krueger, F.; Madeja, Z.; Hemberger, M.; McMahon, M.; Cook, S.J.; Gaunt, S.J. Down-regulation of Cdx2 in colorectal carcinoma cells by the Raf-MEK-ERK 1/2 pathway. *Cell. Signal.* **2009**, *21*, 1846–1856. [CrossRef] [PubMed]

28. Li, J.; Wang, G.W.; Wang, C.Y.; Zhao, Y.; Zhang, H.; Tan, Z.J.; Song, Z.H.; Ding, M.X.; Deng, H.K. MEK/ERK signaling contributes to the maintenance of human embryonic stem cell self-renewal. *Differentiation* **2007**, *75*, 299–307. [CrossRef] [PubMed]

29. Wang, L.F.; Li, X.; Gao, Y.B.; Wang, S.M.; Zhao, L.; Dong, J.; Yao, B.W.; Xu, X.P.; Chang, G.M.; Zhou, H.M.; et al. Activation of VEGF/Flk-1-ERK Pathway Induced Blood-Brain Barrier Injury after Microwave Exposure. *Mol. Neurobiol.* **2015**, *52*, 478–491. [CrossRef] [PubMed]

30. Guo, R.J.; Suh, E.R.; Lynch, J.P. The role of Cdx proteins in intestinal development and cancer. *Cancer Biol. Ther.* **2004**, *3*, 593–601. [CrossRef] [PubMed]

31. Coskun, M.; Troelsen, J.T.; Nielsen, O.H. The role of CDX2 in intestinal homeostasis and inflammation. *BBA Mol. Basis Dis.* **2011**, *1812*, 283–289. [CrossRef] [PubMed]

32. Lemieux, E.; Boucher, M.J.; Mongrain, S.; Boudreau, F.; Asselin, C.; Rivard, N. Constitutive activation of the MEK/ERK pathway inhibits intestinal epithelial cell differentiation. *Am. J. Physiol. Gastrointest. Liv. Physiol.* **2011**, *301*, G719–G730. [CrossRef] [PubMed]
33. Suzuki, T.; Yoshinaga, N.; Tanabe, S. Interleukin-6 (IL-6) Regulates Claudin-2 Expression and Tight Junction Permeability in Intestinal Epithelium. *J. Biol. Chem.* **2011**, *286*, 31263–31271. [CrossRef] [PubMed]
34. Kiatsurayanon, C.; Niyonsaba, F.; Smithrithee, R.; Akiyama, T.; Ushio, H.; Hara, M.; Okumura, K.; Ikeda, S.; Ogawa, H. Host Defense (Antimicrobial) Peptide, Human beta-Defensin-3, Improves the Function of the Epithelial Tight-Junction Barrier in Human Keratinocytes. *J. Investig. Dermatol.* **2014**, *134*, 2163–2173. [CrossRef] [PubMed]
35. Ren, W.K.; Yin, J.; Xiao, H.; Chen, S.; Liu, G.; Tan, B.; Li, N.Z.; Peng, Y.Y.; Li, T.J.; Zeng, B.H.; et al. Intestinal Microbiota-Derived GABA Mediates Interleukin-17 Expression during Enterotoxigenic *Escherichia coli* Infection. *Front. Immunol.* **2017**, *7*, 685. [CrossRef] [PubMed]

 International Journal of
Molecular Sciences

Article

Anti-Inflammatory and Gastroprotective Roles of *Rabdosia inflexa* through Downregulation of Pro-Inflammatory Cytokines and MAPK/NF-κB Signaling Pathways

Md Rashedunnabi Akanda [1,2], In-Shik Kim [1], Dongchoon Ahn [1], Hyun-Jin Tae [1], Hyeon-Hwa Nam [3], Byung-Kil Choo [3], Kyunghwa Kim [4] and Byung-Yong Park [1,*]

[1] College of Veterinary Medicine and Bio-safety Research Institute, Chonbuk National University, Iksan 54596, Korea; rashed.mvd@gmail.com (M.R.A.); iskim@jbnu.ac.kr (I.-S.K.); ahndc@jbnu.ac.kr (D.A.); hjtae@jbnu.ac.kr (H.-J.T.)
[2] Department of Pharmacology and Toxicology, Sylhet Agricultural University, Sylhet 3100, Bangladesh
[3] Department of Crop Science and Biotechnology, Chonbuk National University, Jeonju 54896, Korea; hh_hh@jbnu.ac.kr (H.-H.N.); bkchoo@jbnu.ac.kr (B.-K.C.)
[4] Department of Cardiothoracic Surgery, Research Institute of Clinical Medicine, Chonbuk National University, Jeonju 54907, Korea; tcskim@jbnu.ac.kr
* Correspondence: parkb@jbnu.ac.kr; Tel.: +82-63-850-0961

Received: 23 January 2018; Accepted: 13 February 2018; Published: 14 February 2018

Abstract: Globally, gastric ulcer is a vital health hazard for a human. *Rabdosia inflexa* (RI) has been used in traditional medicine for inflammatory diseases. The present study aimed to investigate the protective effect and related molecular mechanism of RI using lipopolysaccharide (LPS)-induced inflammation in RAW 246.7 cells and HCl/EtOH-induced gastric ulcer in mice. We applied 3-(4,5-dimethyl-thiazol-2-yl)-2,5-diphenyltetrazolium bromide (MTT), nitric oxide (NO), reactive oxygen species (ROS), histopathology, malondialdehyde (MDA), quantitative real-time polymerase chain reaction (qPCR), immunohistochemistry (IHC), and Western blot analyses to evaluate the protective role of RI. Study revealed that RI effectively attenuated LPS-promoted NO and ROS production in RAW 246.7 cells. In addition, RI mitigated gastric oxidative stress by inhibiting lipid peroxidation, elevating NO, and decreasing gastric inflammation. RI significantly halted elevated gene expression of pro-inflammatory cytokines such as tumor necrosis factor-α (*TNF-α*), interleukin-1β (*IL-1β*), interleukin-6 (*IL-6*), inducible nitric oxide synthetase (*iNOS*), and cyclooxygenase-2 (*COX-2*) in gastric tissue. Likewise, RI markedly attenuated the mitogen-activated protein kinases (MAPKs) phosphorylation, COX-2 expression, phosphorylation and degradation of inhibitor kappa B (IκBα) and activation of nuclear factor kappa B (NF-κB). Thus, experimental findings suggested that the anti-inflammatory and gastroprotective activities of RI might contribute to regulating pro-inflammatory cytokines and MAPK/NF-κB signaling pathways.

Keywords: *Rabdosia inflexa*; inflammation; gastric ulcer; cytokines; MAPK; NF-κB

---

1. Introduction

Alcohol consumption is a recognized risk factor for human health. The most common diseases include infectious diseases, gastric ulcer, cancer, diabetes, and liver and pancreas disease caused by alcohol consumption either partially or entirely [1]. The pathogenesis of gastric ulcer is complicated and multifactorial; it is usually caused by an acute imbalance between gastric mucosal integrity and mucosal immunity [2]. Usually, ethanol is absorbed through the intestinal wall and metabolized in the liver in different ways: oxidation by alcohol dehydrogenase (ADH), cytochrome P450 2E1

(CYP2E1), and catalase enzymes. All the processes intensify to form acetaldehyde and then acetate by aldehyde dehydrogenase (ALDH). Alcohol metabolism with ADH enhances the generation of reduced forms of nicotinamide adenine dinucleotide (NADH), but production of CYP2E1 continues to produce free radical. Acetaldehyde and free radicals combine with cell compounds and disturb cell physiology [3]. Consequently, oxidative stress plays a crucial role in the pathogenesis of alcoholic tissue damage and increases lipid peroxidation, which injures capillary endothelial cells and increases cellular permeability [4] that are involved in the DNA damage of gastric mucosal epithelial cells [5]. Although the complete mechanism of alcohol-induced gastric mucosal damage has not been fully disclosed, evidence shows that oxidative stress and neutrophil infiltration are associated with the development of acute gastritis [6,7].

Lipopolysaccharide (LPS), a bacterial endotoxin, is commonly used as an inducer of the macrophage cell lineage, acting through Toll-like receptor 4 (TLR4), which activates the mitogen-activated protein kinases (MAPKs) signaling cascades and the pathway that triggers nuclear factor kappa B (NF-κB) [8,9]. MAPKs are the important signaling pathway and play a crucial regulatory role in both adapted and innate immune response [10]. Ethanol-induced oxidative stress stimulates the release of reactive oxygen species (ROS). ROS are recognized as the second messenger to initiate the redox-sensitive signal-transduction pathway with MAPK cascade and are linked with downstream transcription factor: NF-κB [11]. ROS mediate stimulation of inhibitor kappa B (IκB) kinase, which induces proteasomal breakdown of IκBα and activates NF-κB. NF-κB is a transcription factor that binds to κ-β motifs in the promoters of target genes and triggers transcription of inflammatory cytokines and chemokines [12].

The therapeutic and biological activities of indigenous plants and their active compounds have potential importance for their capability to manage and treat many inflammatory and immunomodulatory diseases [13]. *Rabdosia inflexa* (RI), a perennial shrub, is a member of the lamiaceae family, which is cosmopolitan and cultivated throughout Northeast China, the Korean peninsula, and Japan. In South Korea, RI, locally known as "sanbakha", has been used as folk medicine for treating gastrointestinal inflammation and pain. Previously, RI and its active compounds such as inflexin and inflexinol have been reported for pancreatitis and anti-cancer effect [14–16]. Based on its traditional uses and biological activities, the study investigated its anti-inflammatory and gastroprotective activity and its possible molecular mechanisms in both RAW 264.7 cells and HCl/EtOH-induced gastric ulcer in mice.

2. Results

2.1. Analysis of Total Phenolic and Flavonoid Contents of Rabdosia inflexa (RI)

Phenolic and flavonoid contents are the secondary metabolites of a plant, having a wide range of biological activities and usually antioxidant properties. The total phenolic and flavonoid content of RI were investigated and presented in Table 1. The total phenolic and flavonoid content of RI were 143.288 ± 1.68 mg/g gallic acid and 256.301 ± 1.40 mg/g rutin equivalent, respectively.

Table 1. Total phenolic and flavonoid content of *Rabdosia inflexa* (RI).

Plant Extract	Total Phenolic (mg GAE/g Extract)	Total Flavonoid (mg RU/g Extract)	Total Yield (%)
RI	143.288 ± 1.68	256.301 ± 1.40	27.13

Note: Gallic acid and rutin were used as standards. Results are expressed in milligrams of gallic acid equivalent per gram of extract sample (mg GAE/g) and mg of rutin equivalent per gram of extract sample (mg RU/g).

2.2. Effect of RI on Viability and Morphology of RAW 264.7 Cells

The present study measured the anti-inflammatory ability of RI extract using RAW 264.7 cells on LPS-induced inflammation using MTT assay. To investigate the cytotoxicity and cell viability of RI,

RAW 264.7 cells were treated with different concentrations of RI (50, 100, 200, 400, and 800 µg/mL) for 24 h. Among the concentrations, RI (800 µg/mL) significantly reduced the cell viability (Figure 1a). However, the cell viability did not significantly alter after co-treatment with LPS (0.5 µg/mL) and RI (100, 200, and 400 µg/mL) for 24 h (Figure 1b). As shown in Figure 1c, LPS markedly induced morphological changes of RAW 264.7 cells after 24 h of treatment, which was consequently improved by the treatment with RI. Thus, results proposed that RI has not affected the viability and morphology of RAW 264.7 cells and it could be due to the anti-inflammatory effect of RI.

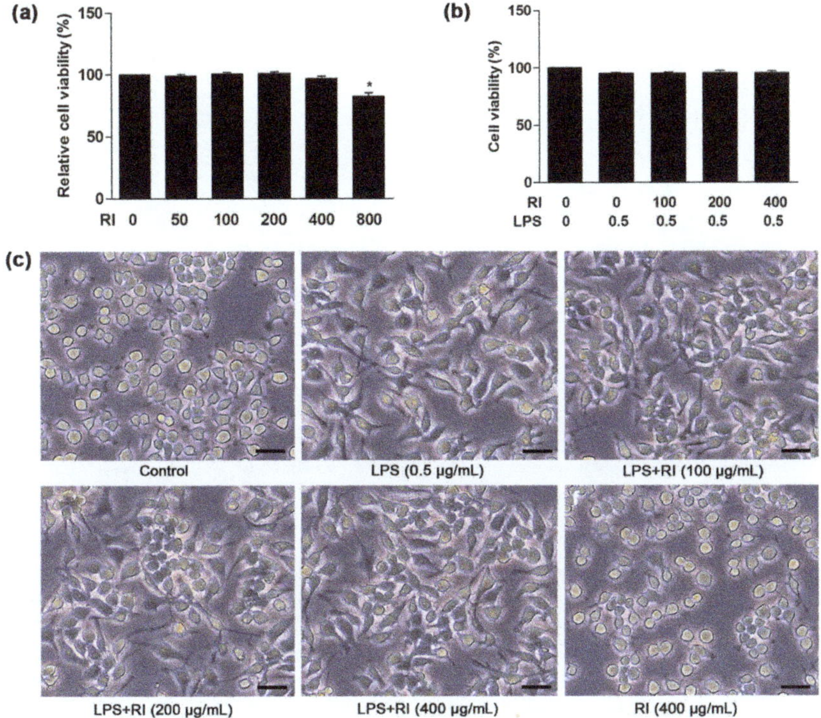

Figure 1. Protective role of *Rabdosia inflexa* (RI) on: (**a**) cytotoxicity; (**b**) cell viability; and (**c**) morphological alteration in RAW 264.7 cells were measured by MTT assay. Cells were pretreated with various concentration of RI (100, 200, and 400 µg/mL) for 1 h, followed by co-treatment with RI and LPS (0.5 µg/mL) for another 24 h. Cell morphology was visualized by optical microscopy (scale bar 200 µm). * $p < 0.05$ when compared with the control. Data are expressed as mean $\pm$ SEM of three independent experiments.

2.3. RI Attenuated the LPS-Induced NO and ROS Production in RAW 264.7 Cells

In LPS-treated cells, there was a marked increase ($p < 0.05$) in NO and ROS production as compared to the control. Conversely, co-treatment with the RI significantly reduced ($p < 0.05$) the NO and ROS production in a dose-dependent manner (Figure 2a,b). Together, RI suppressed the LPS-induced inflammatory response by preventing NO and intracellular ROS production in RAW 264.7 cells.

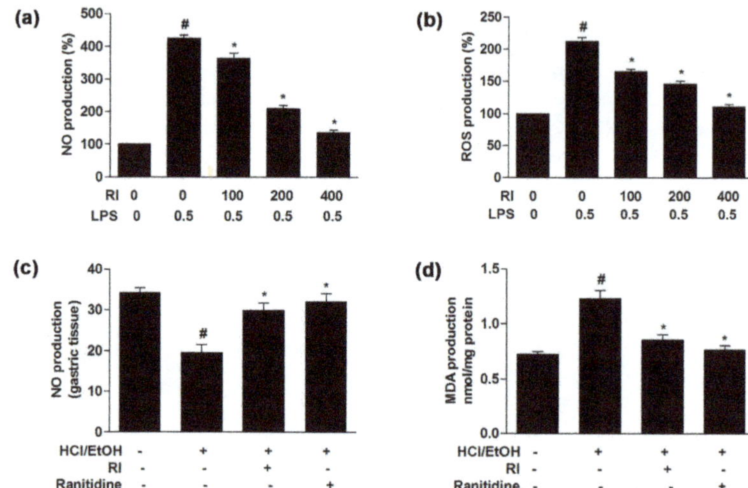

Figure 2. Protective role of RI on NO, intracellular ROS and MDA production in RAW 264.7 cells and gastric tissue. (**a**) NO; and (**b**) ROS production was measured by Griess and ROS-Glo H_2O_2 assays in RAW 264.7 cells (upper panel). Cells were pretreated with various concentration of RI (100, 200, and 400 μg/mL) for 1 h, followed by co-treatment with RI and LPS (0.5 μg/mL) for another 24 h. (**c**) NO; and (**d**) MDA production in gastric tissue were measured by Griess and TBARS assays (lower panel). Mice were pretreated for 1 h with RI (400 mg/kg) and Ranitidine (40 mg/kg). After 1 h, HCl/EtOH (10 μL/g) was given orally. # $p < 0.05$ when compared with the control and * $p < 0.05$ when compared with LPS and HCl/EtOH. Data are expressed as mean ± SEM.

2.4. RI Improved the Gross and Histopathology of Gastric Tissue

The recent study investigated the gastroprotective effect of RI in HCl/EtOH-induced gastric ulcer in mice. HCl/EtOH-induced severe gastric damage, which was notably attenuated by RI pretreatment (Figure 3a, upper panel). In addition, the histological study confirmed that the stomach had the normal structure of mucosa in control group. Besides, in the HCl/EtOH-treated mice, epithelial destruction and inflammatory cells infiltration were found in the mucosa and submucosal area. However, RI and ranitidine-treated groups markedly improved the histopathological changes as compared to HCl/EtOH-treated group (Figure 3a, lower panel). These data are well correlated with the protective abilities of RI against gastric ulcer. Likewise, the gross and histological lesions index of gastric tissue was significantly reduced ($p < 0.05$) by pretreatment with RI and ranitidine treated groups than in HCl/EtOH-induced gastric ulcer mice (Figure 3b,c).

2.5. RI Regulated the NO and MDA Production in Gastric Tissue

To evaluate the oxidative stress level, the NO and MDA production was measured in gastric tissue. After inducing gastric injury, HCl/EtOH significantly ($p < 0.05$) decreased NO and increased MDA production. Overall, pretreatment with the RI effectively ($p < 0.05$) increased and decreased the NO and MDA production as related to standard drug ranitidine, respectively (Figure 2c,d). Therefore, results evidently reveal that RI reduced the oxidative stress in HCl/EtOH-stimulated gastric ulcer for its strong anti-oxidant and anti-inflammatory capacity.

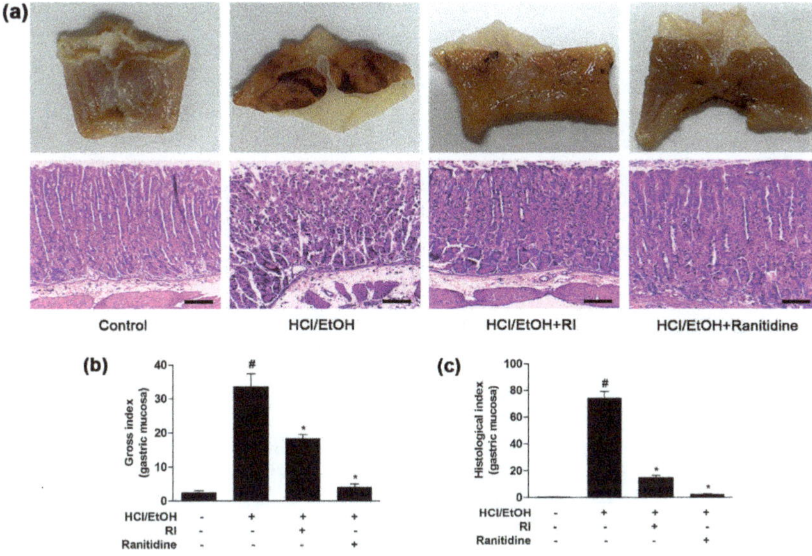

Figure 3. Protective role of RI on HCl/EtOH-induced gastric damage in mice: (**a**) gross lesion (upper panel) and histological lesion (lower panel) (scale bar. 200 µm); (**b**) gross lesion index; and (**c**) histological index. # $p < 0.05$ when compared with the control and * $p < 0.05$ when compared with HCl/EtOH. Data are expressed as mean ± SEM.

2.6. RI Suppressed the Activation Pro-Inflammatory Cytokines in Gastric Tissue

Pro-inflammatory cytokines play a fundamental role in various types of inflammation. To elucidate the protective role of RI, the gene expression of pro-inflammatory cytokines was examined in the glandular stomach samples by qPCR analysis. The gene expression level of *TNF-α*, *IL-1β*, *IL-6*, *iNOS*, and *COX-2* were gradually upregulated ($p < 0.05$) in the HCl/EtOH-treated group as compared to the control, whereas pretreatment with RI and ranitidine groups significantly ($p < 0.05$) downregulated the cytokines expression level than in the HCl/EtOH-treated group (Figure 4a–e). Thus, data suggest that RI inhibited the gene expression of pro-inflammatory cytokines in gastric tissue and thereby mitigated the gastric inflammation.

2.7. RI Inhibited the COX-2 Expression in Gastric Tissue

It is recognized that elevated expression of COX-2 plays a vital role in the inflammatory process and previous study has revealed that HCl/EtOH strongly activates COX-2 expression in gastric tissue [17]. COX-2 expression in the gastric mucosal epithelial cells was revealed by immunohistochemical staining analysis. As observed, COX-2 was slightly expressed in the normal control gastric mucosal epithelial cells; in contrast, HCl/EtOH increased the COX-2 expression of gastric mucosal epithelial cells, which was mostly observed in the gastric mucosal inflammatory area (Figure 5a). The expression of COX-2 was markedly ($p < 0.05$) blocked by the pretreatment of RI as related to the standard drug ranitidine (Figure 5b). Thus, RI significantly blocked the activation of COX-2 expression in the gastric mucosal inflammatory area and reduced the inflammatory activity.

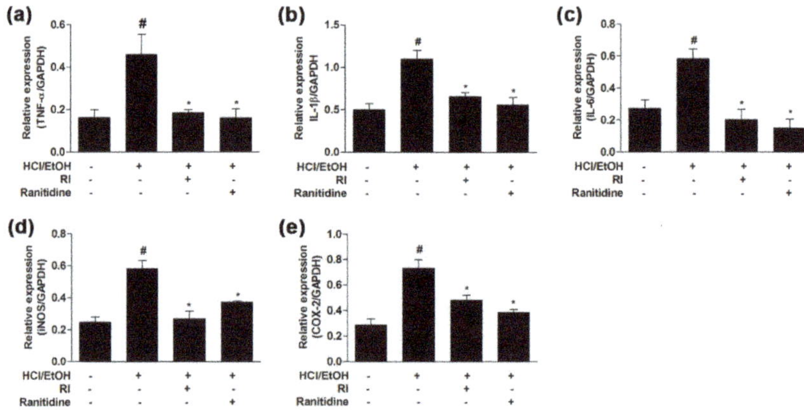

Figure 4. Protective role of RI on gene expression of pro-inflammatory cytokines in gastric tissue. In HCl/EtOH-treated mice, gene expression level of: (**a**) *TNF-α*; (**b**) *IL-1β*; (**c**) *IL-6*; (**d**) *iNOS*; and (**e**) *COX-2* were significantly upregulated, whereas pretreatment with the RI markedly downregulated the gene expression level as related to ranitidine. # $p < 0.05$ when compared with control and * $p < 0.05$ when compared with HCl/EtOH. Data are expressed as mean ± SEM.

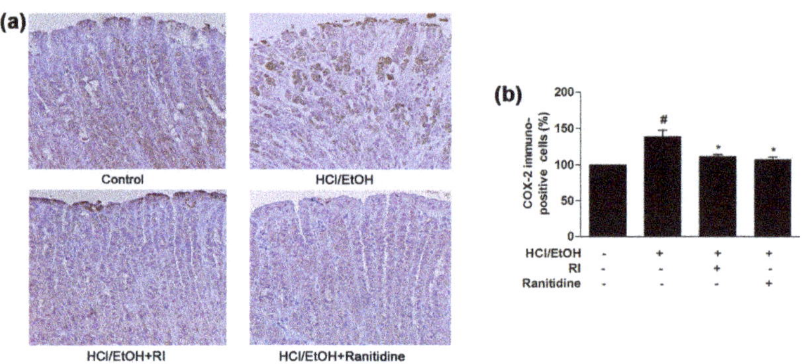

Figure 5. Protective role of RI on COX-2 immunoreactivity in the gastric tissue: (**a**) COX-2 expression in gastric mucosal epithelial cells; and (**b**) COX-2 positive immune-stained cells. Scale bar, 200 μm. # $p < 0.05$ when compared with the control and * $p < 0.05$ when compared with HCl/EtOH. Data are expressed as mean ± SEM.

2.8. RI Blocked the MAPK Cascade, COX-2, and NF-κB Activation

To find the possible molecular mechanisms of the anti-inflammatory and gastroprotective role of RI, the protein expression related to anti-inflammation signaling pathways was evaluated. The present data showed that LPS treatment remarkably elevated the phosphorylation of MAPK family protein (ERK1/2, JNK, and p38) in RAW 264.7 cells, whereas RI pretreatment notably ($p < 0.05$) attenuated the phosphorylation of MAPK proteins (Figure 6, upper panel). Meanwhile, LPS and HCl/EtOH treatment increased COX-2 expression in RAW 264.7 cells and gastric tissues were markedly ($p < 0.05$) blocked by RI pretreatment (Figure 6, middle panel). After LPS and HCl/EtOH stimulation, IκBα and NF-κB phosphorylation were noticeably ($p < 0.05$) increased, indicating the activation of NF-κB. However, IκBα phosphorylation and the nuclear translocation of NF-κB (p65) were gradually reduced ($p < 0.05$) by RI pretreatment (Figure 6, middle and lower panels). Moreover, RI alone does not seem to involve in the signal pathways in vitro study. Together, these results demonstrate that RI significantly inhibited the phosphorylation of MAPK cascade in RAW 264.7 cells as well as activation of COX-2, IκBα, and NF-κB in RAW 264.7 cells and gastric tissues, simultaneously.

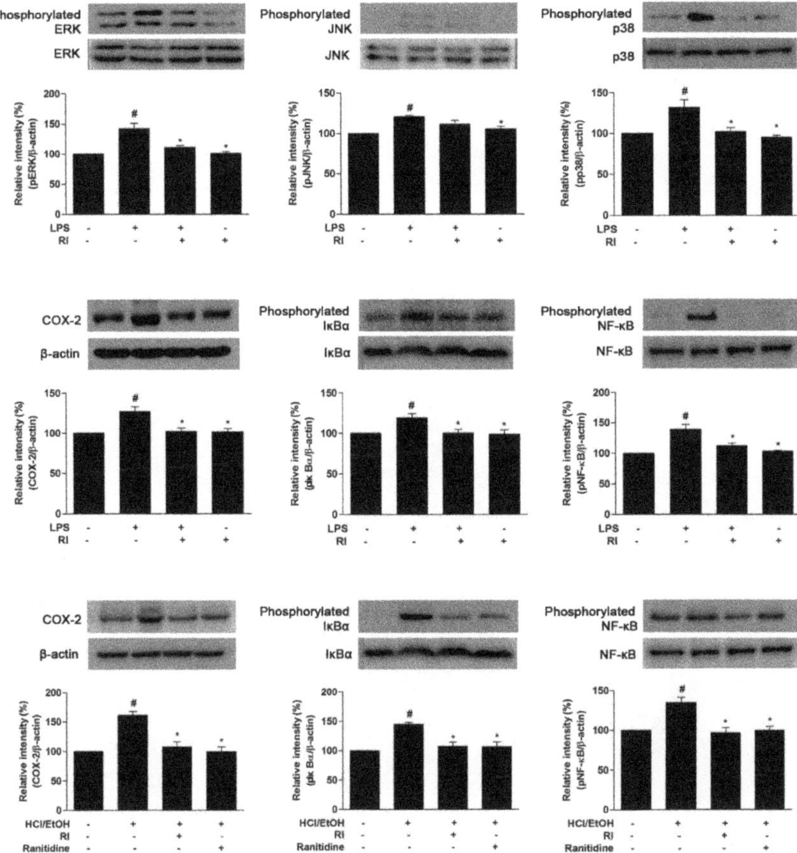

Figure 6. Protective role of RI on the MAPK cascades, COX-2 expression, and activation of IκBα, NF-κB in RAW 264.7 cells and gastric tissue. Here, upper and middle panels represent the MAPKs (pERK1/2, pJNK, and pp38), COX-2, IκBα and NF-κB expression in RAW 264.7 cells and the lower panel represents the COX-2, IκBα and NF-κB expression in the gastric tissue. The relative band intensity of target protein was measured as compared with total protein and β-actin. LPS-induced the phosphorylation of MAPK cascade, whereas pretreatment with the RI reduced the phosphorylation of MAPK cascade. LPS and HCl/EtOH increased the COX-2 expression, kinetic phosphorylation, and degradation of IκBα and phosphorylation of NF-κB. However, pretreatment with the RI notably decreased the COX-2 expression, IκBα phosphorylation, and degradation, NF-κB translocation as related to standard drug ranitidine. # $p < 0.05$ when compared with the control and * $p < 0.05$ when compared with LPS and HCl/EtOH. Data are expressed as mean ± SEM.

3. Discussion

Gastrointestinal disorders are a global health problem affecting millions of people. Inflammation is a defensive biological response to harmful stimuli and infection that promotes the production of inflammatory mediators. Oxidative stress plays a vital role in alterations related to the pathophysiology of inflammation. Modulation of the inflammatory mediators is considered a promising strategy for facing inflammatory disease. Although RI has been traditionally used to treat inflammatory diseases, the underlying molecular mechanism of anti-inflammation properties is still not understood. The study investigated the anti-inflammatory and gastroprotective activity of RI in a model of LPS-induced inflammation in RAW 264.7 cells and HCl/EtOH-induced experimental gastric ulcer.

Antioxidants play an important role in redox mechanisms in a biological system, protecting it against inflammation and apoptosis. Phenolic and flavonoid are the most important plant secondary metabolites and have the strong antioxidant capacity [18,19]. Antioxidants act as oxygen scavenger capable of catalyzing the oxidative process [20]. Significant amounts of phenolic and flavonoid content were found in RI (Table 2) that may be the major donor for the anti-oxidative as well as anti-inflammatory role against gastric damage. For the possible mechanism by which RI protects against LPS-induced macrophage cell damage, these results showed that it may act through its anti-oxidative and anti-inflammatory effects (Figure 1). Treatment with the RI inhibited the LPS-induced intracellular oxidative stress (ROS) and NO production (Figure 2, upper panel). These data strongly suggest that RI could cure various inflammatory symptoms based on its anti-inflammatory properties. Tissue-related macrophages play an important role in the loss of physiological functions of the organ by releasing toxic and inflammatory molecules, such as NO and ROS [21]. The synthesis of these inflammatory molecules is responsible for the progression of various inflammatory diseases, such as gastric ulcer [22,23].

Table 2. The nucleotide sequence of the primers for qPCR.

Gene	Primers Sequence (5′–3′)	Genebank Accession No.
TNF-α	TTGACCTCAGCGCTGAGTTG CCTGTAGCCCACGTCGTAGC	NM_013693
IL-1β	CAGGATGAGGACATGAGCACC CTCTGCAGACTCAAACTCCAC	XM_006498795
IL-6	GTACTCCAGAAGACCAGAGG TGCTGGTGACAACCACGGCC	NM_001314054
iNOS	CCCTTCCGAAGTTTCTGGCAGCAGC GGCTGTCAGAGCCTCGTGGCTTTGG	XM_006532446
COX-2	CACTACATCCTGACCCACTT ATGCTCCTGCTTGAGTATGT	NM_011198
GAPDH	CACTCACGGCAAATTCAACGGCAC GACTCCACGACATACTCAGCAC	XM_017321385

The pathogenesis of ethanol-induced gastric injury is very complex and related to oxidative stress that has been confirmed by recent studies [24,25]. Gastric tissue damage is caused by an imbalance between the damage of the gastric tissue and protective factors. In addition, NO plays a complex role in gastric mucosal integrity and its synthesized independently [26]. NO level declines in patients suffering from gastric distress considerably. By increasing gastric mucosal blood flow, normal production of NO could retain the integrity of gastric mucosa and contribute to the defense and healing of mucosal damage, while preventing chemotaxis and adhesion of inflammatory cells to guard gastric mucosa [27]. Following gastric damage, the gastric tissue may be partially oxidized due to injury. Lipid peroxidation is the result of ROS reaction against cell membranes and produces a significant level of pro-oxidant such as MDA, which leads to oxidative gastric damage [28]. In the present study, RI markedly increased NO and decreased MDA levels in gastric mucosa, demonstrating the anti-inflammatory and antioxidant potential of RI (Figure 2, lower panel). This finding is consistent with an earlier report [27]. In this study, RI attenuated the macroscopic and histopathologic lesions in gastric mucosa and inflammatory cells influx that signifies its prospective anti-gastric ulcer activities (Figure 3), as also shown for the standard compound ranitidine hydrochloride, a histamine-2 receptor antagonist clinically recommended for the treatment of gastric ulcer [29]. The ethanol-induced gastric lesion is a key experimental model commonly used for likely anti-gastric ulcer activity since ethanol is thought to be a leading cause of gastric ulcer [30,31]. Ethanol has been revealed to cause hemorrhagic gastric lesions characterized by mucosal friability and infiltration of inflammatory cells [32]. Reduction of the infiltration of inflammatory cells (neutrophil) has been considered to be a vital anti-inflammatory mechanism by which effective anti-gastric-ulcer medicine protects against mucosal injuries [33].

Pro-inflammatory cytokines and enzymes such as *TNF-α*, *IL-1β*, *IL-6*, *iNOS* and *COX-2* genes in the tissue may be used as biomarkers of gastric visceral damage. Following inflammatory stimuli, inflammatory mediators have been elevated to prompt deleterious effect in the stomach. In this study, increased production of inflammatory cytokines in ulcerated gastric tissue can be attributed to the damaging effect of ethanol. A high level of cytokines triggers neutrophils, lymphocytes, and monocytes at the inflammatory site; these, in turn, start a different oxidative disturbance, toxic metabolites, and lysosomal enzymes liable for local tissue damage in gastric ulcer [34]. Present study observed that RI remarkably suppressed the pro-inflammatory cytokine production in gastric tissue (Figure 4). The inhibition of NF-κB is a primary mechanism for RI suppression of gastric ulcer since the expression of various pro-inflammatory cytokines including *TNF-α* is mainly regulated by the transcription of NF-κB [35].

Inflammation is a physiological and immunological response triggered by both cell and tissue injury that is primarily controlled by a MAPK cascade signaling pathway [36]. MAPKs are kinases that are responsible for most cellular responses to inflammatory cytokines and external stress signals, and these kinases are essential for the regulation of the production of various inflammation mediators [37,38]. MAPK cascade pathway is activated by inflammatory stimuli such as LPS. MAPK comprising ERK, JNK, and p38, is controlled in response to the triggering of extracellular signal cytoskeletal proteins, nuclear transcription factors and the stabilization of cytokines gene. Experimental results revealed that LPS gradually upregulated the phosphorylation of MAPKs, and these phosphorylated MAPKs was notably inhibited by pretreatment with RI, resulting in the elevation of antioxidant response element of RI regulated phase II enzymes, which are involved in cellular protection mechanism (Figure 6, upper panel). Similarly, COX-2 is an important factor which plays key roles in the pathogenesis of inflammation. The protein expression of COX-2 significantly increased after LPS and HCl/EtOH treatment. However, the expression of COX-2 was decreased markedly by RI treatment (Figure 6, middle and lower panels). The immunohistochemical data showed that RI pretreatment markedly blocked the COX-2 localization in the gastric mucosal epithelium cells of the inflammatory area indicating that COX-2 was involved in the inhibition of inflammatory cells activation and mitigates the oxidative stress, and improve the healing process in the gastric mucosa (Figure 5).

It is known that pro-inflammatory cytokines are regulated by NF-κB signaling pathway [39]. The activation of NF-κB is induced by ROS and NO produced from macrophages following exposure of LPS and it depends on the phosphorylation and degradation of the corresponding upstream factor IκBα. Similarly, gastric damages lead to the production of free radicals that prompt the migration and accumulation of macrophages and leukocytes in the damaged sites and the release of pro-inflammatory mediators [40]. The NF-κB dimers are normally sequestered in the cytosol by binding to the IκB inhibitory protein. The previous study reported that ethanol stimulation activates NF-κB, which leads to phosphorylation of IκBα and p50/p65 heterodimer [27]. These p50/p65 dimers enhanced by phosphorylated IκBα promote gene expression of pro-inflammatory cytokines, which translocate to the nucleus [41]. In the present study, the results showed that RI significantly inhibited phosphorylation of IκBα and NF-κB p65 in RAW 264.7 cells and gastric tissue after LPS and HCl/EtOH stimulation, respectively (Figure 6, middle and lower panels). The study indicates that RI may inhibit early steps of inflammation and modulate upregulation of pro-inflammatory cytokines through suppression of NF-κB translocation.

4. Materials and Methods

4.1. Chemicals and Antibodies

LPS, MTT, penicillin/streptomycin, trypsin-EDTA, hematoxylin, eosin, and protease inhibitor were purchased from Sigma-Aldrich (St. Louis, MO, USA). DMEM, FBS, and other cell culture reagents were supplied by Gibco (Carlsbad, CA, USA). DMSO was obtained from Bioshop (Burlington, ON, Canada). RNA extraction kit (RiboEx and Hybrid-R) was bought from Gene All (Seoul, Korea). Griess reagent, cDNA synthesis kit (ReverTra Ace qPCR RT Kit), T-PER, and BCA protein assay kit were purchased from Thermo Scientific (Waltham, MA, USA). SYBR Green qPCR Kit obtained from TOYOBO (Tokyo, Japan). Primary antibodies ERK1/2, JNK, p38, COX-2, IκBα, NF-κB, and β-actin were supplied by Cell Signaling (Danvers, MA, USA). Secondary antibody (goat anti-rabbit immunoglobulin g horseradish peroxidase) was provided by Santa Cruz (Dallas, TX, USA). WESTSAVE gold ECL detection kit was obtained from Abfrontire (Seoul, Korea). TBARS assay kit by Cayman (Ann Arbor, MI, USA). ROS-Glo H_2O_2 assay kit was supplied by Promega (Madison, WI, USA). Zoletil 50 was bought from Virbac (Carros, France).

4.2. Collection and Preparation of Rabdosia Inflexa Extract

The aerial part of RI was collected from the Jiri mountain area in the southern part of Korea and was authenticated from its microscopic and macroscopic features by the Korean Institute of Oriental Medicine (KIOM). We prepared RI extract according to the previously described method [42]. Briefly, the aerial parts were chopped and dried completely. The extract was prepared by maceration of the sample with 70% ethanol (twice for 2 h reflux), and then the filtered extract was concentrated under vacuum centrifuge and dehydrated with a lyophilizer. The powder extract was liquefied in dimethyl sulfoxide (DMSO) and was sterilized using a 0.22 μm syringe filter. RI was dissolved in 0.04% DMSO in media for cell culture experiments and 0.1% DMSO in saline for oral gavage. The dried extract was kept at −20 °C. The study was conducted using a single batch of extract to avoid batch-to-batch variation and maximize the product constancy.

4.3. Phytochemical Analysis

Total phenolic and flavonoid content of RI extract was measured using Folin–Ciocalteu (FC) method according to the previously described method [43].

4.4. RAW 264.7 Cells Culture

Mouse macrophage RAW 264.7 cells were cultured in Dulbecco modified Eagle medium (DMEM) enriched with 10% fetal bovine serum (FBS) and 1% penicillin and streptomycin in a 5% CO_2 humidified incubator at 37 °C. Cells were maintained as a monolayer and subcultured once cells reached about 90% confluency in the culture flask.

4.5. Cell Viability and Morphological Study

Cell viability was detected using a 3-(4,5-dimethyl-thiazol-2-yl)-2,5-diphenyltetrazolium bromide (MTT) assay, following a prescribed method [34]. Briefly, RAW 264.7 cells (1×10^6 cells/mL) were cultured overnight. To determine the cytotoxicity, cells were treated with RI (50, 100, 200, 400, and 800 μg/mL) for 24 h. In contrast, measuring the cell viability, cells were pretreated with RI (100, 200, and 400 μg/mL) for 1 h and co-treated with LPS (0.5 μg/mL) and RI for another 24 h. Moreover, for morphological evaluation of RAW 264.7 cells, the image of the cells was acquired by an inverted microscope (CKX41, Olympus, Tokyo, Japan).

4.6. Measurement of NO and ROS in RAW 264.7 Cells

Mouse macrophage RAW 264.7 cells (1×10^6 cells/mL) were cultured in 96 well plates. After overnight culture, cells were pretreated with RI (100, 200. and 400 µg/mL) for 1 h and were then co-treated with LPS (0.5 µg/mL) and RI for 24 h. NO, and ROS activity was measured by Griess reagent and a ROS-Glo H_2O_2 assay kit according to the manufacturer's recommendation. Intracellular ROS and NO levels were measured at 570 nm by the tunable Versa max microplate reader.

4.7. Mice Management and HCl/EtOH-Induced Gastric Ulcer Model

Six-week-old ICR mice were handled in accordance with the published method [34] and in accordance with the guide for the care and use of laboratory animals (Eighth Edition, 2011, published by The National Academies Press, Washington, DC, USA) and Institutional Animal Care and Use Committee (IACUC; CBNU 2017-0126) of the Chonbuk National University Laboratory Animal Center in Korea. An experimental gastric ulcer model in mice was induced using HCl/EtOH [44]. Forty ICR mice were randomly divided into four groups. Mice fasted overnight before the experiment. Fasted mice were orally given RI (400 mg/kg) and ranitidine (40 mg/kg) in respected groups for 1 h. After 1 h, 60% EtOH in 150 mM HCl (10 µL/g) was given orally. Control mice were treated with normal saline. Mice were anesthetized with Zoletil 50 (10 mg/kg) 1 h after administration of HCl/EtOH and samples were collected for experimental analysis.

4.8. Gross and Histopathology of Gastric Mucosal Tissue

To evaluate the gross and histological changes in glandular stomach tissue, we followed previous study [34]. For quantifying the degree of gross and histological lesions index, we followed a prescribed method with slight modification [45]. Briefly, the gross lesions of gastric mucosa were measured using following formula; (number of lesions of type I) + (number of lesions of type II) $\times$ 2 + (number of lesions of type III) $\times$ 3. Here, the gross lesions were characterized by: the presence of single submucosal punctiform hemorrhage, edema, type I; the presence of submucosal hemorrhagic lesions with slight erosions, type II; and the presence of deep ulcer with erosions and invasive lesions, type III. Besides, the histological lesions of gastric mucosa were measured using following formula; (% type I lesion) $\times$ 1 + (% type II lesion) $\times$ 2 + (% type III lesion) $\times$ 3. Here, the histological lesions were characterized by: gastric mucosal cells appeared intact and had a normal shape, type 0 lesion; surface epithelial cells and the uppermost 2 or 3 cells lining the glands were damaged, type I damage; damage greater than I but involving <50% of the thickness of the gastric mucosa, type II damage; and damage involving >50% of the thickness of the gastric mucosa, type III damage.

4.9. Analysis of Lipid Peroxidation and NO Production

Malondialdehyde (MDA) and NO are important indicators of oxidative stress. The gastric tissue was homogenized and centrifuged at 10,000 rpm at 4 °C for 10 min. The supernatant was collected and kept at −80 °C for experimental analysis. MDA and NO concentration were measured in the gastric tissue samples according to the commercial kit instructions.

4.10. RNA Extraction and Quantitative Real-Time Polymerase Chain Reaction (qPCR)

The RNA was extracted from a gastric tissue according to the manufacturer's instructions. The concentration of total RNA was quantified with the BioSpec-nano spectrophotometer (Shimadzu Biotech, Tokyo, Japan) at a 260/280 nm ratio. For complementary DNA (cDNA) synthesis, total RNA (3 µg) was used, and cDNA synthesis was maintained according to the manufacturer's instructions. qPCR was performed SYBR Green Real-Time PCR master mix according to Roche LightCycler™. Relative expression of target genes was normalized to the reference gene: glyceraldehyde 3-phosphate dehydrogenase (GAPDH). The sequences of the primers (Bioneer, Daejeon, Korea) used are shown in Table 2 [46].

4.11. Immunohistochemical (IHC) Analysis

COX-2 immunopositive cells expression in the gastric tissue was performed using according to Vectastain ABC kit recommendations. Briefly, paraffin section was deparaffinized in xylene and hydrated in ethanol. Citrate buffer was used for antigen retrieval and 3% hydrogen peroxide (H_2O_2) was used for inactivating the endogenous peroxidase activity. Tissue was blocked with normal serum for 1 h. Anti-rabbit monoclonal COX-2 antibody (dilution 1:200) was incubated overnight at 4 °C. Subsequently, the section was incubated with biotinylated secondary antibody for 1 h and Vectastain ABC reagent for 30 min at room temperature. The sections were incubated with diaminobenzidine (DAB) in the dark until brown color development. After counterstain, the section was dehydrated in ethanol, cleared in xylene and mounted on a glass slide. The section was imaged at a fixed 100× magnification using Leica DM2500 microscope (Leica Microsystems, Wetzlar, Germany).

4.12. Western Blot Analysis

RAW 264.7 cells and gastric tissue were harvested and washed twice with ice-cold PBS. Cells and tissues were lysed by the lysis buffer; radioimmunoprecipitation assay buffer (RIPA), and/or tissue protein extraction reagent (T-PER), phenylmethanesulfonyl fluoride (PMSF), sodium orthovanadate (Na_3VO_4), and protease inhibitor cocktail. The total concentration of protein of lysate cells and tissues were measured with a bicinchoninic acid (BCA) protein assay protein kit. An equal amount of protein was separated by 10–12% sodium dodecyl sulfate-polyacrylamide gel electrophoresis (SDS-PAGE) and transferred to a nitrocellulose membrane. The membrane was incubated with blocking serum; 5% bovine serum albumin (BSA) in Tris-buffered saline with tween twenty (TBST) for 2 h at room temperature and by primary antibodies for overnight at 4 °C. Then, the blot was washed and incubated with secondary antibodies for 2 h. Bands were detected using an enhanced chemiluminescence (ECL) detection kit, and bands images were taken by a LAS-400 image system, (GE Healthcare, Little Chalfont, UK); β-actin was used as the reference antibody.

4.13. Statistical Analysis

Data were analyzed with Graph Pad Prism 5.0 (Graph Pad Software, Inc., San Diego, CA, USA) and are expressed as mean ± standard error (SEM). Statistical analyses were assessed by analysis of variance (ANOVA) followed by Bonferroni post-hoc tests. The minimum statistical significance was considered to be $p < 0.05$ for all analyses.

5. Conclusions

RI mitigates inflammation and maintains normal gastric mucosal integrity. The present study validates that RI protects against inflammation and gastric ulcer by mitigating the inflammation response and oxidative stress via downregulation of the pro-inflammatory cytokines mediated by MAPK/NF-κB signaling pathways. Therefore, RI could be a promising phytomedicine and has advantages for prospective clinical applications in the future for an oxidative stress-mediated gastric ulcer.

Acknowledgments: This research was supported by Basic Science Research Program through the National Research Foundation of Korea funded by the Ministry of Education (2017R1D1A1B03035765) and Biomedical Research Institute, Chonbuk National University Hospital.

Author Contributions: Md Rashedunnabi Akanda and Byung-Yong Park conceived and designed the study. Md Rashedunnabi Akanda performed the experiment, analyzed the data and wrote the original manuscript. Md Rashedunnabi Akanda, In-Shik Kim, Dongchoon Ahn, Hyun-Jin Tae, Hyeon-Hwa Nam, Byung-Kil Choo, Kyunghwa Kim and Byung-Yong Park read and approved the final version of the manuscript.

Conflicts of Interest: The authors declare no conflict of interest.

References

1. Rehm, J. The risks associated with alcohol use and alcoholism. *Alcohol Res. Health* **2011**, *34*, 135–143. [PubMed]
2. Tarnawski, A.S.; Ahluwalia, A.; Jones, M.K. Increased susceptibility of aging gastric mucosa to injury: The mechanisms and clinical implications. *World J. Gastroenterol.* **2014**, *20*, 4467–4482. [CrossRef] [PubMed]
3. Orywal, K.; Szmitkowski, M. Alcohol dehydrogenase and aldehyde dehydrogenase in malignant neoplasms. *Clin. Exp. Med.* **2017**, *17*, 131–139. [CrossRef] [PubMed]
4. Kwiecien, S.; Jasnos, K.; Magierowski, M.; Sliwowski, Z.; Pajdo, R.; Brzozowski, B.; Mach, T.; Wojcik, D.; Brzozowski, T. Lipid peroxidation, reactive oxygen species and antioxidative factors in the pathogenesis of gastric mucosal lesions and mechanism of protection against oxidative stress—Induced gastric injury. *J. Physiol. Pharmacol.* **2014**, *65*, 613–622. [PubMed]
5. Singh, P.; Vishwakarma, S.P.; Singh, R.L. Antioxidant, oxidative DNA damage protective and antimicrobial activities of the plant Trigonella foenum-graecum. *J. Sci. Food Agric.* **2014**, *94*, 2497–2504. [CrossRef] [PubMed]
6. Allavena, P.; Garlanda, C.; Borrello, M.G.; Sica, A.; Mantovani, A. Pathways connecting inflammation and cancer. *Curr. Opin. Genet. Dev.* **2008**, *18*, 3–10. [CrossRef] [PubMed]
7. El-Maraghy, S.A.; Rizk, S.M.; Shahin, N.N. Gastroprotective effect of crocin in ethanol-induced gastric injury in rats. *Chem. Biol. Interact.* **2015**, *229*, 26–35. [CrossRef] [PubMed]
8. Akira, S. Toll-like receptors: Lessons from knockout mice. *Biochem. Soc. Trans.* **2000**, *28*, 551–556. [CrossRef] [PubMed]
9. Song, X.; Zhang, W.; Wang, T.; Jiang, H.; Zhang, Z.; Fu, Y.; Yang, Z.; Cao, Y.; Zhang, N. Geniposide plays an anti-inflammatory role via regulating TLR4 and downstream signaling pathways in lipopolysaccharide-induced mastitis in mice. *Inflammation* **2014**, *37*, 1588–1598. [CrossRef] [PubMed]
10. Zhang, Y.L.; Dong, C. MAP kinases in immune responses. *Cell. Mol. Immunol.* **2005**, *2*, 20–27. [PubMed]
11. Ryan, S.; McNicholas, W.T.; Taylor, C.T. A critical role for p38 map kinase in NF-kappaB signaling during intermittent hypoxia/reoxygenation. *Biochem. Biophys. Res. Commun.* **2007**, *355*, 728–733. [CrossRef] [PubMed]
12. Dambrova, M.; Zvejniece, L.; Skapare, E.; Vilskersts, R.; Svalbe, B.; Baumane, L.; Muceniece, R.; Liepinsh, E. The anti-inflammatory and antinociceptive effects of NF-kappaB inhibitory guanidine derivative ME10092. *Int. Immunopharmacol.* **2010**, *10*, 455–460. [CrossRef] [PubMed]
13. Nakajima, H.; Itokawa, H.; Ikuta, A. [Studies on the constituents of the flower of Camellia japonica (2)]. *Yakugaku Zasshi* **1984**, *104*, 157–161. [CrossRef] [PubMed]
14. Fujita, T.; Takeda, Y.; Yuasa, E.; Okamura, A.; Shingu, T.; Yokoi, T. Structure of Inflexinol, a New Cyto-Toxic Diterpene from Rabdosia-Inflexa. *Phytochemistry* **1982**, *21*, 903–905. [CrossRef]
15. Kubo, I.; Nakanishi, K.; Kamikawa, T.; Isobe, T.; Kubota, T. Structure of Inflexin. *Chem. Lett.* **1977**, *6*, 99–102. [CrossRef]
16. Ahn, D.W.; Ryu, J.K.; Kim, J.; Kim, Y.T.; Yoon, Y.B.; Lee, K.; Hong, J.T. Inflexinol reduces severity of acute pancreatitis by inhibiting nuclear factor-kappaB activation in cerulein-induced pancreatitis. *Pancreas* **2013**, *42*, 279–284. [CrossRef] [PubMed]
17. Salama, S.M.; Gwaram, N.S.; AlRashdi, A.S.; Khalifa, S.A.; Abdulla, M.A.; Ali, H.M.; El-Seedi, H.R. A Zinc Morpholine Complex Prevents HCl/Ethanol-Induced Gastric Ulcers in a Rat Model. *Sci. Rep.* **2016**, *6*, 29646. [CrossRef] [PubMed]
18. Lin, D.R.; Xiao, M.S.; Zhao, J.J.; Li, Z.H.; Xing, B.S.; Li, X.D.; Kong, M.Z.; Li, L.Y.; Zhang, Q.; Liu, Y.W.; et al. An Overview of Plant Phenolic Compounds and Their Importance in Human Nutrition and Management of Type 2 Diabetes. *Molecules* **2016**, *21*, 1374. [CrossRef] [PubMed]
19. Mierziak, J.; Kostyn, K.; Kulma, A. Flavonoids as Important Molecules of Plant Interactions with the Environment. *Molecules* **2014**, *19*, 16240–16265. [CrossRef] [PubMed]
20. Dzoyem, J.P.; Eloff, J.N. Anti-inflammatory, anticholinesterase and antioxidant activity of leaf extracts of twelve plants used traditionally to alleviate pain and inflammation in South Africa. *J. Ethnopharmacol.* **2015**, *160*, 194–201. [CrossRef] [PubMed]

21. Kim, M.J.; Kadayat, T.; Kim, D.E.; Lee, E.S.; Park, P.H. TI-I-174, a Synthetic Chalcone Derivative, Suppresses Nitric Oxide Production in Murine Macrophages via Heme Oxygenase-1 Induction and Inhibition of AP-1. *Biomol. Ther.* **2014**, *22*, 390–399. [CrossRef] [PubMed]

22. Li, B.; Alli, R.; Vogel, P.; Geiger, T.L. IL-10 modulates DSS-induced colitis through a macrophage-ROS-NO axis. *Mucosal Immunol.* **2014**, *7*, 869–878. [CrossRef] [PubMed]

23. Yang, W.S.; Jeong, D.; Yi, Y.S.; Lee, B.H.; Kim, T.W.; Htwe, K.M.; Kim, Y.D.; Yoon, K.D.; Hong, S.; Lee, W.S.; et al. Myrsine seguinii ethanolic extract and its active component quercetin inhibit macrophage activation and peritonitis induced by LPS by targeting to Syk/Src/IRAK-1. *J. Ethnopharmacol.* **2014**, *151*, 1165–1174. [CrossRef] [PubMed]

24. Xie, M.; Chen, H.; Nie, S.; Tong, W.; Yin, J.; Xie, M. Gastroprotective effect of gamma-aminobutyric acid against ethanol-induced gastric mucosal injury. *Chem. Biol. Interact.* **2017**, *272*, 125–134. [CrossRef] [PubMed]

25. Lee, J.S.; Oh, T.Y.; Kim, Y.K.; Baik, J.H.; So, S.; Hahm, K.B.; Surh, Y.J. Protective effects of green tea polyphenol extracts against ethanol-induced gastric mucosal damages in rats: Stress-responsive transcription factors and MAP kinases as potential targets. *Mutat. Res. Fundam. Mol. Mech. Mutagen.* **2005**, *579*, 214–224. [CrossRef] [PubMed]

26. Calatayud, S.; Barrachina, D.; Esplugues, J.V. Nitric oxide: Relation to integrity, injury, and healing of the gastric mucosa. *Microsc. Res. Tech.* **2001**, *53*, 325–335. [CrossRef] [PubMed]

27. Chen, S.; Zhao, X.; Sun, P.; Qian, J.; Shi, Y.; Wang, R. Preventive effect of Gardenia jasminoides on HCl/ethanol induced gastric injury in mice. *J. Pharmacol. Sci.* **2017**, *133*, 1–8. [CrossRef] [PubMed]

28. Yu, T.; Yang, Y.; Kwak, Y.S.; Song, G.G.; Kim, M.Y.; Rhee, M.H.; Cho, J.Y. Ginsenoside Rc from Panax ginseng exerts anti-inflammatory activity by targeting TANK-binding kinase 1/interferon regulatory factor-3 and p38/ATF-2. *J. Ginseng Res.* **2017**, *41*, 127–133. [CrossRef] [PubMed]

29. Park, J.G.; Kim, S.C.; Kim, Y.H.; Yang, W.S.; Kim, Y.; Hong, S.; Kim, K.H.; Yoo, B.C.; Kim, S.H.; Kim, J.H.; et al. Anti-Inflammatory and Antinociceptive Activities of Anthraquinone-2-Carboxylic Acid. *Mediat. Inflamm.* **2016**, *2016*, 1903849. [CrossRef] [PubMed]

30. Liu, Y.; Tian, X.; Gou, L.; Fu, X.; Li, S.; Lan, N.; Yin, X. Protective effect of l-citrulline against ethanol-induced gastric ulcer in rats. *Environ. Toxicol. Pharmacol.* **2012**, *34*, 280–287. [CrossRef] [PubMed]

31. Boligon, A.A.; de Freitas, R.B.; de Brum, T.F.; Waczuk, E.P.; Klimaczewski, C.V.; de Avila, D.S.; Athayde, M.L.; de Freitas Bauermann, L. Antiulcerogenic activity of Scutia buxifolia on gastric ulcers induced by ethanol in rats. *Acta Pharm. Sin. B* **2014**, *4*, 358–367. [CrossRef] [PubMed]

32. Park, S.W.; Oh, T.Y.; Kim, Y.S.; Sim, H.; Park, S.J.; Jang, E.J.; Park, J.S.; Baik, H.W.; Hahm, K.B. Artemisia asiatica extracts protect against ethanol-induced injury in gastric mucosa of rats. *J. Gastroenterol. Hepatol.* **2008**, *23*, 976–984. [CrossRef] [PubMed]

33. Mei, X.; Xu, D.; Xu, S.; Zheng, Y.; Xu, S. Novel role of Zn(II)-curcumin in enhancing cell proliferation and adjusting proinflammatory cytokine-mediated oxidative damage of ethanol-induced acute gastric ulcers. *Chem. Biol. Interact.* **2012**, *197*, 31–39. [CrossRef] [PubMed]

34. Akanda, M.R.; Park, B.Y. Involvement of MAPK/NF-kappaB signal transduction pathways: Camellia japonica mitigates inflammation and gastric ulcer. *Biomed. Pharmacother.* **2017**, *95*, 1139–1146. [CrossRef] [PubMed]

35. Li, W.; Huang, H.; Niu, X.; Fan, T.; Mu, Q.; Li, H. Protective effect of tetrahydrocoptisine against ethanol-induced gastric ulcer in mice. *Toxicol. Appl. Pharmacol.* **2013**, *272*, 21–29. [CrossRef] [PubMed]

36. Ci, X.; Ren, R.; Xu, K.; Li, H.; Yu, Q.; Song, Y.; Wang, D.; Li, R.; Deng, X. Schisantherin A exhibits anti-inflammatory properties by down-regulating NF-kappaB and MAPK signaling pathways in lipopolysaccharide-treated RAW 264.7 cells. *Inflammation* **2010**, *33*, 126–136. [CrossRef] [PubMed]

37. Wei, S.G.; Yu, Y.; Weiss, R.M.; Felder, R.B. Endoplasmic reticulum stress increases brain MAPK signaling, inflammation and renin-angiotensin system activity and sympathetic nerve activity in heart failure. *Am. J. Physiol. Heart Circ. Physiol.* **2016**, *311*, H871–H880. [CrossRef] [PubMed]

38. Jin, X.; Han, J.; Yang, S.; Hu, Y.; Liu, H.; Zhao, F. 11-*O*-acetylcyathatriol inhibits MAPK/p38-mediated inflammation in LPS-activated RAW 264.7 macrophages and has a protective effect on ethanol-induced gastric injury. *Mol. Med. Rep.* **2016**, *14*, 874–880. [CrossRef] [PubMed]

39. Liang, Y.; Zhou, Y.; Shen, P. NF-kappaB and its regulation on the immune system. *Cell. Mol. Immunol.* **2004**, *1*, 343–350. [PubMed]

40. Kang, J.W.; Yun, N.; Han, H.J.; Kim, J.Y.; Kim, J.Y.; Lee, S.M. Protective Effect of Flos Lonicerae against Experimental Gastric Ulcers in Rats: Mechanisms of Antioxidant and Anti-Inflammatory Action. *Evid. Based Complement. Altern. Med.* **2014**, *2014*. [CrossRef] [PubMed]

41. Hinz, M.; Scheidereit, C. The IkappaB kinase complex in NF-kappaB regulation and beyond. *EMBO Rep.* **2014**, *15*, 46–61. [CrossRef] [PubMed]

42. Akanda, M.; Tae, H.-J.; Kim, I.-S.; Ahn, D.; Tian, W.; Islam, A.; Nam, H.-H.; Choo, B.-K.; Park, B.-Y. Hepatoprotective Role of Hydrangea macrophylla against Sodium Arsenite-Induced Mitochondrial-Dependent Oxidative Stress via the Inhibition of MAPK/Caspase-3 Pathways. *Int. J. Mol. Sci.* **2017**, *18*, 1482. [CrossRef] [PubMed]

43. Jing, L.; Ma, H.; Fan, P.; Gao, R.; Jia, Z. Antioxidant potential, total phenolic and total flavonoid contents of Rhododendron anthopogonoides and its protective effect on hypoxia-induced injury in PC12 cells. *BMC Complement. Altern. Med.* **2015**, *15*, 287. [CrossRef] [PubMed]

44. Yang, Y.; Yu, T.; Lee, Y.G.; Yang, W.S.; Oh, J.; Jeong, D.; Lee, S.; Kim, T.W.; Park, Y.C.; Sung, G.H.; et al. Methanol extract of Hopea odorata suppresses inflammatory responses via the direct inhibition of multiple kinases. *J. Ethnopharmacol.* **2013**, *145*, 598–607. [CrossRef] [PubMed]

45. Nam, S.Y.; Kim, N.; Lee, C.S.; Choi, K.D.; Lee, H.S.; Jung, H.C.; Song, I.S. Gastric mucosal protection via enhancement of MUC5AC and MUC6 by geranylgeranylacetone. *Dig. Dis. Sci.* **2005**, *50*, 2110–2120. [CrossRef] [PubMed]

46. Yang, Y.; Yu, T.; Jang, H.J.; Byeon, S.E.; Song, S.Y.; Lee, B.H.; Rhee, M.H.; Kim, T.W.; Lee, J.; Hong, S.; et al. In vitro and in vivo anti-inflammatory activities of Polygonum hydropiper methanol extract. *J. Ethnopharmacol.* **2012**, *139*, 616–625. [CrossRef] [PubMed]

International Journal of
Molecular Sciences

Review

Role of p38 MAPK in Atherosclerosis and Aortic Valve Sclerosis

Anna Reustle [1,2] **and Michael Torzewski** [3,*]

1   Dr. Margarete-Fischer-Bosch-Institute of Clinical Pharmacology, 70376 Stuttgart, Germany; anna.reustle@ikp-stuttgart.de
2   University of Tuebingen, 72074 Tuebingen, Germany
3   Department of Laboratory Medicine and Hospital Hygiene, Robert Bosch-Hospital, 70376 Stuttgart, Germany
*   Correspondence: michael.torzewski@rbk.de; Tel.: +49-711-8101-3500

Received: 18 October 2018; Accepted: 22 November 2018; Published: 27 November 2018

Abstract: Atherosclerosis and aortic valve sclerosis are cardiovascular diseases with an increasing prevalence in western societies. Statins are widely applied in atherosclerosis therapy, whereas no pharmacological interventions are available for the treatment of aortic valve sclerosis. Therefore, valve replacement surgery to prevent acute heart failure is the only option for patients with severe aortic stenosis. Both atherosclerosis and aortic valve sclerosis are not simply the consequence of degenerative processes, but rather diseases driven by inflammatory processes in response to lipid-deposition in the blood vessel wall and the aortic valve, respectively. The p38 mitogen-activated protein kinase (MAPK) is involved in inflammatory signaling and activated in response to various intracellular and extracellular stimuli, including oxidative stress, cytokines, and growth factors, all of which are abundantly present in atherosclerotic and aortic valve sclerotic lesions. The responses generated by p38 MAPK signaling in different cell types present in the lesions are diverse and might support the progression of the diseases. This review summarizes experimental findings relating to p38 MAPK in atherosclerosis and aortic valve sclerosis and discusses potential functions of p38 MAPK in the diseases with the aim of clarifying its eligibility as a pharmacological target.

Keywords: atherosclerosis; aortic valve sclerosis; aortic valve stenosis; p38; MAPK

---

1. Introduction

Cardiovascular diseases are the leading cause of death worldwide [1]. Among the diseases, atherosclerosis is the one with the highest mortality in the western world [2]. Risk factors are associated with western lifestyle and include smoking, hypertension, and high blood glucose, lipid, and cholesterol levels. Atherosclerosis develops in arterial blood vessel walls and most commonly occurs in coronary arteries, in branch points of the carotid artery, and the big leg arteries. The vessel wall consists of three tissue layers: the intima at the luminal side, the media, and the adventitia that is in contact with the surrounding perivascular tissue. The intima is composed of a single layer of endothelial cells and subendothelial connective tissue, providing a barrier between the blood flow and the underlying tissue. The media, mainly composed of vascular smooth muscle cells (SMCs) and elastic connective tissue, is the thickest layer and confers stability and elasticity to the vessel wall. Finally, the adventitia represents the most complex layer, pervaded with nerves and small blood vessels (vaso vasorum) that supply the larger vessel with nerve signals and nutrients to regulate vessel wall function. In atherosclerosis, the diameter of the intima layer increases locally due to proliferation and growth of SMCs, connective tissue deposition, and accumulation of lipids from the blood stream, together forming an atherosclerotic plaque (Figure 1; upper right). As a consequence, the lumen of the vessel narrows, impairing the perfusion of the adjacent tissues. If the plaque is instable and ruptures, the thrombogenic lipid core gets into contact with circulating blood, leading to coagulation and thrombus formation. Patients with atherosclerosis and risk of plaque rupture are treated with statins, which lower blood cholesterol levels and stabilize the plaque, possibly by inhibition of inflammatory processes and modulation of plaque composition [3,4].

Aortic valve stenosis (AVS), not as prevalent as atherosclerosis, is one of the most common indications for cardiac surgery and affects around 12% of the elderly population above the age of 74 [5]. Aortic valve sclerosis, or calcific aortic valve disease (CAVD), is an early stage of AVS and is marked by thickening and calcification of the valve tissue. Common risk factors to develop aortic valve sclerosis are high blood pressure, high blood lipid and cholesterol levels, obesity, diabetes mellitus, smoking, and chronic kidney disease [6–11]. The healthy human aortic valve is composed of three thin leaflets (<1 mm), each made up of three layers—the fibrosa on the aortic side, the spongiosa, and the ventricularis on the ventricular side of the valve (Figure 1; upper left). Valve interstitial cells (VICs) are scattered throughout the layers, and endothelial cells line the leaflets on both sides. The layers differ in their extracellular matrix (ECM) composition, providing the leaflets with the stability and flexibility needed to open and close with every contraction of the left ventricle, to allow the blood to enter the aorta and supply the body with oxygen-rich blood. In aortic valve sclerosis, or CAVD, the leaflets are obstructed by fibrosis and calcium deposition, mainly in the fibrosa layer, which impairs their ability to smoothly open and close the passage from the heart to the aorta. As a consequence, the valve narrows and the pressure in the left ventricle rises, leading to increased stress and eventually heart failure. To date, no treatments exist to prevent the development of aortic valve sclerosis, or to halt its progression to AVS. Therefore, valve replacement surgery is the only therapeutic option for patients with severe AVS.

Although atherosclerosis and CAVD are distinct diseases, with differing prevalence and disease manifestations, they share common risk factors such as smoking, obesity, high blood pressure, and elevated blood low density lipoprotein (LDL)-cholesterol levels. Independent of lifestyle choices, individuals may be predisposed to some of these risk factors by genetic mutations. In fact, genome-wide association studies (GWAS) have revealed distinct genomic loci which are frequently affected in individuals with cardiovascular diseases, including genes involved in blood coagulation, inflammation, endothelial cell adhesion, and lipid metabolism and transport [12,13]. The *LPA* gene, encoding the lipoprotein a (Lp(a)), a known cardiovascular risk factor, was also identified in GWAS of aortic valve sclerosis [14,15] and shown to be present in elevated levels in plasma and aortic valves of patients [16–18]. The presence of common risk factors and genetic dispositions of atherosclerosis and CAVD highlight the existence of shared disease initiation mechanisms [19]. In both diseases, endothelial damage, followed by lipid insudation and accumulation in the intima or fibrosa layers, respectively, are thought to represent the initiating events. To dispose of excess lipids, macrophages are recruited to the sites by damage-activated endothelial cells. If the lipid burden is too high, macrophages accumulate and transform to lipid-laden foam cells. So called fatty streaks, or intimal xanthoma in the vessel walls are thought to be the signs of such early lesions, although they might as well regress without progression into atherosclerotic plaques [20,21]. During progression however, further immune cells are recruited to the lesions by pro-inflammatory cytokines that are secreted by macrophages, endothelial cells, and lesion smooth muscle cells (SMCs) or VICs. Fibrosis occurs due to cell proliferation and ECM remodeling, leading to thickening of the tissues. The chronic inflammatory environment is thought to furthermore promote the tissue calcification that is seen in both pathologies [22–24]. Since immune cell infiltration is an early event and chronic inflammation a suspected driver in both pathologies, therapeutic targeting of inflammatory signaling could represent an instrument to intervene with progression of atherosclerosis as well as aortic valve sclerosis and to avoid the fatal consequences of both diseases.

In the context of chronic inflammation, the p38 mitogen-activated protein kinase (MAPK) pathway has gained attention in the field of both atherosclerosis and CAVD research. p38 MAPK signaling is implicated in diverse biological processes, such as tissue development, cell proliferation, apoptosis, inflammation, and cancer (reviewed in [25]). p38 MAPK is activated by various extracellular inducers of inflammation, which are highly abundant in atherosclerotic and CAVD lesions. To illuminate the role of p38 MAPK signaling in atherosclerosis and aortic valve sclerosis, in this review we summarize relevant experimental findings related to p38 MAPK in both pathologies. To acknowledge the tissue complexity of the diseases, we dissected the findings into the different cell types that make up the lesions and influence disease progression. The aim of this review is to give an overview of p38 MAPK signaling in atherosclerosis and aortic valve sclerosis, and to discuss potential therapeutic implications.

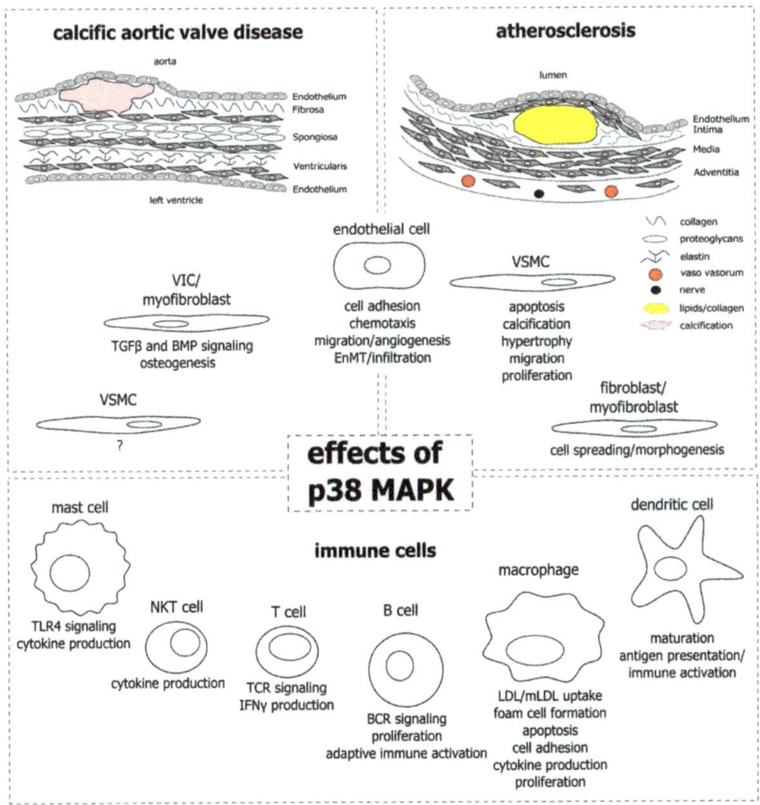

Figure 1. Functional involvement of p38 mitogen-activated protein kinase (MAPK) signaling in calcific aortic valve disease (CAVD) and atherosclerosis. Upper left: CAVD lesion. Schematic cross-section of an aortic valve leaflet composed of the fibrosa, spongiosa, and ventricularis tissue layers. The layers are dispersed by matrix producing valve interstitial cells (VICs) and lined by endothelial cells on both sides which face the aorta or the left ventricle. Lipids accumulate mainly in the collagen-rich fibrosa layer, which is also where the calcification develops. Upper right: atherosclerotic lesion. Schematic cross-section of the vessel wall containing an atherosclerotic plaque. The vessel wall consists of a collagen-rich intima layer lined by endothelial cells that are in direct contact with the blood flow. The underlying media layer contains vascular smooth muscle cells (VSMCs) that contract and dilate in response to nerve signals from the adventitia layer, thereby regulating local blood pressure. The adventitia contains nerves and blood vessels that supply the VSMCs. Atherosclerotic plaques develop in the intima layer and are stabilized by VSMCs from the media. Lower panel: functions attributed to p38 MAPK activity in different immune cells present in CAVD and atherosclerotic lesions. TGFβ: transforming growth factor β; BMP: bone morphogenic protein; EnMT: endothelial to mesenchymal transition; NKT: natural killer T cells; TLR4: toll-like receptor 4; TCR: T cell receptor; IFNγ: interferon γ; BCR: B cell receptor; mLDL: modified LDL.

2. p38 MAPK Signaling

The p38 MAPKs are members of the mitogen-activated serine/threonine kinase family, together with the extracellular signal-regulated kinases (ERKs) and the c-Jun N-terminal kinases (JNKs). The p38 MAPKs are activated in the presence of certain pathogenic stimuli, such as lipopolysaccharides (LPS), by pro-inflammatory cytokines, or when cells experience extracellular stress, such as ultraviolet radiation, heat shock, or hypoxia. Intracellular stress triggered by miss-folded proteins in the endoplasmic reticulum (ER) or DNA damage can also lead to p38 MAPK activation. Common for all extracellular and intracellular inducers of MAPKs is that binding of the associated ligands to their respective receptors sets in motion a cascade of successive phosphorylation events, where MAPK kinase kinases (MAPKKKs/MEKKs) phosphorylate MAPK kinases (MAPKKs/MKKs/MEKs), which in turn phosphorylate and activate MAPKs. MEK3 and MEK6 are the primary MAPK kinases that phosphorylate the p38 MAPKs, whereas different sets of MEKs mainly activate ERKs and JNKs. MAPK-activated protein kinase 2 (MAPKAPK2/MK2) and the heat shock protein 27 (HSP27) are important downstream targets of activated p38 MAPK, functioning to protect cells from heat shock and osmotic stress [26]. A multitude of other p38 MAPK downstream targets are known today, which execute the cellular responses upon p38 MAPK activation. The cell type and cellular context seem to impact the generated response, which can be as diverse as pro-apoptotic, pro-inflammatory, or anti-proliferative. Comprehensive and exhaustive reviews of the p38 MAPK signaling pathway are provided by others, e.g. Cargnello and Roux, or Coulthard et al. [25,27,28].

Since atherosclerosis and aortic valve sclerosis have become recognized as active, inflammation-driven processes, the p38 MAPK has gained attention in this research field. An increasing number of studies investigating p38 MAPK signaling in different cell types associated with these two cardiovascular diseases have been published in recent years. In the following sections, the experimental findings are summarized and subdivided into the cell types that are present in atherosclerotic and CAVD lesions and that are expected to be involved in disease pathogenesis.

3. Endothelial Cells

Endothelial cells (ECs) line the luminal surface of blood vessels and both sides of the aortic valve leaflets. The endothelium builds a protective barrier for the underlying tissue and has important functions in regulating the composition of the associated tissue layers, as well as in adhesion and invasion of inflammatory cells. Vascular endothelium is, furthermore, involved in regulation of the vascular tone, by production of nitric oxide (NO) and other vasomodulators. Healthy endothelium acts atheroprotectively by controlling local blood pressure and by inhibition of inflammation and thrombosis. When the endothelium is damaged, for instance by increased shear stress due to high blood pressure, the endothelial barrier breaks and allows the ingress of blood components into the tissue. As a consequence, ECs become activated to initiate repair of the damaged tissue and disposal of intruded cells and molecules. In the process, ECs express adhesion molecules that allow attachment and invasion of inflammatory cells. Atherosclerosis or CAVD develop if early lesions cannot be resolved and progress into chronically inflamed sites. The presence of excess LDL, and its modification in the damaged tissue, seems to be a crucial factor in the initiation of pathologic lesions [29].

Due to its presence in early atherosclerotic and CAVD lesions, native LDL and/or its modification products might represent important inducers of p38 MAPK signaling at early disease stages. Indeed, LDL has been demonstrated to induce p38 MAPK signaling in ECs [30], and various functions have been identified for the p38 kinase, including upregulation of the cell adhesion molecules E-selectin [30] and vascular cell adhesion protein 1 (VCAM-1) [31] and the chemokine monocyte-chemoattractant protein-1 (MCP-1) [32], all involved in pro-inflammatory signaling and local recruitment of immune cells. Lp(a), consisting of LDL covalently bound to apolipoprotein a, is a known risk factor for cardiovascular diseases and has been shown to increase phosphorylation of p38 MAPK and other kinases in human umbilical-vein endothelial cells (HUVEC), inducing cell growth and migration [33]. Other studies showed that p38 MAPK might be involved in EC migration associated with angiogenesis [34,35], which is observed in atherosclerotic and CAVD lesions [36,37]. p38 MAPK has also been shown to be involved in regulation of EC permeability [38]. Interestingly, high density lipoprotein (HDL), which is inversely correlated with the risk of cardiovascular disease development [39], has been shown to inhibit p38 MAPK activity in HUVEC, leading to decreased interleukin (IL)-6 secretion [40]. In contrast, HDL and oxidized HDL (oxHDL) have been shown to activate p38 MAPK signaling in HUVEC, macrophages, and vascular SMCs in other studies [41–44]. Which of the described functions of p38 MAPK in ECs are associated with atherosclerosis and/or CAVD disease development and progression, or might even represent drivers or inhibitors of the diseases, warrants further investigation.

Both vascular and valve ECs may undergo endothelial-to-mesenchymal transition (EnMT) accompanied by upregulation of alpha smooth muscle actin expression, the acquisition of contractile properties, and the ability to infiltrate the underlying tissue layers. EnMT of vascular and valve ECs is required during development and maintenance of adult tissue homeostasis by replenishing pools of tissue-resident VICs and SMCs. In the pathogenesis of atherosclerosis and CAVD, however, EnMT allows ECs to transition to VICs or SMCs and to further acquire osteoblast-like properties [45]. Osteoblast-like cells are observed in the lesions, and osteogenic processes are thought to drive tissue calcification in advanced disease stages [36].

The initiation events of EC EnMT are not entirely solved to date. However, mechanical strain has been demonstrated to cause EnMT in valve ECs [46]. The strain increases continually with the rise in tissue calcification, representing a self-replenishing process. In the same study, transforming growth factor beta (TGFβ) and Wnt/β-catenin signaling were demonstrated to drive the strain-induced transformation of ECs to VICs on the cellular level. Interestingly, in other studies, the p38 MAPK has also been shown to be activated when vascular ECs are exposed to shear stress, with impact on actin dynamics and cell-cell alignment [47–49]. Whether p38 MAPK impacts EnMT of valve and/or vascular ECs has not been investigated so far.

Despite the diverse p38 MAPK functions that point to a pro-inflammatory and generally disease-promoting character of p38 MAPK signaling in ECs, an in vivo study with an ApoE$^{-/-}$ atherogenic mouse model did not identify an impact of EC-specific knockout of p38 MAPK expression on disease progression and outcome [50]. Although in vivo models better account for the complex environment of a disease, their informative value is compromised by differences between organisms species. Especially when inflammatory signaling and immune processes are involved, mouse models might deviate from the human organism [51,52]. The discrepancy between the in vitro and in vivo findings highlights the need for suitable atherosclerosis and CAVD model systems that account for the complex environment and signaling networks of the diseases.

4. Smooth Muscle Cells

Vascular smooth muscle cells (SMCs) are the most abundant cell type in the blood vessel wall, and small numbers of SMCs are also present in healthy valve leaflets, although here, the predominant cell type are VICs (see below). SMCs produce matrix proteins, including collagen, providing the tissue with the required stability and flexibility. In atherosclerosis, vascular SMCs are stimulated to grow, proliferate, and migrate and are highly involved in the thickening of the intima layer of the blood vessel wall. SMCs seem to play a dual role in atherosclerosis progression, as they form protective layers around lipid cores, thereby protecting them from luminal stresses that might provoke plaque rupture and, as a consequence, thrombosis. On the other hand, vascular SMCs are drivers of atherosclerotic progression. In advanced atherosclerotic lesions, SMCs have been shown to undergo enhanced apoptosis, leading to plaque destabilization accompanied by an increased risk of plaque rupture [53,54]. In addition, apoptotic SMCs release matrix vesicles, which serve as nucleation sites for calcium crystals and thereby support plaque calcification [55].

The reason for the increased apoptosis of vascular SMCs in atherosclerotic plaques is not entirely understood. Several lines of evidence, however, point to a potential role of LDL and its modification products in the process [56–58]. LDL molecules have also been shown to induce the p38 MAPK pathway in SMCs. Treatment of rat vascular SMCs with oxidized LDL (oxLDL) induced p38 MAPK phosphorylation and its nuclear translocation in a pathway that includes G-protein coupled receptors and the phospholipase C [59]. In this study, activation of p38 MAPK led to increased cytotoxicity in vascular SMCs. Pro-calcific and pro-apoptotic effects of p38 MAPK activation in vascular SMCs, as a consequence of oxLDL-induced cellular ceramide levels, have been reported in other studies [60,61]. At this point we want to note that although oxLDL in this review is used as a general term to refer to oxidized LDL particles, several methods of oxLDL generation do exist. Depending on the method, extensively or minimally oxidized LDL particles are generated that differ by the extent and type of phospholipid and protein modifications within the particles [62]. Importantly, extensively and minimally oxidized LDL particles interact with different pattern recognition receptors, inducing distinct or even opposed cellular processes [63,64]. The studies cited above used extensively oxLDL [59,60] and minimally oxLDL [61], respectively. All studies cited in the following sections used extensively oxLDL in their experiments.

Hypertrophy of vascular SMCs is a common feature of atherosclerosis and is one of the underlying mechanisms of intima thickening observed in pre-atherosclerotic lesions [65]. Angiotensin II (Ang II), an inducer of vascular SMC hypertrophy [66], has been shown to activate the p38 kinase by augmenting intracellular oxidative stress [67,68]. In an alternative pathway, Ang II induces p38 MAPK phosphorylation via epidermal growth factor receptor (EGFR) and TGFβ signaling, with effects on peroxisome proliferator-activated receptor gamma (PPARγ) expression and vascular SMC hypertrophy [69,70]. Furthermore, Ang II-mediated stimulation of p38 MAPK signaling has been shown to increase vascular SMCs migration [71] and proliferation [72], all processes involved in the aberrant growth of the intima in atherosclerosis.

As already mentioned above, the role of vascular SMCs in atherosclerosis is dual. In the development of atherosclerotic lesions, proliferation, migration, and cell growth of SMCs seem to be driving factors, whereas later in disease, when the atherosclerotic plaque has already formed, SMC apoptosis gains more importance. The summarized studies show that p38 MAPK signaling is involved in all disease stages by activating different cellular responses that seem to be dependent on the presence of signaling molecules such as modified LDL (mLDL), ceramide, TGFβ, or Ang II, and others that are not discussed in this review. When it comes to LDL, the exact nature of modification and the local concentration further impact the generated response. The matter is actively debated and reviewed elsewhere [29]. In order to dissect the function of p38 MAPK activation in vascular SMCs, more studies are needed. Especially, the activation of p38 kinase in early-to-late atherosclerotic lesions needs to be further assessed, as well as the local context of expression, such as the presence of mLDL and other signaling molecules.

Since SMCs constitute only a minor fraction of cells in the healthy aortic valve, their contribution to CAVD development has not been extensively studied [73,74]. In calcific aortic valves, however, the numbers of SMCs increase and they have been shown to co-localize with calcified regions [75]. The authors of the latter study speculate that in CAVD, TGFβ might be involved in the trans-differentiation of SMCs from cell types present in valve tissues, such as ECs, VICs, or myofibroblasts (MFBs; see below), leading to the observed increase in SMC numbers. Whether p38 MAPK is expressed in valve SMCs, similar to that in vascular SMCs, has not been studied so far.

5. Aortic Valve Interstitial Cells, Myofibroblasts, and Vascular Fibroblasts

Valve interstitial cells (VICs) are the dominant cell type in aortic valves. They are found in the fibrosa, spongiosa, and ventricularis layer, and maintain tissue homeostasis by layer-specific ECM deposition. During development and progression of aortic valve sclerosis, VICs take an active part. It has been shown that VICs transform to activated, smooth muscle actin-expressing myofibroblasts (MFBs) in early valve lesions [76]. A subset of MFBs further differentiate into osteoblast-like cells, which express osteogenic factors and produce a calcium-rich bone matrix. It has been furthermore suggested that apoptotic VICs provide initiation sites for calcific nodule formation, additionally supporting aortic valve calcification [77]. Many signaling networks probably interact to drive the pathologic transformation of VICs in aortic valve sclerosis, and so far, the detailed molecular processes and their chronology are not completely understood. The cytokine TGFβ1, however, is thought to play a major role in driving the processes. Through interaction with its cognate receptors expressed on VICs and MFBs, TGFβ1 activates the small mothers against decapentaplegic (SMAD) signaling cascade, which induces the transcription of osteogenic genes, promoting osteogenesis and calcification in the aortic valve [78,79]. In porcine aortic valves, it was shown that TGFβ1 induces calcium nodule formation, generation of reactive oxygen species, and VIC senescence through SMAD, extracellular signal regulated kinase (ERK)1/2, and p38 MAPK signaling [80]. The p38 kinase has also been implicated in TGFβ-mediated induction of the osteogenic transcription factor RUNX2, and seems to be necessary for the osteoblastic differentiation [81,82]. In addition, p38 MAPK was shown to be involved in bone morphogenic protein 2 (BMP2) signaling, which is another potent inducer of SMAD signaling and RUNX2 expression [81,83]. Independent of TGFβ1, oxLDL may induce p38 MAPK as well as JNK phosphorylation in VICs through the pro-osteogenic receptor for advanced glycosylation end-products (RAGE) [84]. In addition, p38 MAPK has been shown to be activated by sphingosine 1-phosphate and LPS in VICs, inducing pro-inflammatory, pro-angiogenic, and osteogenic processes [85]. Therefore, in aortic valve VICs and MFBs, p38 MAPK seems to be involved in major osteogenic signaling pathways and to support the osteogenic transformation of cells that drives tissue calcification.

In atherosclerosis, fibroblasts in the adventitia layer are also activated at early time points and have the ability to transform to MFBs [86]. Whether they contribute to calcification of plaques similar to VICs in aortic valve sclerosis is not known to date. Experimentally, it has been shown that native LDL can induce p38 MAPK signaling in fibroblasts with consequences for cell spreading and morphology [87]. Here, cholesterol was the component of LDL that induced p38 MAPK most potently [88].

6. Monocytes and Macrophages

In both atherosclerosis and aortic valve sclerosis, monocytes are recruited to early lesions and differentiate into macrophages once they enter the tissue [89]. Macrophages promote lesion progression in different ways. For one, they support remodeling of the ECM by secretion of proteolytic enzymes such as matrix metalloproteinases (MMPs) and cathepsins that degrade collagens and elastins, providing initiation sites for calcium crystallization [90–92]. Secondly, in both pathologies macrophages are transformed to foam cells by the uptake of mLDL. The exact nature of the LDL modification that induces foam cell formation in vivo has not been definitely clarified to date, and different modifications, including oxLDL and enzymatically modified LDL (eLDL), are used to model the process in vitro. Importantly, the uptake of mLDL seems to be required for p38 MAPK activation and foam cell formation in macrophages [93]. Activation of the p38 MAPK pathway has been demonstrated in macrophages associated with atherosclerotic plaques via immunohistochemistry in patient-derived tissues [94] as well as in animal models [95]. Several in vitro studies have investigated the biological consequences of p38 MAPK activation in atherosclerosis-associated macrophages. First of all, p38 MAPK has been demonstrated to be part of a positive feedback mechanism that drives foam cell formation. Here, oxLDL and eLDL induce p38 MAPK activation in macrophages, which in turn enhances LDL uptake by PPARγ-mediated upregulation of LDL uptake receptors such as CD36 [93,94]. In another study, oxLDL has been shown to enhance the adhesive capacity of monocytes by p38 MAPK-mediated upregulation of the chemokine receptor CXCR2 [96]. Furthermore, p38 MAPK activation in macrophages has been shown to induce the expression of pro-inflammatory cytokines in response to mLDL incubation [97,98]. In addition, Senokuchi et al. demonstrated that oxLDL-mediated induction of p38 MAPK is important for the production of the cytokine granulocyte-macrophage colony-stimulating factor (GM-CSF) and proliferation of macrophages [99]. Taken together, the in vitro studies provide concordant evidence for a pro-inflammatory role of p38 MAPK expression in atherosclerosis-associated macrophages, driving disease progression by promotion of macrophage proliferation and chronic inflammation.

In vivo studies in atherosclerotic mouse models on the other hand achieved conflicting results. In a study by Seimon et al. conditional p38 MAPK-deficiency in macrophages of ApoE$^{-/-}$ mice led to increased macrophage apoptosis and atherosclerotic plaque progression induced by ER stress and unfolded protein response [100]. A subsequent study with the same mouse model could not detect any effects of macrophage-specific p38 MAPK depletion on atherosclerotic plaque progression [50]. In an earlier study, systemic depletion of the p38 MAPK downstream kinase MAPK-activated protein kinase-2 (MAPKAPK2) in hypercholesteremic mice resulted in decreased foam cell formation and inflammatory signaling [101].

Similar to atherosclerosis, macrophages infiltrate early valve lesions [76]. To our knowledge, valve-associated macrophages have not been studied so far and therefore their role in disease progression is not clear. In aortic valve sclerosis, lipid-laden macrophages (foam cells) have been detected immunohistologically [76,102], and thus, a similar pro-inflammatory and disease-promoting function as in atherosclerosis might be suspected. Whether p38 MAPK signaling plays a role in valve-associated macrophages and disease progression remains to be investigated.

7. Other Immune Cells: Mast Cells, T Cells, Natural Killer T Cells, B Cells, and Dendritic Cells

Besides macrophages, other immune cells are present in atherosclerotic and aortic valve lesions. In the following section, we summarize major findings around other immune cells that are present in atherosclerotic and/or calcified aortic valve lesions, with a focus on p38 MAPK signaling. The immunologic landscape of these lesions is highly diverse and complex, and a detailed description is beyond the scope of this review. Comprehensive reviews that give detailed insight into the relationship of cardiovascular diseases and inflammation have been published in recent years (e.g., [103–106]).

Mast cells, a leukocyte population that is involved in allergic reactions and wound healing, have been detected in both atherosclerotic and aortic valve lesions [107,108] and shown to induce p38 MAPK phosphorylation in response to oxLDL stimulation in vitro [109]. The authors of the latter study suspected p38 MAPK, together with other MAPKs and NF-κB, to act downstream of toll-like receptor 4 (TLR4), inducing the expression of pro-inflammatory cytokines in the presence of oxLDL, thereby promoting disease progression by recruitment of inflammatory cells and, as a consequence, atherosclerotic plaque destabilization.

T lymphocytes are associated with both calcific nodules in aortic valves [110] and the fibrous cap and plaque in atherosclerosis [111]. The lymphocytes infiltrate early in lesion development, most likely recruited from the blood stream by pro-inflammatory cytokines secreted by macrophages, SMCs and/or VICs. Their entry into the tissue is facilitated by the expression of adhesion molecules, such as VCAM-1, intercellular adhesion molecule 1 (ICAM-1), and P-selectin, by activated endothelial cells. In later disease stages, neo-angiogenesis in the transformed tissue might provide additional access routes for T lymphocytes [112]. T cells are thought to promote lesion progression by maintaining the chronic inflammatory environment that supports tissue remodeling and destabilization. In addition, cytotoxic T cells induce apoptosis in target cells, producing nucleation sites for calcium crystallization [55]. In CAVD, Nagy et al. furthermore showed that activated and clonally expanded cytotoxic CD8^{+} T cells specifically target and kill osteoclasts, a cell type that is usually found in bone tissues and mediates calcium resorption and bone turnover [113]. Although p38 MAPK has been shown to be involved in T cell receptor (TCR) signaling and is important for interferon γ (IFNγ) production in T cells [114,115], to our knowledge, the role of the p38 MAPK signaling pathway in T cells has never been investigated in the context of atherosclerosis or CAVD.

Natural killer T (NKT) cells are a type of T cell that recognizes lipid antigens presented by antigen presenting cells in conjunction with the CD1 surface molecule. In atherosclerosis, NKT cells are present in regions with CD1 expressing foam cells and are suspected to promote the local inflammation by secretion of a variety of cytokines [116]. In agreement with this, NKT cell stimulation exacerbated atherosclerosis in the presence of the CD1 antigen in an ApoE$^{-/-}$ mouse model [117]. The pro-atherogenic function of NKT cells is thought to be a consequence of granzyme B and perforin secretion, two molecules with cytolytic activity [118]. Interestingly, apart from their cytolytic function, NKT cells have been implicated in neo-angiogenesis, which is frequently observed in atherosclerotic plaques and thought to contribute to plaque destabilization [119]. The pro-angiogenic function is thought to be mediated by IL-8 that is secreted by lipid-antigen stimulated NKT cells, which induces epidermal growth factor receptor (EGFR) expression in endothelial cells [120]. NKT cells are also present in sclerotic aortic valves, and have been shown to associate with disease progression in a mouse model [121] and in human valves [122]. The cytokine IL-2 has been shown to upregulate p38 MAPK in NKT cells, resulting in the production of pro- and anti-inflammatory cytokines [123]. Of note, the increased cytokine production mediated by p38 MAPK has been shown to be regulated translationally, but not on the gene expression level [124]. In contrast to these results, in another study, p38 MAPK has been shown to inhibit secretion of IL-2 and IL-4 by NKT cells, and inhibition of p38 MAPK rescued cytokine secretion [125].

B lymphocytes have been detected in atherosclerotic lesions [126,127] and also in calcified heart valves [128]. They gained more attention as antibodies against atherosclerosis-associated epitopes were detected in the lesions and in the blood circulation of patients [129,130]. Generally, two different types of antibodies exist: so called natural antibodies produced by innate-like B-1 cells [131] and antibodies produced by conventional B-2 cells as part of an adaptive immune response. The concept is emerging that natural antibodies targeting autoantigens, including products from oxidative metabolism such as oxLDL, are atheroprotective [132,133], whereas the adaptive immune response mediated by B-2 cells fuels local inflammation and is rather atherogenic [134] (reviewed in [135]). p38 MAPK signaling has not been investigated in the context of atherosclerosis- or CAVD-associated B cells. It has been shown, however, that p38 MAPK is activated upon B cell receptor stimulation, leading to B cell proliferation [136]. The authors of the study moreover showed that p38 MAPK acts in collaboration with the transcription factor MEF2C specifically in B cells that mediate an adaptive immune response. In this regard, p38 MAPK could represent a potential target to reduce atherogenic B cell-mediated inflammatory signaling in atherosclerosis and possibly also in CAVD.

Finally, dendritic cells (DCs) are present in healthy aorta and aortic valves [137]. The endogenous function of DCs is to present antigens to T cells in order to activate an adaptive immune response in the presence of foreign antigens. DCs have also been detected in atherosclerotic lesions [138], especially at sites that are prone to rupture [139]. The role DCs play in atherosclerosis, however, is not well understood. In a review by Koltsova and Ley, studies investigating the impact of DCs for atherosclerosis are summarized, with most studies suggesting a disease-promoting function of DCs, in which they accumulate lipids similar to macrophages and drive local inflammation [140]. No studies investigating the function of DCs in CAVD exist. Experimental findings concerning the importance of the p38 MAPK pathway for DC function are two sided. On the one hand, p38 MAPK activity is required for the maturation of immature DCs [141]. On the other hand, p38 MAPK inhibition in DC progenitor cells leads to enhanced antigen presentation and immune activation [142]. Whether the p38 MAPK pathway is relevant for DC function in atherosclerotic and/or aortic valve lesions, and whether it has an effect on disease development or progression, has not been studied to date.

8. Conclusions

Despite the similarities, atherosclerosis and aortic valve sclerosis are distinct in their pathogenesis. This is highlighted by the fact that, although they share common risk factors, not all patients with atherosclerosis are affected by aortic valve sclerosis, and vice versa. In addition, statins that are effectively applied in atherosclerosis therapy show no clinical benefit for patients with aortic valve sclerosis [143–145]. Both the blood vessel wall and the aortic valve are complex structures comprised of endothelial cells and underlying connective tissue layers, that are dispersed by ECM-producing SMCs and VICs, respectively. Immune cells are present in healthy tissues, however, their numbers rise dramatically as they are recruited to early lesions and establish sites of chronic inflammation in the courses of the diseases. The p38 MAPK is involved in inflammatory signaling in different settings and cell types and has gained interest in atherosclerosis and CAVD research, especially since these diseases have become recognized as inflammation-driven. With this review we aimed at elucidating the role of p38 MAPK in the development and progression of atherosclerosis and CAVD by outlining its functions in the different cell types that constitute the lesions and impact disease progression. Figure 1 provides an overview of the cell types and the corresponding functions that have been attributed to p38 MAPK activity.

One of the biggest challenges of p38 MAPK research is the multitude of different stimuli that induce its phosphorylation, such as cytokines, growth factors, and osmotic, oxidative, and mechanical stresses (reviewed in [146]), and the difficulty in dissecting the most relevant factors in certain (patho)physiological conditions. In atherosclerosis and CAVD, combinations of such stimulants are present in a spatiotemporal distribution, most likely leading to a variable p38 MAPK activation status throughout lesions. In addition, other cellular pathways, including the c-Jun terminal kinase (JNK), extracellular signal regulated kinase (ERK), and TGFβ signaling crosstalk and interact with components of p38 MAPK signaling, further modulating intracellular signal transduction and eventually the generated response. Finally, p38 MAPK activation leads to different responses in different cell types, which again influence each other, producing dynamic interconnected networks. To date, in vitro experimental models cannot capture the entire complexity of the p38 MAPK signaling network of pathological conditions such as atherosclerosis or CAVD. Nevertheless, they are suitable for the investigation of simplified processes under a controlled environment. Here, well defined experimental conditions and the use of highly specific p38 MAPK inhibitors are requisite for the generation of meaningful results. Studies in animals better account for the complexity of lesions, although certainly none of the models used in atherosclerosis or CAVD research today perfectly resemble the human situation [147]. Therefore, as for in vitro experiments, results obtained with animal models should always be reviewed critically before they are translated to the human organism.

In the end, the question should be answered whether p38 MAPK activation is a driver of human atherosclerosis and/or CAVD or merely a consequence of the pro-inflammatory, stress-laden microenvironment of the lesion. Depending on the outcome, pharmacologic targeting of p38 MAPK with highly specific inhibitors such as skepinone-L [94,148], or targeting of components of the corresponding signaling cascade could become an option for a therapeutic intervention in atherosclerosis and/or CAVD in the future. Indeed, clinical trials with the p38 MAPK inhibitors dilmapimod (SB681323, GalaxoSmithKline, London, UK) and losmapimod (GW856553, GalaxoSmithKline, London, UK) were carried out with atherosclerotic patients and shown to reduce inflammation in atherosclerotic lesions [149–151]. However, patients with acute myocardial infarction did not benefit from the treatment with losmapimod in a subsequent phase III trial [152]. Whether patients might benefit from p38 MAPK inhibition at earlier disease stages has, to our knowledge, never been investigated. Especially in the case of CAVD, therapies that slow, halt, or even reverse disease progression are desperately needed to provide an alternative for otherwise inevitable valve replacement procedures.

Author Contributions: Conceptualization, A.R. and M.T.; writing—original draft preparation, A.R.; writing—review and editing, M.T.; visualization, A.R.; supervision, M.T.; funding acquisition, M.T.

Funding: This work was funded by the Robert-Bosch Foundation, Stuttgart, Germany.

Conflicts of Interest: The authors declare no conflict of interest.

Abbreviations

Ang II	angiotensin II
ApoE	apolipoprotein E
AVS	aortic valve stenosis
BMP2	bone morphogenic protein 2
CAVD	calcific aortic valve disease
DC	dendritic cell
EC	endothelial cell
ECM	extracellular matrix
EGFR	epidermal growth factor receptor
eLDL	enzymatically modified LDL
EnMT	endothelial-to-mesenchymal transition
ER	endoplasmic reticulum
ERK	extracellular signal regulated kinase
GM-CSF	granulocyte-macrophage colony-stimulating factor
GWAS	genome-wide association study
HDL	high-density lipoprotein
HUVEC	human umbilical-vein endothelial cell
ICAM-1	intercellular adhesion molecule 1
IL	interleukin
JNK	c-Jun N-terminal kinase
IFN	interferon
LDL	low-density lipoprotein
Lp(a)	lipoprotein a
LPS	lipopolysaccharide
MAPK	mitogen-activated protein kinase
MAPKAPK2	MAPK-activated protein kinase 2
MAPKK	MAPK kinase
MAPKKK	MAPK kinase kinase
MCP-1	monocyte-chemoattractant protein 1
MFB	myofibroblast
mLDL	modified LDL
MMP	matrix metalloproteinase
NKT	natural killer T cell
NO	nitric oxide
oxHDL	oxidized HDL
oxLDL	oxidized LDL
PPARγ	peroxisome proliferator-activated receptor gamma
SMAD	small mothers against decapentaplegic
SMC	smooth muscle cell
TCR	T cell receptor
TGFβ	transforming growth factor beta
TLR4	toll-like receptor 4
VCAM-1	vascular cell adhesion protein 1
VIC	valve interstitial cell
VSMC	vascular smooth muscle cell

References

1. World Health Organization. Cardiovascular Diseases. Available online: http://www.euro.who.int/en/health-topics/noncommunicable-diseases/cardiovascular-diseases (accessed on 10 September 2018).
2. Roth, G.A.; Johnson, C.; Abajobir, A.; Abd-Allah, F.; Abera, S.F.; Abyu, G.; Ahmed, M.; Aksut, B.; Alam, T.; Alam, K.; et al. Global, Regional, and National Burden of Cardiovascular Diseases for 10 Causes, 1990 to 2015. *J. Am. Coll. Cardiol.* **2017**, *70*, 1–25. [CrossRef] [PubMed]

3. Banach, M.; Serban, C.; Sahebkar, A.; Mikhailidis, D.P.; Ursoniu, S.; Ray, K.K.; Rysz, J.; Toth, P.P.; Muntner, P.; Mosteoru, S.; et al. Impact of statin therapy on coronary plaque composition: A systematic review and meta-analysis of virtual histology intravascular ultrasound studies. *BMC Med.* **2015**, *13*, 229. [CrossRef] [PubMed]

4. Nakamura, K.; Sasaki, T.; Cheng, X.W.; Iguchi, A.; Sato, K.; Kuzuya, M. Statin prevents plaque disruption in apoE-knockout mouse model through pleiotropic effect on acute inflammation. *Atherosclerosis* **2009**, *206*, 355–361. [CrossRef] [PubMed]

5. Osnabrugge, R.L.J.; Mylotte, D.; Head, S.J.; van Mieghem, N.M.; Nkomo, V.T.; LeReun, C.M.; Bogers, A.J.J.C.; Piazza, N.; Kappetein, A.P. Aortic stenosis in the elderly: Disease prevalence and number of candidates for transcatheter aortic valve replacement: A meta-analysis and modeling study. *J. Am. Coll. Cardiol.* **2013**, *62*, 1002–1012. [CrossRef] [PubMed]

6. Larsson, S.C.; Wolk, A.; Bäck, M. Alcohol consumption, cigarette smoking and incidence of aortic valve stenosis. *J. Intern. Med.* **2017**, *282*, 332–339. [CrossRef] [PubMed]

7. Larsson, S.C.; Wolk, A.; Håkansson, N.; Bäck, M. Overall and abdominal obesity and incident aortic valve stenosis: Two prospective cohort studies. *Eur. Heart J.* **2017**, *38*, 2192–2197. [CrossRef] [PubMed]

8. Rahimi, K.; Mohseni, H.; Kiran, A.; Tran, J.; Nazarzadeh, M.; Rahimian, F.; Woodward, M.; Dwyer, T.; MacMahon, S.; Otto, C.M. Elevated blood pressure and risk of aortic valve disease: A cohort analysis of 5.4 million UK adults. *Eur. Heart J.* **2018**, *39*, 3596–3603. [CrossRef] [PubMed]

9. Stewart, B.F.; Siscovick, D.; Lind, B.K.; Gardin, J.M.; Gottdiener, J.S.; Smith, V.E.; Kitzman, D.W.; Otto, C.M. Clinical factors associated with calcific aortic valve disease. Cardiovascular Health Study. *J. Am. Coll. Cardiol.* **1997**, *29*, 630–634. [CrossRef]

10. Guerraty, M.A.; Chai, B.; Hsu, J.Y.; Ojo, A.O.; Gao, Y.; Yang, W.; Keane, M.G.; Budoff, M.J.; Mohler, E.R. Relation of aortic valve calcium to chronic kidney disease (from the Chronic Renal Insufficiency Cohort Study). *Am. J. Cardiol.* **2015**, *115*, 1281–1286. [CrossRef] [PubMed]

11. Katz, R.; Wong, N.D.; Kronmal, R.; Takasu, J.; Shavelle, D.M.; Probstfield, J.L.; Bertoni, A.G.; Budoff, M.J.; O'Brien, K.D. Features of the metabolic syndrome and diabetes mellitus as predictors of aortic valve calcification in the Multi-Ethnic Study of Atherosclerosis. *Circulation* **2006**, *113*, 2113–2119. [CrossRef] [PubMed]

12. Klarin, D.; Zhu, Q.M.; Emdin, C.A.; Chaffin, M.; Horner, S.; McMillan, B.J.; Leed, A.; Weale, M.E.; Spencer, C.C.A.; Aguet, F.; et al. Genetic analysis in UK Biobank links insulin resistance and transendothelial migration pathways to coronary artery disease. *Nat. Genet.* **2017**, *49*, 1392–1397. [CrossRef] [PubMed]

13. Howson, J.M.M.; Zhao, W.; Barnes, D.R.; Ho, W.-K.; Young, R.; Paul, D.S.; Waite, L.L.; Freitag, D.F.; Fauman, E.B.; Salfati, E.L.; et al. Fifteen new risk loci for coronary artery disease highlight arterial-wall-specific mechanisms. *Nat. Genet.* **2017**, *49*, 1113–1119. [CrossRef] [PubMed]

14. Thanassoulis, G.; Campbell, C.Y.; Owens, D.S.; Smith, J.G.; Smith, A.V.; Peloso, G.M.; Kerr, K.F.; Pechlivanis, S.; Budoff, M.J.; Harris, T.B.; et al. Genetic associations with valvular calcification and aortic stenosis. *N. Engl. J. Med.* **2013**, *368*, 503–512. [CrossRef] [PubMed]

15. Helgadottir, A.; Thorleifsson, G.; Gretarsdottir, S.; Stefansson, O.A.; Tragante, V.; Thorolfsdottir, R.B.; Jonsdottir, I.; Bjornsson, T.; Steinthorsdottir, V.; Verweij, N.; et al. Genome-wide analysis yields new loci associating with aortic valve stenosis. *Nat. Commun.* **2018**, *9*, 987. [CrossRef] [PubMed]

16. Torzewski, M.; Ravandi, A.; Yeang, C.; Edel, A.; Bhindi, R.; Kath, S.; Twardowski, L.; Schmid, J.; Yang, X.; Franke, U.F.W.; et al. Lipoprotein(a) Associated Molecules are Prominent Components in Plasma and Valve Leaflets in Calcific Aortic Valve Stenosis. *JACC Basic Transl. Sci.* **2017**, *2*, 229–240. [CrossRef] [PubMed]

17. Bozbas, H.; Yildirir, A.; Atar, I.; Pirat, B.; Eroglu, S.; Aydinalp, A.; Ozin, B.; Muderrisoglu, H. Effects of serum levels of novel atherosclerotic risk factors on aortic valve calcification. *J. Heart Valve Dis.* **2007**, *16*, 387–393. [PubMed]

18. Gotoh, T.; Kuroda, T.; Yamasawa, M.; Nishinaga, M.; Mitsuhashi, T.; Seino, Y.; Nagoh, N.; Kayaba, K.; Yamada, S.; Matsuo, H. Correlation between lipoprotein(a) and aortic valve sclerosis assessed by echocardiography (the JMS Cardiac Echo and Cohort Study). *Am. J. Cardiol.* **1995**, *76*, 928–932. [CrossRef]

19. Rajamannan, N.M. Calcific aortic stenosis: Lessons learned from experimental and clinical studies. *Arterioscler. Thromb. Vasc. Biol.* **2009**, *29*, 162–168. [CrossRef] [PubMed]

20. Virmani, R.; Kolodgie, F.D.; Burke, A.P.; Farb, A.; Schwartz, S.M. Lessons from sudden coronary death: A comprehensive morphological classification scheme for atherosclerotic lesions. *Arterioscler. Thromb. Vasc. Biol.* **2000**, *20*, 1262–1275. [CrossRef] [PubMed]

21. Stary, H.C. Natural history and histological classification of atherosclerotic lesions: An update. *Arterioscler. Thromb. Vasc. Biol.* **2000**, *20*, 1177–1178. [CrossRef] [PubMed]

22. Libby, P. Inflammation and Atherosclerosis. *Circulation* **2002**, *105*, 1135–1143. [CrossRef] [PubMed]

23. Libby, P. Inflammation in atherosclerosis. *Nature* **2002**, *420*, 868–874. [CrossRef] [PubMed]

24. Freeman, R.V.; Otto, C.M. Spectrum of Calcific Aortic Valve Disease: Pathogenesis, Disease Progression, and Treatment Strategies. *Circulation* **2005**, *111*, 3316–3326. [CrossRef] [PubMed]

25. Coulthard, L.R.; White, D.E.; Jones, D.L.; McDermott, M.F.; Burchill, S.A. p38(MAPK): Stress responses from molecular mechanisms to therapeutics. *Trends Mol. Med.* **2009**, *15*, 369–379. [CrossRef] [PubMed]

26. Guay, J.; Lambert, H.; Gingras-Breton, G.; Lavoie, J.N.; Huot, J.; Landry, J. Regulation of actin filament dynamics by p38 map kinase-mediated phosphorylation of heat shock protein 27. *J. Cell Sci.* **1997**, *110 Pt 3*, 357–368.

27. Cargnello, M.; Roux, P.P. Activation and function of the MAPKs and their substrates, the MAPK-activated protein kinases. *Microbiol. Mol. Biol. Rev.* **2011**, *75*, 50–83. [CrossRef] [PubMed]

28. Bonney, E.A. Mapping out p38MAPK. *Am. J. Reprod. Immunol.* **2017**, *77*. [CrossRef] [PubMed]

29. Torzewski, M. Enzymatically modified LDL, atherosclerosis and beyond: Paving the way to acceptance. *Front. Biosci.* **2018**, *23*, 1257–1271. [CrossRef]

30. Zhu, Y.; Liao, H.; Wang, N.; Ma, K.S.; Verna, L.K.; Shyy, J.Y.; Chien, S.; Stemerman, M.B. LDL-activated p38 in endothelial cells is mediated by Ras. *Arterioscler. Thromb. Vasc. Biol.* **2001**, *21*, 1159–1164. [CrossRef] [PubMed]

31. Pietersma, A.; Tilly, B.C.; Gaestel, M.; de Jong, N.; Lee, J.C.; Koster, J.F.; Sluiter, W. p38 mitogen activated protein kinase regulates endothelial VCAM-1 expression at the post-transcriptional level. *Biochem. Biophys. Res. Commun.* **1997**, *230*, 44–48. [CrossRef] [PubMed]

32. Goebeler, M.; Kilian, K.; Gillitzer, R.; Kunz, M.; Yoshimura, T.; Bröcker, E.B.; Rapp, U.R.; Ludwig, S. The MKK6/p38 stress kinase cascade is critical for tumor necrosis factor-alpha-induced expression of monocyte-chemoattractant protein-1 in endothelial cells. *Blood* **1999**, *93*, 857–865. [PubMed]

33. Liu, L.; Craig, A.W.; Meldrum, H.D.; Marcovina, S.M.; Elliott, B.E.; Koschinsky, M.L. Apolipoprotein(a) stimulates vascular endothelial cell growth and migration and signals through integrin alphaVbeta3. *Biochem. J.* **2009**, *418*, 325–336. [CrossRef] [PubMed]

34. Rousseau, S.; Houle, F.; Landry, J.; Huot, J. p38 MAP kinase activation by vascular endothelial growth factor mediates actin reorganization and cell migration in human endothelial cells. *Oncogene* **1997**, *15*, 2169–2177. [CrossRef] [PubMed]

35. McMullen, M.E.; Bryant, P.W.; Glembotski, C.C.; Vincent, P.A.; Pumiglia, K.M. Activation of p38 has opposing effects on the proliferation and migration of endothelial cells. *J. Biol. Chem.* **2005**, *280*, 20995–21003. [CrossRef] [PubMed]

36. Mohler, E.R.; Gannon, F.; Reynolds, C.; Zimmerman, R.; Keane, M.G.; Kaplan, F.S. Bone formation and inflammation in cardiac valves. *Circulation* **2001**, *103*, 1522–1528. [CrossRef] [PubMed]

37. Herrmann, J.; Lerman, L.O.; Mukhopadhyay, D.; Napoli, C.; Lerman, A. Angiogenesis in atherogenesis. *Arterioscler. Thromb. Vasc. Biol.* **2006**, *26*, 1948–1957. [CrossRef] [PubMed]

38. Borbiev, T.; Birukova, A.; Liu, F.; Nurmukhambetova, S.; Gerthoffer, W.T.; Garcia, J.G.N.; Verin, A.D. p38 MAP kinase-dependent regulation of endothelial cell permeability. *Am. J. Physiol. Lung Cell. Mol. Physiol.* **2004**, *287*, L911–L918. [CrossRef] [PubMed]

39. Rader, D.J.; Hovingh, G.K. HDL and cardiovascular disease. *Lancet* **2014**, *384*, 618–625. [CrossRef]

40. Gomaraschi, M.; Basilico, N.; Sisto, F.; Taramelli, D.; Eligini, S.; Colli, S.; Sirtori, C.R.; Franceschini, G.; Calabresi, L. High-density lipoproteins attenuate interleukin-6 production in endothelial cells exposed to pro-inflammatory stimuli. *Biochim. Biophys. Acta* **2005**, *1736*, 136–143. [CrossRef] [PubMed]

41. Martínez-González, J.; Escudero, I.; Badimon, L. Simvastatin potenciates PGI(2) release induced by HDL in human VSMC: Effect on Cox-2 up-regulation and MAPK signalling pathways activated by HDL. *Atherosclerosis* **2004**, *174*, 305–313. [CrossRef] [PubMed]

42. Ren, J.; Jin, W.; Chen, H. oxHDL decreases the expression of CD36 on human macrophages through PPARgamma and p38 MAP kinase dependent mechanisms. *Mol. Cell. Biochem.* **2010**, *342*, 171–181. [CrossRef] [PubMed]

43. Norata, G.D.; Banfi, C.; Pirillo, A.; Tremoli, E.; Hamsten, A.; Catapano, A.L.; Eriksson, P. Oxidised-HDL3 induces the expression of PAI-1 in human endothelial cells. Role of p38MAPK activation and mRNA stabilization. *Br. J. Haematol.* **2004**, *127*, 97–104. [CrossRef] [PubMed]

44. Norata, G.D.; Callegari, E.; Inoue, H.; Catapano, A.L. HDL3 induces cyclooxygenase-2 expression and prostacyclin release in human endothelial cells via a p38 MAPK/CRE-dependent pathway: Effects on COX-2/PGI-synthase coupling. *Arterioscler. Thromb. Vasc. Biol.* **2004**, *24*, 871–877. [CrossRef] [PubMed]

45. Wylie-Sears, J.; Aikawa, E.; Levine, R.A.; Yang, J.-H.; Bischoff, J. Mitral valve endothelial cells with osteogenic differentiation potential. *Arterioscler. Thromb. Vasc. Biol.* **2011**, *31*, 598–607. [CrossRef] [PubMed]

46. Balachandran, K.; Alford, P.W.; Wylie-Sears, J.; Goss, J.A.; Grosberg, A.; Bischoff, J.; Aikawa, E.; Levine, R.A.; Parker, K.K. Cyclic strain induces dual-mode endothelial-mesenchymal transformation of the cardiac valve. *Proc. Natl. Acad. Sci. USA* **2011**, *108*, 19943–19948. [CrossRef] [PubMed]

47. Sumpio, B.E.; Yun, S.; Cordova, A.C.; Haga, M.; Zhang, J.; Koh, Y.; Madri, J.A. MAPKs (ERK1/2, p38) and AKT can be phosphorylated by shear stress independently of platelet endothelial cell adhesion molecule-1 (CD31) in vascular endothelial cells. *J. Biol. Chem.* **2005**, *280*, 11185–11191. [CrossRef] [PubMed]

48. Azuma, N.; Duzgun, S.A.; Ikeda, M.; Kito, H.; Akasaka, N.; Sasajima, T.; Sumpio, B.E. Endothelial cell response to different mechanical forces. *J. Vasc. Surg.* **2000**, *32*, 789–794. [CrossRef] [PubMed]

49. Azuma, N.; Akasaka, N.; Kito, H.; Ikeda, M.; Gahtan, V.; Sasajima, T.; Sumpio, B.E. Role of p38 MAP kinase in endothelial cell alignment induced by fluid shear stress. *Am. J. Physiol. Heart Circ. Physiol.* **2001**, *280*, H189–H197. [CrossRef] [PubMed]

50. Kardakaris, R.; Gareus, R.; Xanthoulea, S.; Pasparakis, M. Endothelial and macrophage-specific deficiency of P38α MAPK does not affect the pathogenesis of atherosclerosis in ApoE$^-$/$^-$ mice. *PLoS ONE* **2011**, *6*, e21055. [CrossRef] [PubMed]

51. Mestas, J.; Hughes, C.C.W. Of mice and not men: Differences between mouse and human immunology. *J. Immunol.* **2004**, *172*, 2731–2738. [CrossRef] [PubMed]

52. von Scheidt, M.; Zhao, Y.; Kurt, Z.; Pan, C.; Zeng, L.; Yang, X.; Schunkert, H.; Lusis, A.J. Applications and Limitations of Mouse Models for Understanding Human Atherosclerosis. *Cell Metab.* **2017**, *25*, 248–261. [CrossRef] [PubMed]

53. Davies, M.J.; Richardson, P.D.; Woolf, N.; Katz, D.R.; Mann, J. Risk of thrombosis in human atherosclerotic plaques: Role of extracellular lipid, macrophage, and smooth muscle cell content. *Br. Heart J.* **1993**, *69*, 377–381. [CrossRef] [PubMed]

54. Kockx, M.M. Apoptosis in the atherosclerotic plaque: Quantitative and qualitative aspects. *Arterioscler. Thromb. Vasc. Biol.* **1998**, *18*, 1519–1522. [CrossRef] [PubMed]

55. Proudfoot, D.; Skepper, J.N.; Hegyi, L.; Bennett, M.R.; Shanahan, C.M.; Weissberg, P.L. Apoptosis regulates human vascular calcification in vitro: Evidence for initiation of vascular calcification by apoptotic bodies. *Circ. Res.* **2000**, *87*, 1055–1062. [CrossRef] [PubMed]

56. Nishio, E.; Arimura, S.; Watanabe, Y. Oxidized LDL induces apoptosis in cultured smooth muscle cells: A possible role for 7-ketocholesterol. *Biochem. Biophys. Res. Commun.* **1996**, *223*, 413–418. [CrossRef] [PubMed]

57. Lee, T.; Chau, L. Fas/Fas ligand-mediated death pathway is involved in oxLDL-induced apoptosis in vascular smooth muscle cells. *Am. J. Physiol. Cell Physiol.* **2001**, *280*, C709–C718. [CrossRef] [PubMed]

58. Jovinge, S.; Crisby, M.; Thyberg, J.; Nilsson, J. DNA fragmentation and ultrastructural changes of degenerating cells in atherosclerotic lesions and smooth muscle cells exposed to oxidized LDL in vitro. *Arterioscler. Thromb. Vasc. Biol.* **1997**, *17*, 2225–2231. [CrossRef] [PubMed]

59. Jing, Q.; Xin, S.M.; Cheng, Z.J.; Zhang, W.B.; Zhang, R.; Qin, Y.W.; Pei, G. Activation of p38 mitogen-activated protein kinase by oxidized LDL in vascular smooth muscle cells: Mediation via pertussis toxin-sensitive G proteins and association with oxidized LDL-induced cytotoxicity. *Circ. Res.* **1999**, *84*, 831–839. [CrossRef] [PubMed]

60. Liao, L.; Zhou, Q.; Song, Y.; Wu, W.; Yu, H.; Wang, S.; Chen, Y.; Ye, M.; Lu, L. Ceramide mediates Ox-LDL-induced human vascular smooth muscle cell calcification via p38 mitogen-activated protein kinase signaling. *PLoS ONE* **2013**, *8*, e82379. [CrossRef] [PubMed]

61. Loidl, A.; Claus, R.; Ingolic, E.; Deigner, H.-P.; Hermetter, A. Role of ceramide in activation of stress-associated MAP kinases by minimally modified LDL in vascular smooth muscle cells. *Biochim. Biophys. Acta* **2004**, *1690*, 150–158. [CrossRef] [PubMed]

62. Watson, A.D.; Leitinger, N.; Navab, M.; Faull, K.F.; Hörkkö, S.; Witztum, J.L.; Palinski, W.; Schwenke, D.; Salomon, R.G.; Sha, W.; et al. Structural identification by mass spectrometry of oxidized phospholipids in minimally oxidized low density lipoprotein that induce monocyte/endothelial interactions and evidence for their presence in vivo. *J. Biol. Chem.* **1997**, *272*, 13597–13607. [CrossRef] [PubMed]

63. Miller, Y.I.; Choi, S.-H.; Wiesner, P.; Fang, L.; Harkewicz, R.; Hartvigsen, K.; Boullier, A.; Gonen, A.; Diehl, C.J.; Que, X.; et al. Oxidation-specific epitopes are danger-associated molecular patterns recognized by pattern recognition receptors of innate immunity. *Circ. Res.* **2011**, *108*, 235–248. [CrossRef] [PubMed]

64. Boullier, A.; Li, Y.; Quehenberger, O.; Palinski, W.; Tabas, I.; Witztum, J.L.; Miller, Y.I. Minimally oxidized LDL offsets the apoptotic effects of extensively oxidized LDL and free cholesterol in macrophages. *Arterioscler. Thromb. Vasc. Biol.* **2006**, *26*, 1169–1176. [CrossRef] [PubMed]

65. Geisterfer, A.A.; Peach, M.J.; Owens, G.K. Angiotensin II induces hypertrophy, not hyperplasia, of cultured rat aortic smooth muscle cells. *Circ. Res.* **1988**, *62*, 749–756. [CrossRef] [PubMed]

66. Berk, B.C.; Vekshtein, V.; Gordon, H.M.; Tsuda, T. Angiotensin II-stimulated protein synthesis in cultured vascular smooth muscle cells. *Hypertension* **1989**, *13*, 305–314. [CrossRef] [PubMed]

67. Ushio-Fukai, M.; Alexander, R.W.; Akers, M.; Griendling, K.K. p38 Mitogen-activated protein kinase is a critical component of the redox-sensitive signaling pathways activated by angiotensin II. Role in vascular smooth muscle cell hypertrophy. *J. Biol. Chem.* **1998**, *273*, 15022–15029. [CrossRef] [PubMed]

68. Kyaw, M.; Yoshizumi, M.; Tsuchiya, K.; Kirima, K.; Tamaki, T. Antioxidants inhibit JNK and p38 MAPK activation but not ERK $\frac{1}{2}$ activation by angiotensin II in rat aortic smooth muscle cells. *Hypertens. Res.* **2001**, *24*, 251–261. [CrossRef] [PubMed]

69. Subramanian, V.; Golledge, J.; Heywood, E.B.; Bruemmer, D.; Daugherty, A. Regulation of peroxisome proliferator-activated receptor-γ by angiotensin II via transforming growth factor-β1-activated p38 mitogen-activated protein kinase in aortic smooth muscle cells. *Arterioscler. Thromb. Vasc. Biol.* **2012**, *32*, 397–405. [CrossRef] [PubMed]

70. Xu, X.; Ha, C.-H.; Wong, C.; Wang, W.; Hausser, A.; Pfizenmaier, K.; Olson, E.N.; McKinsey, T.A.; Jin, Z.-G. Angiotensin II stimulates protein kinase D-dependent histone deacetylase 5 phosphorylation and nuclear export leading to vascular smooth muscle cell hypertrophy. *Arterioscler. Thromb. Vasc. Biol.* **2007**, *27*, 2355–2362. [CrossRef] [PubMed]

71. Lee, H.M.; Lee, C.-K.; Lee, S.H.; Roh, H.Y.; Bae, Y.M.; Lee, K.-Y.; Lim, J.; Park, P.-J.; Park, T.-K.; Lee, Y.L.; et al. p38 mitogen-activated protein kinase contributes to angiotensin II-stimulated migration of rat aortic smooth muscle cells. *J. Pharmacol. Sci.* **2007**, *105*, 74–81. [CrossRef] [PubMed]

72. Won, S.-M.; Park, Y.-H.; Kim, H.-J.; Park, K.-M.; Lee, W.-J. Catechins inhibit angiotensin II-induced vascular smooth muscle cell proliferation via mitogen-activated protein kinase pathway. *Exp. Mol. Med.* **2006**, *38*, 525–534. [CrossRef] [PubMed]

73. Bairati, A.; DeBiasi, S. Presence of a smooth muscle system in aortic valve leaflets. *Anat. Embryol.* **1981**, *161*, 329–340. [CrossRef] [PubMed]

74. Della Rocca, F.; Sartore, S.; Guidolin, D.; Bertiplaglia, B.; Gerosa, G.; Casarotto, D.; Pauletto, P. Cell composition of the human pulmonary valve: A comparative study with the aortic valve—The VESALIO Project. *Ann. Thorac. Surg.* **2000**, *70*, 1594–1600. [CrossRef]

75. Latif, N.; Sarathchandra, P.; Chester, A.H.; Yacoub, M.H. Expression of smooth muscle cell markers and co-activators in calcified aortic valves. *Eur. Heart J.* **2015**, *36*, 1335–1345. [CrossRef] [PubMed]

76. Otto, C.M.; Kuusisto, J.; Reichenbach, D.D.; Gown, A.M.; O'Brien, K.D. Characterization of the early lesion of 'degenerative' valvular aortic stenosis. Histological and immunohistochemical studies. *Circulation* **1994**, *90*, 844–853. [CrossRef] [PubMed]

77. Jian, B.; Narula, N.; Li, Q.-Y.; Mohler, E.R.; Levy, R.J. Progression of aortic valve stenosis: TGF-β1 is present in calcified aortic valve cusps and promotes aortic valve interstitial cell calcification via apoptosis. *Ann. Thorac. Surg.* **2003**, *75*, 457–465. [CrossRef]

78. Osman, L.; Yacoub, M.H.; Latif, N.; Amrani, M.; Chester, A.H. Role of human valve interstitial cells in valve calcification and their response to atorvastatin. *Circulation* **2006**, *114*, I547–I552. [CrossRef] [PubMed]

79. Miyazono, K.; Kusanagi, K.; Inoue, H. Divergence and convergence of TGF-beta/BMP signaling. *J. Cell. Physiol.* **2001**, *187*, 265–276. [CrossRef] [PubMed]

80. Das, D.; Holmes, A.; Murphy, G.A.; Mishra, K.; Rosenkranz, A.C.; Horowitz, J.D.; Kennedy, J.A. TGF-beta1-Induced MAPK activation promotes collagen synthesis, nodule formation, redox stress and cellular senescence in porcine aortic valve interstitial cells. *J. Heart Valve Dis.* **2013**, *22*, 621–630. [PubMed]

81. Lee, K.-S.; Hong, S.-H.; Bae, S.-C. Both the Smad and p38 MAPK pathways play a crucial role in Runx2 expression following induction by transforming growth factor-beta and bone morphogenetic protein. *Oncogene* **2002**, *21*, 7156–7163. [CrossRef] [PubMed]

82. Gallea, S.; Lallemand, F.; Atfi, A.; Rawadi, G.; Ramez, V.; Spinella-Jaegle, S.; Kawai, S.; Faucheu, C.; Huet, L.; Baron, R.; et al. Activation of mitogen-activated protein kinase cascades is involved in regulation of bone morphogenetic protein-2-induced osteoblast differentiation in pluripotent C2C12 cells. *Bone* **2001**, *28*, 491–498. [CrossRef]

83. Song, R.; Zeng, Q.; Ao, L.; Yu, J.A.; Cleveland, J.C.; Zhao, K.-S.; Fullerton, D.A.; Meng, X. Biglycan induces the expression of osteogenic factors in human aortic valve interstitial cells via Toll-like receptor-2. *Arterioscler. Thromb. Vasc. Biol.* **2012**, *32*, 2711–2720. [CrossRef] [PubMed]

84. Li, F.; Zhao, Z.; Cai, Z.; Dong, N.; Liu, Y. Oxidized low-density lipoprotein promotes osteoblastic differentiation of valvular interstitial cells through RAGE/MAPK. *Cardiology* **2015**, *130*, 55–61. [CrossRef] [PubMed]

85. Fernández-Pisonero, I.; López, J.; Onecha, E.; Dueñas, A.I.; Maeso, P.; Crespo, M.S.; San Román, J.A.; García-Rodríguez, C. Synergy between sphingosine 1-phosphate and lipopolysaccharide signaling promotes an inflammatory, angiogenic and osteogenic response in human aortic valve interstitial cells. *PLoS ONE* **2014**, *9*, e109081. [CrossRef] [PubMed]

86. Xu, F.; Ji, J.; Li, L.; Chen, R.; Hu, W.-C. Adventitial fibroblasts are activated in the early stages of atherosclerosis in the apolipoprotein E knockout mouse. *Biochem. Biophys. Res. Commun.* **2007**, *352*, 681–688. [CrossRef] [PubMed]

87. Dobreva, I.; Waeber, G.; Mooser, V.; James, R.W.; Widmann, C. LDLs induce fibroblast spreading independently of the LDL receptor via activation of the p38 MAPK pathway. *J. Lipid Res.* **2003**, *44*, 2382–2390. [CrossRef] [PubMed]

88. Dobreva, I.; Zschörnig, O.; Waeber, G.; James, R.W.; Widmann, C. Cholesterol is the major component of native lipoproteins activating the p38 mitogen-activated protein kinases. *Biol. Chem.* **2005**, *386*, 909–918. [CrossRef] [PubMed]

89. Gerrity, R.G. The role of the monocyte in atherogenesis: I. Transition of blood-borne monocytes into foam cells in fatty lesions. *Am. J. Pathol.* **1981**, *103*, 181–190. [PubMed]

90. Qin, X.; Corriere, M.A.; Matrisian, L.M.; Guzman, R.J. Matrix metalloproteinase inhibition attenuates aortic calcification. *Arterioscler. Thromb. Vasc. Biol.* **2006**, *26*, 1510–1516. [CrossRef] [PubMed]

91. Deguchi, J.-O.; Aikawa, E.; Libby, P.; Vachon, J.R.; Inada, M.; Krane, S.M.; Whittaker, P.; Aikawa, M. Matrix metalloproteinase-13/collagenase-3 deletion promotes collagen accumulation and organization in mouse atherosclerotic plaques. *Circulation* **2005**, *112*, 2708–2715. [CrossRef] [PubMed]

92. Aikawa, E.; Nahrendorf, M.; Sosnovik, D.; Lok, V.M.; Jaffer, F.A.; Aikawa, M.; Weissleder, R. Multimodality molecular imaging identifies proteolytic and osteogenic activities in early aortic valve disease. *Circulation* **2007**, *115*, 377–386. [CrossRef] [PubMed]

93. Zhao, M.; Liu, Y.; Wang, X.; New, L.; Han, J.; Brunk, U.T. Activation of the p38 MAP kinase pathway is required for foam cell formation from macrophages exposed to oxidized LDL. *APMIS* **2002**, *110*, 458–468. [CrossRef] [PubMed]

94. Cheng, F.; Twardowski, L.; Fehr, S.; Aner, C.; Schaeffeler, E.; Joos, T.; Knorpp, T.; Dorweiler, B.; Laufer, S.; Schwab, M.; et al. Selective p38α MAP kinase/MAPK14 inhibition in enzymatically modified LDL-stimulated human monocytes: Implications for atherosclerosis. *FASEB J.* **2017**, *31*, 674–686. [CrossRef] [PubMed]

95. Shafi, S.; Codrington, R.; Gidden, L.M.; Ferns, G.A.A. Increased expression of phosphorylated forms of heat-shock protein-27 and p38MAPK in macrophage-rich regions of fibro-fatty atherosclerotic lesions in the rabbit. *Int. J. Exp. Pathol.* **2016**, *97*, 56–65. [CrossRef] [PubMed]

96. Lei, Z. OxLDL upregulates CXCR2 expression in monocytes via scavenger receptors and activation of p38 mitogen-activated protein kinase. *Cardiovasc. Res.* **2002**, *53*, 524–532. [CrossRef]

97. Wang, S.; Zhou, H.; Feng, T.; Wu, R.; Sun, X.; Guan, N.; Qu, L.; Gao, Z.; Yan, J.; Xu, N.; et al. β-Glucan attenuates inflammatory responses in oxidized LDL-induced THP-1 cells via the p38 MAPK pathway. *Nutr. Metab. Cardiovasc. Dis.* **2014**, *24*, 248–255. [CrossRef] [PubMed]

98. Hakala, J.K.; Lindstedt, K.A.; Kovanen, P.T.; Pentikäinen, M.O. Low-Density Lipoprotein Modified by Macrophage-Derived Lysosomal Hydrolases Induces Expression and Secretion of IL-8 Via p38 MAPK and NF-κB by Human Monocyte-Derived Macrophages. *Arterioscler. Thromb. Vasc. Biol.* **2006**, *26*, 2504–2509. [CrossRef] [PubMed]

99. Senokuchi, T.; Matsumura, T.; Sakai, M.; Matsuo, T.; Yano, M.; Kiritoshi, S.; Sonoda, K.; Kukidome, D.; Nishikawa, T.; Araki, E. Extracellular signal-regulated kinase and p38 mitogen-activated protein kinase mediate macrophage proliferation induced by oxidized low-density lipoprotein. *Atherosclerosis* **2004**, *176*, 233–245. [CrossRef] [PubMed]

100. Seimon, T.A.; Wang, Y.; Han, S.; Senokuchi, T.; Schrijvers, D.M.; Kuriakose, G.; Tall, A.R.; Tabas, I.A. Macrophage deficiency of p38alpha MAPK promotes apoptosis and plaque necrosis in advanced atherosclerotic lesions in mice. *J. Clin. Investig.* **2009**, *119*, 886–898. [CrossRef] [PubMed]

101. Jagavelu, K.; Tietge, U.J.F.; Gaestel, M.; Drexler, H.; Schieffer, B.; Bavendiek, U. Systemic deficiency of the MAP kinase-activated protein kinase 2 reduces atherosclerosis in hypercholesterolemic mice. *Circ. Res.* **2007**, *101*, 1104–1112. [CrossRef] [PubMed]

102. O'Brien, K.D.; Reichenbach, D.D.; Marcovina, S.M.; Kuusisto, J.; Alpers, C.E.; Otto, C.M. Apolipoproteins B, (a), and E accumulate in the morphologically early lesion of 'degenerative' valvular aortic stenosis. *Arterioscler. Thromb. Vasc. Biol.* **1996**, *16*, 523–532. [CrossRef] [PubMed]

103. Wu, M.-Y.; Li, C.-J.; Hou, M.-F.; Chu, P.-Y. New Insights into the Role of Inflammation in the Pathogenesis of Atherosclerosis. *Int. J. Mol. Sci.* **2017**, *18*, 2034. [CrossRef] [PubMed]

104. Moriya, J. Critical roles of inflammation in atherosclerosis. *J. Cardiol.* **2018**. [CrossRef] [PubMed]

105. Pawade, T.A.; Newby, D.E.; Dweck, M.R. Calcification in Aortic Stenosis: The Skeleton Key. *J. Am. Coll. Cardiol.* **2015**, *66*, 561–577. [CrossRef] [PubMed]

106. Towler, D.A. Oxidation, inflammation, and aortic valve calcification peroxide paves an osteogenic path. *J. Am. Coll. Cardiol.* **2008**, *52*, 851–854. [CrossRef] [PubMed]

107. Helske, S.; Lindstedt, K.A.; Laine, M.; Mäyränpää, M.; Werkkala, K.; Lommi, J.; Turto, H.; Kupari, M.; Kovanen, P.T. Induction of local angiotensin II-producing systems in stenotic aortic valves. *J. Am. Coll. Cardiol.* **2004**, *44*, 1859–1866. [CrossRef] [PubMed]

108. Kaartinen, M.; Penttilä, A.; Kovanen, P.T. Accumulation of activated mast cells in the shoulder region of human coronary atheroma, the predilection site of atheromatous rupture. *Circulation* **1994**, *90*, 1669–1678. [CrossRef] [PubMed]

109. Meng, Z.; Yan, C.; Deng, Q.; Dong, X.; Duan, Z.-M.; Gao, D.-F.; Niu, X.-L. Oxidized low-density lipoprotein induces inflammatory responses in cultured human mast cells via Toll-like receptor 4. *Cell. Physiol. Biochem.* **2013**, *31*, 842–853. [CrossRef] [PubMed]

110. Olsson, M.; Dalsgaard, C.J.; Haegerstrand, A.; Rosenqvist, M.; Rydén, L.; Nilsson, J. Accumulation of T lymphocytes and expression of interleukin-2 receptors in nonrheumatic stenotic aortic valves. *J. Am. Coll. Cardiol.* **1994**, *23*, 1162–1170. [CrossRef]

111. Jonasson, L.; Holm, J.; Skalli, O.; Bondjers, G.; Hansson, G.K. Regional accumulations of T cells, macrophages, and smooth muscle cells in the human atherosclerotic plaque. *Arterioscler. Thromb. Vasc. Biol.* **1986**, *6*, 131–138. [CrossRef]

112. Mazzone, A.; Epistolato, M.C.; de Caterina, R.; Storti, S.; Vittorini, S.; Sbrana, S.; Gianetti, J.; Bevilacqua, S.; Glauber, M.; Biagini, A.; et al. Neoangiogenesis, T-lymphocyte infiltration, and heat shock protein-60 are biological hallmarks of an immunomediated inflammatory process in end-stage calcified aortic valve stenosis. *J. Am. Coll. Cardiol.* **2004**, *43*, 1670–1676. [CrossRef] [PubMed]

113. Nagy, E.; Lei, Y.; Martínez-Martínez, E.; Body, S.C.; Schlotter, F.; Creager, M.; Assmann, A.; Khabbaz, K.; Libby, P.; Hansson, G.K.; et al. Interferon-γ Released by Activated CD8+ T Lymphocytes Impairs the Calcium Resorption Potential of Osteoclasts in Calcified Human Aortic Valves. *Am. J. Pathol.* **2017**, *187*, 1413–1425. [CrossRef] [PubMed]

114. Zhang, J.; Salojin, K.V.; Gao, J.X.; Cameron, M.J.; Bergerot, I.; Delovitch, T.L. p38 mitogen-activated protein kinase mediates signal integration of TCR/CD28 costimulation in primary murine T cells. *J. Immunol.* **1999**, *162*, 3819–3829. [PubMed]

115. Rincón, M.; Enslen, H.; Raingeaud, J.; Recht, M.; Zapton, T.; Su, M.S.; Penix, L.A.; Davis, R.J.; Flavell, R.A. Interferon-gamma expression by Th1 effector T cells mediated by the p38 MAP kinase signaling pathway. *EMBO J.* **1998**, *17*, 2817–2829. [CrossRef] [PubMed]

116. Melián, A.; Geng, Y.J.; Sukhova, G.K.; Libby, P.; Porcelli, S.A. CD1 expression in human atherosclerosis. A potential mechanism for T cell activation by foam cells. *Am. J. Pathol.* **1999**, *155*, 775–786. [CrossRef]

117. Tupin, E.; Nicoletti, A.; Elhage, R.; Rudling, M.; Ljunggren, H.-G.; Hansson, G.K.; Berne, G.P. CD1d-dependent activation of NKT cells aggravates atherosclerosis. *J. Exp. Med.* **2004**, *199*, 417–422. [CrossRef] [PubMed]

118. Li, Y.; To, K.; Kanellakis, P.; Hosseini, H.; Deswaerte, V.; Tipping, P.; Smyth, M.J.; Toh, B.-H.; Bobik, A.; Kyaw, T. CD4+ natural killer T cells potently augment aortic root atherosclerosis by perforin- and granzyme B-dependent cytotoxicity. *Circ. Res.* **2015**, *116*, 245–254. [CrossRef] [PubMed]

119. Kyriakakis, E.; Cavallari, M.; Andert, J.; Philippova, M.; Koella, C.; Bochkov, V.; Erne, P.; Wilson, S.B.; Mori, L.; Biedermann, B.C.; et al. Invariant natural killer T cells: Linking inflammation and neovascularization in human atherosclerosis. *Eur. J. Immunol.* **2010**, *40*, 3268–3279. [CrossRef] [PubMed]

120. Kyriakakis, E.; Cavallari, M.; Pfaff, D.; Fabbro, D.; Mestan, J.; Philippova, M.; de Libero, G.; Erne, P.; Resink, T.J. IL-8-mediated angiogenic responses of endothelial cells to lipid antigen activation of iNKT cells depend on EGFR transactivation. *J. Leukoc. Biol.* **2011**, *90*, 929–939. [CrossRef] [PubMed]

121. Shuvy, M.; Ben Ya'acov, A.; Zolotarov, L.; Lotan, C.; Ilan, Y. Beta glycosphingolipids suppress rank expression and inhibit natural killer T cell and CD8+ accumulation in alleviating aortic valve calcification. *Int. J. Immunopathol. Pharmacol.* **2009**, *22*, 911–918. [CrossRef] [PubMed]

122. Mazur, P.; Mielimonka, A.; Natorska, J.; Wypasek, E.; Gawęda, B.; Sobczyk, D.; Kapusta, P.; Malinowski, K.P.; Kapelak, B. Lymphocyte and monocyte subpopulations in severe aortic stenosis at the time of surgical intervention. *Cardiovasc. Pathol.* **2018**, *35*, 1–7. [CrossRef] [PubMed]

123. Bessoles, S.; Fouret, F.; Dudal, S.; Besra, G.S.; Sanchez, F.; Lafont, V. IL-2 triggers specific signaling pathways in human NKT cells leading to the production of pro- and anti-inflammatory cytokines. *J. Leukoc. Biol.* **2008**, *84*, 224–233. [CrossRef] [PubMed]

124. Nagaleekar, V.K.; Sabio, G.; Aktan, I.; Chant, A.; Howe, I.W.; Thornton, T.M.; Benoit, P.J.; Davis, R.J.; Rincon, M.; Boyson, J.E. Translational control of NKT cell cytokine production by p38 MAPK. *J. Immunol.* **2011**, *186*, 4140–4146. [CrossRef] [PubMed]

125. Stuart, J.K.; Bisch, S.P.; Leon-Ponte, M.; Hayatsu, J.; Mazzuca, D.M.; Maleki Vareki, S.; Haeryfar, S.M.M. Negative modulation of invariant natural killer T cell responses to glycolipid antigens by p38 MAP kinase. *Int. Immunopharmacol.* **2010**, *10*, 1068–1076. [CrossRef] [PubMed]

126. Houtkamp, M.A.; de Boer, O.J.; van der Loos, C.M.; van der Wal, A.C.; Becker, A.E. Adventitial infiltrates associated with advanced atherosclerotic plaques: Structural organization suggests generation of local humoral immune responses. *J. Pathol.* **2001**, *193*, 263–269. [CrossRef]

127. Ramshaw, A.L.; Parums, D.V. Immunohistochemical characterization of inflammatory cells associated with advanced atherosclerosis. *Histopathology* **1990**, *17*, 543–552. [CrossRef] [PubMed]

128. Wallby, L.; Steffensen, T.; Jonasson, L.; Broqvist, M. Inflammatory Characteristics of Stenotic Aortic Valves: A Comparison between Rheumatic and Nonrheumatic Aortic Stenosis. *Cardiol. Res. Pract.* **2013**, *2013*, 895215. [CrossRef] [PubMed]

129. Shaw, P.X.; Hörkkö, S.; Chang, M.K.; Curtiss, L.K.; Palinski, W.; Silverman, G.J.; Witztum, J.L. Natural antibodies with the T15 idiotype may act in atherosclerosis, apoptotic clearance, and protective immunity. *J. Clin. Investig.* **2000**, *105*, 1731–1740. [CrossRef] [PubMed]

130. Palinski, W.; Hörkkö, S.; Miller, E.; Steinbrecher, U.P.; Powell, H.C.; Curtiss, L.K.; Witztum, J.L. Cloning of monoclonal autoantibodies to epitopes of oxidized lipoproteins from apolipoprotein E-deficient mice. Demonstration of epitopes of oxidized low density lipoprotein in human plasma. *J. Clin. Investig.* **1996**, *98*, 800–814. [CrossRef] [PubMed]

131. Binder, C.J.; Shaw, P.X.; Chang, M.-K.; Boullier, A.; Hartvigsen, K.; Hörkkö, S.; Miller, Y.I.; Woelkers, D.A.; Corr, M.; Witztum, J.L. The role of natural antibodies in atherogenesis. *J. Lipid Res.* **2005**, *46*, 1353–1363. [CrossRef] [PubMed]

132. Binder, C.J.; Hörkkö, S.; Dewan, A.; Chang, M.-K.; Kieu, E.P.; Goodyear, C.S.; Shaw, P.X.; Palinski, W.; Witztum, J.L.; Silverman, G.J. Pneumococcal vaccination decreases atherosclerotic lesion formation: Molecular mimicry between Streptococcus pneumoniae and oxidized LDL. *Nat. Med.* **2003**, *9*, 736–743. [CrossRef] [PubMed]

133. Chou, M.-Y.; Fogelstrand, L.; Hartvigsen, K.; Hansen, L.F.; Woelkers, D.; Shaw, P.X.; Choi, J.; Perkmann, T.; Bäckhed, F.; Miller, Y.I.; et al. Oxidation-specific epitopes are dominant targets of innate natural antibodies in mice and humans. *J. Clin. Investig.* **2009**, *119*, 1335–1349. [CrossRef] [PubMed]

134. Kyaw, T.; Tay, C.; Khan, A.; Dumouchel, V.; Cao, A.; To, K.; Kehry, M.; Dunn, R.; Agrotis, A.; Tipping, P.; et al. Conventional B2 B cell depletion ameliorates whereas its adoptive transfer aggravates atherosclerosis. *J. Immunol.* **2010**, *185*, 4410–4419. [CrossRef] [PubMed]

135. Tsiantoulas, D.; Diehl, C.J.; Witztum, J.L.; Binder, C.J. B cells and humoral immunity in atherosclerosis. *Circ. Res.* **2014**, *114*, 1743–1756. [CrossRef] [PubMed]

136. Khiem, D.; Cyster, J.G.; Schwarz, J.J.; Black, B.L. A p38 MAPK-MEF2C pathway regulates B-cell proliferation. *Proc. Natl. Acad. Sci. USA* **2008**, *105*, 17067–17072. [CrossRef] [PubMed]

137. Choi, J.-H.; Do, Y.; Cheong, C.; Koh, H.; Boscardin, S.B.; Oh, Y.-S.; Bozzacco, L.; Trumpfheller, C.; Park, C.G.; Steinman, R.M. Identification of antigen-presenting dendritic cells in mouse aorta and cardiac valves. *J. Exp. Med.* **2009**, *206*, 497–505. [CrossRef] [PubMed]

138. Bobryshev, Y.V.; Lord, R.S. S-100 positive cells in human arterial intima and in atherosclerotic lesions. *Cardiovasc. Res.* **1995**, *29*, 689–696. [CrossRef]

139. Yilmaz, A.; Lochno, M.; Traeg, F.; Cicha, I.; Reiss, C.; Stumpf, C.; Raaz, D.; Anger, T.; Amann, K.; Probst, T.; et al. Emergence of dendritic cells in rupture-prone regions of vulnerable carotid plaques. *Atherosclerosis* **2004**, *176*, 101–110. [CrossRef] [PubMed]

140. Koltsova, E.K.; Ley, K. How dendritic cells shape atherosclerosis. *Trends Immunol.* **2011**, *32*, 540–547. [CrossRef] [PubMed]

141. Arrighi, J.F.; Rebsamen, M.; Rousset, F.; Kindler, V.; Hauser, C. A critical role for p38 mitogen-activated protein kinase in the maturation of human blood-derived dendritic cells induced by lipopolysaccharide, TNF-alpha, and contact sensitizers. *J. Immunol.* **2001**, *166*, 3837–3845. [CrossRef] [PubMed]

142. Lu, Y.; Zhang, M.; Wang, S.; Hong, B.; Wang, Z.; Li, H.; Zheng, Y.; Yang, J.; Davis, R.E.; Qian, J.; et al. p38 MAPK-inhibited dendritic cells induce superior antitumour immune responses and overcome regulatory T-cell-mediated immunosuppression. *Nat. Commun.* **2014**, *5*, 4229. [CrossRef] [PubMed]

143. Cowell, S.J.; Newby, D.E.; Prescott, R.J.; Bloomfield, P.; Reid, J.; Northridge, D.B.; Boon, N.A. A randomized trial of intensive lipid-lowering therapy in calcific aortic stenosis. *N. Engl. J. Med.* **2005**, *352*, 2389–2397. [CrossRef] [PubMed]

144. Rossebø, A.B.; Pedersen, T.R.; Boman, K.; Brudi, P.; Chambers, J.B.; Egstrup, K.; Gerdts, E.; Gohlke-Bärwolf, C.; Holme, I.; Kesäniemi, Y.A.; et al. Intensive lipid lowering with simvastatin and ezetimibe in aortic stenosis. *N. Engl. J. Med.* **2008**, *359*, 1343–1356. [CrossRef] [PubMed]

145. Chan, K.L.; Teo, K.; Dumesnil, J.G.; Ni, A.; Tam, J. Effect of Lipid lowering with rosuvastatin on progression of aortic stenosis: Results of the aortic stenosis progression observation: Measuring effects of rosuvastatin (ASTRONOMER) trial. *Circulation* **2010**, *121*, 306–314. [CrossRef] [PubMed]

146. Raman, M.; Chen, W.; Cobb, M.H. Differential regulation and properties of MAPKs. *Oncogene* **2007**, *26*, 3100–3112. [CrossRef] [PubMed]

147. Daugherty, A.; Tall, A.R.; Daemen, M.J.A.P.; Falk, E.; Fisher, E.A.; García-Cardeña, G.; Lusis, A.J.; Owens, A.P.; Rosenfeld, M.E.; Virmani, R. Recommendation on Design, Execution, and Reporting of Animal Atherosclerosis Studies: A Scientific Statement From the American Heart Association. *Circ. Res.* **2017**, *121*, e53–e79. [CrossRef] [PubMed]

148. Koeberle, S.C.; Romir, J.; Fischer, S.; Koeberle, A.; Schattel, V.; Albrecht, W.; Grütter, C.; Werz, O.; Rauh, D.; Stehle, T.; et al. Skepinone-L is a selective p38 mitogen-activated protein kinase inhibitor. *Nat. Chem. Biol.* **2012**, *8*, 141–143. [CrossRef] [PubMed]

149. Sarov-Blat, L.; Morgan, J.M.; Fernandez, P.; James, R.; Fang, Z.; Hurle, M.R.; Baidoo, C.; Willette, R.N.; Lepore, J.J.; Jensen, S.E.; et al. Inhibition of p38 mitogen-activated protein kinase reduces inflammation after coronary vascular injury in humans. *Arterioscler. Thromb. Vasc. Biol.* **2010**, *30*, 2256–2263. [CrossRef] [PubMed]

150. Elkhawad, M.; Rudd, J.H.F.; Sarov-Blat, L.; Cai, G.; Wells, R.; Davies, L.C.; Collier, D.J.; Marber, M.S.; Choudhury, R.P.; Fayad, Z.A.; et al. Effects of p38 mitogen-activated protein kinase inhibition on vascular and systemic inflammation in patients with atherosclerosis. *JACC Cardiovasc. Imaging* **2012**, *5*, 911–922. [CrossRef] [PubMed]

151. Newby, L.K.; Marber, M.S.; Melloni, C.; Sarov-Blat, L.; Aberle, L.H.; Aylward, P.E.; Cai, G.; de Winter, R.J.; Hamm, C.W.; Heitner, J.F.; et al. Losmapimod, a novel p38 mitogen-activated protein kinase inhibitor, in non-ST-segment elevation myocardial infarction: A randomised phase 2 trial. *Lancet* **2014**, *384*, 1187–1195. [CrossRef]

152. O'Donoghue, M.L.; Glaser, R.; Cavender, M.A.; Aylward, P.E.; Bonaca, M.P.; Budaj, A.; Davies, R.Y.; Dellborg, M.; Fox, K.A.A.; Gutierrez, J.A.T.; et al. Effect of Losmapimod on Cardiovascular Outcomes in Patients Hospitalized With Acute Myocardial Infarction: A Randomized Clinical Trial. *JAMA* **2016**, *315*, 1591–1599. [CrossRef] [PubMed]

International Journal of
Molecular Sciences

Review

Roles of Mitogen-Activated Protein Kinases in Osteoclast Biology

Kyunghee Lee, Incheol Seo, Mun Hwan Choi and Daewon Jeong *

Department of Microbiology, Laboratory of Bone Metabolism and Control, Yeungnam University College of Medicine, Daegu 42415, Korea; kyungheelee@ynu.ac.kr (K.L.); htr@daum.net (I.S.); choibak@ynu.ac.kr (M.H.C.)
* Correspondence: dwjeong@ynu.ac.kr; Tel.: +82-53-640-6944

Received: 24 August 2018; Accepted: 27 September 2018; Published: 1 October 2018

Abstract: Bone undergoes continuous remodeling, which is homeostatically regulated by concerted communication between bone-forming osteoblasts and bone-degrading osteoclasts. Multinucleated giant osteoclasts are the only specialized cells that degrade or resorb the organic and inorganic bone components. They secrete proteases (e.g., cathepsin K) that degrade the organic collagenous matrix and establish localized acidosis at the bone-resorbing site through proton-pumping to facilitate the dissolution of inorganic mineral. Osteoporosis, the most common bone disease, is caused by excessive bone resorption, highlighting the crucial role of osteoclasts in intact bone remodeling. Signaling mediated by mitogen-activated protein kinases (MAPKs), including extracellular signal-regulated kinase (ERK), c-Jun N-terminal kinase (JNK), and p38, has been recognized to be critical for normal osteoclast differentiation and activation. Various exogenous (e.g., toll-like receptor agonists) and endogenous (e.g., growth factors and inflammatory cytokines) stimuli contribute to determining whether MAPKs positively or negatively regulate osteoclast adhesion, migration, fusion and survival, and osteoclastic bone resorption. In this review, we delineate the unique roles of MAPKs in osteoclast metabolism and provide an overview of the upstream regulators that activate or inhibit MAPKs and their downstream targets. Furthermore, we discuss the current knowledge about the differential kinetics of ERK, JNK, and p38, and the crosstalk between MAPKs in osteoclast metabolism.

Keywords: mitogen-activated protein kinases (MAPKs); MAPK kinetics; osteoclast differentiation; bone remodeling

---

1. Introduction

Normal bone physiology depends on the coupled processes of removing old bone and replacing it with new bone [1]. Throughout life, bone undergoes continuous remodeling through concerted bone matrix formation and mineralization (anabolic process) by osteoblasts and mineralized bone matrix degradation (catabolic process) by osteoclasts [2]. Osteoclasts are multinucleated cells that are formed from monocyte/macrophage lineage cells, and their differentiation and function are regulated by various cytokines, hormones, and growth factors [3,4]. Especially, macrophage colony stimulating factor (M-CSF) and receptor activator for nuclear factor κ-B ligand (RANKL) are indispensable in the regulation of the sequential processes of osteoclastogenesis, including osteoclast precursor proliferation, adhesion, migration, and cell-cell fusion to form multinucleated cells, as well as in the migration, survival, and bone-resorptive function of mature osteoclasts [5]. Both M-CSF and RANKL act through mitogen-activated protein kinases (MAPKs), extracellular signal-regulated kinase (ERK), c-Jun N-terminal kinase (JNK), and p38 signaling during osteoclast differentiation and bone resorption. MAPK signaling activated by M-CSF is mainly involved in the regulation of osteoclast precursor proliferation, whereas RANKL-induced MAPK activation is primarily implicated in osteoclast differentiation [6,7]. Moreover, a recent, advanced study revealed that the activation of MAPKs by M-CSF or RANKL differs in terms of the extent, duration, and isoform specificity of MAPK phosphorylation, thus determining the distinct cell fates of proliferation or differentiation in osteoclast precursors [8]. In osteoclast precursors, two ERK forms, ERK1/2, and three JNK isoforms, JNK1/2/3, are mainly involved in osteoclast precursor proliferation and osteoclast apoptosis, respectively [4,8,9]. Among the four isoforms of p38 (α, β, γ, and δ), p38α is highly expressed in osteoclast precursors and mature osteoclasts and plays a key role in osteoclast differentiation and bone resorption [10].

MAPKs convert a variety of extracellular stimuli into specific cellular responses, thus acting as signaling hubs, in eukaryotic cells [11]. MAPK pathways are organized into three-tiered cascades comprised of three molecules: MAPK, MAPK kinase (MAPKK or MEK), and MAPKK kinase (MAPKKK or MEKK). In the phosphorelay system, MAPKKKs, which are serine/threonine protein kinases, phosphorylate and activate MAPKKs, which then dually phosphorylate the threonine and tyrosine residues of the conserved TXY motif ("X" stands for glutamic acid, proline, or glycine) of the activation loop of MAPKs, including ERK, JNK, and p38 [12]. The activities of MAPK pathways are properly regulated through dephosphorylation of the threonine or tyrosine residue of the TXY motif by phosphatases, such as dual-specificity phosphatases (DUSPs), which counter the activities of kinases [13]. The activation of MAPK pathways leads to various biological outcomes, including gene induction, cell proliferation and survival, apoptosis, and differentiation, as well as cellular stress and inflammatory responses.

The signaling transmission of extracellular stimuli via MAPK activation to appropriate intracellular molecules has been established to be essential for the regulation of osteoclast differentiation and bone remodeling. Numerous studies exploring the roles of MAPKs in osteoclast metabolism have suggested that ERK, JNK, and p38 are key players in osteoclast differentiation and activation. In this review, we describe the peculiar roles of MAPKs in osteoclast metabolism, as well as various upstream stimulators and inhibitors of MAPKs and their downstream targets. In addition, we discuss current knowledge regarding the distinctive kinetics of MAPKs and the crosstalk between MAPKs in osteoclast metabolism.

2. ERK Signaling in Osteoclasts

The ERK signaling pathway has been implicated in the survival, proliferation, apoptosis, formation, polarity, podosome disassembly, and differentiation of osteoclasts. Combined findings in ERK1 knockout and hematopoietic ERK2 conditional knockout mice showed that ERK1 plays a crucial role in modulating osteoclast differentiation, migration, and bone resorption [14]. A variety of cytokines, growth factors, and hormones positively or negatively regulate ERK signaling in osteoclasts (Figure 1). The ERK signaling cascade consists of a core of three serially phosphorylating protein kinases. The activation of Raf isoforms via Ras-Raf interaction stimulates the MAPKKs MEK1 and MEK2, which then activate ERK1 and ERK2 by dual phosphorylation at the conserved Thr-Glu-Tyr (TEY) motif [15,16], which leads to the phosphorylation of various downstream substrates, including transcription factors.

Recently, the functions of less-well-studied MEK5/ERK5 signaling pathways in bone biology begin to be of interest [17,18]. It was reported that conditional deletion of ERK5 in the mouse prostate using *Nkx3.1-Cre* recombinase expression resulted in a severely deformed and curved spine, with an associated loss of trabecular bone volume [17]. These spinal abnormalities in *Nkx3.1-Cre* ERK5 null mice are associated with increased osteoclast activity. In addition, M-CSF, but not RANKL, induces ERK5 phosphorylation and the consequent M-CSF/MEK5/ERK5 signaling mediates osteoclast differentiation [19].

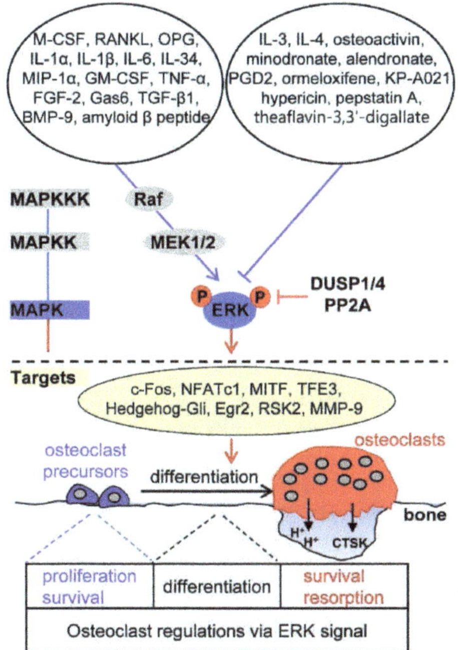

Figure 1. Osteoclastogenic signaling cascade controlled by upstream activators and inhibitors of extracellular signal-regulated kinase (ERK). The marked molecules are described in the text entitled "ERK signaling in osteoclasts". CTSK, cathepsin K; M-CSF, macrophage colony-stimulating factor; RANKL, receptor activator of nuclear factor κ-B ligand; OPG, osteoprotegerin; IL, interleukin; MIP, macrophage inflammatory proteins; GM-CSF, granulocyte-macrophage colony-stimulating factor; TNF, tumor necrosis factor; FGF, fibroblast growth factor; TGF, transforming growth factor; BMP, bone morphogenetic protein; PGD2, prostaglandin D2; MAPK, mitogen-activated protein kinase; MAPKK, mitogen-activated protein kinase kinase; MAPKKK, mitogen-activated protein kinase kinase kinase; DUSP, dual-specificity phosphatase. Arrows indicate activation of the signaling pathways while T bars indicate inhibition of the signaling pathways.

2.1. Upstream Activators of ERK Signaling in Osteoclasts

The osteoclastogenic factors M-CSF and RANKL play a critical role in osteoclast differentiation by inducing the phosphorylation of ERK1 and ERK2 [4]. The binding of M-CSF to its receptor c-Fms results in the phosphorylation of specific tyrosine residues of c-Fms. The phosphorylated site at the intracellular cytosolic tail of c-Fms interacts with growth factor receptor-binding protein-2, a stimulator of the Ras/Raf pathway, which then leads to the activation of ERK1 and ERK2, enhancing osteoclast precursor proliferation and survival [20,21]. Binding of RANKL to its receptor RANK leads to the recruitment of the adaptor protein, TNF receptor-associated factor 6 (TRAF6), to the cytoplasmic tail in a submembrane compartment and then triggers ERK activation. RANKL/RANK/TRAF6/ERK cascades have been shown to regulate osteoclast formation and function [22,23]. Interestingly, osteoprotegerin (OPG), a decoy receptor that binds to RANKL, and thus, suppresses osteoclast differentiation by interrupting the interaction between RANKL and RANK, can also phosphorylate ERK1 and ERK2 and directly induce podosome disassembly in osteoclasts [22,24,25].

Several reports have suggested that ERK activation by inflammatory cytokines positively regulates osteoclastogenesis. Interleukin-1β (IL-1β) acts synergistically with RANKL to increase ERK activation in a Ca^{2+}-dependent manner [26] and IL-1α, IL-6, and IL-34 induce phosphorylation of ERK1 and ERK2, leading to the promotion of osteoclastogenesis [27–29]. Macrophage inflammatory protein-1α (MIP-1α) secreted from multiple myeloma cells induces osteoclast formation by activating the MEK/ERK/c-Fos pathway [30]. Granulocyte-macrophage colony-stimulating factor (GM-CSF)-induced ERK activation promotes the fusion of mononuclear osteoclasts to form multinucleated osteoclasts by inducing the expression of dendritic cell-specific transmembrane protein (DC-STAMP, also known as TM7SF4) via the Ras/ERK pathway [31].

Growth factors, such as fibroblast growth factor-2 (FGF-2), growth arrest-specific gene 6 (Gas6), and tumor necrosis factor-α (TNF-α), stimulate mature osteoclast function and survival through ERK activation [32,33]. ERK is transiently activated during transforming growth factor-β1 (TGF-β1)-induced apoptosis of osteoclasts differentiated from human umbilical cord blood monocytes, via the activation of caspase-9 and upregulation of the pro-apoptotic protein Bim [34]. The binding of bone morphogenetic protein-9 (BMP-9) to its receptor anaplastic lymphoma kinase 1 on the cell surface activates the canonical Smad-1/5/8 pathway and the ERK pathway, and supports the formation, function, and survival of osteoclasts derived from human umbilical cord blood monocytes [35]. Interestingly, in patients with Alzheimer's disease, who have a high risk of osteoporotic hip fracture, amyloid beta peptide, one of the pathological hallmarks of Alzheimer's disease that is abnormally deposited in bone tissues [36], was shown to enhance RANKL-induced ERK and NF-κB activation and to promote osteoclastic bone resorption [37]. Taken together, various upstream stimulators of ERK pathway were found to positively regulate the process of osteoclast differentiation.

2.2. Upstream Inhibitors of ERK Signaling in Osteoclasts

IL-3 and IL-4, known as anti-osteoclastogenic cytokines, suppress osteoclastogenesis and/or osteoclastic bone resorption via inhibition of the ERK pathway and activation of signal transducer and activator of transcription 5 (STAT5) [38–40]. Prostaglandin D2 inactivates ERK signaling during chemoattractant receptor homologous molecule expressed on T-helper type 2 cells (CRTH2)-mediated apoptosis of osteoclasts derived from human peripheral blood mononuclear cells [41]. In osteoactivin-CD44-ERK signal cascades, shedding of the ectodomain of osteoactivin, a heavily glycosylated type I transmembrane protein that is expressed in both osteoclasts and osteoblasts, produces a soluble form of osteoactivin [42] that binds to the CD44 receptor, followed by the inhibition of ERK signaling, and thus, decreased osteoclast differentiation [43].

Several pharmacological compounds, including anti-osteoporotic agents, have been reported to inhibit osteoclastogenesis by suppressing ERK signaling. Nitrogen-containing bisphosphonates, such as minodronate and alendronate, which are used as anti-resorptive drugs for the treatment of metabolic bone diseases [44], have been shown to decrease the phosphorylation of ERK1/2 and Akt, thereby inhibiting osteoclast formation [45]. Ormeloxifene, which is a nonsteroidal selective estrogen receptor modulator that exerts an estrogen-agonistic effect and has anticancer activity in breast cancer [46], has an anti-osteoclastogenic effect, resulting from ERK1/ERK2 and JNK inactivation by inhibiting the generation of RANKL-induced reactive oxygen species (ROS) [47]. KP-A021, a triazole-based compound that exhibits anti-inflammatory, anti-tumor, anti-tubercular, and anti-fungal activities, reportedly inhibits osteoclast differentiation by suppressing RANKL-induced MEK-ERK phosphorylation cascades [48,49]. Hypericin, a naphtodianthrone isolated from *Hypericum perforatum* and a potent and selective inhibitor of protein kinase C that reduces neuropathic pain, attenuates RANKL-induced osteoclastogenesis of bone marrow-derived macrophages via specific inhibition of the ERK signaling pathway without affecting JNK, p38, and NF-κB signaling in vitro, and suppresses titanium particle-induced bone erosion in vivo [50]. Pepstatin A, an inhibitor of aspartic proteinases, such as cathepsins D and E, and theaflavin-3,3′-digallate, a natural active compound derived from black tea, inhibit osteoclast formation and polarization and bone resorption by specifically suppressing RANKL-induced ERK signaling [51,52]. Therefore, inflammatory cytokines and pharmacological agents capable of inhibiting RANKL-induced ERK activation are regarded to negatively regulate osteoclast differentiation and function.

2.3. Downstream Targets of ERK Signaling in Osteoclasts

ERKs phosphorylate numerous downstream target substrates to control osteoclastogenesis. ERKs govern various transcription factors during osteoclastogenesis. c-Fos is phosphorylated at its C-terminal domain on serines 362 and 374 through ERK activation in response to M-CSF or RANKL [53,54]. The expression of c-Fos and nuclear factor of activated T-cells, cytoplasmic 1 (NFATc1), which are crucial osteoclastogenic transcription factors, is regulated via GM-CSF-induced ERK signaling [31]. M-CSF-stimulated ERK1 and ERK2 activation directly phosphorylates microphthalmia-associated transcription factor (MITF), a basic/helix-loop-helix/leucine-zipper transcription factor essential for osteoclast maturation, and its partner protein, transcription factor E3 (TFE3) [55]. Estrogen (17β-estradiol)-induced ERK activation inhibits osteoclastogenesis and promotes osteoclast apoptosis by downregulating signaling through the transcription factor Hedgehog-Gli. This fact implies that estrogen deficiency in postmenopausal osteoporosis induces increased and activated osteoclasts via the activation of Hedgehog-Gli signaling regulated by a MEK/ERK cascade [56]. M-CSF-induced, immediate gene induction of the Krüppel-like zinc finger transcription factor Egr2 in osteoclasts maintains osteoclast survival by inducing the pro-survival Bcl2 family member Mcl1 and proteolytic degradation of the pro-apoptotic Bim [57]. In addition, ribosomal S6 kinase 2 (RSK2), a member of the p90RSK family of serine/threonine kinases, is a downstream target of ERK1/ERK2 that participates in modulating M-CSF-induced PI3K/Akt activation through an ERK/RSK2-mediated negative feedback loop in macrophages [58]. RANKL-induced ERK activation induces the expression and activity of matrix metalloproteinase 9 (MMP-9, also termed gelatinase B/type IV collagenase), which is implicated in osteoclast migration and bone resorption [59], through TRAF6, but not TRAF2, in osteoclast precursors [60].

2.4. Phosphatase Regulation of ERK Signaling in Osteoclasts

DUSPs, protein phosphatases that dephosphorylate both phosphotyrosine and phosphoserine/phosphothreonine protein residues, play an important role in the duration, magnitude, and spatiotemporal regulation of MAPK activities [61]. STAT5, a member of the STAT family of transcription factors essential for cytokine-regulated processes [62], negatively regulates the activity of MAPKs, in particular, ERK1/2, by inducing the expression of *DUSP1* and *DUSP2*, thus suppressing the bone-resorbing activity of osteoclasts [39]. RANKL promotes osteoclastogenesis via sustained ERK activation, whereas RANKL together with the toll-like receptor 9 (TLR9) ligand, oligodeoxynucleotides containing unmethylated CpG dinucleotides (CpG-ODN), induces transient ERK activation by enhanced ERK dephosphorylation, due to the expression of phosphatase PP2A, a serine/threonine phosphatase, thus accelerating the degradation of osteoclastogenic transcription factor c-Fos and thereby inhibiting osteoclastogenesis [53].

3. JNK Signaling in Osteoclasts

JNK signaling plays an important role in the regulation of apoptosis, formation, and differentiation of osteoclasts [9,63,64]. Bone marrow-derived macrophages isolated from mice lacking JNK1 or carrying a mutated form (JunAA/JunAA) of c-Jun that cannot be phosphorylated by the JNKs show reduced osteoclast differentiation and bone resorption activity [63]. Moreover, impairment of JNK signaling by overexpression of dominant-negative JNK1, c-Jun, and c-Fos, or the JNK-specific inhibitor SP600125 abrogates the anti-apoptotic effect of RANKL/RANK/TRAF6 signaling in osteoclasts [9]. This indicates that JNK/c-Jun signaling mediates the RANKL-induced anti-apoptotic process in mature osteoclasts. Blockade of JNK activity at the pre-fusion osteoclast stage results in the reversion of tartrate-resistant acid phosphatase (TRAP)-positive cells (representing pre-osteoclasts at the pre-fusion stage) to TRAP-negative cells (representing osteoclast precursors), even in the continuous presence of RANKL, demonstrating that the JNK pathway is required for maintaining osteoclastic commitment, until fusion [65]. In osteoclasts, JNK signaling pathways are reported to structurally organize as a signaling cascade (Figure 2). MAPKKKs, such as MEKK1, and transforming growth factor beta-activated kinase 1 (TAK1) stimulate the MAPKKs MKK4 and MKK7, which induce dual phosphorylation of JNK at a conserved TPY motif [9,66].

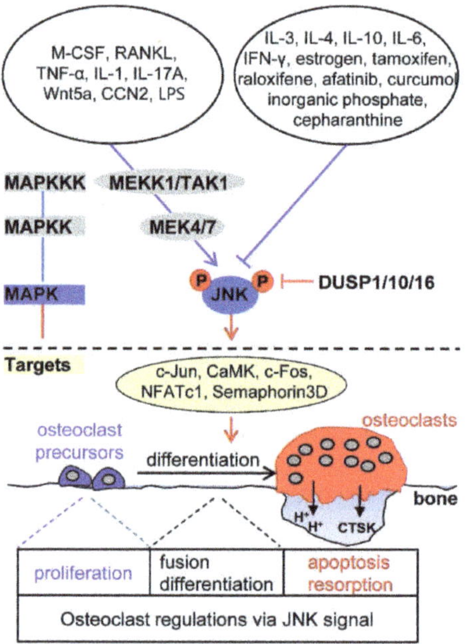

Figure 2. Osteoclastogenic signaling cascade controlled by upstream activators and inhibitors of c-Jun N-terminal kinase (JNK). The indicated molecules are described in the section of "JNK signaling in osteoclasts". Arrows indicate activation of the signaling pathways while T bars indicate inhibition of the signaling pathways.

3.1. Upstream Activators of JNK Signaling in Osteoclasts

The osteoclastogenic factor RANKL activates JNK signaling through TRAF6, thus stimulating osteoclast differentiation [9,64]. M-CSF produces ceramide 1-phosphate, which is reported to be mitogenic for fibroblasts and acts as a lipid second messenger, in murine bone marrow-derived macrophages, and ceramide 1-phosphate from M-CSF-stimulated cells mediates their proliferation via rapid phosphorylation of protein kinase B (also known as Akt) and JNK [67,68].

JNK signaling activated by the inflammatory cytokines TNF-α and IL-1 induces cell-cell fusion to form osteoclasts and enhances osteoclast survival, respectively [69–71]. IL-17A facilitates autophagic activity of osteoclast precursors and promotes osteoclastogenesis via activating the RANKL-JNK pathway [72].

In JNK signaling induced by growth factors and other signals, Wnt5a, a non-canonical Wnt ligand secreted from osteoblasts, binds to its receptor, receptor tyrosine kinase-like orphan receptor (Ror2) expressed in the plasma membrane of osteoclasts. Wnt5a-Ror2 signaling induces RANK expression through JNK activation and recruitment of c-Jun to the promoter of RANK-coding gene in osteoclast precursors, thereby enhancing RANKL-mediated osteoclastogenesis [73], suggesting that Wnt5a-Ror2-JNK signaling between osteoblasts and osteoclast precursors mediates osteoclastogenesis. In addition, CCN2 (connective tissue growth factor, cystein rich protein, and nephroblastoma overexpressed gene), known as a connective tissue growth factor, directly binds to RANK or OPG to enhance osteoclastogenesis via the activation of RANKL-RANK-JNK signaling and removal of the anti-osteoclastogenic effect of OPG [74]. Lipopolysaccharide (LPS), a prominent pathogenic factor in inflammatory bone diseases, induces osteoclast formation by activating the JNK-STAT3-NFATc1 pathway via the generation of ROS as second messengers [75].

3.2. Upstream Inhibitors of JNK Signaling in Osteoclasts

Several inflammatory cytokines that negatively influence osteoclastogenesis via JNK inactivation have been identified. IL-3 induces the irreversible inhibition of RANK expression by downregulating JNK activation in osteoclast precursors, and thus suppresses RANKL-induced osteoclastogenesis [76,77]. IL-4 blocks RANKL-induced activation of NF-κB and JNK depending on STAT6, and IL-10 downregulates RANKL-induced expression of NFATc1, c-Jun, and c-Fos, and JNK phosphorylation, ultimately impairing osteoclastogenesis [40,78]. IL-6 suppresses osteoclast differentiation through inhibition of JNK activation, at least in part by upregulating the expression of *DUSP1* and *DUSP16*, which dephosphorylate JNK [79]. Interferon-γ inhibits RANKL-induced activation of JNK through degradation of TRAF6 in osteoclast precursors or induction of osteoclast inhibitory peptide-1 expression [80,81].

Some therapeutic agents having anti-osteoporosis, anti-tumor, and anti-inflammation activities have been reported to suppress osteoclastogenesis through JNK inactivation. The anti-osteoporotic agent estrogen and the selective estrogen-receptor modulators tamoxifen and raloxifene that mimic the anti-osteoporotic effect of estrogen suppress RANKL-induced JNK-c-Jun axis signaling, resulting in a decrease in osteoclast formation and differentiation [82,83]. The anti-tumor agent afatinib (an ATP-competitive 4-anilinoquinazoline derivative), an irreversible epidermal growth factor receptor tyrosine kinase inhibitor, has proven efficacious in phase III trials in patients with non-small cell lung cancer, which is known as the third most common cause of bone metastases [84]. This inhibitor with anti-tumor effect specifically inhibits RANKL-induced phosphorylation of JNK and Akt, ameliorating the differentiation and bone resorbing activity of osteoclasts [85]. Cepharanthine, a natural alkaloid extracted from *Stephania cepharantha* Hayata, has been used in the clinic for the treatment of tumors and inflammatory diseases [86]. This agent prevents estrogen deficiency-induced bone loss by inhibiting osteoclastogenesis via the attenuation of JNK and PI3K-Akt signaling [87]. Curcumol, a sesquiterpene and one of the major components of the essential oil of *Rhizoma curcumae* with antitumor and anti-inflammatory properties, inhibits osteoclastogenesis by specifically impairing RANKL-induced JNK-activator protein-1 (AP-1) signaling [88]. In pathological conditions, such as uremic disease, elevated serum phosphate levels are closely related with ectopic extraskeletal calcification, especially in vascular calcification [89]. A high concentration of extracellular inorganic phosphate inhibits osteoclast differentiation and bone resorption activity through specific downregulation of RANKL-induced JNK and Akt activation, with no significant changes in p38 and ERK phosphorylation [90].

3.3. Downstream Targets of JNK Signaling in Osteoclasts

RANKL-induced activated JNK phosphorylates the transcription factor c-Jun, which forms a complex with c-Fos, an essential transcription factor for osteoclast formation [9,91]. JNK signaling also induces the expression of the calcium/calmodulin-dependent protein kinase (CaMK), c-Fos, and NFATc1, which are involved in the maintenance of osteoclast lineage commitment [65,75]. Semaphorin 3D is a downstream target of JNK signaling and is involved in stimulating TNF-α-induced osteoclastogenesis [69].

3.4. Phosphatase Regulation of JNK Signaling in Osteoclasts

DUSP10, a member of the MAPK phosphatase family of dual-specificity phosphatases, predominantly dephosphorylates JNK and is induced in osteoclasts by RANKL stimulation. RANKL-induced activation of DUSP10 is thought to limit JNK signaling in order to inhibit osteoclasts from undergoing apoptosis [92]. Results obtained through in-vitro differentiation of osteoclast precursors obtained from DUSP1-deficient mice revealed that DUSP1 removes RANKL-induced phosphorylation of JNK and negatively regulates osteoclast differentiation and activation [13].

4. p38 Signaling in Osteoclasts

The p38 signaling pathway plays a key role in the regulation of osteoclast formation and maturation, and thus, in bone resorption and remodeling [5,93]. A study using conditional p38α knockout mice showed that p38α deficiency induces increased bone mass in young mice, with decreased numbers of osteoclasts and bone resorption [93]. Consistent with these in-vivo data, expression of dominant negative forms of p38α and treatment with a specific p38 inhibitor in osteoclast precursors resulted in complete blockage of RANKL-induced osteoclastogenesis in vitro [94]. Furthermore, p38α plays an important role in coupling osteoclastogenesis and osteoblastogenesis, as demonstrated by the fact that specific ablation of p38α in monocytic osteoclast precursors obtained from p38α-flox; LysM-Cre mice indirectly inhibited osteoblast proliferation and differentiation via a decrease in the expression and secretion of coupling factors, BMP-2 and platelet-derived growth factor AA, which are expressed via the p38 MAPK-Creb axis in osteoclasts [93]. In osteoclastogenesis, osteoclastogenic factors stimulate MAPKKKs, including TAK1 and relay the phosphorylation of MAPKKs, MKK3, and MKK6 (Figure 3) [95,96]. Subsequently, the activated MKK3 and MKK6 induce dual phosphorylation of p38α at a conserved TGY motif, facilitating osteoclastogenesis by the activation of NF-κB signaling and NFATc1 induction [96,97].

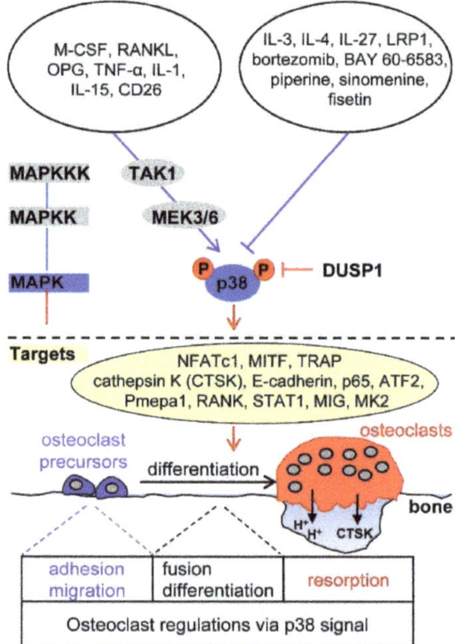

Figure 3. Osteoclastogenic signaling cascade controlled by upstream activators and inhibitors of p38 mitogen-activated protein kinases (MAPK). The presented molecules are described in the part of "p38 signaling in osteoclasts". Arrows indicate activation of the signaling pathways while T bars indicate inhibition of the signaling pathways.

4.1. Upstream Activators of p38 Signaling in Osteoclasts

M-CSF-c-Fms signaling induces p38 activation during macrophage development [8,98,99]. The binding of RANKL to its cognate receptor RANK leads to the phosphorylation of p38 in osteoclast precursors through adaptor protein TRAF6, thus inducing osteoclast differentiation [5,94]. OPG directly activates p38 signaling, thereby potentiating osteoclast function through MMP-9 expression [24] or by retracting osteoclast adhesion structures [25].

TNF-α and IL-1 directly activate p38 in a RANKL-independent manner [100,101]. IL-15 allows a synergistic effect of RANKL-induced osteoclast formation and bone resorption activity through p38 activation [102]. The expression of CD26, a cell-surface glycoprotein with dipeptidyl peptidase IV activity, in osteoclasts is accompanied by increased activation of MKK3/6-p38-MITF signaling, which is essential for early osteoclast differentiation [103]. On the contrary of the positive role of p38 activation in osteoclastogenesis, p38 activation by any stimuli has been reported to negatively regulate osteoclastogenesis. TLR stimulation with the TLR2 agonist Pam3Cys or the TLR4 agonist LPS inhibits the differentiation of human peripheral blood mononuclear cells into osteoclasts by downregulating RANK transcription and cell-surface expression of the M-CSF receptor c-Fms. Especially, it has been shown that TLR2-induced proteolytic cleavage of c-Fms depends on both p38 and ERK activation [104]. Interestingly, serum amyloid A, which is a major acute-phase protein that is secreted from liver cells in response to infection or injury, blocks M-CSF/c-Fms signaling via activation of p38 and ERK, thus inhibiting osteoclast formation by repressing osteoclast-associated genes, such as RANK and TRAF6, and inducing expression of anti-osteoclastogenic genes, such as MafB and the gene encoding interferon regulatory factor 8 [105].

4.2. Upstream Inhibitors of p38 Signaling in Osteoclasts

Various reports have suggested that any stimulus can suppress RANKL-induced p38 activation and thus suppress osteoclastogenesis. IL-3 negatively regulates p38 signaling through the activation of STAT5 in the early stages of RANKL-induced osteoclast differentiation, thus inhibiting osteoclastogenesis via the upregulation of the anti-osteoclastogenic regulators Id1 and Id2 [38]. IL-4 inhibits osteoclastogenesis by specifically blocking the activation of RANKL-induced p38 and NF-κB signaling in a STAT6-dependent manner, but not M-CSF signaling [40]. IL-27 inhibits the differentiation of human peripheral blood mononuclear cells into osteoclasts by downregulating the expression of osteoclastogenic factors RANK, triggering receptor expressed on myeloid cells (TREM-2), and NFATc1, as well as RANKL-induced activation of p38, ERK, and NF-κB signaling [106].

Blockage of osteoclast formation by the TLR9 agonist CpG-ODN is attributed to reduced expression of c-Fos by shifting from RANKL-induced sustained activation of ERK, JNK, and p38 to transient activation resulting from increased expression of PP2A [53]. Ctsk-Cre;Lrp1f/f mice with osteoclast-specific deletion of low-density-lipoprotein receptor-related protein 1 (LRP1) showed dramatically decreased trabecular bone mass with significantly increased osteoclast formation. Consistent herewith, ex-vivo culture experiments revealed that LRP1-deficient bone marrow-derived macrophages from Ctsk-Cre;Lrp1f/f mice more efficiently differentiated into osteoclasts by elevating NF-κB and p38 signaling than LRP1$^{+/+}$ macrophage cells, indicating that LRP1 negatively regulates osteoclastogenesis by blunting p38 and NF-κB signaling [107].

There exist synthetic and natural compounds that regulate osteoclast differentiation through the modulation of p38. Bortezomib, a synthetic proteasome inhibitor approved by the Food and Drug Administration for use in multiple myeloma, inhibits p38-triggered early osteoclast differentiation and thus blocks osteoclastic bone resorption [108]. Stimulation of the A2B adenosine receptor with its specific agonist BAY 60-6583 inhibits RANKL-induced NF-κB and p38 signaling and leads to a decrease in both cell-cell fusion in the late stage of osteoclast differentiation by notably reducing osteoclast fusion factors (Atp6v0d2 and DC-STAMP) and osteoclastic bone resorption [109]. Natural compounds, such as piperine and sinomenine, which are plant alkaloids, and fisetin, a flavonoid found in the smoke tree, not only show anti-angiogenic, anti-inflammatory, and anti-tumor activities, but also suppress RANKL-induced osteoclast differentiation via the downregulation of p38 activity [110–112].

4.3. Downstream Targets of p38 in Osteoclasts

Activated p38 directly phosphorylates and stimulates NFATc1 and MITF, transcription factors essential for osteoclastogenesis, inducing gene expression of osteoclastic proteins, such as TRAP, cathepsin K, and E-cadherin [113,114]. p38 activated by RANKL-TAK1-MKK6 signaling induces the phosphorylation of the NF-κB p65 subunit on Ser-536, resulting in increased transcription of NF-κB and NFATc1 [96]. In addition, RANKL/RANK/TRAF6/MKK3/6 signaling induces p38 activation followed by phosphorylation of activating transcription factor 2 (ATF2), stimulating RANKL-induced osteoclast differentiation, but not osteoclast function [115]. RANKL-stimulated active p38 strongly induces the expression of prostate transmembrane protein androgen induced 1 (Pmepa1), which subsequently upregulates cell-surface expression of RANK on osteoclasts [116]. RANKL-induced p38 activation induces STAT1 phosphorylation at Ser727 and promotes the expression and secretion of monokine induced by interferon-γ (MIG), which stimulates the adhesion and migration of osteoclast precursors and differentiated osteoclasts [117]. RANKL-stimulated active p38 directly phosphorylates MAPK-activated protein kinase-2 (MK2), which is critical for regulating the expression of osteoclastic fusion genes, DC-STAMP and osteoclast stimulatory transmembrane protein (OC-STAMP) [118].

4.4. Phosphatase Regulation of p38 Signaling in Osteoclasts

Mice lacking *DUSP1* exhibit drastic osteoclast activation in response to local LPS injection, and osteoclast precursors derived from $DUSP1^{-/-}$ mice show increased cell-cell fusion to multinucleated osteoclasts and osteoclastic bone-resorptive activity [119]. Further, DUSP1 is expressed by RANKL stimulation, is localized into the nucleus, and preferentially dephosphorylates the threonine and tyrosine residues of activated p38 and JNK over those of ERK, thus inducing transient p38 and JNK activation in response to stress and negatively regulating osteoclast formation and function by inactivating p38 MAPK-dependent signaling [13,119].

5. Distinct Kinetics of MAPKs and Crosstalk between MAPKs

The strength and duration of the response of MAPKs to different exogenous stimuli determines the biological outcome of the response. For instance, epidermal growth factor-induced transient ERK activation via the Raf-MEK-ERK axis induces proliferation in PC12 neuroendocrine cells, whereas nerve growth factor-induced prolonged ERK activation via the Raf-MEK-ERK axis leads to the differentiation of PC12 cells into sympathetic neuron-like cells [120,121]. Interestingly, prolonged ERK activation by epidermal growth factor in PC12 cells overexpressing the epidermal growth factor receptor switches cell fate from proliferation to differentiation [120]. Thus, cell fate decision is regarded a consequence of the duration of ERK activation. A recent report indicated that in osteoclast precursors, MAPK signaling induced by M-CSF or RANKL differed in terms of the extent and duration of ERK, p38, and JNK phosphorylation, as well as the selective phosphorylation of JNK isoforms [8]: (i) M-CSF induced more pronounced and sustained ERK phosphorylation than RANKL, (ii) RANKL induced more and longer p38 phosphorylation than M-CSF, (iii) M-CSF favorably phosphorylated JNK1 rather than JNK2 or JNK3, whereas RANKL had no such preference, and iv) M-CSF induced immediate monophasic activation (5 to 20 min) of MAPKs, whereas RANKL induced biphasic immediate (5 to 20 min) and delayed activation (8 to 24 h) of MAPKs. The different kinetics of MAPK activation by M-CSF or RANKL are considered to be related to cell fate decision of osteoclast precursor proliferation or differentiation [8].

5.1. ERK Kinetics in Osteoclast Metabolism

RANKL induces sustained ERK phosphorylation and maintains increased protein levels of c-Fos, resulting in enhanced osteoclastogenesis [53]. In contrast, the TLR9 ligand CpG-ODN leads to the transition of RANKL-induced sustained ERK activation into transient ERK activation by enhancing the expression of phosphatase PP2A, thereby decreasing the c-Fos level by degrading c-Fos mRNA and protein, and consequently inhibiting osteoclastogenesis. Moreover, when osteoclast precursors were pretreated with okadaic acid, a phosphatase inhibitor, RANKL-induced transient ERK activation by CpG-ODN was reverted to persistent activation, consequently inducing c-Fos expression. Additionally, 17β-estradiol triggers osteoclast apoptosis via transient ERK activation, peaking at 5 min after estrogen administration and returning to the basal level by 30 min, but blocks osteoblast apoptosis via long-lasting ERK phosphorylation for at least 24 h [122,123]. The opposite effects of ERK activation on apoptosis were accounted as a result of the differential duration of ERK phosphorylation in the osteoclasts and osteoblasts [122,123]. Collectively, the switch between RANKL-induced transient and persistent ERK activation is able to be regulated by tuning the activity of phosphatase and TLR-mediated signaling in the osteoclast precursors or mature osteoclasts.

5.2. JNK Kinetics in Osteoclast Metabolism

Lissencephaly-1 (LIS1)-flox;LysM-Cre mice, in which LIS1, a key regulator of microtubules and the cytoplasmic dynein motor complex, is specifically deleted in myeloid cells relevant to osteoclast precursors, exhibit increased bone mass, due to defective osteoclast formation and bone resorption [124]. Consistent with these findings in vivo, osteoclast precursors derived from LIS1 conditional knockout mice exhibited impaired osteoclast formation and accelerated apoptotic cell death through the suppression of M-CSF-induced prolonged ERK activation and the induction of RANKL-induced prolonged JNK activation. Moreover, the ablation of RelA, a component of NF-κB, induced strong activation of JNK by RANKL and resulted in JNK-Bid-mediated apoptosis of osteoclast precursors [92]. Together, these results indicate that changes in the activities of ERK and JNK during M-CSF- and RANKL-mediated osteoclastogenic signaling regulate the apoptosis of osteoclast precursors.

5.3. p38 Kinetics in Osteoclast Metabolism

p38α-flox;LysM-Cre mice exhibit bone defects in an age-dependent manner, displaying osteopetrosis at 2.5 months and osteoporosis at 6 months of age [93]. When compared with the differentiation of osteoclast precursors obtained from 2.5-month-old wild-type mice, osteoclast precursors isolated from age-matched p38α-deficient mice showed increased osteoclast formation at low cell density, but decreased osteoclast formation at high cell density. Hotokezaka et al. suggested that ERK inactivation induces RANKL-induced strong p38 activation and positively regulates osteoclastogenesis via the inhibition of ERK-mediated osteoclast precursor proliferation [125]. We also reported that p38 activation via the RANKL-RANK-TRAF6 axis leads to a shift from proliferation to differentiation in osteoclast precursors [8]. Therefore, the positive role of p38 in RANKL-, but not M-CSF-induced osteoclastogenesis seems to be differential in osteoclast formation and bone remodeling, according to spatial conditions of cell-cell confluency and physiological development stage, respectively.

5.4. Crosstalk between ERK and p38 in Osteoclast Metabolism

ERK inactivation by PD98059, a specific MEK inhibitor, suppressed serum-stimulated proliferation of SaOS-2 human osteosarcoma cells, but stimulated the osteogenic differentiation of these cells via accelerated p38 activation [126]. This phenomenon could be explained by a competition and balancing of p38-induced cell differentiation and MEK/ERK-mediated cell proliferation. In accordance herewith, ERK inactivation in osteoclast precursors by treatment with MEK inhibitors (U0126 and PD98059) elevated RANKL-induced p38 activation and resulted in enhanced osteoclast [125]. In addition, treatment of osteoclast precursors with p38 inhibitors (SB203580 and PD169316) induced an increase in RANKL-induced ERK activation and led to decreased osteoclast differentiation. These results suggest that MEK/ERK and p38 pathway may be involved in the suppression and induction of osteoclastogenesis, respectively, by regulating a seesaw-like crosstalk between ERK and p38 MAPK signaling. Of note, PD98059 and U0126 were found to show off-target effects on the MEK5/ERK5 pathway at higher concentrations, suggesting the possibility that ERK5 may contribute to some of the roles ascribed to ERK1/2 in osteoclastogenesis [127].

6. Conclusions

M-CSF and RANKL act as osteoclastogenic key regulators in normal osteoclast metabolism and share ERK, JNK, and p38 as signal mediators, but exhibit differences in the extent and duration of activation and MAPK isoform specificity. Moreover, M-CSF induces monophasic activation with an immediate phosphorylation (5 to 20 min) of MAPKs; distinctively, RANKL leads to biphasic activation with both immediate (5 to 20 min) and delayed phosphorylation (8 to 24 h) of MAPKs. The timing of RANKL-induced delayed MAPK activation coincided with the onset of osteoclast differentiation. Thus, the differential MAPK signaling induced by M-CSF and RANKL is recognized to determine the osteoclast precursor proliferation and osteoclast differentiation, respectively (Figure 4) [8]. JNK and p38 activated via RANKL-RANK signaling predominantly mediate osteoclastic apoptosis and promote osteoclast differentiation and function, respectively, whereas ERK activation via M-CSF/c-Fms axis preferentially potentiates osteoclast precursor proliferation [4,6–9,128,129]. Because p38 signaling is more tightly connected to the control of osteoclast metabolism than ERK and JNK signaling, researchers have tried to apply p38 inhibitors to prevent periopathogen-induced periodontal and active alveolar bone loss with degradation of mineralized and non-mineralized tooth tissues [130–132] and to treat rheumatoid arthritis with synovial inflammation, overactive osteoclast function, cartilage degradation, and bone erosion [133]. Although p38 is currently considered as a potential therapeutic target for inflammation-mediated bone loss [134], osteoclast-specific indirect regulators of p38 rather than direct p38 inhibitors should be developed to avoid side effects to other tissues and cells. Further studies are needed to clarify the detailed molecular mechanism underlying the crosstalk between MAPKs and the regulation of MAPKs by the balancing of kinases and phosphatases, and to explore the roles of specific isoforms of JNK1/2/3 and co-modulators capable of tuning p38 MAPK cascades in osteoclast metabolism.

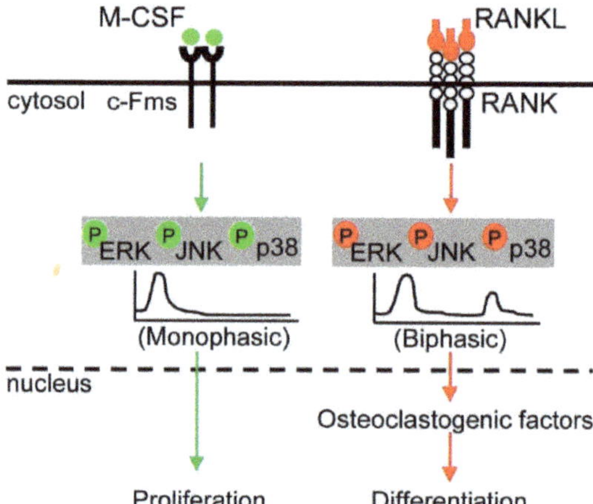

Figure 4. Osteoclast precursor proliferation by macrophage colony stimulating factor (M-CSF)/c-Fms-mediated monophasic activation of MAPKs and osteoclast differentiation by receptor activator for nuclear factor κ-B ligand (RANKL)/RANK-mediated biphasic activation of MAPKs. Arrows indicate activation of the signaling pathways, a solid line indicates the plasma membrane, and a dotted line indicates the nuclear membrane of osteoclast precursors. Green color means the signaling pathway induced by M-CSF and red color means the signaling pathway induced by RANKL.

Author Contributions: All authors contributed equally to this work. K.L., I.S., M.H.C., and D.J. formulated the theme, outline of the review, and wrote the manuscript.

Funding: This work was supported by grants from the National Research Foundation of Korea (Nos. 2016R1A2B2012108, 2015R1A5A2009124, and 2016R1A6A3A11930818).

Conflicts of Interest: The authors declare no conflict of interest.

References

1. Datta, H.K.; Ng, W.F.; Walker, J.A.; Tuck, S.P.; Varanasi, S.S. The cell biology of bone metabolism. *J. Clin. Pathol.* **2008**, *61*, 577–587. [CrossRef] [PubMed]
2. Zaidi, M. Skeletal remodeling in health and disease. *Nat. Med.* **2007**, *13*, 791–801. [CrossRef] [PubMed]
3. Chambers, T.J. Regulation of the differentiation and function of osteoclasts. *J. Pathol.* **2000**, *192*, 4–13. [CrossRef]
4. Teitelbaum, S.L.; Ross, F.P. Genetic regulation of osteoclast development and function. *Nat. Rev. Genet.* **2003**, *4*, 638–649. [CrossRef] [PubMed]
5. Boyle, W.J.; Simonet, W.S.; Lacey, D.L. Osteoclast differentiation and activation. *Nature* **2003**, *423*, 337–342. [CrossRef] [PubMed]
6. Ross, F.P. M-csf, c-fms, and signaling in osteoclasts and their precursors. *Ann. N. Y. Acad. Sci.* **2006**, *1068*, 110–116. [CrossRef] [PubMed]
7. Wada, T.; Nakashima, T.; Hiroshi, N.; Penninger, J.M. Rankl-rank signaling in osteoclastogenesis and bone disease. *Trends Mol. Med.* **2006**, *12*, 17–25. [CrossRef] [PubMed]
8. Lee, K.; Chung, Y.H.; Ahn, H.; Kim, H.; Rho, J.; Jeong, D. Selective regulation of mapk signaling mediates rankl-dependent osteoclast differentiation. *Int. J. Biol. Sci.* **2016**, *12*, 235–245. [CrossRef] [PubMed]
9. Ikeda, F.; Matsubara, T.; Tsurukai, T.; Hata, K.; Nishimura, R.; Yoneda, T. Jnk/c-jun signaling mediates an anti-apoptotic effect of rankl in osteoclasts. *J. Bone Miner. Res.* **2008**, *23*, 907–914. [CrossRef] [PubMed]
10. Bohm, C.; Hayer, S.; Kilian, A.; Zaiss, M.M.; Finger, S.; Hess, A.; Engelke, K.; Kollias, G.; Kronke, G.; Zwerina, J.; et al. The alpha-isoform of p38 mapk specifically regulates arthritic bone loss. *J. Immunol.* **2009**, *183*, 5938–5947. [CrossRef] [PubMed]
11. Chang, L.; Karin, M. Mammalian map kinase signalling cascades. *Nature* **2001**, *410*, 37–40. [CrossRef] [PubMed]
12. Kyriakis, J.M.; Avruch, J. Mammalian mitogen-activated protein kinase signal transduction pathways activated by stress and inflammation. *Physiol. Rev.* **2001**, *81*, 807–869. [CrossRef] [PubMed]
13. Carlson, J.; Cui, W.; Zhang, Q.; Xu, X.; Mercan, F.; Bennett, A.M.; Vignery, A. Role of mkp-1 in osteoclasts and bone homeostasis. *Am. J. Pathol.* **2009**, *175*, 1564–1573. [CrossRef] [PubMed]
14. He, Y.; Staser, K.; Rhodes, S.D.; Liu, Y.; Wu, X.; Park, S.J.; Yuan, J.; Yang, X.; Li, X.; Jiang, L.; et al. Erk1 positively regulates osteoclast differentiation and bone resorptive activity. *PLoS ONE* **2011**, *6*, e24780. [CrossRef] [PubMed]
15. Kinoshita, T.; Shirouzu, M.; Kamiya, A.; Hashimoto, K.; Yokoyama, S.; Miyajima, A. Raf/mapk and rapamycin-sensitive pathways mediate the anti-apoptotic function of p21ras in il-3-dependent hematopoietic cells. *Oncogene* **1997**, *15*, 619–627. [CrossRef] [PubMed]
16. Raman, M.; Chen, W.; Cobb, M.H. Differential regulation and properties of mapks. *Oncogene* **2007**, *26*, 3100–3112. [CrossRef] [PubMed]
17. Loveridge, C.J.; van't Hof, R.J.; Charlesworth, G.; King, A.; Tan, E.H.; Rose, L.; Daroszewska, A.; Prior, A.; Ahmad, I.; Welsh, M.; et al. Analysis of nkx3.1:Cre-driven erk5 deletion reveals a profound spinal deformity which is linked to increased osteoclast activity. *Sci. Rep.* **2017**, *7*, 13241. [CrossRef] [PubMed]
18. Adam, C.; Gluck, L.; Ebert, R.; Goebeler, M.; Jakob, F.; Schmidt, M. The mek5/erk5 mitogen-activated protein kinase cascade is an effector pathway of bone-sustaining bisphosphonates that regulates osteogenic differentiation and mineralization. *Bone* **2018**, *111*, 49–58. [CrossRef] [PubMed]
19. Amano, S.; Chang, Y.T.; Fukui, Y. Erk5 activation is essential for osteoclast differentiation. *PLoS ONE* **2015**, *10*, e0125054. [CrossRef] [PubMed]
20. Sherr, C.J. Colony-stimulating factor-1 receptor. *Blood* **1990**, *75*, 1–12. [PubMed]
21. Mancini, A.; Niedenthal, R.; Joos, H.; Koch, A.; Trouliaris, S.; Niemann, H.; Tamura, T. Identification of a second grb2 binding site in the v-fms tyrosine kinase. *Oncogene* **1997**, *15*, 1565–1572. [CrossRef] [PubMed]
22. Lacey, D.L.; Timms, E.; Tan, H.L.; Kelley, M.J.; Dunstan, C.R.; Burgess, T.; Elliott, R.; Colombero, A.; Elliott, G.; Scully, S.; et al. Osteoprotegerin ligand is a cytokine that regulates osteoclast differentiation and activation. *Cell* **1998**, *93*, 165–176. [CrossRef]

23. Kong, Y.Y.; Yoshida, H.; Sarosi, I.; Tan, H.L.; Timms, E.; Capparelli, C.; Morony, S.; Oliveira-dos-Santos, A.J.; Van, G.; Itie, A.; et al. Opgl is a key regulator of osteoclastogenesis, lymphocyte development and lymph-node organogenesis. *Nature* **1999**, *397*, 315–323. [CrossRef] [PubMed]

24. Theoleyre, S.; Wittrant, Y.; Couillaud, S.; Vusio, P.; Berreur, M.; Dunstan, C.; Blanchard, F.; Redini, F.; Heymann, D. Cellular activity and signaling induced by osteoprotegerin in osteoclasts: Involvement of receptor activator of nuclear factor kappab ligand and mapk. *Biochim. Biophys. Acta* **2004**, *1644*, 1–7. [CrossRef] [PubMed]

25. Zhao, H.; Liu, X.; Zou, H.; Dai, N.; Yao, L.; Gao, Q.; Liu, W.; Gu, J.; Yuan, Y.; Bian, J.; et al. Osteoprotegerin induces podosome disassembly in osteoclasts through calcium, erk, and p38 mapk signaling pathways. *Cytokine* **2015**, *71*, 199–206. [CrossRef] [PubMed]

26. Chung, Y.H.; Choi, B.; Song, D.H.; Song, Y.; Kang, S.W.; Yoon, S.Y.; Kim, S.W.; Lee, H.K.; Chang, E.J. Interleukin-1beta promotes the lc3-mediated secretory function of osteoclast precursors by stimulating the ca(2)(+)-dependent activation of erk. *Int. J. Biochem. Cell Biol.* **2014**, *54*, 198–207. [CrossRef] [PubMed]

27. Lee, Z.H.; Lee, S.E.; Kim, C.W.; Lee, S.H.; Kim, S.W.; Kwack, K.; Walsh, K.; Kim, H.H. Il-1alpha stimulation of osteoclast survival through the pi 3-kinase/akt and erk pathways. *J. Biochem.* **2002**, *131*, 161–166. [CrossRef] [PubMed]

28. Feng, W.; Liu, H.; Luo, T.; Liu, D.; Du, J.; Sun, J.; Wang, W.; Han, X.; Yang, K.; Guo, J.; et al. Combination of il-6 and sil-6r differentially regulate varying levels of rankl-induced osteoclastogenesis through nf-kappab, erk and jnk signaling pathways. *Sci. Rep.* **2017**, *7*, 41411. [CrossRef] [PubMed]

29. Baud'huin, M.; Renault, R.; Charrier, C.; Riet, A.; Moreau, A.; Brion, R.; Gouin, F.; Duplomb, L.; Heymann, D. Interleukin-34 is expressed by giant cell tumours of bone and plays a key role in rankl-induced osteoclastogenesis. *J. Pathol.* **2010**, *221*, 77–86. [CrossRef] [PubMed]

30. Tsubaki, M.; Kato, C.; Isono, A.; Kaneko, J.; Isozaki, M.; Satou, T.; Itoh, T.; Kidera, Y.; Tanimori, Y.; Yanae, M.; et al. Macrophage inflammatory protein-1alpha induces osteoclast formation by activation of the mek/erk/c-fos pathway and inhibition of the p38mapk/irf-3/ifn-beta pathway. *J. Cell. Biochem.* **2010**, *111*, 1661–1672. [CrossRef] [PubMed]

31. Lee, M.S.; Kim, H.S.; Yeon, J.T.; Choi, S.W.; Chun, C.H.; Kwak, H.B.; Oh, J. Gm-csf regulates fusion of mononuclear osteoclasts into bone-resorbing osteoclasts by activating the ras/erk pathway. *J. Immunol.* **2009**, *183*, 3390–3399. [CrossRef] [PubMed]

32. Kawaguchi, H.; Katagiri, M.; Chikazu, D. Osteoclastic bone resorption through receptor tyrosine kinase and extracellular signal-regulated kinase signaling in mature osteoclasts. *Mod. Rheumatol.* **2004**, *14*, 1–5. [CrossRef] [PubMed]

33. Lee, S.E.; Chung, W.J.; Kwak, H.B.; Chung, C.H.; Kwack, K.B.; Lee, Z.H.; Kim, H.H. Tumor necrosis factor-alpha supports the survival of osteoclasts through the activation of akt and erk. *J. Biol. Chem.* **2001**, *276*, 49343–49349. [CrossRef] [PubMed]

34. Houde, N.; Chamoux, E.; Bisson, M.; Roux, S. Transforming growth factor-beta1 (tgf-beta1) induces human osteoclast apoptosis by up-regulating bim. *J. Biol. Chem.* **2009**, *284*, 23397–23404. [CrossRef] [PubMed]

35. Fong, D.; Bisson, M.; Laberge, G.; McManus, S.; Grenier, G.; Faucheux, N.; Roux, S. Bone morphogenetic protein-9 activates smad and erk pathways and supports human osteoclast function and survival in vitro. *Cell Signal* **2013**, *25*, 717–728. [CrossRef] [PubMed]

36. Li, S.; Liu, B.; Zhang, L.; Rong, L. Amyloid beta peptide is elevated in osteoporotic bone tissues and enhances osteoclast function. *Bone* **2014**, *61*, 164–175. [CrossRef] [PubMed]

37. Li, S.; Yang, B.; Teguh, D.; Zhou, L.; Xu, J.; Rong, L. Amyloid beta peptide enhances rankl-induced osteoclast activation through nf-kappab, erk, and calcium oscillation signaling. *Int. J. Mol. Sci.* **2016**, *17*, 1683. [CrossRef] [PubMed]

38. Lee, J.; Seong, S.; Kim, J.H.; Kim, K.; Kim, I.; Jeong, B.C.; Nam, K.I.; Kim, K.K.; Hennighausen, L.; Kim, N. Stat5 is a key transcription factor for il-3-mediated inhibition of rankl-induced osteoclastogenesis. *Sci. Rep.* **2016**, *6*, 30977. [CrossRef] [PubMed]

39. Hirose, J.; Masuda, H.; Tokuyama, N.; Omata, Y.; Matsumoto, T.; Yasui, T.; Kadono, Y.; Hennighausen, L.; Tanaka, S. Bone resorption is regulated by cell-autonomous negative feedback loop of stat5-dusp axis in the osteoclast. *J. Exp. Med.* **2014**, *211*, 153–163. [CrossRef] [PubMed]

40. Wei, S.; Wang, M.W.; Teitelbaum, S.L.; Ross, F.P. Interleukin-4 reversibly inhibits osteoclastogenesis via inhibition of nf-kappa b and mitogen-activated protein kinase signaling. *J. Biol. Chem.* **2002**, *277*, 6622–6630. [CrossRef] [PubMed]

41. Yue, L.; Haroun, S.; Parent, J.L.; de Brum-Fernandes, A.J. Prostaglandin d(2) induces apoptosis of human osteoclasts through erk1/2 and akt signaling pathways. *Bone* **2014**, *60*, 112–121. [CrossRef] [PubMed]

42. Rose, A.A.; Annis, M.G.; Dong, Z.; Pepin, F.; Hallett, M.; Park, M.; Siegel, P.M. Adam10 releases a soluble form of the gpnmb/osteoactivin extracellular domain with angiogenic properties. *PLoS ONE* **2010**, *5*, e12093. [CrossRef] [PubMed]

43. Sondag, G.R.; Mbimba, T.S.; Moussa, F.M.; Novak, K.; Yu, B.; Jaber, F.A.; Abdelmagid, S.M.; Geldenhuys, W.J.; Safadi, F.F. Osteoactivin inhibition of osteoclastogenesis is mediated through cd44-erk signaling. *Exp. Mol. Med.* **2016**, *48*, e257. [CrossRef] [PubMed]

44. Rogers, M.J.; Gordon, S.; Benford, H.L.; Coxon, F.P.; Luckman, S.P.; Monkkonen, J.; Frith, J.C. Cellular and molecular mechanisms of action of bisphosphonates. *Cancer* **2000**, *88*, 2961–2978. [CrossRef]

45. Tsubaki, M.; Komai, M.; Itoh, T.; Imano, M.; Sakamoto, K.; Shimaoka, H.; Takeda, T.; Ogawa, N.; Mashimo, K.; Fujiwara, D.; et al. Nitrogen-containing bisphosphonates inhibit rankl- and m-csf-induced osteoclast formation through the inhibition of erk1/2 and akt activation. *J. Biomed. Sci.* **2014**, *21*, 10. [CrossRef] [PubMed]

46. Misra, N.C.; Nigam, P.K.; Gupta, R.; Agarwal, A.K.; Kamboj, V.P. Centchroman–a non-steroidal anti-cancer agent for advanced breast cancer: Phase-ii study. *Int. J. Cancer* **1989**, *43*, 781–783. [CrossRef] [PubMed]

47. Kharkwal, G.; Chandra, V.; Fatima, I.; Dwivedi, A. Ormeloxifene inhibits osteoclast differentiation in parallel to downregulating rankl-induced ros generation and suppressing the activation of erk and jnk in murine raw264.7 cells. *J. Mol. Endocrinol.* **2012**, *48*, 261–270. [CrossRef] [PubMed]

48. Zhou, C.H.; Wang, Y. Recent researches in triazole compounds as medicinal drugs. *Curr. Med. Chem.* **2012**, *19*, 239–280. [CrossRef] [PubMed]

49. Ihn, H.J.; Lee, D.; Lee, T.; Shin, H.I.; Bae, Y.C.; Kim, S.H.; Park, E.K. The 1,2,3-triazole derivative kp-a021 suppresses osteoclast differentiation and function by inhibiting rankl-mediated mek-erk signaling pathway. *Exp. Biol. Med. (Maywood)* **2015**, *240*, 1690–1697. [CrossRef] [PubMed]

50. Ouyang, Z.; Zhai, Z.; Li, H.; Liu, X.; Qu, X.; Li, X.; Fan, Q.; Tang, T.; Qin, A.; Dai, K. Hypericin suppresses osteoclast formation and wear particle-induced osteolysis via modulating erk signalling pathway. *Biochem. Pharmacol.* **2014**, *90*, 276–287. [CrossRef] [PubMed]

51. Hu, X.; Ping, Z.; Gan, M.; Tao, Y.; Wang, L.; Shi, J.; Wu, X.; Zhang, W.; Yang, H.; Xu, Y.; et al. Theaflavin-3,3′-digallate represses osteoclastogenesis and prevents wear debris-induced osteolysis via suppression of erk pathway. *Acta Biomater.* **2017**, *48*, 479–488. [CrossRef] [PubMed]

52. Yoshida, H.; Okamoto, K.; Iwamoto, T.; Sakai, E.; Kanaoka, K.; Hu, J.P.; Shibata, M.; Hotokezaka, H.; Nishishita, K.; Mizuno, A.; et al. Pepstatin a, an aspartic proteinase inhibitor, suppresses rankl-induced osteoclast differentiation. *J. Biochem.* **2006**, *139*, 583–590. [CrossRef] [PubMed]

53. Amcheslavsky, A.; Bar-Shavit, Z. Toll-like receptor 9 ligand blocks osteoclast differentiation through induction of phosphatase. *J. Bone Miner. Res.* **2007**, *22*, 1301–1310. [CrossRef] [PubMed]

54. Wagner, E.F.; Matsuo, K. Signalling in osteoclasts and the role of fos/ap1 proteins. *Ann. Rheum. Dis.* **2003**, *62* Suppl. 2, ii83–85. [CrossRef]

55. Weilbaecher, K.N.; Motyckova, G.; Huber, W.E.; Takemoto, C.M.; Hemesath, T.J.; Xu, Y.; Hershey, C.L.; Dowland, N.R.; Wells, A.G.; Fisher, D.E. Linkage of m-csf signaling to mitf, tfe3, and the osteoclast defect in mitf(mi/mi) mice. *Mol. Cell* **2001**, *8*, 749–758. [CrossRef]

56. Li, X.; Jie, Q.; Zhang, H.; Zhao, Y.; Lin, Y.; Du, J.; Shi, J.; Wang, L.; Guo, K.; Li, Y.; et al. Disturbed mek/erk signaling increases osteoclast activity via the hedgehog-gli pathway in postmenopausal osteoporosis. *Prog. Biophys. Mol. Biol.* **2016**, *122*, 101–111. [CrossRef] [PubMed]

57. Bradley, E.W.; Ruan, M.M.; Oursler, M.J. Novel pro-survival functions of the kruppel-like transcription factor egr2 in promotion of macrophage colony-stimulating factor-mediated osteoclast survival downstream of the mek/erk pathway. *J. Biol. Chem.* **2008**, *283*, 8055–8064. [CrossRef] [PubMed]

58. Wang, L.; Iorio, C.; Yan, K.; Yang, H.; Takeshita, S.; Kang, S.; Neel, B.G.; Yang, W. A erk/rsk-mediated negative feedback loop regulates m-csf-evoked pi3k/akt activation in macrophages. *FASEB J.* **2018**, *32*, 875–887. [CrossRef] [PubMed]

59. Engsig, M.T.; Chen, Q.J.; Vu, T.H.; Pedersen, A.C.; Therkidsen, B.; Lund, L.R.; Henriksen, K.; Lenhard, T.; Foged, N.T.; Werb, Z.; et al. Matrix metalloproteinase 9 and vascular endothelial growth factor are essential for osteoclast recruitment into developing long bones. *J. Cell Biol.* **2000**, *151*, 879–889. [CrossRef] [PubMed]

60. Sundaram, K.; Nishimura, R.; Senn, J.; Youssef, R.F.; London, S.D.; Reddy, S.V. Rank ligand signaling modulates the matrix metalloproteinase-9 gene expression during osteoclast differentiation. *Exp. Cell Res.* **2007**, *313*, 168–178. [CrossRef] [PubMed]

61. Caunt, C.J.; Keyse, S.M. Dual-specificity map kinase phosphatases (mkps): Shaping the outcome of map kinase signalling. *FEBS J.* **2013**, *280*, 489–504. [CrossRef] [PubMed]

62. Akira, S. Functional roles of stat family proteins: Lessons from knockout mice. *Stem. Cells* **1999**, *17*, 138–146. [CrossRef] [PubMed]

63. David, J.P.; Sabapathy, K.; Hoffmann, O.; Idarraga, M.H.; Wagner, E.F. Jnk1 modulates osteoclastogenesis through both c-jun phosphorylation-dependent and -independent mechanisms. *J. Cell Sci.* **2002**, *115*, 4317–4325. [CrossRef] [PubMed]

64. Otero, J.E.; Dai, S.; Foglia, D.; Alhawagri, M.; Vacher, J.; Pasparakis, M.; Abu-Amer, Y. Defective osteoclastogenesis by ikkbeta-null precursors is a result of receptor activator of nf-kappab ligand (rankl)-induced jnk-dependent apoptosis and impaired differentiation. *J. Biol. Chem.* **2008**, *283*, 24546–24553. [CrossRef] [PubMed]

65. Chang, E.J.; Ha, J.; Huang, H.; Kim, H.J.; Woo, J.H.; Lee, Y.; Lee, Z.H.; Kim, J.H.; Kim, H.H. The jnk-dependent camk pathway restrains the reversion of committed cells during osteoclast differentiation. *J. Cell Sci.* **2008**, *121*, 2555–2564. [CrossRef] [PubMed]

66. Qi, B.; Cong, Q.; Li, P.; Ma, G.; Guo, X.; Yeh, J.; Xie, M.; Schneider, M.D.; Liu, H.; Li, B. Ablation of tak1 in osteoclast progenitor leads to defects in skeletal growth and bone remodeling in mice. *Sci. Rep.* **2014**, *4*, 7158. [CrossRef] [PubMed]

67. Stanley, E.R.; Chitu, V. Csf-1 receptor signaling in myeloid cells. *Cold Spring Harb. Perspect. Biol.* **2014**, *6*, a021857. [CrossRef] [PubMed]

68. Gangoiti, P.; Granado, M.H.; Wang, S.W.; Kong, J.Y.; Steinbrecher, U.P.; Gomez-Munoz, A. Ceramide 1-phosphate stimulates macrophage proliferation through activation of the pi3-kinase/pkb, jnk and erk1/2 pathways. *Cell Signal.* **2008**, *20*, 726–736. [CrossRef] [PubMed]

69. Sang, C.; Zhang, J.; Zhang, Y.; Chen, F.; Cao, X.; Guo, L. Tnf-alpha promotes osteoclastogenesis through jnk signaling-dependent induction of semaphorin3d expression in estrogen-deficiency induced osteoporosis. *J. Cell. Physiol.* **2017**, *232*, 3396–3408. [CrossRef] [PubMed]

70. Hotokezaka, H.; Sakai, E.; Ohara, N.; Hotokezaka, Y.; Gonzales, C.; Matsuo, K.; Fujimura, Y.; Yoshida, N.; Nakayama, K. Molecular analysis of rankl-independent cell fusion of osteoclast-like cells induced by tnf-alpha, lipopolysaccharide, or peptidoglycan. *J. Cell Biochem.* **2007**, *101*, 122–134. [CrossRef] [PubMed]

71. Jimi, E.; Akiyama, S.; Tsurukai, T.; Okahashi, N.; Kobayashi, K.; Udagawa, N.; Nishihara, T.; Takahashi, N.; Suda, T. Osteoclast differentiation factor acts as a multifunctional regulator in murine osteoclast differentiation and function. *J. Immunol.* **1999**, *163*, 434–442. [PubMed]

72. Ke, D.; Fu, X.; Xue, Y.; Wu, H.; Zhang, Y.; Chen, X.; Hou, J. Il-17a regulates the autophagic activity of osteoclast precursors through rankl-jnk1 signaling during osteoclastogenesis in vitro. *Biochem. Biophys. Res. Commun.* **2018**, *497*, 890–896. [CrossRef] [PubMed]

73. Maeda, K.; Kobayashi, Y.; Udagawa, N.; Uehara, S.; Ishihara, A.; Mizoguchi, T.; Kikuchi, Y.; Takada, I.; Kato, S.; Kani, S.; et al. Wnt5a-ror2 signaling between osteoblast-lineage cells and osteoclast precursors enhances osteoclastogenesis. *Nat. Med.* **2012**, *18*, 405–412. [CrossRef] [PubMed]

74. Aoyama, E.; Kubota, S.; Khattab, H.M.; Nishida, T.; Takigawa, M. Ccn2 enhances rankl-induced osteoclast differentiation via direct binding to rank and opg. *Bone* **2015**, *73*, 242–248. [CrossRef] [PubMed]

75. Park, H.; Noh, A.L.; Kang, J.H.; Sim, J.S.; Lee, D.S.; Yim, M. Peroxiredoxin ii negatively regulates lipopolysaccharide-induced osteoclast formation and bone loss via jnk and stat3. *Antioxid. Redox Signal.* **2015**, *22*, 63–77. [CrossRef] [PubMed]

76. Khapli, S.M.; Mangashetti, L.S.; Yogesha, S.D.; Wani, M.R. Il-3 acts directly on osteoclast precursors and irreversibly inhibits receptor activator of nf-kappa b ligand-induced osteoclast differentiation by diverting the cells to macrophage lineage. *J. Immunol.* **2003**, *171*, 142–151. [CrossRef] [PubMed]

77. Khapli, S.M.; Tomar, G.B.; Barhanpurkar, A.P.; Gupta, N.; Yogesha, S.D.; Pote, S.T.; Wani, M.R. Irreversible inhibition of rank expression as a possible mechanism for il-3 inhibition of rankl-induced osteoclastogenesis. *Biochem. Biophys. Res. Commun.* **2010**, *399*, 688–693. [CrossRef] [PubMed]

78. Mohamed, S.G.; Sugiyama, E.; Shinoda, K.; Taki, H.; Hounoki, H.; Abdel-Aziz, H.O.; Maruyama, M.; Kobayashi, M.; Ogawa, H.; Miyahara, T. Interleukin-10 inhibits rankl-mediated expression of nfatc1 in part via suppression of c-fos and c-jun in raw264.7 cells and mouse bone marrow cells. *Bone* **2007**, *41*, 592–602. [CrossRef] [PubMed]

79. Yoshitake, F.; Itoh, S.; Narita, H.; Ishihara, K.; Ebisu, S. Interleukin-6 directly inhibits osteoclast differentiation by suppressing receptor activator of nf-kappab signaling pathways. *J. Biol. Chem.* **2008**, *283*, 11535–11540. [CrossRef] [PubMed]

80. Takayanagi, H.; Ogasawara, K.; Hida, S.; Chiba, T.; Murata, S.; Sato, K.; Takaoka, A.; Yokochi, T.; Oda, H.; Tanaka, K.; et al. T-cell-mediated regulation of osteoclastogenesis by signalling cross-talk between rankl and ifn-gamma. *Nature* **2000**, *408*, 600–605. [CrossRef] [PubMed]

81. Koide, M.; Maeda, H.; Roccisana, J.L.; Kawanabe, N.; Reddy, S.V. Cytokineregulation and the signaling mechanism of osteoclast inhibitory peptide-1 (oip-1/hsca) to inhibit osteoclast formation. *J. Bone Miner. Res.* **2003**, *18*, 458–465. [CrossRef] [PubMed]

82. Shevde, N.K.; Bendixen, A.C.; Dienger, K.M.; Pike, J.W. Estrogens suppress rank ligand-induced osteoclast differentiation via a stromal cell independent mechanism involving c-jun repression. *Proc. Natl. Acad. Sci. USA* **2000**, *97*, 7829–7834. [CrossRef] [PubMed]

83. Srivastava, S.; Toraldo, G.; Weitzmann, M.N.; Cenci, S.; Ross, F.P.; Pacifici, R. Estrogen decreases osteoclast formation by down-regulating receptor activator of nf-kappa b ligand (rankl)-induced jnk activation. *J. Biol. Chem.* **2001**, *276*, 8836–8840. [CrossRef] [PubMed]

84. Wind, S.; Schnell, D.; Ebner, T.; Freiwald, M.; Stopfer, P. Clinical pharmacokinetics and pharmacodynamics of afatinib. *Clin. Pharmacokinet.* **2017**, *56*, 235–250. [CrossRef] [PubMed]

85. Ihn, H.J.; Kim, J.A.; Bae, Y.C.; Shin, H.I.; Baek, M.C.; Park, E.K. Afatinib ameliorates osteoclast differentiation and function through downregulation of rank signaling pathways. *BMB Rep.* **2017**, *50*, 150–155. [CrossRef] [PubMed]

86. Rogosnitzky, M.; Danks, R. Therapeutic potential of the biscoclaurine alkaloid, cepharanthine, for a range of clinical conditions. *Pharmacol. Rep.* **2011**, *63*, 337–347. [CrossRef]

87. Zhou, C.H.; Meng, J.H.; Yang, Y.T.; Hu, B.; Hong, J.Q.; Lv, Z.T.; Chen, K.; Heng, B.C.; Jiang, G.Y.; Zhu, J.; et al. Cepharanthine prevents estrogen deficiency-induced bone loss by inhibiting bone resorption. *Front. Pharmacol.* **2018**, *9*, 210. [CrossRef] [PubMed]

88. Yu, M.; Chen, X.; Lv, C.; Yi, X.; Zhang, Y.; Xue, M.; He, S.; Zhu, G.; Wang, H. Curcumol suppresses rankl-induced osteoclast formation by attenuating the jnk signaling pathway. *Biochem. Biophys. Res. Commun.* **2014**, *447*, 364–370. [CrossRef] [PubMed]

89. Giachelli, C.M. Vascular calcification: In vitro evidence for the role of inorganic phosphate. *J. Am. Soc. Nephrol.* **2003**, *14*, S300–S304. [CrossRef]

90. Mozar, A.; Haren, N.; Chasseraud, M.; Louvet, L.; Maziere, C.; Wattel, A.; Mentaverri, R.; Morliere, P.; Kamel, S.; Brazier, M.; et al. High extracellular inorganic phosphate concentration inhibits rank-rankl signaling in osteoclast-like cells. *J. Cell. Physiol.* **2008**, *215*, 47–54. [CrossRef] [PubMed]

91. Grigoriadis, A.E.; Wang, Z.Q.; Cecchini, M.G.; Hofstetter, W.; Felix, R.; Fleisch, H.A.; Wagner, E.F. C-fos: A key regulator of osteoclast-macrophage lineage determination and bone remodeling. *Science* **1994**, *266*, 443–448. [CrossRef] [PubMed]

92. Vaira, S.; Alhawagri, M.; Anwisye, I.; Kitaura, H.; Faccio, R.; Novack, D.V. Rela/p65 promotes osteoclast differentiation by blocking a rankl-induced apoptotic jnk pathway in mice. *J. Clin. Investig.* **2008**, *118*, 2088–2097. [CrossRef] [PubMed]

93. Cong, Q.; Jia, H.; Li, P.; Qiu, S.; Yeh, J.; Wang, Y.; Zhang, Z.L.; Ao, J.; Li, B.; Liu, H. P38alpha mapk regulates proliferation and differentiation of osteoclast progenitors and bone remodeling in an aging-dependent manner. *Sci. Rep.* **2017**, *7*, 45964. [CrossRef] [PubMed]

94. Matsumoto, M.; Sudo, T.; Saito, T.; Osada, H.; Tsujimoto, M. Involvement of p38 mitogen-activated protein kinase signaling pathway in osteoclastogenesis mediated by receptor activator of nf-kappa b ligand (rankl). *J. Biol. Chem.* **2000**, *275*, 31155–31161. [CrossRef] [PubMed]

95. Lamothe, B.; Lai, Y.; Xie, M.; Schneider, M.D.; Darnay, B.G. Tak1 is essential for osteoclast differentiation and is an important modulator of cell death by apoptosis and necroptosis. *Mol. Cell. Biol.* **2013**, *33*, 582–595. [CrossRef] [PubMed]

96. Huang, H.; Ryu, J.; Ha, J.; Chang, E.J.; Kim, H.J.; Kim, H.M.; Kitamura, T.; Lee, Z.H.; Kim, H.H. Osteoclast differentiation requires tak1 and mkk6 for nfatc1 induction and nf-kappab transactivation by rankl. *Cell Death Differ.* **2006**, *13*, 1879–1891. [CrossRef] [PubMed]

97. Boyle, D.L.; Hammaker, D.; Edgar, M.; Zaiss, M.M.; Teufel, S.; David, J.P.; Schett, G.; Firestein, G.S. Differential roles of mapk kinases mkk3 and mkk6 in osteoclastogenesis and bone loss. *PLoS ONE* **2014**, *9*, e84818. [CrossRef]

98. Zhu, N.; Cui, J.; Qiao, C.; Li, Y.; Ma, Y.; Zhang, J.; Shen, B. Camp modulates macrophage development by suppressing m-csf-induced mapks activation. *Cell Mol. Immunol.* **2008**, *5*, 153–157. [CrossRef] [PubMed]

99. Chung, Y.H.; Jang, Y.; Choi, B.; Song, D.H.; Lee, E.J.; Kim, S.M.; Song, Y.; Kang, S.W.; Yoon, S.Y.; Chang, E.J. Beclin-1 is required for rankl-induced osteoclast differentiation. *J. Cell. Physiol.* **2014**, *229*, 1963–1971. [CrossRef] [PubMed]

100. Matsumoto, M.; Sudo, T.; Maruyama, M.; Osada, H.; Tsujimoto, M. Activation of p38 mitogen-activated protein kinase is crucial in osteoclastogenesis induced by tumor necrosis factor. *FEBS Lett.* **2000**, *486*, 23–28. [CrossRef]

101. Kim, J.H.; Jin, H.M.; Kim, K.; Song, I.; Youn, B.U.; Matsuo, K.; Kim, N. The mechanism of osteoclast differentiation induced by il-1. *J. Immunol.* **2009**, *183*, 1862–1870. [CrossRef] [PubMed]

102. Okabe, I.; Kikuchi, T.; Mogi, M.; Takeda, H.; Aino, M.; Kamiya, Y.; Fujimura, T.; Goto, H.; Okada, K.; Hasegawa, Y.; et al. Il-15 and rankl play a synergistically important role in osteoclastogenesis. *J. Cell Biochem.* **2017**, *118*, 739–747. [CrossRef] [PubMed]

103. Nishida, H.; Suzuki, H.; Madokoro, H.; Hayashi, M.; Morimoto, C.; Sakamoto, M.; Yamada, T. Blockade of cd26 signaling inhibits human osteoclast development. *J. Bone Miner. Res.* **2014**, *29*, 2439–2455. [CrossRef] [PubMed]

104. Ji, J.D.; Park-Min, K.H.; Shen, Z.; Fajardo, R.J.; Goldring, S.R.; McHugh, K.P.; Ivashkiv, L.B. Inhibition of rank expression and osteoclastogenesis by tlrs and ifn-gamma in human osteoclast precursors. *J. Immunol.* **2009**, *183*, 7223–7233. [CrossRef] [PubMed]

105. Oh, E.; Lee, H.Y.; Kim, H.J.; Park, Y.J.; Seo, J.K.; Park, J.S.; Bae, Y.S. Serum amyloid a inhibits rankl-induced osteoclast formation. *Exp. Mol. Med.* **2015**, *47*, e194. [CrossRef] [PubMed]

106. Kalliolias, G.D.; Zhao, B.; Triantafyllopoulou, A.; Park-Min, K.H.; Ivashkiv, L.B. Interleukin-27 inhibits human osteoclastogenesis by abrogating rankl-mediated induction of nuclear factor of activated t cells c1 and suppressing proximal rank signaling. *Arthritis Rheum.* **2010**, *62*, 402–413. [PubMed]

107. Lu, D.; Li, J.; Liu, H.; Foxa, G.E.; Weaver, K.; Li, J.; Williams, B.O.; Yang, T. Lrp1 suppresses bone resorption in mice by inhibiting the rankl-stimulated nf-kappab and p38 pathways during osteoclastogenesis. *J Bone Miner. Res.* **2018**, in press. [CrossRef] [PubMed]

108. von Metzler, I.; Krebbel, H.; Hecht, M.; Manz, R.A.; Fleissner, C.; Mieth, M.; Kaiser, M.; Jakob, C.; Sterz, J.; Kleeberg, L.; et al. Bortezomib inhibits human osteoclastogenesis. *Leukemia* **2007**, *21*, 2025–2034. [CrossRef] [PubMed]

109. Kim, B.H.; Oh, J.H.; Lee, N.K. The inactivation of erk1/2, p38 and nf-kb is involved in the down-regulation of osteoclastogenesis and function by a2b adenosine receptor stimulation. *Mol. Cells* **2017**, *40*, 752–760. [PubMed]

110. Choi, S.W.; Son, Y.J.; Yun, J.M.; Kim, S.H. Fisetin inhibits osteoclast differentiation via downregulation of p38 and c-fos-nfatc1 signaling pathways. *Evid. Based Complement. Alternat. Med.* **2012**, *2012*, 810563. [CrossRef] [PubMed]

111. Deepak, V.; Kruger, M.C.; Joubert, A.; Coetzee, M. Piperine alleviates osteoclast formation through the p38/c-fos/nfatc1 signaling axis. *Biofactors* **2015**, *41*, 403–413. [CrossRef] [PubMed]

112. Li, X.; He, L.; Hu, Y.; Duan, H.; Li, X.; Tan, S.; Zou, M.; Gu, C.; Zeng, X.; Yu, L.; et al. Sinomenine suppresses osteoclast formation and mycobacterium tuberculosis h37ra-induced bone loss by modulating rankl signaling pathways. *PLoS ONE* **2013**, *8*, e74274. [CrossRef] [PubMed]

113. Mansky, K.C.; Sankar, U.; Han, J.; Ostrowski, M.C. Microphthalmia transcription factor is a target of the p38 mapk pathway in response to receptor activator of nf-kappa b ligand signaling. *J. Biol. Chem.* **2002**, *277*, 11077–11083. [CrossRef] [PubMed]

114. Matsumoto, M.; Kogawa, M.; Wada, S.; Takayanagi, H.; Tsujimoto, M.; Katayama, S.; Hisatake, K.; Nogi, Y. Essential role of p38 mitogen-activated protein kinase in cathepsin k gene expression during osteoclastogenesis through association of nfatc1 and pu.1. *J. Biol. Chem.* **2004**, *279*, 45969–45979. [CrossRef] [PubMed]

115. Li, X.; Udagawa, N.; Itoh, K.; Suda, K.; Murase, Y.; Nishihara, T.; Suda, T.; Takahashi, N. P38 mapk-mediated signals are required for inducing osteoclast differentiation but not for osteoclast function. *Endocrinology* **2002**, *143*, 3105–3113. [CrossRef] [PubMed]

116. Funakubo, N.; Xu, X.; Kukita, T.; Nakamura, S.; Miyamoto, H.; Kukita, A. Pmepa1 induced by rankl-p38 mapk pathway has a novel role in osteoclastogenesis. *J. Cell. Physiol.* **2018**, *233*, 3105–3118. [CrossRef] [PubMed]

117. Kwak, H.B.; Lee, S.W.; Jin, H.M.; Ha, H.; Lee, S.H.; Takeshita, S.; Tanaka, S.; Kim, H.M.; Kim, H.H.; Lee, Z.H. Monokine induced by interferon-gamma is induced by receptor activator of nuclear factor kappa b ligand and is involved in osteoclast adhesion and migration. *Blood* **2005**, *105*, 2963–2969. [CrossRef] [PubMed]

118. Herbert, B.A.; Valerio, M.S.; Gaestel, M.; Kirkwood, K.L. Sexual dimorphism in mapk-activated protein kinase-2 (mk2) regulation of rankl-induced osteoclastogenesis in osteoclast progenitor subpopulations. *PLoS ONE* **2015**, *10*, e0125387. [CrossRef] [PubMed]

119. Sartori, R.; Li, F.; Kirkwood, K.L. Map kinase phosphatase-1 protects against inflammatory bone loss. *J. Dent. Res.* **2009**, *88*, 1125–1130. [CrossRef] [PubMed]

120. Marshall, C.J. Specificity of receptor tyrosine kinase signaling: Transient versus sustained extracellular signal-regulated kinase activation. *Cell* **1995**, *80*, 179–185. [CrossRef]

121. Murphy, L.O.; Smith, S.; Chen, R.H.; Fingar, D.C.; Blenis, J. Molecular interpretation of erk signal duration by immediate early gene products. *Nat. Cell Biol.* **2002**, *4*, 556–564. [CrossRef] [PubMed]

122. Kameda, T.; Mano, H.; Yuasa, T.; Mori, Y.; Miyazawa, K.; Shiokawa, M.; Nakamaru, Y.; Hiroi, E.; Hiura, K.; Kameda, A.; et al. Estrogen inhibits bone resorption by directly inducing apoptosis of the bone-resorbing osteoclasts. *J. Exp. Med.* **1997**, *186*, 489–495. [CrossRef] [PubMed]

123. Chen, J.R.; Plotkin, L.I.; Aguirre, J.I.; Han, L.; Jilka, R.L.; Kousteni, S.; Bellido, T.; Manolagas, S.C. Transient versus sustained phosphorylation and nuclear accumulation of erks underlie anti-versus pro-apoptotic effects of estrogens. *J. Biol. Chem.* **2005**, *280*, 4632–4638. [CrossRef] [PubMed]

124. Ye, S.; Fujiwara, T.; Zhou, J.; Varughese, K.I.; Zhao, H. Lis1 regulates osteoclastogenesis through modulation of m-scf and rankl signaling pathways and cdc42. *Int. J. Biol. Sci.* **2016**, *12*, 1488–1499. [CrossRef] [PubMed]

125. Hotokezaka, H.; Sakai, E.; Kanaoka, K.; Saito, K.; Matsuo, K.; Kitaura, H.; Yoshida, N.; Nakayama, K. U0126 and pd98059, specific inhibitors of mek, accelerate differentiation of raw264.7 cells into osteoclast-like cells. *J. Biol. Chem.* **2002**, *277*, 47366–47372. [CrossRef] [PubMed]

126. Shimo, T.; Matsumura, S.; Ibaragi, S.; Isowa, S.; Kishimoto, K.; Mese, H.; Nishiyama, A.; Sasaki, A. Specific inhibitor of mek-mediated cross-talk between erk and p38 mapk during differentiation of human osteosarcoma cells. *J. Cell Commun. Signal.* **2007**, *1*, 103–111. [CrossRef] [PubMed]

127. Mody, N.; Leitch, J.; Armstrong, C.; Dixon, J.; Cohen, P. Effects of map kinase cascade inhibitors on the mkk5/erk5 pathway. *FEBS Lett.* **2001**, *502*, 21–24. [CrossRef]

128. Jaworowski, A.; Wilson, N.J.; Christy, E.; Byrne, R.; Hamilton, J.A. Roles of the mitogen-activated protein kinase family in macrophage responses to colony stimulating factor-1 addition and withdrawal. *J. Biol. Chem.* **1999**, *274*, 15127–15133. [CrossRef] [PubMed]

129. Valledor, A.F.; Comalada, M.; Xaus, J.; Celada, A. The differential time-course of extracellular-regulated kinase activity correlates with the macrophage response toward proliferation or activation. *J. Biol. Chem.* **2000**, *275*, 7403–7409. [CrossRef] [PubMed]

130. Kirkwood, K.L.; Li, F.; Rogers, J.E.; Otremba, J.; Coatney, D.D.; Kreider, J.M.; D'Silva, N.J.; Chakravarty, S.; Dugar, S.; Higgins, L.S.; et al. A p38alpha selective mitogen-activated protein kinase inhibitor prevents periodontal bone loss. *J. Pharmacol. Exp. Ther.* **2007**, *320*, 56–63. [CrossRef] [PubMed]

131. Rogers, J.E.; Li, F.; Coatney, D.D.; Otremba, J.; Kriegl, J.M.; Protter, T.A.; Higgins, L.S.; Medicherla, S.; Kirkwood, K.L. A p38 mitogen-activated protein kinase inhibitor arrests active alveolar bone loss in a rat periodontitis model. *J. Periodontol.* **2007**, *78*, 1992–1998. [CrossRef] [PubMed]

132. Kirkwood, K.L.; Rossa, C., Jr. The potential of p38 mapk inhibitors to modulate periodontal infections. *Curr. Drug Metab.* **2009**, *10*, 55–67. [CrossRef] [PubMed]

133. Clark, A.R.; Dean, J.L. The p38 mapk pathway in rheumatoid arthritis: A. sideways look. *Open Rheumatol. J.* **2012**, *6*, 209–219. [CrossRef] [PubMed]
134. Wei, S.; Siegal, G.P. P38 mapk as a potential therapeutic target for inflammatory osteolysis. *Adv. Anat. Pathol.* **2007**, *14*, 42–45. [CrossRef] [PubMed]

International Journal of
Molecular Sciences

MDPI

Communication

ERK5 Phosphorylates $K_V4.2$ and Inhibits Inactivation of the A-Type Current in PC12 Cells

Yurina Kashino [1], Yutaro Obara [1,*], Yosuke Okamoto [1], Takeo Saneyoshi [2], Yasunori Hayashi [2] and Kuniaki Ishii [1]

1   Department of Pharmacology, Yamagata University School of Medicine, Yamagata 990-9585, Japan; yurikamo_chronicle@yahoo.co.jp (Y.K.); okamoto@med.akita-u.ac.jp (Y.O.); kuishii@med.id.yamagata-u.ac.jp (K.I.)
2   Department of Pharmacology, Kyoto University Graduate School of Medicine, Kyoto 606-8501, Japan; saneyoshi.takeo.3v@kyoto-u.ac.jp (T.S.); yhayashi-tky@umin.ac.jp (Y.H.)
*   Correspondence: obaray@med.id.yamagata-u.ac.jp; Tel.: +81-22-628-5234

Received: 14 June 2018; Accepted: 28 June 2018; Published: 10 July 2018

Abstract: Extracellular signal-regulated kinase 5 (ERK5) regulates diverse physiological responses such as proliferation, differentiation, and gene expression. Previously, we demonstrated that ERK5 is essential for neurite outgrowth and catecholamine biosynthesis in PC12 cells and sympathetic neurons. However, it remains unclear how ERK5 regulates the activity of ion channels, which are important for membrane excitability. Thus, we examined the effect of ERK5 on the ion channel activity in the PC12 cells that overexpress both ERK5 and the constitutively active MEK5 mutant. The gene and protein expression levels of voltage-dependent Ca^{2+} and K^+ channels were determined by RT-qPCR or Western blotting. The A-type K^+ current was recorded using the whole-cell patch clamp method. In these ERK5-activated cells, the gene expression levels of voltage-dependent L- and P/Q-type Ca^{2+} channels did not alter, but the N-type Ca^{2+} channel was slightly reduced. In contrast, those of $K_V4.2$ and $K_V4.3$, which are components of the A-type current, were significantly enhanced. Unexpectedly, the protein levels of $K_V4.2$ were not elevated by ERK5 activation, but the phosphorylation levels were increased by ERK5 activation. By electrophysiological analysis, the inactivation time constant of the A-type current was prolonged by ERK5 activation, without changes in the peak current. Taken together, ERK5 inhibits an inactivation of the A-type current by phosphorylation of $K_V4.2$, which may contribute to the neuronal differentiation process.

Keywords: extracellular signal-regulated kinase 5 (ERK5); $K_V4.2$; PC12 cells

---

1. Introduction

Conventional mitogen-activated protein kinases (MAPKs) involve extracellular signal-regulated kinases (ERKs) 1, 2, and 5, c-Jun N-terminal kinase and p38 MAPKs, and atypical MAPKs include ERK3, 4, and 7 and nemo-like kinase [1]. In response to growth factors or neurotrophic factors, ERKs are strongly activated and regulate diverse physiological responses, such as proliferation, differentiation, and gene expression. The signal transduction leading to ERK1/2 activation and the involvement of ERK1/2 in cellular responses are the best studied among the MAPK family members.

ERK5 shares homology in the amino acid sequence in the kinase-domain with ERK1/2, and possesses a unique long C-terminal domain [2,3]. In the past 10 years, specific inhibitors of ERK5 signaling, such as BIX02189 [4,5] and XMD8-92 [6], have been developed. Using these pharmacological inhibitors, the role of ERK5 in tumor genesis and metastatic progression has been especially well understood [7,8]. We have shown that the levels of ERK5 and tyrosine hydroxylase, a rate-limiting enzyme for catecholamine biosynthesis, are co-related in normal human adrenal medulla, but this correlation is disrupted in pheochromocytomas [9]. However, signaling pathways for ERK5 activation and physiological roles of ERK5 in neuronal development are relatively unclear. For example, involvements of small G-proteins in ERK5 activation are vague [10], whereas it has been established that ERK1/2 is activated through Ras and Rap1 [11,12]. Some limited studies suggest that ERK5 is necessary and sufficient for neuronal differentiation of progenitor cells [13], and is essential for adult hippocampal neurogenesis [14,15]. ERK5 promotes neuronal survival in sympathetic or sensory neurons [16,17]. We have shown that ERK5 is essential for neurite/axon outgrowth and catecholamine biosynthesis in PC12 cells and sympathetic neurons [5,9]. Thus, ERK5 plays important roles in neuronal survival, as well as morphological and functional differentiation. However, although it is well known that ERK1/2 regulates membrane excitability (i.e., neuronal activity) [18–21], ERK5 regulation of membrane excitability has been poorly understood. Therefore, in the present study, we attempted to clarify the effect of ERK5 signaling on ion channel activity, which is important for regulating membrane excitability.

2. Results

To examine the effect of ERK5 signaling, we attempted to activate ERK5 selectively by the overexpression of ERK5 wildtype and a constitutively active mutant of MAPK/ERK kinase (MEK) 5 (MEK5S311D/T315D, or MEK5D for short). To confirm that ERK signaling is activated by transfection with these DNA constructs, we measured the myocyte-enhancer factor (MEF) 2 transcriptional activity by reporter gene assay as an index of ERK5 activation. It has been well-established that ERK5 phosphorylates MEF2C directly and the transcriptional activity increases [22]. In human embryonic kidney 293 cells (HEK293 cells), overexpression of MEK5D and ERK5 resulted in a dramatic enhancement of MEF2C activity (Figure 1).

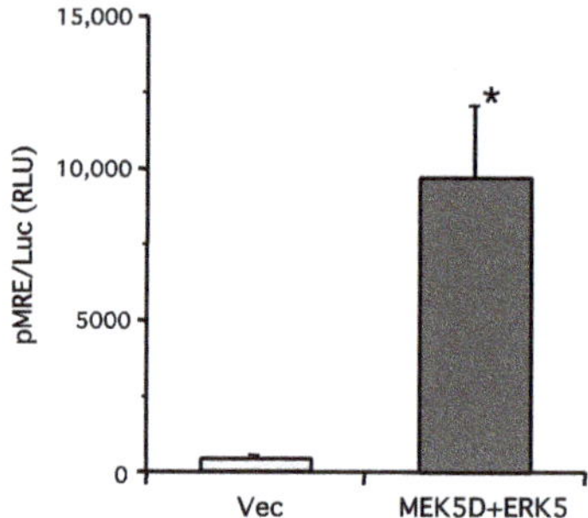

Figure 1. Overexpression of constitutively active mitogen-activated protein and extracellular signal-regulated (MAPK/ERK) kinase (MEK) 5 mutant and ERK5 causes activation of ERK5 signaling. Human embryonic kidney 293 cells (HEK293 cells) were transfected with tandem myocyte-enhancer factor (MEF) 2 response element (MRE)-luciferase reporter gene and empty vector (Vec) or MEK5D and ERK5. Two days after transfection, the luciferase activity resulting from MEF2 activation was measured. ERK5 significantly increased MEF2C activity (one experiment in triplicate ($n = 3$), * p <0.05, unpaired Student's *t*-test).

We previously demonstrated that overexpression of MEK5D and ERK5 strengthens ERK5 signaling, accompanied by the phosphorylation of the Thr-Glu-Tyr (TEY) activation motif and auto-phosphorylation sites on ERK5, but the ERK1/2 TEY phosphorylation site is not affected [23]. Next, PC12 cells were co-transfected with MEK5D and ERK5, and the messenger RNA (mRNA) expression levels of the major voltage-dependent Ca^{2+} and K^+ channels were measured by RT-qPCR (Figure 2). There were no significant changes in the expression levels of $Ca_V1.2$ (L-type) and $Ca_V2.1$ (P/Q-type), but a significant reduction of $Ca_V2.2$ (N-type) was observed. In contrast, the $K_V4.2$ and $K_V4.3$ expression levels, which are responsible for the transient outward I_{to} current

(A type-current), were significantly promoted by ERK5 signaling. K^+ channel-interacting proteins (KChIPs) are β-subunit for $K_v4.2$, and the A-type current is influenced by the expression of KChiPs [24]. We previously performed RNA-sequencing to examine the gene expression levels comprehensively in PC12 cells [25]. The RPKM values for KChIPs 1, 2, 3, and 4 were 0.021284, 0, 1.21248, and 0.067198, respectively ($n = 3$). Because KChIP3 is a major β-subunit for $K_v4.2$ in PC12 cells, we examined the KChIP3 expression levels. But, there was no significant change in the expression levels. It has been shown that overexpression of the RasG12V (RasV12) oncogenic mutant can strongly activate ERK1/2 signaling without affecting the phosphorylation status of the ERK5 TEY motif [5,23]. Constitutive ERK1/2 activation by the overexpression of RasV12 did not elevate the significant expression of $K_v4.2$ and $K_v4.3$ in our condition (0.791-fold, $n = 6$, $p = 0.603$ for $K_v4.2$ and 0.522-fold, $n = 3$, $p = 0.454$ for $K_v4.3$). Because $K_v4.2$ mediates the majority of the A-type current and is a critical molecule for the modulation of neuronal excitability in many types of neurons, including the cornu ammonis (CA) 1 pyramidal neurons of the hippocampus and the dorsal horn neurons [20,24], we focused on ERK5 regulation of $K_v4.2$ for further study.

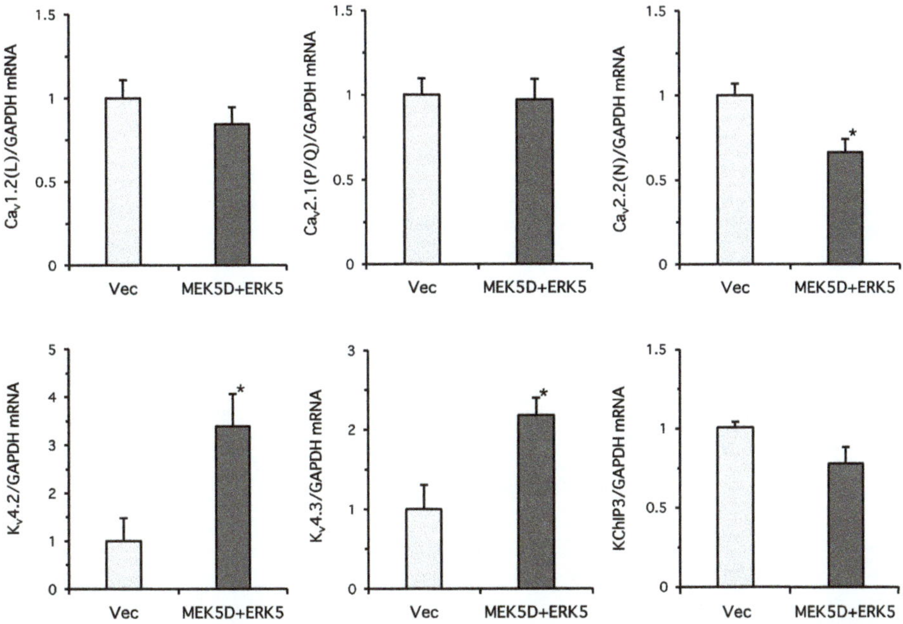

Figure 2. ERK5 promotes gene expression of $K_v4.2$ and $K_v4.3$ in PC12 cells. PC12 cells were transfected with empty vector (Vec) or MEK5 and ERK5. Two days after transfection, the total RNA was isolated from the cell lysates and RT-qPCR was performed using specific primers for $Ca_v1.2$, $Ca_v2.1$, $Ca_v2.2$, $K_v4.2$, $K_v4.3$, and KChIP3. ERK5 significantly promoted the gene expression of $K_v4.2$ and $K_v4.3$, and attenuated $Ca_v2.2$ expression (two independent experiments in triplicate ($n = 6$), * $p < 0.05$, unpaired Student's *t*-test).

We next examined the protein levels of $K_v4.2$ after th eERK5 activation in PC12 cells. Surprisingly, although the mRNA levels were increased, the protein levels were not altered significantly (Figure 3a). In addition to expression levels, we investigated the phosphorylation status of $K_v4.2$, because it has been reported that ERK1/2 phosphorylates at least three Ser/Thr residues at the C-terminus of $K_v4.2$ and both ERK5 and ERK1/2 preferentially phosphorylate Ser/Thr residues that have a similar minimum consensus sequence (i.e., Ser/Thr-Pro) [24,26,27]. We used a Phos-tag reagent, which tightly binds phosphorylated amino acids in the presence of Mn^{2+} or Zn^{2+}. In principle, the Phos-tag-mixed sodium dodecyl sulfate-polyacrylamide gel electrophoresis (SDS-PAGE) causes a band-shift of phosphorylated proteins, and they can be clearly distinguished from unphosphorylated proteins. The overexpression of MEK5D and ERK5 caused the band-shift of $K_v4.2$, which was significantly diminished by the dominant-negative ERK5 kinase-dead mutant (ERK5K83M, or ERK5KD for short), suggesting that ERK5 signaling promoted phosphorylation levels of $K_v4.2$ (Figure 3b).

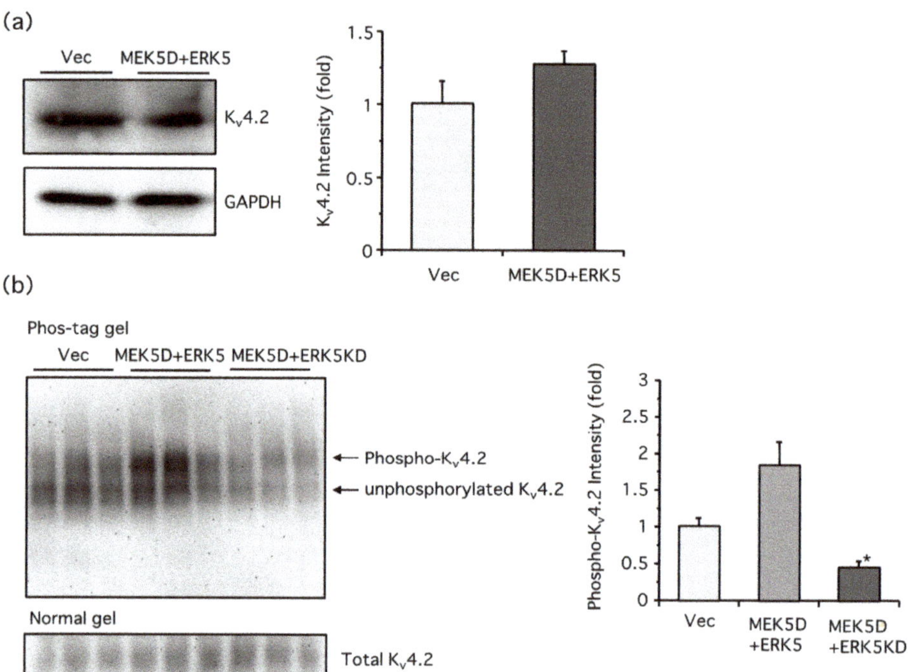

Figure 3. ERK5 did not alter the protein expression of $K_v4.2$, but did promote the phosphorylation of $K_v4.2$ in PC12 cells. (**a**) PC12 cells were transfected with empty vector (Vec) or MEK5D and ERK5. Two days after transfection, Western blotting using $K_v4.2$ and glyceraldehyde-3-phosphate dehydrogenase (GAPDH) antibodies was performed. The density of the $K_v4.2$ bands was expressed as a fold of control cells (Vec) (two independent experiments in triplicate (n = 6), unpaired Student's *t*-test); (**b**) PC12 cells were transfected with empty vector (Vec), MEK5D and ERK5, or MEK5D and ERK5KD. Two days after transfection, Western blotting using Phos-tag-contained polyacrylamide gels or Phos-tag-free gels was carried out with a $K_v4.2$ antibody. The density of the phospho-$K_v4.2$ bands was expressed as a fold of the control cells (Vec). Overexpression of MEK5D and ERK5KD significantly attenuated the MEK5D and ERK5-induced $K_v4.2$ phosphorylation (three similar independent experiments in triplicate (n = 3), * p <0.05, Tukey's method).

We next examined the effect of ERK5 signaling on the A-type current in PC12 cells. The peak current was unchanged in the PC12 cells overexpressing MEK5D and ERK5, but there was a significant slowing of inactivation (Figure 4). The time constant (τ) at the fast and slow phases was 6.663 and 213.0 (ms), respectively, in the control cells, and 16.69 and 334.7 (ms), respectively, in the cells co-transfected with MEK5D and ERK5. These results suggest that the A-type current inactivation was inhibited by the ERK5 activation, regulating membrane excitability.

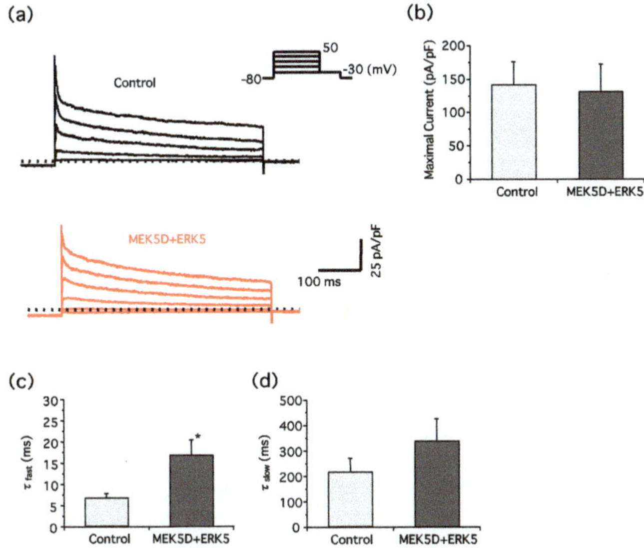

Figure 4. ERK5 inhibits inactivation of the A-type current in PC12 cells. PC12 cells were co-transfected with EGFP, MEK5D, and ERK5. Two days after transfection, the A-type current was recorded. (**a**) Representative traces and step-pulse protocol are shown; (**b**) maximal peak current was measured at +50 mV. The ERK5 did not significantly change amplitude levels (data from three independent experiments ($n = 4$), unpaired Student's *t*-test); (c,d) the time constant (τ) at fast (**c**) and slow (**d**) phases at +50 mV was calculated. ERK5 significantly changed the time constant (τ) at the fast phase (data from three independent experiments ($n = 4$), * $p < 0.05$, unpaired Student's *t*-test).

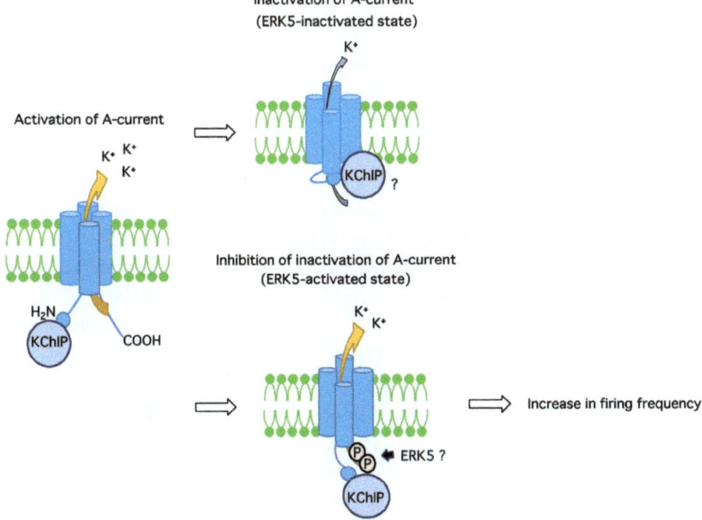

Figure 5. Putative mechanism of regulation of $K_v4.2$ channels by ERK5. The ERK5 phosphorylates unidentified Ser/Thr residue(s) on $K_v4.2$, resulting in the inhibition of the A-type current inactivation. This mechanism may contribute to rapid repolarization toward resting potential, which is necessary for causing the next firing.

3. Discussion

In the present study, we found that ERK5 signaling promoted the mRNA expression of the $K_v4.2$ primary subunits that underlie the transient A-type current in PC12 cells. However, its protein levels were not reflected by this mRNA up-regulation. Instead, the phosphorylation of endogenous $K_v4.2$ proteins was promoted by ERK5 and the inactivation rate of the A-type current decreased in these cells. This putative mechanism is shown in Figure 5.

It has been demonstrated that the ERK5 knock-down by antisense oligonucleotides suppressed levels of transient receptor potential (TRP) V1 and A1 in dorsal root ganglion neurons [28]. In the study above, ERK5 regulated the TRPV1 and TRPA1 expression by an unknown mechanism. It has been shown that various transcription factors bind to the $K_v4.2$ promoter and regulate the transcription. For example, GATA4 and 6, as well as FOG2 enhance $K_v4.2$ transcription in PC12 cells, although there is a possibility that these transcription factors influence indirectly, as the GATA-binding consensus sequence is lacking in the minimum $K_v4.2$ promoter [29]. Another study shows that the calcineurin/nuclear factor of the activated T cells (NFAT) pathway increases the $K_v4.2$ mRNA and protein expression and promoter activity, without affecting the KChIP2 and $K_v4.3$ levels in rat neonatal ventricular myocytes [30]. Furthermore, neuritin increases the A-type current density accompanied by the up-regulation of $K_v4.2$ mRNA and protein via the Ca^{2+}/calmodulin/calcineurin/NFATc4 and ERK/NFATc4 pathways in the central neurons, and affects neuronal excitability with increased dendritic spine formation [21]. Because there are NFAT binding sites in the $K_v4.2$ promoter [21], ERK5 may phosphorylate NFAT to promote $K_v4.2$ transcription, as ERK1/2 activation resulted in phosphorylation of Ser676 on NFATc4 [31]. However, the reason remains unknown as to why the $K_v4.2$ protein levels were not reflected by its mRNA expression in this study. In contrast, ERK5 phosphorylation of $K_v4.2$ was promoted without changes in the protein expression levels. This may reflect the results obtained by electrophysiological experiments that the peak current was not altered by the ERK5 activation. Furthermore, it is also reasonable that the change in the time constant of the A-type current was affected by the changes in $K_v4.2$ phosphorylation status, but not the protein levels.

It has been shown that ERK1/2 directly phosphorylates Thr602, Thr607, and Ser616 residues at the C-terminal cytoplasmic domain of $K_v4.2$ [26]. These amino acids are entirely preserved among human and rat $K_v4.2$. In this study, epidermal growth factor enhanced phosphorylation of $K_v4.2$ at these three sites in COS7 cells overexpressing $K_v4.2$. Although this phosphorylation was attenuated by U0126, which blocks ERK1/2 signaling, the remaining phosphorylated band was still observed. Because epidermal growth factor can activate both ERK1/2 and ERK5 [5,23], the remaining U0126-resistant $K_v4.2$ phosphorylation component may result from ERK5 activity. This group further examined the effects of these three phosphorylated amino acids on the A-type current [24]. The mutation of these three amino acids to Asp caused the activation curve to shift toward more depolarized membrane potentials, whereas the mutation of these three amino acids to Ala showed no effect. Interestingly, the site-directed T607D mutant caused a rightward shift of the activation curve only in the presence of KChIP3, as observed in the case of the triple D mutant, but the S616D mutant caused a leftward shift, which is the totally opposite effect. It has been shown that ERK1/2 also phosphorylates $K_v4.2$, reducing its conductance in neurons [18]. The minimum consensus sequence of ERK5 and ERK1/2 is similar, but ERK5 may preferentially phosphorylate the Ser616 residue, which results in rapid repolarization to increase the firing frequency, as described below. In contrast, the pituitary adenylate cyclase-activating polypeptide (PACAP) down-regulates the A-type current density without influencing the voltage-dependence of the Kv4.2 channel currents by ERK1/2 phosphorylation of $K_v4.2$ in rat hippocampal neurons [19]. However, the characteristics of the site-directed mutants of the three amino acids above (T602A, T607A, and S616A) are different from the results found by the group mentioned above. The $K_v4.2$ S616A mutant did not show any pituitary adenylate cyclase-activating polypeptide (PACAP) induced reduction in the channel current density, whereas the overexpression of T607A mutants partially blocked the inhibitory effect of PACAP. Additionally, the mutational analysis of $K_V4.2$ indicates that Ser616 is the functionally relevant ERK1/2 phosphorylation site for the modulation of the $K_V4.2$-mediated currents in neurons derived from spinal cord dorsal horns [20]. Therefore, the roles of the ERK phosphorylation site at the $K_V4.2$ C-terminus are still controversial. Further study is necessary to identify the ERK5 phosphorylation site on $K_V4.2$, and to examine the effect on the A-type current.

Adjusting the classical Hodgkin–Huxley models, Rush and Rinzel studied the effects of the A-current on the steady firing rate of neurons. They showed that the number of spikes per burst increases as the conductance of the A-current decreases and as inactivation decreases [32]. When ERK5 was activated, the A-current inactivation rate was reduced in our results (Figure 4). According to their model, we assume that ERK5 may contribute to more rapid repolarization toward the resting potential for responding to the next firing. Therefore, ERK5 may increase the firing frequency through the phosphorylation of $K_V4.2$.

In conclusion, this study revealed, for the first time, that ERK5 signaling promotes phosphorylation of $K_V4.2$ and inhibits the inactivation of the A-type current for the enhancement of membrane excitability in PC12 cells. ERK5 promotes neurite outgrowth and catecholamine biosynthesis. In addition to these roles, the regulation of membrane excitability may be essential for the differentiation process toward mature neurons. Future directions are examining the role of the ERK5-enhanced A-type current in neuronal morphological changes and functions in primary cultured neurons, using ERK5 conditional knockout mice.

4. Materials and Methods

Materials: HRP-conjugated anti-glyceraldehyde-3-phosphate dehydrogenase (GAPDH) antibody was purchased from Cell Signaling Technology (Beverly, MA, USA). Anti-$K_V4.2$ antibody was purchased from UC Davis/NIH NeuroMab (Davis CA, USA), and HRP-conjugated anti-mouse IgG secondary antibody was purchased from GE Healthcare (Buckinghamshire, UK). Enhanced chemiluminescence (ECL) assay kits were purchased from either GE Healthcare, PerkinElmer (Waltham, MA, USA) or Nacalai Tesque (Kyoto, Japan). Lipofectamine 2000 was purchased from Invitrogen (Grand Island, NY, USA). Mn^{2+}-Phos-tag was purchased from Wako Pure Chemicals (Osaka, Japan). TriPure Isolation Reagent for the total RNA extraction, and the FastStart Essential DNA Green Master for real-time PCR were purchased from Roche (Indianapolis, IN, USA), and a Reverse Transcription kit was purchased from Toyobo (Osaka, Japan). A DNA plasmid encoding enhanced green fluorescent protein (EGFP) was purchased from Takara (Tokyo, Japan). The DNA plasmid encoding a tandem MRE-driven firefly luciferase was kindly given by Ron Prywes (Columbia University, NY, USA), and MEK5D (S311D/T315D) was kindly given by Eisuke Nishida (Kyoto University, Japan). DNA plasmid encoding oncogenic RasG12V mutant was used to activate ERK1/2. It was kindly given from Philip J.S. Stork (Vollum Institute, Oregon Health Sciences University, OR, USA). Because these DNA plasmids were kind gifts, as described above, there is restriction for the availability of these plasmids. ERK5KD (K83M) mutant was created from wildtype ERK5 as a template, as described previously [5].

Cell lines: The HEK293 cells and PC12 cells are provided by the Department of Cellular Signaling, Graduate School of Pharmaceutical Sciences, Tohoku University, Japan. Results using these cell lines have been published [9,23]. The HEK293 cells were grown in Dulbecco's modified Eagle's medium (DMEM), supplemented with 10% fetal bovine serum, penicillin (50 units/mL), and streptomycin (50 µg/mL), in a 5% CO_2 incubator at 37 °C. The PC12 cells were grown in DMEM, supplemented with 10% FBS, 5% horse serum, penicillin (50 units/mL), and streptomycin (50 µg/mL) in a 5% CO_2 incubator at 37 °C.

qRT-PCR: The total RNA from the PC12 cells was extracted using TriPure isolation reagent according to the manufacturer's protocol. The RNA was then reverse transcribed using a RT-PCR kit, and real-time PCR was performed using a LightCycler Nano thermal cycler (Roche), as described previously [33]. The PCR primers used in the PC12 cell experiments were as follows: $Ca_V1.2$ (5'-TGT

TTC CAG ATG AGA CCC GC-3' and 5'-GAG GCC CTT CGA CCT AGA GA-3'), $Ca_v2.1$ (5'- CTG CTT TGA AGA GGG GAC AG-3' and 5'-GGA AAA CAG TGA GCA CAG CA-3'), $Ca_v2.2$ (5'-TCA TTG TGG TCT TCG CTC TG-3' and 5'-CCT TTG CTG ACT CCT CCT TG-3'), $K_v4.2$ (5'-TTG GCG ACT GCT GTT ATG AG-3' and 5'-TGA CTG AGA CGG CAA TGA AG-3'), $K_v4.3$ (5'-GGC TAC ACC CTG AAG AG CTG-3' and 5'-GCC AAA TAT CTT CCC AGC AA-3'), KChIP3 (5'-GCC TTC GAT GCT GAT GGG AA-3' and 5'-AGA GGT GCG TCC TTT CGC AG-3'), and GAPDH (5'-ACC ACA GTC CAT GCC ATC AC-3' and 5'-TCC ACC ACC CTG TTG CTG TA-3'). PCR products were quantified and normalized to the GAPDH control before finally being presented as a fold change.

Reporter gene assay: Reporter gene assays were performed similarly, as described previously [33]. MEK5D, ERK5, and MRE-luciferase reporter genes were co-transfected into HEK293 cells in 24-well plates using Lipofectamine 2000. Two days after transfection, the lysates were collected and the luciferase activity was measured using a luminometer (Lumat LB9507, Berthold Japan K.K., Tokyo, Japan).

SDS-PAGE with or without Phos-tag and Western blotting: The proteins were separated by electrophoresis using 10–11% polyacrylamide gels. The proteins were then transferred from the gel onto a polyvinylidene difluoride membrane (GE Healthcare), according to standard protocols. The membranes were blocked for 0.5 h at room temperature in 5% skim milk in Tris-buffered saline containing 0.1% tween-20 (TBST), then incubated with the indicated primary antibodies overnight at 4 °C. The antibodies were dissolved in the blocking buffer, and used at the following dilutions: anti-$K_v4.2$ (1:500 or 1:1000), and HRP-conjugated anti-GAPDH (1:1000). The membranes were washed several times with TBST before being incubated with HRP-conjugated anti-mouse IgG secondary antibodies (diluted 1:5000 in blocking buffer) at room temperature for 1–2 h. The membranes were then washed with TBST, developed using an ECL chemiluminescence assay kit, and visualized using a ChemiDoc XRS imaging system (BioRad, Hercules, CA, USA) or LAS1000 (Fuji Film, Tokyo, Japan). The relative intensities of the bands corresponding to $K_v4.2$ and the internal control GAPDH were determined using Image-J densitometry software (National Institute of Health, Bethesda, MD, USA).

For electrophoresis using Phos-tag, the proteins were separated with 5% polyacrylamide gels containing 30 μM Phos-tag and 60 μM $MnCl_2$. After the gels were washed twice for 10 min with transfer buffer containing 10 mM EDTA to remove Mn^{2+}, the proteins were then transferred from the gel onto a polyvinylidene difluoride membrane at 30 V for 16 h. The further procedure is performed similarly, as described above.

Electrophysiology by patch-clamping: The PC12 cells were co-transfected with EGFP, MEK5D, and ERK5. EGFP was used as a marker for the transfected cells. The whole-cell patch clamp method was used for recording the membrane currents (patch-clamp amplifier Axopatch 200B, Molecular Devices, Chicago, IL, USA), as described previously [34]. Borosilicate glass electrodes had tip resistances between 2.5 and 4.5 MΩ when filled with internal solution composed of (mM) KOH 120, aspartic acid 80, Mg-ATP 5, KCl 20, HEPES 5, EGTA 5, and GTP-Na_2 0.1 (pH 7.2 with aspartic acid). The composition of the external solution (mM) was: NaCl 136.9, KCl 5.4, $CaCl_2$ 1.8, $MgCl_2$ 0.5, NaH_2PO_4 0.33, HEPES 5.0, and glucose 5.5 (pH 7.4 with NaOH). To evoke membrane currents, the cells were held at a potential of −80 mM and depolarized for 500 ms to various potentials, ranging from −30 to +50 mV in 20 mV increments at 37 ± 0.5 °C. The pulse protocol and data acquisition and storage were accomplished with Clampex 9.2 (Molecular Devices). The sampling frequency was 10 kHz and low-pass filtering was performed at 5 kHz. The cell membrane capacitance (C_m) was determined by integrating the area under the capacitive transient elicited, by applying a 50 ms hyperpolarizing voltage-step from a potential of −40 to −45 mV. All membrane currents (I_m) were normalized by Cm, then analyzed using IGOR software (Wavemetrics, Portland, OR, USA). The time-course of inactivation at 50 mV was fitted with a first order biexponential function, as follows:

$$I_m(t) = y_0 + y_1\left\{1 - \exp\left(-\frac{t}{\tau_{fast}}\right)\right\} + y_2\left\{1 - \exp\left(-\frac{t}{\tau_{slow}}\right)\right\}$$

where τ_{fast} and τ_{slow} are fast and slow time constants, respectively.

Statistics: Data are expressed as means $\pm$ S.E.M., and the statistical significance of the differences between groups was analyzed using the unpaired Student's *t*-test or one-way ANOVA, with post hoc test using Tukey's test for multiple comparisons.

Author Contributions: Y.O. (Yutaro Obara), Y.H., and K.I. conceived and designed the experiments; Y.K., Y.O. (Yutaro Obara), Y.O. (Yosuke Okamoto), and T.S. performed the experiments; Y.O. (Yutaro Obara), Y.O. (Yosuke Okamoto), and K.I. wrote the paper.

Funding: This work was supported in part by Grants-in-Aid from the Japan Society for the Promotion of Science (No. 15K07963 to Y.O. [Yutaro Obara]) and the Takeda Science Foundation (Y.O. [Yutaro Obara]). These grants cover the costs to publish in open access. This work was also supported by RIKEN, Grant-in-Aid for Scientific Research (A) and Grant-in-Aid for Scientific Research on Innovative Area "Principles of memory dynamism elucidated from a diversity of learning systems", Human Frontier Science Foundation, The Naito Foundation, Research Foundation for Opto-Science and Technology, Novartis Foundation to Y.H. and Grant-in-Aid for Scientific Research (B) from the MEXT, Japan (T.S.). The founding sponsors had no role in the design of the study; in the collection, analyses, or interpretation of data; in the writing of the manuscript; and in the decision to publish the results.

Acknowledgments: We thank Philip J.S. Stork (Vollum Institute, Oregon Health Sciences University, Portland, OR, USA), Ron Prywes (Columbia University, New York, NY, USA), and Eisuke Nishida (Kyoto University, Kyoto, Japan) for kindly providing DNA plasmids.

Conflicts of Interest: Y.H. is partly supported by Fujitsu Laboratories and Dwango. The other authors declare no conflict of interest.

Abbreviations

MAPK	Mitogen-activated protein kinase
ERK	Extracellular signal-regulated kinase
MEK	MAPK/ERK kinase
MEF	Myocyte-enhancer factor
MRE HEK293 cells	MEF2 response elementhuman embryonic kidney 293 cells
SDS-PAGE	Sodium dodecyl sulfate-polyacrylamide gel electrophoresis
TRP	Transient receptor potential
NFAT	Nuclear factor of activated T cells
KChIPCA	K^+ channel-interacting proteincornu ammonis
PACAP	Pituitary adenylate cyclase-activating polypeptide
GAPDH	Glyceraldehyde-3-phosphate dehydrogenase
ECL	Enhanced chemiluminescence
EGFP	Enhanced green fluorescent protein
DMEM	Dulbecco's modified Eagle's medium
TBST	Tris-buffered saline containing 0.1% tween-20

References

1. Coulombe, P.; Meloche, S. Atypical mitogen-activated protein kinase: Structure, regulation and functions. *Biochim. Biophys. Acta* **2007**, *1773*, 1376–1387. [CrossRef] [PubMed]
2. Drew, B.A.; Burow, M.E.; Beckman, B.S. MEK5/ERK5 pathway: The first fifteen years. *Biochim. Biophys. Acta* **2012**, *1825*, 37–48. [CrossRef] [PubMed]
3. Wang, X.; Tournier, C. Regulation of cellular functions by the ERK5 signalling pathway. *Cell. Signal.* **2006**, *18*, 753–760. [CrossRef] [PubMed]
4. Tatake, R.J.; O'Neill, M.M.; Kennedy, C.A.; Wayne, A.L.; Jakes, S.; Wu, D.; Kugler, S.Z., Jr.; Kashem, M.A.; Kaplita, P.; Snow, R.J. Identification of pharmacological inhibitors of the MEK5/ERK5 pathway. *Biochem. Biophys. Res. Commun.* **2008**, *377*, 120–125. [CrossRef] [PubMed]
5. Obara, Y.; Yamauchi, A.; Takehara, S.; Nemoto, W.; Takahashi, M.; Stork, P.J.; Nakahata, N. ERK5 Activity Is Required for Nerve Growth Factor-induced Neurite Outgrowth and Stabilization of Tyrosine Hydroxylase in PC12 Cells. *J. Biol. Chem.* **2009**, *284*, 23564–23573. [CrossRef] [PubMed]

6. Yang, Q.; Deng, X.; Lu, B.; Cameron, M.; Fearns, C.; Patricelli, M.P.; Yates, J.R., 3rd; Gray, N.S.; Lee, J.D. Pharmacological inhibition of BMK1 suppresses tumor growth through promyelocytic leukemia protein. *Cancer Cell.* **2010**, *18*, 258–267. [CrossRef] [PubMed]

7. Hoang, V.T.; Yan, T.J.; Cavanaugh, J.E.; Flaherty, P.T.; Beckman, B.S.; Burow, M.E. MEK5-ERK5 signaling in cancer: Implications for targeted therapy. *Cancer Lett.* **2017**, *392*, 51. [CrossRef] [PubMed]

8. Simoes, A.E.; Rodrigues, C.M.; Borralho, P.M. The MEK5/ERK5 signalling pathway in cancer: A promising novel therapeutic target. *Drug Discov. Today* **2016**, *21*, 1654–1663. [CrossRef] [PubMed]

9. Obara, Y.; Nagasawa, R.; Nemoto, W.; Pellegrino, M.J.; Takahashi, M.; Habecker, B.A.; Stork, P.J.; Ichiyanagi, O.; Ito, H.; Tomita, Y.; et al. ERK5 induces ankrd1 for catecholamine biosynthesis and homeostasis in adrenal medullary cells. *Cell. Signal.* **2016**, *28*, 177–189. [CrossRef] [PubMed]

10. Obara, Y.; Nakahata, N. The signaling pathway leading to extracellular signal-regulated kinase 5 (ERK5) activation via G-proteins and ERK5-dependent neurotrophic effects. *Mol. Pharmacol.* **2010**, *77*, 10–16. [CrossRef] [PubMed]

11. Obara, Y.; Labudda, K.; Dillon, T.J.; Stork, P.J. PKA phosphorylation of Src mediates Rap1 activation in NGF and cAMP signaling in PC12 cells. *J. Cell. Sci.* **2004**, *117*, 6085–6094. [CrossRef] [PubMed]

12. York, R.D.; Yao, H.; Dillon, T.; Ellig, C.L.; Eckert, S.P.; McCleskey, E.W.; Stork, P.J. Rap1 mediates sustained MAP kinase activation induced by nerve growth factor. *Nature* **1998**, *392*, 622–626. [CrossRef] [PubMed]

13. Liu, L.; Cundiff, P.; Abel, G.; Wang, Y.; Faigle, R.; Sakagami, H.; Xu, M.; Xia, Z. Extracellular signal-regulated kinase (ERK) 5 is necessary and sufficient to specify cortical neuronal fate. *Proc. Natl. Acad. Sci. USA* **2006**, *103*, 9697–9702. [CrossRef] [PubMed]

14. Pan, Y.W.; Chan, G.C.; Kuo, C.T.; Storm, D.R.; Xia, Z. Inhibition of adult neurogenesis by inducible and targeted deletion of ERK5 mitogen-activated protein kinase specifically in adult neurogenic regions impairs contextual fear extinction and remote fear memory. *J. Neurosci.* **2012**, *32*, 6444–6455. [CrossRef] [PubMed]

15. Pan, Y.W.; Zou, J.; Wang, W.; Sakagami, H.; Garelick, M.G.; Abel, G.; Kuo, C.T.; Storm, D.R.; Xia, Z. Inducible and conditional deletion of extracellular signal-regulated kinase 5 disrupts adult hippocampal neurogenesis. *J. Biol. Chem.* **2012**, *287*, 23306–23317. [CrossRef] [PubMed]

16. Finegan, K.G.; Wang, X.; Lee, E.J.; Robinson, A.C.; Tournier, C. Regulation of neuronal survival by the extracellular signal-regulated protein kinase 5. *Cell. Death Differ.* **2009**, *16*, 674–683. [CrossRef] [PubMed]

17. Watson, F.L.; Heerssen, H.M.; Bhattacharyya, A.; Klesse, L.; Lin, M.Z.; Segal, R.A. Neurotrophins use the Erk5 pathway to mediate a retrograde survival response. *Nat. Neurosci.* **2001**, *4*, 981–988. [CrossRef] [PubMed]

18. Morozov, A.; Muzzio, I.A.; Bourtchouladze, R.; Van-Strien, N.; Lapidus, K.; Yin, D.; Winder, D.G.; Adams, J.P.; Sweatt, J.D.; Kandel, E.R. Rap1 couples cAMP signaling to a distinct pool of p42/44MAPK regulating excitability, synaptic plasticity, learning, and memory. *Neuron* **2003**, *39*, 309–325. [CrossRef]

19. Gupte, R.P.; Kadunganattil, S.; Shepherd, A.J.; Merrill, R.; Planer, W.; Bruchas, M.R.; Strack, S.; Mohapatra, D.P. Convergent phosphomodulation of the major neuronal dendritic potassium channel Kv4.2 by pituitary adenylate cyclase-activating polypeptide. *Neuropharmacology* **2016**, *101*, 291–308. [CrossRef] [PubMed]

20. Hu, H.J.; Carrasquillo, Y.; Karim, F.; Jung, W.E.; Nerbonne, J.M.; Schwarz, T.L.; Gereau, R.W.T. The kv4.2 potassium channel subunit is required for pain plasticity. *Neuron* **2006**, *50*, 89–100. [CrossRef] [PubMed]

21. Yao, J.J.; Zhao, Q.R.; Liu, D.D.; Chow, C.W.; Mei, Y.A. Neuritin Up-regulates Kv4.2 alpha-Subunit of Potassium Channel Expression and Affects Neuronal Excitability by Regulating the Calcium-Calcineurin-NFATc4 Signaling Pathway. *J. Biol. Chem.* **2016**, *291*, 17369–17381. [CrossRef] [PubMed]

22. Kato, Y.; Kravchenko, V.V.; Tapping, R.I.; Han, J.; Ulevitch, R.J.; Lee, J.D. BMK1/ERK5 regulates serum-induced early gene expression through transcription factor MEF2C. *Embo. J.* **1997**, *16*, 7054–7066. [CrossRef] [PubMed]

23. Honda, T.; Obara, Y.; Yamauchi, A.; Couvillon, A.D.; Mason, J.J.; Ishii, K.; Nakahata, N. Phosphorylation of ERK5 on Thr732 is associated with ERK5 nuclear localization and ERK5-dependent transcription. *PLoS ONE* **2015**, *10*, e0117914. [CrossRef] [PubMed]

24. Schrader, L.A.; Birnbaum, S.G.; Nadin, B.M.; Ren, Y.; Bui, D.; Anderson, A.E.; Sweatt, J.D. ERK/MAPK regulates the Kv4.2 potassium channel by direct phosphorylation of the pore-forming subunit. *Am. J. Physiol. Cell. Physiol* **2006**, *290*, C852-861. [CrossRef] [PubMed]

25. Obara, Y.; Ishii, K. Transcriptome Analysis Reveals That Midnolin Regulates mRNA Expression Levels of Multiple Parkinson's Disease Causative Genes. *Biol. Pharm. Bull.* **2018**, *41*, 20–23. [CrossRef] [PubMed]

26. Adams, J.P.; Anderson, A.E.; Varga, A.W.; Dineley, K.T.; Cook, R.G.; Pfaffinger, P.J.; Sweatt, J.D. The A-type potassium channel Kv4.2 is a substrate for the mitogen-activated protein kinase ERK. *J. Neurochem.* **2000**, *75*, 2277–2287. [CrossRef] [PubMed]

27. Lugo, J.N.; Barnwell, L.F.; Ren, Y.; Lee, W.L.; Johnston, L.D.; Kim, R.; Hrachovy, R.A.; Sweatt, J.D.; Anderson, A.E. Altered phosphorylation and localization of the A-type channel, Kv4.2 in status epilepticus. *J. Neurochem.* **2008**, *106*, 1929–1940. [CrossRef] [PubMed]

28. Katsura, H.; Obata, K.; Mizushima, T.; Sakurai, J.; Kobayashi, K.; Yamanaka, H.; Dai, Y.; Fukuoka, T.; Sakagami, M.; Noguchi, K. Activation of extracellular signal-regulated protein kinases 5 in primary afferent neurons contributes to heat and cold hyperalgesia after inflammation. *J. Neurochem.* **2007**, *102*, 1614–1624. [CrossRef] [PubMed]

29. Jia, Y.; Takimoto, K. GATA and FOG2 transcription factors differentially regulate the promoter for Kv4.2 K(+) channel gene in cardiac myocytes and PC12 cells. *Cardiovasc. Res.* **2003**, *60*, 278–287. [CrossRef]

30. Gong, N.; Bodi, I.; Zobel, C.; Schwartz, A.; Molkentin, J.D.; Backx, P.H. Calcineurin increases cardiac transient outward K+ currents via transcriptional up-regulation of Kv4.2 channel subunits. *J. Biol. Chem.* **2006**, *281*, 38498–38506. [CrossRef] [PubMed]

31. Yang, T.T.; Xiong, Q.; Graef, I.A.; Crabtree, G.R.; Chow, C.W. Recruitment of the extracellular signal-regulated kinase/ribosomal S6 kinase signaling pathway to the NFATc4 transcription activation complex. *Mol. Cell. Biol.* **2005**, *25*, 907–920. [CrossRef] [PubMed]

32. Rush, M.E.; Rinzel, J. The potassium A-current, low firing rates and rebound excitation in Hodgkin-Huxley models. *Bull. Math. Biol.* **1995**, *57*, 899–929. [CrossRef] [PubMed]

33. Obara, Y.; Imai, T.; Sato, H.; Takeda, Y.; Kato, T.; Ishii, K. Midnolin is a novel regulator of parkin expression and is associated with Parkinson's Disease. *Sci. Rep.* **2017**, *7*, 5885. [CrossRef] [PubMed]

34. Okamoto, Y.; Kawamura, K.; Nakamura, Y.; Ono, K. Pathological impact of hyperpolarization-activated chloride current peculiar to rat pulmonary vein cardiomyocytes. *J. Mol. Cell. Cardiol.* **2014**, *66*, 53–62. [CrossRef] [PubMed]

International Journal of
Molecular Sciences

Review

Role of Mitogen Activated Protein Kinase Signaling in Parkinson's Disease

Anastasiia Bohush, Grazyna Niewiadomska and Anna Filipek *

Nencki Institute of Experimental Biology, Polish Academy of Sciences, 3 Pasteur Street, 02-093 Warsaw, Poland;
a.bohush@nencki.gov.pl (A.B.); g.niewiadomska@nencki.gov.pl (G.N.)
* Correspondence: a.filipek@nencki.gov.pl; Tel.: +48-22-5892-332; Fax: +48-22-822-53-42

Received: 30 August 2018; Accepted: 26 September 2018; Published: 29 September 2018

Abstract: Parkinson's disease (PD) is a neurodegenerative disorder caused by insufficient dopamine production due to the loss of 50% to 70% of dopaminergic neurons. A shortage of dopamine, which is predominantly produced by the dopaminergic neurons within the substantia nigra, causes clinical symptoms such as reduction of muscle mass, impaired body balance, akinesia, bradykinesia, tremors, postural instability, etc. Lastly, this can lead to a total loss of physical movement and death. Since no cure for PD has been developed up to now, researchers using cell cultures and animal models focus their work on searching for potential therapeutic targets in order to develop effective treatments. In recent years, genetic studies have prominently advocated for the role of improper protein phosphorylation caused by a dysfunction in kinases and/or phosphatases as an important player in progression and pathogenesis of PD. Thus, in this review, we focus on the role of selected MAP kinases such as JNKs, ERK1/2, and p38 MAP kinases in PD pathology.

Keywords: apoptosis; ERK1/2; JNKs; mitochondrial dysfunction; neurodegeneration; neuro-inflammation; oxidative stress; p38 MAPKs; Parkinson's disease

---

1. Introduction

Parkinson's disease (PD) is a neurodegenerative disorder of the central nervous system (CNS), which affects about 1% of human population over the age of 60 [1] around the world and, for which, up to now, no cure has been developed [2]. Resting tremor, rigidity, hypokinesia, and postural instability are the four cardinal motor symptoms of PD resulting from the loss of dopaminergic neurons in the substantia nigra pars compacta, which is a key regulatory structure of basal ganglia circuitry. As the disease progresses, patients frequently develop cognitive impairment and depression. Most motor symptoms can be attributed to the degeneration of dopaminergic neurons within the substantia nigra pars compacta [3]. Nonetheless, in recent years, it has become increasingly appreciated that several other non-dopaminergic neuronal populations also degenerate (Figure 1). These include various autonomic nuclei and the locus coeruleus as well as glutamatergic neurons throughout the cerebral cortex. PD is characterized by the formation of specific inclusions called Lewy bodies (LBs) in neurons of several brain structures. LBs consist mostly of misfolded proteins such as α-synuclein, tubulin and microtubule associated proteins, ubiquitin, amyloid precursor protein, synaptic vesicle proteins, various enzymes, and chaperons/co-chaperons [4].

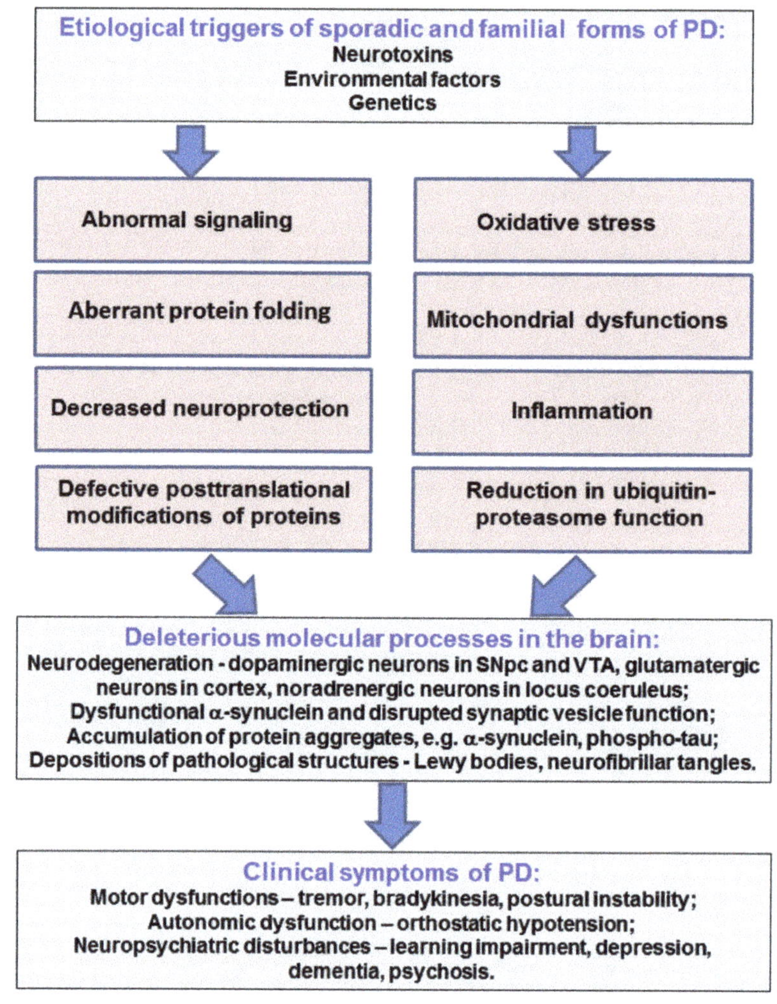

Figure 1. Principal pathological processes in PD etiology and clinical hallmarks of the disease. SNpc—substantia nigra pars compacta, VTA—ventral tegmental area.

The MAP (mitogen-activated protein) kinase family is one of the oldest and evolutionarily conserved family of serine/threonine protein kinases responsible for intracellular signaling in *Eukaryota* [5]. MAPKs (MAP kinases) regulate many physiological processes such as gene expression, mitosis, metabolism, cell differentiation and motility, stress response, survival, or cell death [6]. In mammalian cells, there are four main groups of conventional MAPKs: ERK1/2 (called also MAPK3 and MAPK1, respectively), ERK5, JNKs (JNK1, JNK2, and JNK3 called MAPK8, MAPK9, and MAPK10, respectively) and p38 MAPKs (p38α, p38β, p38γ, and p38δ called also MAPK14, MAPK11, MAPK12, and MAPK13, respectively). All these isoforms share sequence similarities but their cellular targets/substrates differ substantially. In addition, atypical MAPKs including NLK (Nemo-like kinase), ERK3/4, and ERK7/8 classified into a separate group have been described [6]. All these kinases collaborate in transmitting signals from numerous extracellular stimuli and control intracellular processes triggered by them. Thus, in consequence, MAPKs are capable of phosphorylating and altering the activities of countless substrates in different subcellular compartments. MAPK substrates have been found not only in the cytoplasm but also in mitochondria, the Golgi apparatus, the endoplasmic reticulum, and the nucleus [7,8].

1.1. MAPK Signalling

The MAPK signalling cascade provides a mechanism for cells to respond to a catalogue of external signals. In fact, the diversity and specificity of cellular responses is facilitated through a linear cascade of events, which is comprised of a sequentially operating set of three evolutionarily conserved groups of protein kinases known as: MAPK, MAPK kinase (MAP2K), and MAPK kinase kinase (MAP3K). MAP3Ks are serine/threonine kinases, which are activated either via phosphorylation and/or due to the interaction with a small GTP-binding protein of the Ras/Rho family in response to extracellular stimulus. MAP3Ks activation results in phosphorylation and activation of MAP2Ks, which consequently stimulate MAPKs activity through dual phosphorylation of threonine and tyrosine residues positioned in the activation loop of kinase subdomain VIII. The activated MAPKs then phosphorylate target substrates specifically on serine or threonine residues followed by a proline residue. MAP2Ks such as MEK3 and MEK6 are activated by a wide range of MAP3Ks (MEKK1–3, MLK2/3, ASK1, Tpl2, TAK1, and TAO1/2), which become activated in response to oxidative stress, UV irradiation, hypoxia, ischemia, and cytokines including IL-1 (interleukin-1) and TNF-α (tumor necrosis factor alpha). Lastly, these events lead to altered gene expression and modulate crucial cellular functions under normal and pathological conditions such as Parkinson's disease [9].

1.2. JNK Signaling

JNKs (c-Jun N-terminal kinases) are a family of protein kinases activated in response to cytokines, growth factors, pathogens, and stress. JNK-mediated signaling pathways affect gene expression, neuronal plasticity, regeneration, apoptosis, or cellular senescence [10]. JNKs are activated through a dual phosphorylation of threonine and tyrosine residues within a threonine-proline-tyrosine (Thr-Pro-Tyr) motif by two MAP kinase kinases: MKK4 and MKK7. These two MAP kinase kinases can be inactivated by serine/threonine and tyrosine protein phosphatases [9]. In addition to the regulation by upstream kinases, the JNK signaling pathways are modulated by various scaffolding proteins including JNK-interacting protein 1, 2, and 3 (JIP1-3). The JNK family consists of 10 isoforms derived from three genes: JNK1 (four isoforms), JNK2 (four isoforms), and JNK3 (two isoforms). In mammalian cells, JNK1 and JNK2 are ubiquitously expressed while JNK3 is found mainly in the brain, heart, and testis [11]. In order to understand the biological function of JNKs, gene knockout studies were performed. It was found that mice deficient in JNK1, JNK2, JNK3, and JNK1/JNK3 or JNK2/JNK3 survived normally. Compound mutants lacking genes encoding JNK1 and JNK2 were embryonically lethal and had severe dysregulation of apoptosis of brain cells [12]. Under normal conditions, JNKs phosphorylate a variety of substrates. Examples of these substrates include a diverse assortment of nuclear transcription factors (Jun, ATF2, Myc, Elk1), cytoplasmic proteins involved in cytoskeleton regulation (DCX, Tau, WDR62), cell membrane receptors (e.g., BMPR2), mitochondrial proteins (e.g., Mcl1 and Bim), or proteins involved in vesicular transport (e.g., JIP1 and JIP3) [13].

In mammalian brains, JNK transcripts have been detected at levels similar to those in peripheral organs. However, JNK activity is noticeably higher in CNS than in peripheral organs. This activity can be increased by noninvasive environmental stimuli, which underlines the important role of JNKs in the brain [14] under norm and pathology and suggests that it may be implicated in neurodegenerative disorders such as Parkinson's disease [15]. In this respect, it should be stressed that JNKs can be activated by a number of factors implicated in PD such as toxicants [16] and unfolded/misfolded proteins [17]. Some studies have demonstrated that JNKs are significantly activated in several common animal models of PD induced by neurotoxins such as MPTP (1-methyl-4-phenyl-1,2,3,6-tetrahydropyridine), 6-OHDA (6-hydroxydopamine) or LPS (lipopolysaccharide) [18–21]. Genetic deletion of JNK2 and JNK3 protects against MPTP-induced neurodegeneration in mice [22]. Moreover, some other studies have indicated that antioxidant and anti-inflammatory compounds provide neuroprotection in the MPTP and 6-OHDA model of PD, at least in part, through the inhibition of JNK activation [23,24]. In addition, it was found that inhibition of JNKs with the SP-600125 inhibitor protects dopaminergic neurons both from MPP+ (1-methyl-4-phenylpyridinium)-induced neuronal apoptosis in vitro and in MPTP and 6-OHDA models of PD [15]. Another inhibitor of JNKs, SR-3306, was found to reduce the loss of dopaminergic cell bodies in the substantia nigra and their terminals in the striatum [25]. SR-3306 was also shown to have a therapeutic effect in Alzheimer's disease. A marked improvement of cognitive deficits, a significant decrease in the amount of β-amyloid plaques, and a decrease in tau phosphorylation in inflammatory responses were observed in transgenic animals treated for 12 weeks with the JNK inhibitor SP-600125 [26]. Recently, it has been reported that instant activation of JNK phosphorylation following treatment of cells with the HMGB1 (high mobility group box 1) protein cause an increase in the expression of tyrosine hydroxylase. The imbalance of this reduces dopamine synthesis and induces PD [27].

JNKs are not only implicated in the survival of dopaminergic neurons but also in dopamine transmission, which is, among the pathways, most impaired during the course of Parkinson's disease [28]. Dopamine plays a central role in motor and cognitive functions as well as in reward processing by regulating glutamatergic inputs in the striatum. Release of dopamine rapidly exerts its influence on synaptic transmission and regulates both AMPA (α-amino-3-hydroxy-5-methyl-4-isoxazolepropionic acid) and NMDA (*N*-methyl-D-aspartate) receptors [29]. JNKs were found to be downstream targets of postsynaptic NMDA receptors and, moreover, NMDA activity is linked to the presence of a JNK scaffolding protein, JIP1. It was shown that NMDA-evoked glutamate release is controlled by presynaptic JNK-JIP1 interaction. Using JNK2 knock-out mice, it was proven that this kinase is essential in mediating glutamate release [30]. Activation of the glutamatergic pathway together with the dopaminergic one is responsible for synaptic plasticity, long-term potentiation (LTP), and long-term depression (LTD), which underlie motor learning. Accordingly, it has been found that LTP and LTD are altered in animal models of PD [31]. In addition, it has been found that the JNK1-Rac1 signaling pathway mediates phosphorylation of serine 295 in the PSD-95 (postsynaptic density protein 95) protein and, thus, enhances its synaptic accumulation and capability to recruit surface AMPA receptors and, lastly, potentiates excitatory postsynaptic currents [32]. It is worth mentioning that AMPA receptors, which are extremely relevant for synaptic plasticity, are physiological substrates of JNKs [33]. Lately, it has been found that, in a mouse model of PD, the JNK pathway is required for dopamine D1 receptor (D1R)-dependent modulation of corticostriatal synaptic plasticity. Pharmacological activation of D1R evokes a large increase in JNK phosphorylation. Electrophysiological experiments on brain slices from PD mice show that inhibition of JNK signaling in the pathway of striatal projection neurons prevents the increase in synaptic strength caused by activation of D1Rs [28].

It should be stressed that JNKs are implicated in other processes essential for neuronal homeostasis that seem to be severely dysregulated during Parkinson's disease. For example, JNKs seem to play a role in protein transport in the brain. A study on *Caenorhabditis elegans* provides evidence that components of the JNK pathway are necessary for normal protein transport [34] and JNKs have been shown to modulate the interaction of kinesin with microtubules [35]. Therefore, inadequate JNK activity may be at the root of the impairment in the axonal transport frequently observed in PD and a number of many other neurodegenerative disorders [36,37].

JNK signaling is also linked to the apoptosis in neurons [38]. There is a study showing that, in cultured neurons, c-Jun activation is required for NGF (nerve growth factor) withdrawal-induced apoptosis and inhibition of c-Jun protects neurons from induced cell death. For instance, NGF deprivation-induced apoptosis is associated with the activation of the GTPase Cdc42 and JNKs in primary superior cervical ganglion sympathetic neurons. In addition, overexpression of the MAP3K apoptosis signal-regulated kinase 1 (ASK1), has been found to activate JNKs and to induce apoptosis in NGF-differentiated pheochromocytoma PC12 cells and primary rat sympathetic neurons [39]. On the other hand, inhibition of JNK signaling has been shown to reduce apoptosis of many other cells [40–43]. Therefore, JNKs may be critical for pathological cell death observed in Parkinson's disease [44].

Numerous studies have implicated JNKs in oxidative stress, which is known to play an important role in Parkinson's disease and other neurodegenerative disorders [45–47]. Research on *Drosophila melanogaster* showed that flies with mutations that accelerate JNK signaling accumulate less oxidative damage and live longer than wild-type flies [48]. JNKs activation has been also linked to stress evoked by misfolded proteins. Neuropathogenic forms of the huntingtin receptor and the androgen receptor were shown to inhibit axonal transport [49] and subsequent studies showed that this inhibition is mediated by JNK [37].

It should be noted that several other studies provide new insights into the role of JNK-mediated pathways in the control of the balance of autophagy in response to genotoxic stress i.e., the process that plays an important role in neurodegeneration including Parkinson's disease [50–52].

1.3. ERK1/2 Signaling

The ERK1/2 (extracellular signal–regulated kinases 1 and 2) signaling cascade is a central MAPK pathway that plays a role in the regulation of various cellular processes such as proliferation, differentiation, development, learning, survival, apoptosis etc. [6]. This pathway is activated by growth factors [53], insulin [54], ligands of G protein-coupled receptors [55], or stress factors [56]. All these activators trigger a signal transmission by interacting with specific receptors such as receptors with tyrosine kinase activity (RTKs) or G protein-coupled receptors (GPCRs) [57].

Under a normal condition, the ERK1/2 pathway plays an essential role in the regulation of transcription. It phosphorylates and activates different transcription factors such as Elk1, c-Fos, p53, Ets1/2, c-Myc, and NFAT, which, in turn, activate numerous genes that encode proteins involved in proliferation [58]. Activation of ERK1/2 is crucial for efficient G_1/S phase progression in a normal cell cycle. As mentioned above, ERK1/2 directly phosphorylate Elk1, which is involved in the expression of immediate-early genes. In addition, through direct phosphorylation of c-Fos, ERK1/2 promotes its association with c-Jun and the formation of a transcriptionally active AP-1 (activator protein 1) complex. The expression of cyclin D1, which is a protein that interacts with CDKs (cyclin-dependent kinases), permits G_1/S transition depending on AP-1 activity [9]. In addition, ERK1/2 extend the MAPK cascade by phosphorylating and activating MAPKAPK (MAPK activated protein kinase) family members including RSKs (ribosomal S6 kinases), MSKs (mitogen and stress activated protein kinases), and MNKs (MAPK interacting protein kinases) [8]. Additionally, ERK1/2 phosphorylate members of the STAT transcription factor family (STAT1, STAT3, STAT4, and STAT5), which are known to mediate many aspects of survival, proliferation, differentiation [59], cytokine dependent inflammation [60], and apoptosis [61]. Genetic studies highlight differences between ERK1 and ERK2 isoforms. Some of them show that ERK1-null mice are neurologically normal with no impairment in the ability to learn, which may suggest that ERK2 compensates for the loss of ERK1 [62]. In contrast, it was shown that ERK2 knock-out mice are embryonically lethal [63].

In an adult mammalian central nervous system, ERK1/2 are expressed at a higher level in post-mitotic neurons. Immunohistochemical studies have demonstrated that ERK2 is localized in the soma and dendritic trees of neurons of the neocortex, the hippocampus, the striatum, and the cerebellum [64].

The ERK1/2 pathway mediates dopaminergic and glutamatergic signaling in the central nervous system and maintains normal activity of striatal neurons. Activation of ERK1/2 is important for associative learning, memory, visual cortical plasticity, etc. [65]. In addition, it is involved in the activation of D1 and D2 dopamine receptors in the striatum and ionotropic or metabotropic glutamatergic receptors in the dentate gyrus [66]. Moreover, it has been shown that the ERK1/2 signaling pathway plays an important role in the maintenance of spatial memory and long-term fear memory [67]. Another study has revealed the importance of ERK1/2 phosphorylation in neuronal development. Transiently repressed ERK1/2 phosphorylation in mice during the neonatal stage by intraperitoneal injection of MEK1/2 inhibitor, SL327, caused apoptosis of brain cells and had an effect on brain functioning: reduced LTP, impaired memory, and deficits in social behavior [68].

ERK1/2 signaling is involved in neuronal death, which is a major phenomenon in all neurodegenerative diseases including Parkinson's disease [69]. Using the oligodendroglial CG4 cell line, it was shown that H_2O_2-induced cell death is prevented by the application of the ERK1/2 pathway inhibitor, PD98059 [70]. Additionally, it has been shown that nitric oxide produced by glial cells induces neuronal degeneration through ERK1/2 activation [71] and that this degeneration might be blocked by applying, PD98059 [72]. Application of another inhibitor of this pathway, U0126, also indicated that death of striatal neurons induced by dopamine was associated with ERK1/2 activation [73].

There are several processes that link ERK1/2 and Parkinson's disease. In particular, these include: oxidative stress and mitochondria dysfunction, cell survival and apoptosis, neuroprotection, and inflammation. Regarding mitochondria dysfunction, it was found that, in the substantia nigra of PD patients, there is a mild deficiency in mitochondrial complex I [74]. Moreover, using confocal microscopy, it was established that phosphorylated ERK1/2 (p-ERK1/2) immune-reactivity was associated with mitochondrial proteins called MsSOPs and that some vesicular-appearing p-ERK1/2 granules enveloped enlarged mitochondria. In addition, p-ERK1/2 were found within the mitochondria of degenerating neurons derived from Parkinson's disease patients and patients with Lewy body dementia [75]. There are also some other studies that support an idea that ERK1/2 inhibition activates both apoptotic and necrotic cell death-inducing pathways [76]. ERK1/2 directly phosphorylates mitochondrial transcription factor A (TFAM) on serine 117, which affects TFAM-DNA binding and, in consequence, leads to mitochondrial dysfunction. In addition, it was found that TFAM, which is downregulated by ERK1/2 in cells chronically treated with a complex 1 inhibitor, MPP+, regulates mitochondrial biogenesis [77]. Regarding oxidative stress, Wang et al. [78] have shown that the DJ-1 transcription factor interacted with ERK1/2 and was required for the nuclear translocation of ERK1/2. This translocation was suppressed in DJ-1 knock-down cells and DJ-1 null mice treated with an oxidative insult. Additionally, endoplasmic reticulum (ER) stress seems to play a critical role in the progression of Parkinson's disease. Results obtained by Cai et al. [79] indicate that ER stress-induced apoptosis in PD might be inhibited by a basic fibroblast growth factor (bFGF). Administration of bFGF improved motor function recovery, increased tyrosine hydroxylase positive neuron survival, and upregulated the levels of neurotransmitters in the brain of a rat model of Parkinson's disease. Another study has shown that a redox protein, thioredoxin-1, protects neurons from injuries and attenuates symptoms of Parkinson's disease [80]. Additionally, it has been shown that the ERK1/2- and JNK1/2-c-Jun systems are linked with L-DOPA-induced neurotoxicity of dopaminergic neurons in a cellular model of PD [81] and that PI3K/Akt and ERK1/2 signaling pathways are involved in the protection of dopaminergic neurons against MPTP/MPP+-induced neurotoxicity [82]. In addition to that, it was found that, in LPS-induced PD models in vivo and in vitro, a flavonoid known as licochalcone A (Lico.A) significantly inhibited the production of pro-inflammatory mediators and microglial activation by blocking phosphorylation of ERK1/2 [83].

Lastly, it should be mentioned that some of the functions attributed to ERK1/2 in neuronal survival might be carried out by ERK5 since PD98059 and U0126 known as MEK1/2 inhibitors might inhibit the ERK5 pathway as well [84,85]. There is also a study that sheds light on the distinct roles of ERK1/2 and ERK5 in the survival of dopaminergic neurons under physiological conditions and acute oxidative stress. The latter condition is extensively linked to the molecular pathogenesis of Parkinson's disease. The interaction between ERK5 and ERK1/2 pathways was found to promote basal survival of dopaminergic neurons when exposed to oxidative stress. When both pathways were inhibited, the decline in basal survival of MN9D dopaminergic cells after exposure to a toxic agent, 6-OHDA, was observed. In addition, it was found that ERK5 and ERK1/2 have different roles in neuronal metabolism. Activation of ERK5 promoted the survival of MN9D cells but had no influence on the toxic effect of 6-OHDA on these cells [86].

1.4. p38 MAPK Signaling

The p38 MAPKs are strongly activated by extracellular stimuli such as UV light, heat shock, osmotic shock, inflammatory cytokines (e.g., TNF-α, IL-1β), or growth factors (e.g., CSF-1). Thus, these kinases are also known as stress-activated ones [87]. There are four isoforms of p38 MAPKs known as α, β, γ, and δ. All of them share up to 60% sequence similarities and 40% to 45% with other MAP kinase family members [88]. p38 MAPK isoforms have a different expression pattern. p38α MAPK is ubiquitously expressed in most cell types. p38β MAPK is mainly expressed in the brain while p38γ MAPK—in skeletal muscle and p38δ MAPK—is expressed in endocrine glands [89].

Regarding the role of p38α MAPK, it was found that the knockout of the gene encoding this protein is lethal [90] while mice lacking the *p38β* gene were viable and exhibited no apparent health problems. When embryonic fibroblasts from *p38β*$^{-/-}$ mice were analyzed, expression and activation of p38α MAPK, ERK1/2, and JNKs in response to cellular stress remained unchanged, which suggests that the α isoform of p38 MAPK is the main one responsible for controlling all of the detrimental consequences of the p38 MAPK activation such as microglia activation, neuro-inflammation, oxidative stress due to reactive oxygen species (ROS) accumulation, nitric oxide activity, and neuronal apoptosis [91–93].

It is worthy to note that several lines of evidence suggest that p38 MAPKs play a role in neuronal apoptosis, which is linked to Parkinson's disease [94,95]. For instance, it has been reported that they induce apoptosis by phosphorylating Bcl-2 (B-cell lymphoma 2) family members [96]. Interestingly, phosphorylation of one such member, BimEL, on serine 65 may be a common regulatory point for cell death induced by both p38 MAPK and JNK pathways [97]. In addition, oxidative stress in dopaminergic neurons has been shown to trigger the p38 MAPK pathway which, in consequence, may lead to uncontrolled activation of apoptosis in cellular and animal models of Parkinson's disease [98–100]. Together these data suggest that both oxidative stress and p38 MAPK operate to balance the pro-apoptotic and anti-apoptotic phenotypes of dopaminergic neurons. Some other studies show the link between the generation of ROS, initiation of the p38 MAPK/JNK signaling, and apoptosis of neuronal cells in different models of Parkinson's disease [101–105]. Interestingly, it has been shown that the exacerbating effects of deletion of *Park2* gene (encoding parkin protein) on ethanol-induced ROS generation, mitophagy, mitochondrial dysfunction, and cell death were reduced by p38 MAPK inhibitor, SB203580, in vitro and in vivo. In the case of dopaminergic neurons it has been shown that deletion of this gene exacerbates ethanol-induced damage through p38 MAPK dependent inhibition of autophagy and mitochondrial function [106]. Similar data revealed that the p38 MAPK-parkin signaling pathway regulates mitochondrial homeostasis and neuronal degeneration in the A53T α-synuclein mutant model of Parkinson's disease [107].

Attention to p38 MAPKs in terms of neurodegeneration is driven by the fact that these kinases are involved in dopaminergic signaling, which is a pathway known to be disrupted during Parkinson's disease [93]. There is a study by Wu et al. [108] showing that degeneration of nigral dopaminergic neurons was accompanied by an increase in the level of p38 MAPKs and their phosphorylated forms. In agreement are the results published by Yoon et al. [109] showing that phosphorylation of p38 MAPKs by the LRRK2-ASK1 pathway regulated neuronal toxicity and apoptosis. Pharmacological inhibition of this kinase with SB203580 blunted MPTP neurotoxin induced cell apoptosis [110]. Similarly, neuronal protection was observed by applying another p38 MAPK inhibitor, SB239063 [111], or celastrol in rotenone-evoked neuroblastoma SH-SY5Y cellular model of Parkinson's disease [112].

An important issue in Parkinson's disease is neuro-inflammation that can be associated with alterations in glial cells including astrocytes and microglia. The response of neurons to activation of microglia promotes oxidative stress, inflammation, and cytokine-receptor-mediated apoptosis, which eventually contribute to the death of dopaminergic neurons and to the progression of the disease [113]. Rotenone, which is an inhibitor of the mitochondrial complex I, can directly activate microglial cells through the p38 MAPK pathway and initiate dopaminergic neuronal damage in substantia nigra, which ultimately results in parkinsonism. Unfortunately, the exact mechanism behind the selective degeneration of nigral dopaminergic neurons is not fully understood [103]. Moreover, it has been reported that montelukas, which is a cysteinyl leukotriene receptor antagonist, exerted neuroprotective effects in the rotenone-induced PD animal model through the attenuation of microglial cell activation and p38 MAPK expression [114]. It has been suggested that degeneration of nigral dopaminergic neurons was followed by an increase in the expression of p38 MAPKs, p53, and Bax (Bcl-2-associated X protein). Neurotoxins exhibited a similar effect on the level of these proteins in cultured pheochromocytoma PC12 cells, which shows that this phenomenon occurs both in vitro and in vivo. When activated, Bax is exported into the mitochondrial membrane where it oligomerizes and triggers mitochondrial apoptotic signaling. This observation strongly indicates that p38 MAPK/p53 stimulation of Bax can certainly contribute to rotenone's neurotoxicity in models of Parkinson's disease [108]. p38 MAPK also plays a role in neurotoxicity induced by MPTP [115].

Inflammation and autophagy are highly interdependent cellular processes. Autophagy plays an anti-inflammatory role and suppresses pro-inflammatory process by regulating innate immune signalling pathways and inflammasome activity [116]. Inflammatory signals also function to reciprocally control autophagy [117]. However, the mechanism of mutual regulation of both processes is not yet explained. A recent study has shown that the α isoform of p38 MAPK plays a direct and essential role in relieving autophagic control in response to an inflammatory signal by direct phosphorylation of UNC51-like kinase-1, which is the serine/threonine kinase involved in the autophagic cascade in microglia. Moreover, phosphorylation of UNC51-like kinase-1 by p38α MAPK inhibited activity of this kinase, disrupted its interaction with autophagy-related protein 13, ATG13, and, thus, reduced the level of autophagy [118]. Because autophagy disorders are more commonly associated with neurodegenerative diseases, the role of p38 MAPKs in autophagy was studied in a human neuroblastoma SK-N-SH cellular model of Parkinson's disease. The studies revealed that microRNA (miR)-181a regulated apoptosis and autophagy by inhibiting the p38 MAPK/JNK pathway [119].

2. Conclusions

Consistent with the critical role of MAPKs in key cellular activities including cell proliferation, differentiation, and survival or death. The MAPK signaling pathways have been implicated in the pathogenesis of many human diseases. Various observations suggest that they contribute to Parkinson's disease-related pathological processes such as oxidative stress, neuro-inflammation, autophagy, and neuronal death (Figure 2). In addition, MAPK inhibitors demonstrate important neuroprotective properties upstream of the execution of apoptosis in dopaminergic neurons. Therefore, discovering the relationship between MAPK pathways and the prominent pathological processes observed during Parkinson's disease progression may not only aid us to understand the etiology of this disease but also lend insight into molecular targets for the development of therapeutic drugs.

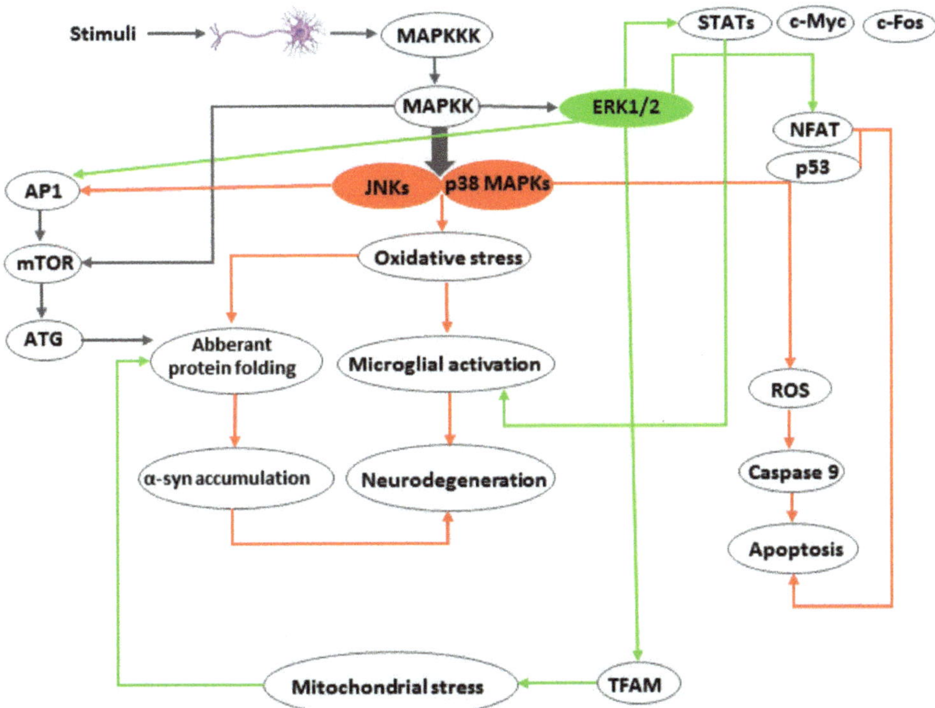

Figure 2. Involvement of MAPKs in processes that lead to Parkinson's disease pathology. Different stimuli (stress stimuli, growth factors, cytokines, mitogens, pathogens, toxins) induce activation of MAPK pathways including activation of MAPKKK and MAPKK followed by phosphorylation of downstream targets such as JNKs, p38 MAPKs, and ERK1/2. JNKs and p38 MAPKs are grouped together due to their involvement in the "death pathway" (marked in red) while ERK1/2 is believed to promote cell growth and differentiation (marked in green). Activation of JNKs and p38 MAPKs promote oxidative stress and apoptosis, which are main contributors to PD pathogenesis. Oxidative stress may also cause microglial activation and chronic inflammation, which are toxic for brain cells and leads to PD pathology. In addition, ERK1/2 contribute to apoptosis of brain cells through the activation of NFAT and p53 and to neuronal inflammation through the activation of STATs. Both ERK1/2 and JNKs activate mTOR signaling, which promotes neurodegeneration such as that observed in PD.

Author Contributions: Each author has participated sufficiently in the work to take public responsibility for appropriate portions of the article content. (1) Authors who made substantial contributions to the concept and design of the review: A.B., A.F., and G.N. (2) Authors who participated in drafting the article: A.B., G.N., and A.F. (3) All Authors gave the final approval of the version to be submitted.

Acknowledgments: Authors would like to thank W. Leśniak for a critical reading of the manuscript. This work has been supported by the European Union's Horizon 2020 research and innovation program under the Marie Sklodowska-Curie grant agreement no 665735 (Bio4Med) and by the funding from the Polish Ministry of Science and Higher Education within 2016–2020 funds for the implementation of international projects (agreement no 3548/H2020/COFUND/2016/2) by NCN grant 2014/15/B/NZ4/05041 to GN and by statutory funds from the Nencki Institute of Experimental Biology PAS.

Conflicts of Interest: The authors declare no conflict of interest.

Abbreviations

AMPA	α-amino-3-hydroxy-5-methyl-4-isoxazolepropionic acid (agonist of the AMPA receptor)
ASK1	Apoptosis signal-regulating kinase 1
AKT	protein kinase B
ATG13	autophagy-related protein 13
bFGF	basic fibroblast growth factor
Cdc42	cell division control protein 42 homolog
CNS	central nervous system
CSF-1	colony stimulating factor 1
ERK1/2	extracellular regulated kinase 1 and 2
EGFR	epidermal growth factor receptor
GPCR	G protein-coupled receptors
HMGB1	high mobility group protein 1
IL-1β	interleukin 1β
JIP1	JNK interacting protein 1
JNK	c-Jun N-terminal kinase
LBs	Lewy bodies
L-DOPA	*L*-3,4-dihydroxyphenylalanine
LRRK2	leucine-rich repeat kinase 2
LTD	long-term depression
MAP kinase	mitogen activated protein kinase
MEK3 and 6	mitogen-activated protein kinase kinase 3 and 6
MEKK1–3	mitogen-activated protein kinase kinase kinase 1 to 3
MLK2/3	mixed lineage kinases 2/3
MN9D	cell line used as a model of dopaminergic neurons
MNK	MAPK interacting protein kinase
MPTP	1-methyl-4-phenyl-1,2,3,6-tetrahydropyridine
MPP+	1-methyl-4-phenylpyridinium
MSK	mitogen and stress activated protein kinase
NGF	neurite growth factor
NMDA	*N*-Methyl-D-aspartic acid
NO	nitric oxide
6-OHDA	6-hydroxydopamine
PD	Parkinson's disease
LTP	long-term potentiation
TH	tyrosine hydroxylase
Rac1	GTPase, member of the Rho family
ROS	reactive oxygen species
RTKs	receptor tyrosine kinases
RSK	ribosomal S6 kinase
SN	substantia nigra
STAT	signal transducer and activator of transcription
TAK1	transforming growth factor beta-activated kinase 1
TAO1/2	thousand and one amino acid kinases 1/2
TFAM	mitochondrial transcription factor A
TNF-α	tumor necrosis factor α
Tpl2	tumor progression locus 2 kinase
UNC51-like kinase-1	serine/threonine kinase involved in the autophagic cascade
NFAT	nuclear factor of activated T-cells
PI3K	Phosphatidylinositol-4,5-bisphosphate 3-kinase

References

1. Nussbaum, R.L.; Ellis, C.E. Alzheimer's disease and Parkinson's disease. *J. N. Engl. Med.* **2003**, *348*, 1356–1364. [CrossRef] [PubMed]
2. Penney, E.B.; Mccabe, B.D. Parkinson's Disease: Insights from Invertebrates. In *Parkinson's Disease Molecular and Therapeutic Insights from Model. Systems*; Nass, R., Przedborski, S., Eds.; Elsevier Inc.: Amsterdam, The Netherlands, 2008; pp. 323–333, ISBN 978-0123740281.
3. Lindenbach, D.; Dupre, K.B.; Eskow Jaunarajs, K.L.; Ostock, C.Y.; Goldenberg, A.A.; Bishop, C. Effects of 5-HT1A receptor stimulation on striatal and cortical M1 pERK induction by L-DOPA and a D1 receptor agonist in a rat model of Parkinson's disease. *Brain Res.* **2013**, *1537*, 327–339. [CrossRef] [PubMed]
4. Beyer, K.; Domingo-Sàbat, M.; Ariza, A. Molecular pathology of Lewy body diseases. *Int. J. Mol. Sci.* **2009**, *10*, 724–745. [CrossRef] [PubMed]
5. Barr, R.K.; Bogoyevitch, M.A. The c-Jun N-terminal protein kinase family of mitogen-activated protein kinases (JNK MAPKs). *Int. J. Biochem. Cell Biol.* **2001**, *33*, 1047–1063. [CrossRef]
6. Plotnikov, A.; Zehorai, E.; Procaccia, S.; Seger, R. The MAPK cascades: Signaling components, nuclear roles and mechanisms of nuclear translocation. *Biochim. Biophys. Acta* **2011**, *1813*, 1619–1633. [CrossRef] [PubMed]
7. Yao, Z.; Seger, R. The ERK signaling cascade—Views from different subcellular compartments. *Biofactors* **2009**, *35*, 407–416. [CrossRef] [PubMed]
8. Yoon, S.; Seger, R. The extracellular signal-regulated kinase: Multiple substrates regulate diverse cellular functions. *Growth Factors* **2006**, *24*, 21–44. [CrossRef] [PubMed]
9. Cargnello, M.; Roux, P.P. Activation and function of the MAPKs and their substrates, the MAPK-activated protein kinases. *Microbiol. Mol. Biol. Rev.* **2011**, *75*, 50–83. [CrossRef] [PubMed]
10. Yarza, R.; Vela, S.; Solas, M.; Ramirez, M.J. c-Jun N-terminal Kinase (JNK) Signaling as a Therapeutic Target for Alzheimer's Disease. *Front. Pharmacol.* **2016**, *6*, 321. [CrossRef] [PubMed]
11. Yamasaki, T.; Kawasaki, H.; Nishina, H. Diverse Roles of JNK and MKK Pathways in the Brain. *J. Signal Transduct.* **2012**, *2012*, 459265. [CrossRef] [PubMed]
12. Kuan, C.Y.; Yang, D.D.; Samanta Roy, D.R.; Davis, R.J.; Rakic, P.; Flavell, R.A. The Jnk1 and Jnk2 protein kinases are required for regional specific apoptosis during early brain development. *Neuron* **1999**, *22*, 667–676. [CrossRef]
13. Zeke, A.; Misheva, M.; Reményi, A.; Bogoyevitch, M.A. JNK Signaling: Regulation and Functions Based on Complex Protein-Protein Partnerships. *Microbiol. Mol. Biol. Rev.* **2016**, *80*, 793–835. [CrossRef] [PubMed]
14. Xu, X.; Raber, J.; Yang, D.; Su, B.; Mucke, L. Dynamic regulation of c-Jun N-terminal kinase activity in mouse brain by environmental stimuli. *Proc. Natl. Acad. Sci. USA* **1997**, *94*, 12655–12660. [CrossRef] [PubMed]
15. Wang, W.; Ma, C.; Mao, Z.; Li, M. JNK inhibition as a potential strategy in treating Parkinson's disease. *Drug News Perspect.* **2004**, *10*, 646. [CrossRef]
16. Kulich, S.M.; Chu, C.T. Sustained extracellular signal-regulated kinase activation by 6-hydroxydopamine: Implications for Parkinson's disease. *J. Neurochem.* **2001**, *77*, 1058–1066. [CrossRef] [PubMed]
17. Brown, M.; Strudwick, N.; Suwara, M.; Sutcliffe, L.K.; Mihai, A.D.; Ali, A.A.; Watson, J.N.; Schröder, M. An initial phase of JNK activation inhibits cell death early in the endoplasmic reticulum stress response. *J. Cell Sci.* **2016**, *129*, 2317–2328. [CrossRef] [PubMed]
18. Pan, J.; Qian, J.; Zhang, Y.; Ma, J.; Wang, G.; Xiao, Q.; Chen, S.; Ding, J. Small peptide inhibitor of JNKs protects against MPTP-induced nigral dopaminergic injury via inhibiting the JNK-signaling pathway. *Lab. Investig.* **2010**, *90*, 156–167. [CrossRef]
19. Zhang, S.; Gui, X.H.; Huang, L.P.; Deng, M.Z.; Fang, R.M.; Ke, X.H.; He, Y.P.; Li, L.; Fang, Y.Q. Neuroprotective Effects of β-Asarone Against 6-Hydroxy Dopamine-Induced Parkinsonism via JNK/Bcl-2/Beclin-1 Pathway. *Mol. Neurobiol.* **2016**, *53*, 83–94. [CrossRef] [PubMed]
20. Badshah, H.; Ali, T.; Rehman, S.-U.; Amin, F.-U.; Ullah, F.; Kim, T.H.; Kim, M.O. Protective Effect of Lupeol Against Lipopolysaccharide-Induced Neuroinflammation via the p38/c-Jun N-Terminal Kinase Pathway in the Adult Mouse Brain. *J. Neuroimmune Pharmacol.* **2016**, *11*, 48–60. [CrossRef] [PubMed]
21. Tao, L.; Zhang, F.; Hao, L.; Wu, J.; Jia, J.; Liu, J.Y.; Zheng, L.T.; Zhen, X. 1-O-tigloyl-1-O-deacetyl-nimbolinin B inhibits LPS-stimulated inflammatory responses by suppressing NF-κB and JNK activation in microglia cells. *J. Pharmacol. Sci.* **2014**, *125*, 364–374. [CrossRef] [PubMed]

22. Brecht, S.; Kirchhof, R.; Chromik, A.; Willesen, M.; Nicolaus, T.; Raivich, G.; Wessig, J.; Waetzig, V.; Goetz, M.; Claussen, M.; et al. Specific pathophysiological functions of JNK isoforms in the brain. *Eur. J. Neurosci.* **2005**, *21*, 363–377. [CrossRef] [PubMed]

23. Lee, Y.; Chun, H.J.; Lee, K.M.; Jung, Y.S.; Lee, J. Silibinin suppresses astroglial activation in a mouse model of acute Parkinson's disease by modulating the ERK and JNK signaling pathways. *Brain Res.* **2015**, *1627*, 233–242. [CrossRef] [PubMed]

24. Pan, J.; Xiao, Q.; Sheng, C.Y.; Hong, Z.; Yang, H.Q.; Wang, G.; Ding, J.Q.; Chen, S.D. Blockade of the translocation and activation of c-Jun N-terminal kinase 3 (JNK3) attenuates dopaminergic neuronal damage in mouse model of Parkinson's disease. *Neurochem. Int.* **2009**, *54*, 418–425. [CrossRef] [PubMed]

25. Crocker, C.E.; Khan, S.; Cameron, M.D.; Robertson, H.A.; Robertson, G.S.; Lograsso, P. JNK Inhibition Protects Dopamine Neurons and Provides Behavioral Improvement in a Rat 6-hydroxydopamine Model of Parkinson's Disease. *ACS Chem. Neurosci.* **2011**, *2*, 207–212. [CrossRef] [PubMed]

26. Zhou, Q.; Wang, M.; Du, Y.; Zhang, W.; Bai, M.; Zhang, Z.; Li, Z.; Miao, J. Inhibition of c-Jun N-terminal kinase activation reverses Alzheimer disease phenotypes in APPswe/PS1dE9 mice. *Ann. Neurol.* **2015**, *77*, 637–654. [CrossRef] [PubMed]

27. Kim, S.J.; Ryu, M.J.; Han, J.; Jang, Y.; Kim, J.; Lee, M.J.; Ryu, I.; Ju, X.; Oh, E.; Chung, W.; et al. Activation of the HMGB1-RAGE axis upregulates TH expression in dopaminergic neurons via JNK phosphorylation. *Biochem. Biophys. Res. Commun.* **2017**, *493*, 358–364. [CrossRef] [PubMed]

28. Spigolon, G.; Cavaccini, A.; Trusel, M.; Tonini, R.; Fisone, G. cJun N-terminal kinase (JNK) mediates cortico-striatal signaling in a model of Parkinson's disease. *Neurobiol. Dis.* **2018**, *110*, 37–46. [CrossRef] [PubMed]

29. Gardoni, F.; Bellone, C. Modulation of the glutamatergic transmission by Dopamine: A focus on Parkinson, Huntington and Addiction diseases. *Front. Cell Neurosci.* **2015**, *9*, 25. [CrossRef] [PubMed]

30. Nisticò, R.; Florenzano, F.; Mango, D.; Ferraina, C.; Grilli, M.; Di Prisco, S.; Nobili, A.; Saccucci, S.; D'Amelio, M.; Morbin, M.; et al. Presynaptic c-Jun N-terminal Kinase 2 regulates NMDA receptor-dependent glutamate release. *Sci. Rep.* **2015**, *12*, 9035. [CrossRef] [PubMed]

31. Calabresi, P.; Galletti, F.; Saggese, E.; Ghiglieri, V.; Picconi, B. Neuronal networks and synaptic plasticity in Parkinson's disease: Beyond motor deficits. *Parkinsonism Relat. Disord.* **2007**, *13* (Suppl. 3), 259–262. [CrossRef]

32. Kim, M.J.; Futai, K.; Jo, J.; Hayashi, Y.; Cho, K.; Sheng, M. Synaptic accumulation of PSD-95 and synaptic function regulated by phosphorylation of serine-295 of PSD-95. *Neuron* **2007**, *56*, 488–502. [CrossRef] [PubMed]

33. Thomas, G.M.; Lin, D.T.; Nuriya, M.; Huganir, R.L. Rapid and bi-directional regulation of AMPA receptor phosphorylation and trafficking by JNK. *EMBO J.* **2008**, *27*, 361–372. [CrossRef] [PubMed]

34. Byrd, D.T.; Kawasaki, M.; Walcoff, M.; Hisamoto, N.; Matsumoto, K.; Jin, Y. UNC-16, a JNK-signaling scaffold protein, regulates vesicle transport in *C. elegans*. *Neuron* **2001**, *32*, 787–800. [CrossRef]

35. Daire, V.; Giustiniani, J.; Leroy-Gori, I.; Quesnoit, M.; Drevensek, S.; Dimitrov, A.; Perez, F.; Poüs, C. Kinesin-1 regulates microtubule dynamics via a c-Jun N-terminal kinase-dependent mechanism. *J. Biol. Chem.* **2009**, *284*, 31992–32001. [CrossRef] [PubMed]

36. Roy, S.; Zhang, B.; Lee, V.M.; Trojanowski, J.Q. Axonal transport defects: A common theme in neurodegenerative diseases. *Acta Neuropathol.* **2005**, *109*, 5–13. [CrossRef] [PubMed]

37. Morfini, G.A.; Burns, M.; Binder, L.I.; Kanaan, N.M.; LaPointe, N.; Bosco, D.A.; Brown, R.H., Jr.; Brown, H.; Tiwari, A.; Hayward, L.; et al. Axonal transport defects in neurodegenerative diseases. *J. Neurosci.* **2009**, *29*, 12776–12786. [CrossRef] [PubMed]

38. Li, D.; Li, X.; Wu, J.; Li, J.; Zhang, L.; Xiong, T.; Tang, J.; Qu, Y.; Mu, D. Involvement of the JNK/FOXO3a/Bim Pathway in Neuronal Apoptosis after Hypoxic-Ischemic Brain Damage in Neonatal Rats. *PLoS ONE* **2015**, *10*, e0132998. [CrossRef] [PubMed]

39. Kanamoto, T.; Mota, M.; Takeda, K.; Rubin, L.L.; Miyazono, K.; Ichijo, H.; Bazenet, C.E. Role of apoptosis signal-regulating kinase in regulation of the c-Jun N-terminal kinase pathway and apoptosis in sympathetic neurons. *Mol. Cell. Biol.* **2000**, *20*, 196–204. [CrossRef] [PubMed]

40. Venugopal, S.K.; Chen, J.; Zhang, Y.; Clemens, D.; Follenzi, A.; Zern, M.A. Role of MAPK phosphatase-1 in sustained activation of JNK during ethanol-induced apoptosis in hepatocyte-like VL-17A cells. *J. Biol. Chem.* **2007**, *282*, 31900–31908. [CrossRef] [PubMed]

41. Ferrandi, C.; Ballerio, R.; Gaillard, P.; Giachetti, C.; Carboni, S.; Vitte, P.A.; Gotteland, J.P.; Cirillo, R. Inhibition of c-Jun N-terminal kinase decreases cardiomyocyte apoptosis and infarct size after myocardial ischemia and reperfusion in anaesthetized rats. *Br. J. Pharmacol.* **2004**, *142*, 953–960. [CrossRef] [PubMed]

42. Son, E.W.; Mo, S.J.; Rhee, D.K.; Pyo, S. Inhibition of ICAM-1 expression by garlic component, allicin, in gamma-irradiated human vascular endothelial cells via downregulation of the JNK signaling pathway. *Int. Immunopharmacol.* **2006**, *6*, 1788–1795. [CrossRef] [PubMed]

43. Shklover, J.; Mishnaevski, K.; Levy-Adam, F.; Kurant, E. JNK pathway activation is able to synchronize neuronal death and glial phagocytosis in Drosophila. *Cell. Death. Dis.* **2015**, *6*, e1649. [CrossRef] [PubMed]

44. Peng, J.; Andersen, J.K. The role of c-Jun N-terminal kinase (JNK) in Parkinson's disease. *IUBMB Life* **2003**, *55*, 267–271. [CrossRef] [PubMed]

45. Hashimoto, M.; Hsu, L.J.; Rockenstein, E.; Takenouchi, T.; Mallory, M.; Masliah, E. Alpha-Synuclein protects against oxidative stress via inactivation of the c-Jun N-terminal kinase stress-signaling pathway in neuronal cells. *J. Biol. Chem.* **2002**, *277*, 11465–11472. [CrossRef] [PubMed]

46. Yan, L.; Liu, S.; Wang, C.; Wang, F.; Song, Y.; Yan, N.; Xi, S.; Liu, Z.; Sun, G. JNK and NADPH oxidase involved in fluoride-induced oxidative stress in BV-2 microglia cells. *Mediat. Inflamm.* **2013**, *2013*, 895975. [CrossRef] [PubMed]

47. Zhan, L.; Xie, Q.; Tibbetts, R.S. Opposing roles of p38 and JNK in a Drosophila model of TDP-43 proteinopathy reveal oxidative stress and innate immunity as pathogenic components of neurodegeneration. *Hum. Mol. Genet.* **2015**, *24*, 757–772. [CrossRef] [PubMed]

48. Wang, M.C.; Bohmann, D.; Jasper, H. JNK signaling confers tolerance to oxidative stress and extends lifespan in Drosophila. *Dev. Cell* **2003**, *5*, 811–816. [CrossRef]

49. Szebenyi, G.; Morfini, G.A.; Babcock, A.; Gould, M.; Selkoe, K.; Stenoien, D.L.; Young, M.; Faber, P.W.; MacDonald, M.E.; McPhaul, M.J.; et al. Neuropathogenic forms of huntingtin and androgen receptor inhibit fast axonal transport. *Neuron* **2003**, *40*, 41–52. [CrossRef]

50. Ren, H.; Fu, K.; Mu, C.; Li, B.; Wang, D.; Wang, G. DJ-1, a cancer and Parkinson's disease associated protein, regulates autophagy through JNK pathway in cancer cells. *Cancer Lett.* **2010**, *297*, 101–108. [CrossRef] [PubMed]

51. Sui, X.; Kong, N.; Ye, L.; Han, W.; Zhou, J.; Zhang, Q.; He, C.; Pan, H. p38 and JNK MAPK pathways control the balance of apoptosis and autophagy in response to chemotherapeutic agents. *Cancer Lett.* **2014**, *344*, 174–179. [CrossRef] [PubMed]

52. Suzuki, M.; Bandoski, C.; Bartlett, J.D. Fluoride induces oxidative damage and SIRT1/autophagy through ROS-mediated JNK signaling. *Free Radic. Biol. Med.* **2015**, *89*, 369–378. [CrossRef] [PubMed]

53. Takahashi-Niki, K.; Kato-Ose, I.; Murata, H.; Maita, H.; Iguchi-Ariga, S.M.; Ariga, H. Epidermal Growth Factor-dependent Activation of the Extracellular Signal-regulated Kinase Pathway by DJ-1 Protein through Its Direct Binding to c-Raf Protein. *J. Biol. Chem.* **2015**, *29*, 17838–17847. [CrossRef] [PubMed]

54. Wang, H.Y.; Doronin, S.; Malbon, C.C. Insulin activation of mitogen-activated protein kinases Erk1,2 is amplified via beta-adrenergic receptor expression and requires the integrity of the Tyr350 of the receptor. *J. Biol. Chem.* **2000**, *275*, 36086–36093. [CrossRef] [PubMed]

55. Naor, Z.; Benard, O.; Seger, R. Activation of MAPK cascades by G-protein-coupled receptors: The case of gonadotropin-releasing hormone receptor. *Trends Endocrinol. Metab.* **2000**, *11*, 91–99. [CrossRef]

56. Chiri, S.; Bogliolo, S.; Ehrenfeld, J.; Ciapa, B. Activation of extracellular signal-regulated kinase ERK after hypo-osmotic stress in renal epithelial A6 cells. *Biochim. Biophys. Acta* **2004**, *1664*, 224–229. [CrossRef] [PubMed]

57. Roskoski, R., Jr. ERK1/2 MAP kinases: Structure, function, and regulation. *Pharmacol. Res.* **2012**, *66*, 105–143. [CrossRef] [PubMed]

58. Sun, Y.; Liu, W.Z.; Liu, T.; Feng, X.; Yang, N.; Zhou, H.F. Signaling pathway of MAPK/ERK in cell proliferation, differentiation, migration, senescence and apoptosis. *Recept Signal. Transduct. Res.* **2015**, *35*, 600–604. [CrossRef] [PubMed]

59. Mitchell, T.J.; John, S. Signal transducer and activator of transcription (STAT) signalling and T-cell lymphomas. *Immunology* **2005**, *114*, 301–312. [CrossRef] [PubMed]

60. Yu, H.; Pardoll, D.; Jove, R. STATs in cancer inflammation and immunity: A leading role for STAT3. *Nat. Rev. Cancer* **2009**, *9*, 798–809. [CrossRef] [PubMed]

61. Yao, K.; Chen, Q.; Wu, Y.; Liu, F.; Chen, X.; Zhang, Y. Unphosphorylated STAT1 represses apoptosis in macrophages during *Mycobacterium tuberculosis* infection. *J. Cell. Sci.* **2017**, *130*, 1740–1751. [CrossRef] [PubMed]

62. Selcher, J.C.; Nekrasova, T.; Paylor, R.; Landreth, G.E.; Sweatt, J.D. Mice lacking the ERK1 isoform of MAP kinase are unimpaired in emotional learning. *Learn. Mem.* **2001**, *8*, 11–19. [CrossRef] [PubMed]

63. Scholl, F.A.; Dumesic, P.A.; Barragan, D.I.; Harada, K.; Bissonauth, V.; Charron, J.; Khavari, P.A. Mek1/2 MAPK kinases are essential for Mammalian development, homeostasis, and Raf-induced hyperplasia. *Dev. Cell* **2007**, *12*, 615–629. [CrossRef] [PubMed]

64. Fiore, R.S.; Bayer, V.E.; Pelech, S.L.; Posada, J.; Cooper, J.A.; Baraban, J.M. p42 mitogen-activated protein kinase in brain: Prominent localization in neuronal cell bodies and dendrites. *Neuroscience* **1993**, *55*, 463–472. [CrossRef]

65. Di Cristo, G.; Berardi, N.; Cancedda, L.; Pizzorusso, T.; Putignano, E.; Ratto, G.M.; Maffei, L. Requirement of ERK activation for visual cortical plasticity. *Science* **2001**, *292*, 2337–2340. [CrossRef] [PubMed]

66. Fieblinger, T.; Sebastianutto, I.; Alcacer, C.; Bimpisidis, Z.; Maslava, N.; Sandberg, S.; Engblom, D.; Cenci, M.A. Mechanisms of dopamine D1 receptor-mediated ERK1/2 activation in the parkinsonian striatum and their modulation by metabotropic glutamate receptor type 5. *J. Neurosci.* **2014**, *34*, 4728–4740. [CrossRef] [PubMed]

67. Hebert, A.E.; Dash, P.K. Extracellular signal-regulated kinase activity in the entorhinal cortex is necessary for long-term spatial memory. *Learn. Mem.* **2002**, *9*, 156–166. [CrossRef] [PubMed]

68. Yufune, S.; Satoh, Y.; Takamatsu, I.; Ohta, H.; Kobayashi, Y.; Takaenoki, Y.; Pagès, G.; Pouysségur, J.; Endo, S.; Kazama, T. Transient Blockade of ERK Phosphorylation in the Critical Period Causes Autistic Phenotypes as an Adult in Mice. *Sci. Rep.* **2015**, *5*, 10252. [CrossRef] [PubMed]

69. Cheung, E.C.; Slack, R.S. Emerging role for ERK as a key regulator of neuronal apoptosis. *Sci. STKE* **2004**, *251*, PE45. [CrossRef] [PubMed]

70. Bhat, N.R.; Zhang, P. Hydrogen peroxide activation of multiple mitogen-activated protein kinases in an oligodendrocyte cell line: Role of extracellular signal-regulated kinase in hydrogen peroxide-induced cell death. *J. Neurochem.* **1999**, *72*, 112–119. [CrossRef] [PubMed]

71. Canals, S.; Casarejos, M.J.; de Bernardo, S.; Solano, R.M.; Mena, M.A. Selective and persistent activation of extracellular signal-regulated protein kinase by nitric oxide in glial cells induces neuronal degeneration in glutathione-depleted midbrain cultures. *Mol. Cell. Neurosci.* **2003**, *24*, 1012–1026. [CrossRef] [PubMed]

72. Teng, L.; Kou, C.; Lu, C.; Xu, J.; Xie, J.; Lu, J.; Liu, Y.; Wang, Z.; Wang, D. Involvement of the ERK pathway in the protective effects of glycyrrhizic acid against the MPP+-induced apoptosis of dopaminergic neuronal cells. *Int. J. Mol. Med.* **2014**, *34*, 742–748. [CrossRef] [PubMed]

73. Chen, J.; Rusnak, M.; Lombroso, P.J.; Sidhu, A. Dopamine promotes striatal neuronal apoptotic death via ERK signaling cascades. *Eur. J. Neurosci.* **2009**, *29*, 287–306. [CrossRef] [PubMed]

74. Schapira, A.H. Mitochondrial complex I deficiency in Parkinson's disease. *Adv. Neurol.* **1993**, *60*, 288–291. [PubMed]

75. Zhu, J.H.; Guo, F.; Shelburne, J.; Watkins, S.; Chu, C.T. Localization of phosphorylated ERK/MAP kinases to mitochondria and autophagosomes in Lewy body diseases. *Brain Pathol.* **2003**, *13*, 473–481. [CrossRef] [PubMed]

76. Monick, M.M.; Powers, L.S.; Barrett, C.W.; Hinde, S.; Ashare, A.; Groskreutz, D.J.; Nyunoya, T.; Coleman, M.; Spitz, D.R.; Hunninghake, G.W. Constitutive ERK MAPK activity regulates macrophage ATP production and mitochondrial integrity. *J. Immunol.* **2008**, *180*, 7485–7496. [CrossRef] [PubMed]

77. Wang, K.Z.; Zhu, J.; Dagda, R.K.; Uechi, G.; Cherra, S.J.; Gusdon, A.M.; Balasubramani, M.; Chu, C.T. ERK-mediated phosphorylation of TFAM downregulates mitochondrial transcription: Implications for Parkinson's disease. *Mitochondrion* **2014**, *17*, 132–140. [CrossRef] [PubMed]

78. Wang, Z.; Liu, J.; Chen, S.; Wang, Y.; Cao, L.; Zhang, Y.; Kang, W.; Li, H.; Gui, Y.; Chen, S.; et al. DJ-1 modulates the expression of Cu/Zn-superoxide dismutase-1 through the Erk1/2-Elk1 pathway in neuroprotection. *Ann. Neurol.* **2011**, *70*, 591–599. [CrossRef] [PubMed]

79. Cai, P.; Ye, J.; Zhu, J.; Liu, D.; Chen, D.; Wei, X.; Johnson, N.R.; Wang, Z.; Zhang, H.; Cao, G.; et al. Inhibition of Endoplasmic Reticulum Stress is Involved in the Neuroprotective Effect of bFGF in the 6-OHDA-Induced Parkinson's Disease Model. *Aging Dis.* **2016**, *7*, 336–449. [CrossRef] [PubMed]

80. Zhang, X.; Bai, L.; Zhang, S.; Zhou, X.; Li, Y.; Bai, J. Trx-1 ameliorates learning and memory deficits in MPTP-induced Parkinson's disease model in mice. *Free Radic. Biol. Med.* **2018**, *124*, 380–387. [CrossRef] [PubMed]

81. Park, K.H.; Shin, K.S.; Zhao, T.T.; Park, H.; Lee, K.E.; Lee, M.K. L-DOPA modulates cell viability through the ERK-c-Jun system in PC12 and dopaminergic neuronal cells. *Neuropharmacology* **2016**, *101*, 87–97. [CrossRef] [PubMed]

82. Cao, Q.; Qin, L.; Huang, F.; Wang, X.; Yang, L.; Shi, H.; Wu, H.; Zhang, B.; Chen, Z.; Wu, X. Amentoflavone protects dopaminergic neurons in MPTP-induced Parkinson's disease model mice through PI3K/Akt and ERK signaling pathways. *Toxicol. Appl. Pharmacol.* **2017**, *319*, 80–90. [CrossRef] [PubMed]

83. Huang, B.; Liu, J.; Ju, C.; Chen, G.; Xu, S.; Zeng, Y.; Yan, X.; Wang, W.; Liu, D.; Fu, S. Licochalcone A Prevents the Loss of Dopaminergic Neurons by Inhibiting Microglial Activation in Lipopolysaccharide (LPS)-Induced Parkinson's Disease Models. *Int. J. Mol. Sci.* **2017**, *18*, 2043. [CrossRef] [PubMed]

84. Kamakura, S.; Moriguchi, T.; Nishida, E. Activation of the protein kinase ERK5/BMK1 by receptor tyrosine kinases. Identification and characterization of a signaling pathway to the nucleus. *J. Biol. Chem.* **1999**, *274*, 26563–26571. [CrossRef] [PubMed]

85. Mody, N.; Leitch, J.; Armstrong, C.; Dixon, J.; Cohen, P. Effects of MAP kinase cascade inhibitors on the MKK5/ERK5 pathway. *FEBS Lett.* **2001**, *502*, 21–24. [CrossRef]

86. Cavanaugh, J.E.; Jaumotte, L.D.; Lakoski, J.M.; Zigmond, M.J. Neuroprotective Role of ERK1/2 and ERK5 in a Dopaminergic Cell Line Under Basal Conditions and in Response to Oxidative Stress. *J. Neurosci. Res.* **2006**, *84*, 1367–1375. [CrossRef] [PubMed]

87. Zarubin, T.; Han, J. Activation and signaling of the p38 MAP kinase pathway. *Cell Res.* **2005**, *15*, 11–18. [CrossRef] [PubMed]

88. Jiang, Y.; Gram, H. Characterization of the structure and function of the fourth member of p38 group mitogen-activated protein kinases, p38delta. *J. Biol. Chem.* **1997**, *272*, 30122–30128. [CrossRef] [PubMed]

89. Cuadrado, A.; Nebreda, A.R. Mechanisms and functions of p38 MAPK signalling. *Biochem. J.* **2010**, *429*, 403–417. [CrossRef] [PubMed]

90. Allen, M.; Svensson, L.; Roach, M.; Hambor, J.; McNeish, J.; Gabel, C.A. Deficiency of the stress kinase p38alpha results in embryonic lethality: Characterization of the kinase dependence of stress responses of enzyme-deficient embryonic stem cells. *J. Exp. Med.* **2000**, *191*, 859–870. [CrossRef] [PubMed]

91. Beardmore, V.A.; Hinton, H.J.; Eftychi, C.; Apostolaki, M.; Armaka, M.; Darragh, J.; McIlrath, J.; Carr, J.M.; Armit, L.J.; Clacher, C.; et al. Generation and characterization of p38beta (MAPK11) gene-targeted mice. *Mol. Cell. Biol.* **2005**, *25*, 10454–10464. [CrossRef] [PubMed]

92. Schonhoff, C.M.; Park, S.W.; Webster, C.R.L.; Anwer, M.S. p38 MAPK α and β isoforms differentially regulate plasma membrane localization of MRP2. *Am. J. Physiol. Gastrointest. Liver Physiol.* **2016**, *310*, G999–G1005. [CrossRef] [PubMed]

93. Jha, S.K.; Jha, N.K.; Kar, R.; Ambasta, R.K.; Kumar, P. p38 MAPK and PI3K/AKT Signalling Cascades in Parkinson's Disease. *Int. J. Mol. Cell Med.* **2015**, *4*, 67–86. [PubMed]

94. Harper, S.J.; LoGrasso, P. Signalling for survival and death in neurones: The role of stress-activated kinases, JNK and p38. *Cell Signal.* **2001**, *13*, 299–310. [CrossRef]

95. Mielke, K.; Herdegen, T. JNK and p38 stresskinases—Degenerative effectors of signal-transduction-cascades in the nervous system. *Prog. Neurobiol.* **2000**, *61*, 45–60. [CrossRef]

96. De Chiara, G.; Marcocci, M.E.; Torcia, M.; Lucibello, M.; Rosini, P.; Bonini, P.; Higashimoto, Y.; Damonte, G.; Armirotti, A.; Amodei, S.; et al. Bcl-2 Phosphorylation by p38 MAPK: Identification of target sites and biologic consequences. *J. Biol. Chem.* **2006**, *281*, 21353–21361. [CrossRef] [PubMed]

97. Cai, B.; Chang, S.H.; Becker, E.B.; Bonni, A.; Xia, Z. p38 MAP kinase mediates apoptosis through phosphorylation of BimEL at Ser-65. *J. Biol. Chem.* **2006**, *281*, 25215–25222. [CrossRef] [PubMed]

98. Choi, W.S.; Eom, D.S.; Han, B.S.; Kim, W.K.; Han, B.H.; Choi, E.J.; Oh, T.H.; Markelonis, G.J.; Cho, J.W.; Oh, Y.J. Phosphorylation of p38 MAPK induced by oxidative stress is linked to activation of both caspase-8- and -9-mediated apoptotic pathways in dopaminergic neurons. *J. Biol. Chem.* **2004**, *279*, 20451–20460. [CrossRef] [PubMed]

99. Tong, H.; Zhang, X.; Meng, X.; Lu, L.; Mai, D.; Qu, S. Simvastatin Inhibits Activation of NADPH Oxidase/p38 MAPK Pathway and Enhances Expression of Antioxidant Protein in Parkinson Disease Models. *Front. Mol. Neurosci.* **2018**, *11*, 165. [CrossRef] [PubMed]

100. Jiang, G.; Hu, Y.; Liu, L.; Cai, J.; Peng, C.; Li, Q. Gastrodin protects against MPP(+)-induced oxidative stress by up regulates heme oxygenase-1 expression through p38 MAPK/Nrf2 pathway in human dopaminergic cells. *Neurochem. Int.* **2014**, *75*, 79–88. [CrossRef] [PubMed]

101. Liedhegner, E.A.; Steller, K.M.; Mieyal, J.J. Levodopa activates apoptosis signaling kinase 1 (ASK1) and promotes apoptosis in a neuronal model: Implications for the treatment of Parkinson's disease. *Chem. Res. Toxicol.* **2011**, *24*, 1644–1652. [CrossRef] [PubMed]

102. Ouyang, M.; Shen, X. Critical role of ASK1 in the 6-hydroxydopamine-induced apoptosis in human neuroblastoma SH-SY5Y cells. *J. Neurochem.* **2006**, *97*, 234–244. [CrossRef] [PubMed]

103. Gao, F.; Chen, D.; Hu, Q.; Wang, G. Rotenone directly induces BV2 cell activation via the p38 MAPK pathway. *PLoS ONE* **2013**, *8*, e72046. [CrossRef] [PubMed]

104. Oh, C.K.; Han, B.S.; Choi, W.S.; Youdim, M.B.; Oh, Y.J. Translocation and oligomerization of Bax is regulated independently by activation of p38 MAPK and caspase-2 during MN9D dopaminergic neurodegeneration. *Apoptosis* **2011**, *16*, 1087–1100. [CrossRef] [PubMed]

105. Chien, W.L.; Lee, T.R.; Hung, S.Y.; Kang, K.H.; Lee, M.J.; Fu, W.M. Impairment of oxidative stress-induced heme oxygenase-1 expression by the defect of Parkinson-related gene of PINK1. *J. Neurochem.* **2011**, *17*, 643–653. [CrossRef]

106. Hwang, C.J.; Kim, Y.E.; Son, D.J.; Park, M.H.; Choi, D.Y.; Park, P.H.; Hellström, M.; Han, S.B.; Oh, K.W.; Park, E.K.; et al. Parkin deficiency exacerbate ethanol-induced dopaminergic neurodegeneration by P38 pathway dependent inhibition of autophagy and mitochondrial function. *Redox Biol.* **2017**, *11*, 456–468. [CrossRef] [PubMed]

107. Chen, J.; Ren, Y.; Gui, C.; Zhao, M.; Wu, X.; Mao, K.; Li, W.; Zou, F. Phosphorylation of Parkin at serine 131 by p38 MAPK promotes mitochondrial dysfunction and neuronal death in mutant A53T α-synuclein model of Parkinson's disease. *Cell Death Dis.* **2018**, *9*, 700. [CrossRef] [PubMed]

108. Wu, F.; Wang, Z.; Gu, J.H.; Ge, J.B.; Liang, Z.Q.; Qin, Z.H. p38(MAPK)/p53-Mediated Bax induction contributes to neurons degeneration in rotenone-induced cellular and rat models of Parkinson's disease. *Neurochem. Int.* **2013**, *63*, 133–140. [CrossRef] [PubMed]

109. Yoon, J.H.; Mo, J.S.; Kim, M.Y.; Ann, E.J.; Ahn, J.S.; Jo, E.H.; Lee, H.J.; Lee, Y.C.; Seol, W.; Yarmoluk, S.M.; et al. LRRK2 functions as a scaffolding kinase of ASK1-mediated neuronal cell death. *Biochim. Biophys. Acta* **2017**, *1864*, 2356–2368. [CrossRef] [PubMed]

110. Wu, R.; Chen, H.; Ma, J.; He, Q.; Huang, Q.; Liu, Q.; Li, M.; Yuan, Z. c-Abl-p38α signaling plays an important role in MPTP-induced neuronal death. *Cell Death Differ.* **2016**, *23*, 542–552. [CrossRef] [PubMed]

111. Karunakaran, S.; Saeed, U.; Mishra, M.; Valli, R.K.; Joshi, S.D.; Meka, D.P.; Seth, P.; Ravindranath, V. Selective activation of p38 mitogen-activated protein kinase in dopaminergic neurons of substantia nigra leads to nuclear translocation of p53 in 1-methyl-4-phenyl-1,2,3,6-tetrahydropyridine-treated mice. *J. Neurosci.* **2008**, *28*, 12500–12509. [CrossRef] [PubMed]

112. Choi, B.S.; Kim, H.; Lee, H.J.; Sapkota, K.; Park, S.E.; Kim, S.; Kim, S.J. Celastrol from 'Thunder God Vine' protects SH-SY5Y cells through the preservation of mitochondrial function and inhibition of p38 MAPK in a rotenone model of Parkinson's disease. *Neurochem. Res.* **2014**, *39*, 84–96. [CrossRef] [PubMed]

113. Chen, W.W.; Zhang, X.; Huang, W.J. Role of neuroinflammation in neurodegenerative diseases (Review). *Mol. Med. Rep.* **2016**, *13*, 3391–3396. [CrossRef] [PubMed]

114. Mansour, R.M.; Ahmed, M.A.E.; El-Sahar, A.E.; El Sayed, N.S. Montelukast attenuates rotenone-induced microglial activation/p38 MAPK expression in rats: Possible role of its antioxidant, anti-inflammatory and antiapoptotic effects. *Toxicol. Appl. Pharmacol.* **2018**. [CrossRef] [PubMed]

115. Thomas, T.; Timmer, M.; Cesnulevicius, K.; Hitti, E.; Kotlyarov, A.; Gaestel, M. MAPKAP kinase 2-deficiency prevents neurons from cell death by reducing neuroinflammation–relevance in a mouse model of Parkinson's disease. *J. Neurochem.* **2008**, *105*, 2039–2052. [CrossRef] [PubMed]

116. Deretic, V.; Saitoh, T.; Akira, S. Autophagy in infection, inflammation and immunity. *Nat. Rev. Immunol.* **2013**, *13*, 722–737. [CrossRef] [PubMed]

117. Shi, C.S.; Shenderov, K.; Huang, N.N.; Kabat, J.; Abu-Asab, M.; Fitzgerald, K.A.; Sher, A.; Kehrl, J.H. Activation of autophagy by inflammatory signals limits IL-1β production by targeting ubiquitinated inflammasomes for destruction. *Nat. Immunol.* **2012**, *13*, 255–263. [CrossRef] [PubMed]

118. He, Y.; She, H.; Zhang, T.; Xu, H.; Cheng, L.; Yepes, M.; Zhao, Y.; Mao, Z. p38 MAPK inhibits autophagy and promotes microglial inflammatory responses by phosphorylating ULK1. *J. Cell. Biol.* **2018**, *217*, 315–328. [CrossRef] [PubMed]
119. Liu, Y.; Song, Y.; Zhu, X. MicroRNA-181a Regulates Apoptosis and Autophagy Process in Parkinson's Disease by Inhibiting p38 Mitogen-Activated Protein Kinase (MAPK)/c-Jun N-Terminal Kinases (JNK) Signaling Pathways. *Med. Sci. Monit.* **2017**, *23*, 1597–1606. [CrossRef] [PubMed]

International Journal of
Molecular Sciences

Article

A Ruthenium(II) *N*-Heterocyclic Carbene (NHC) Complex with Naphthalimide Ligand Triggers Apoptosis in Colorectal Cancer Cells via Activating the ROS-p38 MAPK Pathway

Yasamin Dabiri [1], Alice Schmid [1], Jannick Theobald [1], Biljana Blagojevic [1], Wojciech Streciwilk [2], Ingo Ott [2], Stefan Wölfl [1] and Xinlai Cheng [1,*]

1   Institute of Pharmacy and Molecular Biotechnology, Heidelberg University, Im Neuenheimer Feld 364, 69120 Heidelberg, Germany; dabiri@stud.uni-heidelberg.de (Y.D.); alice.schmid@stud.uni-heidelberg.de (A.S.); jannick.theobald@gmail.com (J.T.); blagojevic@stud.uni-heidelberg.de (B.B.); wolfl@uni-hd.de (S.W.)
2   Institute of Medicinal and Pharmaceutical Chemistry, Technische Universität Braunschweig, Beethovenstraße 55, 38106 Braunschweig, Germany; w.streciwilk@tu-bs.de (W.S.); ingo.ott@tu-bs.de (I.O.)
*   Correspondence: x.cheng@uni-heidelberg.de; Tel.: +49-6221-54-6431

Received: 27 November 2018; Accepted: 6 December 2018; Published: 9 December 2018

Abstract: The p38 MAPK pathway is known to influence the anti-tumor effects of several chemotherapeutics, including that of organometallic drugs. Previous studies have demonstrated the important role of p38 both as a regulator and a sensor of cellular reactive oxygen species (ROS) levels. Investigating the anti-cancer properties of novel 1,8-naphthalimide derivatives containing Rh(I) and Ru(II) *N*-heterocyclic carbene (NHC) ligands, we observed a profound induction of ROS by the complexes, which is most likely generated from mitochondria (mtROS). Further analyses revealed a rapid and consistent activation of p38 signaling by the naphthalimide-NHC conjugates, with the Ru(II) analogue—termed MC6—showing the strongest effect. In view of this, genetic as well as pharmacological inhibition of p38α, attenuated the anti-proliferative and pro-apoptotic effects of MC6 in HCT116 colon cancer cells, highlighting the involvement of this signaling molecule in the compound's toxicity. Furthermore, the influence of MC6 on p38 signaling appeared to be dependent on ROS levels as treatment with general- and mitochondria-targeted anti-oxidants abrogated p38 activation in response to MC6 as well as the molecule's cytotoxic- and apoptogenic response in HCT116 cells. Altogether, our results provide new insight into the molecular mechanisms of naphthalimide-metal NHC analogues via the ROS-induced activation of p38 MAPK, which may have therapeutic interest for the treatment of various cancer types.

Keywords: naphthalimide-metal complex conjugates; *N*-heterocyclic carbene; mitochondria; ROS; p38 MAPK; apoptosis; cancer

---

1. Introduction

Increasing evidence has proven 1,8-naphthalimides as promising candidates for the treatment of cancer, with several such derivatives (e.g., amonafide and mitonafide) being tested in clinical trials against various solid and soft tumors [1]. Despite their potent anti-cancer activity, the clinical application of most of these compounds is hampered due to the toxic side effects [1,2]. Accordingly, several strategies have been developed to modify the naphthalimide ring in order to improve the anti-tumor effects and lower its toxicity [2]. This has led to the synthesis of various naphthalimide-based conjugates, such as metal complexes with naphthalimide ligands [2]. In this regard, naphthalimide-gold(I) phosphine complexes, whose synthesis was inspired by pervious observations on the lead compound auranofin, have shown an increase in the overall cellular uptake and in the nuclear accumulation of gold(I) as compared with the naphthalimide-free analogues [3]. Other interesting examples are the naphthalimide-based ruthenium(II) arene complexes, showing enhanced cancer cell selectivity, which is possibly achieved by the simultaneous action of naphthalimide as a DNA intercalator along with the ability of ruthenium(II) in binding proteins [4]. More recently, 1,8-naphthalimides containing a metal *N*-heterocyclic carbene (NHC) moiety have been synthesized [5,6]. These conjugates are shown to act via both interaction with DNA—related to the naphthalimide structure—as well as metal-based mechanisms, such as the inhibition of the thioredoxin reductase (TrxR) [5,6].

Among the molecular mechanisms that are involved in the anti-cancer efficacy of organometallic drugs is the mitogen activated protein kinase (MAPK) pathway. MAPKs encompass three signaling cascades; (i) extracellular signal-related kinases (ERKs), (ii) the c-Jun N-terminal kinase (JNK), and (iii) p38 MAPK, all of which have key roles in cellular proliferation and survival [7]. p38 MAPK has been repeatedly implicated in cancer therapy and its activation is shown to be necessary for cancer cell death triggered by a number of chemotherapeutic agents [8]. We and others have shown a determinant role for the activation of p38 signaling in the pro-apoptotic effects of metal-based drugs, such as cisplatin [9–11], auranofin [12] as well as gold-containing NHC complexes [13]. However, reports regarding the influence of naphthalimide derivatives on MAPKs are sparse. As an example, a novel amonafide analogue has been shown to down-regulate ERK1/2 and p38 via TAK1 inhibition, leading to its anti-inflammatory effects [14].

In continuation to the aforementioned studies, a series of 4-ethylthio-1,8-naphthalimide conjugates has been recently synthesized, carrying rhodium(I)- and ruthenium(II) NHC fragments as metal units [15]. The compounds were found to interact with DNA via an intercalation mechanism and they were able to trigger strong anti-proliferative effects in MCF-7 breast cancer and HT-29 colon carcinoma cells [15]. In this article, we describe more detailed analyses on the molecular mechanisms underlying the anti-tumor effects of Rh(I)- and Ru(II) naphthalimide-NHC compounds, as well as the metal-free ligand, designated as MC7, MC6, and MC5, respectively. All of the complexes showed potent anti-proliferative effects against various breast- and colorectal cancer (Colorectal cancer (CRC)) cell lines, but exhibited mild toxicity in human foreskin fibroblasts (HFFs). Using HCT116 CRC cells, we have assessed the involvement of reactive oxygen species (ROS) and p38 MAPK signaling in the mode of action of these molecules. We observed elevated intracellular- and mitochondrial ROS production, and a remarkable activation of p38 MAPK in response to naphthalimide-NHC analogues, with no clear regulation of other members of the MAPK family. Additionally, the modulation of ROS and p38α by anti-oxidants and either chemical inhibitors or siRNA, respectively, led to a significant reduction in the pro-apoptotic and growth inhibitory functions of the Ru(II) derivative. Our findings propose p38 signaling as a novel anti-cancer target of organometallic complexes with naphthalimide ligands.

2. Results

2.1. MC5, MC6, and MC7 Exhibit Cytotoxic Effects in Tumor Cells of Different Tissue Origins

To determine the cytotoxicity of the naphthalimide-NHC complexes (Figure 1A) in different cancer models, the cell lines of breast-(MCF-7 and MDA-MB-231) and colorectal (HCT116) tissue origins were treated with increasing concentrations of the compounds for three incubation periods (24, 48, and 72 h), after which Sulforhodamine B (SRB) assay was performed. The metal-free naphthalimide ligand—MC5—as well as the rapid apoptosis inducer, raptinal [16] were included as references. In all of the investigated cell lines, cellular survival was found to decrease time- and concentration-dependently in response to the three compounds, with HCT116 cells showing the highest sensitivity to metal-containing analogues in the first 24 h (Figure 1B). After 48 and 72 h of incubation with MC6 and MC7, cell viabilities became comparable among MCF-7 and HCT116 cells, while they remained substantially higher in the case of MDA-MB-231 at most of the tested concentrations (Figure 1B). Moreover, MC5 was found to be particularly active against MCF-7 breast cancer cells, but not in HCT116 and MDA-MB-231. This is in good agreement with the previous report, showing significantly lower IC_{50} values of metal-free naphthalimide species in MCF-7 as compared to that of HT-29 CRC cells [15]. Notably, the compounds exhibited extremely low cytotoxic effects in HFF cells, even at the highest tested concentrations, which might suggest the preferential toxicity of the naphthalimide-NHC conjugates towards cancer cells (Figure S1).

We also tested the cytotoxicity of the three analogues in the context of p53-deficiency or mutant p53 using the p53-null HCT116 and HT-29 cell lines, respectively. As shown in Figure S2, mutant p53 harboring HT-29 cells are less sensitive to the cytotoxic effects of the compounds at most tested concentrations and time points.

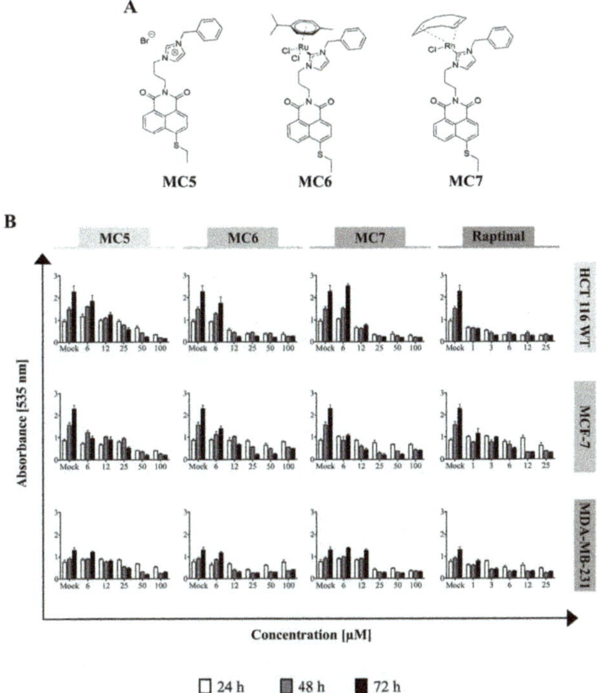

Figure 1. Naphthalimide-*N*-heterocyclic carbene (NHC) analogues exhibit cytotoxic effects against human breast- and colon cancer cells. (**A**) Chemical structures of the compounds; (**B**) Increasing concentrations of each of the complexes, as well as the rapid apoptosis inducer, raptinal [16] (as positive control) were applied to the different cell lines and Sulforhodamine B (SRB) assay was performed after 24, 48, and 72 h of treatment. The Ru(II)- and Rh(I)-containing complexes show the highest and the least efficacy against HCT116 and MDA-MB-231, respectively, in most tested concentrations. 0.1% DMSO-treated cells served as mock. Data represent mean ± SD of three independent experiments, each was done in quadruplicates.

2.2. MC5, MC6, and MC7 Inhibit Cell Cycle Progression in HCT116 CRC Cells

To shed light on the mechanism that is responsible for the inhibitory activity of the compounds on cellular viability, we sought to assess changes in the cell cycle regulation. HCT116 cell line was selected for further investigation based on its higher susceptibility to the MC6- and MC7-mediated cytotoxic effects (Figure 1B). A 24 h post-treatment analysis of the DNA content revealed a G1 arrest in response to treatments (Figure 2A). Although all three compounds caused a significant increase in the G1 phase cell population as compared to mock (0.1% DMSO), this effect was found to be more pronounced in the case of the Rh(I) analogue (MC7), followed by the metal-free ligand (MC5), and finally the Ru(II) complex (MC6) (Figure 2B).

Additionally, we examined the mRNA levels and protein expression of p21, an inhibitor of the complexes of cyclin D and cyclin-dependent kinases (CDKs), which have a key role in the G1 to S phase transition [17]. As expected, p21 mRNA and protein levels were induced after 24 h of treatment, with no statistically significant difference being found across the three compounds (Figure 2C–E).

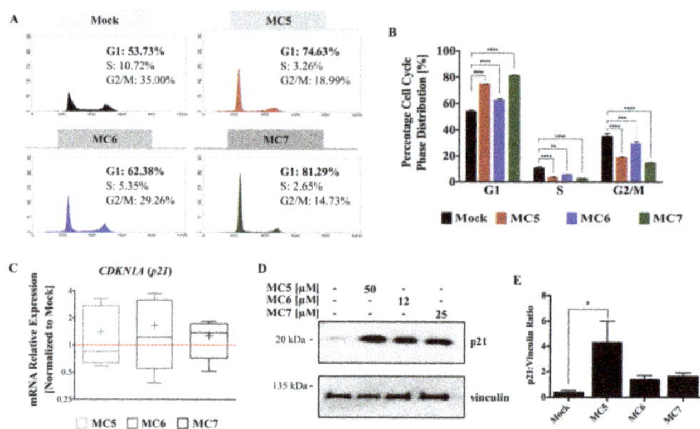

Figure 2. Naphthalimide-NHC analogues induce cell cycle arrest and p21 expression in HCT116 CRC cells. (**A**) Representative histogram plots show the distribution of cell cycle phases in HCT116 cells treated with either 0.1% DMSO (as mock) or the three complexes, MC5, MC6, and MC7 at a concentration of 50, 12, and 25 μM, respectively for 24 h; (**B**) All of the analogues were found to induce a G1 phase arrest as compared to mock treatment. Comparison of the percentage cell population of G1, S, and G2/M phases between mock and each of the three complexes was performed by two-tailed student's *t*-test. Error bars represent the SD of two biological replicates, one of which is depicted in (**A**); (**C**) p21 mRNA levels are up-regulated in response to 24 h of treatments, analyzed by qRT-PCR. Relative expression was calculated using the ΔΔ Ct method where the Ct values of p21 were normalized to those of the housekeeping gene (vinculin). Lower and upper ends of the bars indicate the minimum and maximum values, respectively, and the "+" in the middle represents the mean. Error bars ± SD; *n* = 4; (**D**) p21 protein levels upon 24 h of treatment with the three complexes at the indicated concentrations, determined by immunoblotting; (**E**) Densitometric quantification of p21 bands normalized to those of the loading control (vinculin). Error bars indicate the SEM of two biological replicates, one of which is presented in (**D**). Multiple comparisons were made using one-way ANOVA test and a post-hoc Tukey test. *, **, ***, and **** denote *p*-values less than or equal to 0.05, 0.01, 0.001, and 0.0001, respectively.

2.3. Intracellular- and Mitochondrial ROS Levels of HCT116 Cells Are Differentially Induced by the Three Naphthalimide-NHC Complexes

The involvement of excessive ROS generation has been repeatedly mentioned in the anti-cancer mechanisms of organometallic drugs, including that of metal NHC complexes [18]. We therefore sought to evaluate the influence of the three compounds on intracellular ROS formation in HCT116 cells using dihydroethidium (DHE) staining. After 24 h of treatment with various concentrations of each compound, we found that all of the complexes produced a modest but consistent increase in ROS levels, which occurred concentration-dependently (Figure 3A,B). When comparing the three molecules, the highest fold change (1.7) was observed after treatment with 50 μM of the metal-free ligand (MC5), followed by Ru(II) (MC6) and Rh(I) (MC7) compounds which caused a 1.2-fold increase in ROS levels at the highest used concentrations, 12 and 50 μM, respectively (Figure 3A). Pre-treatment with the ROS scavengers, *N*-acetyl-L-cysteine (NAC) and reduced glutathione (GSH) clearly prevented the

MC6-triggered ROS production (Figure 3C). To gain further insight into the source of ROS, we sought to analyze the role of mitochondrial respiration and included co-treatment with either the complex I inhibitor, rotenone, or the uncoupling agent, CCCP. As shown in Figure 3C, blocking complex I was found to significantly decrease MC6-induced ROS, whereas co-treatment with CCCP (2.5 µM, 2 h) caused a slight increase in the total cellular ROS levels.

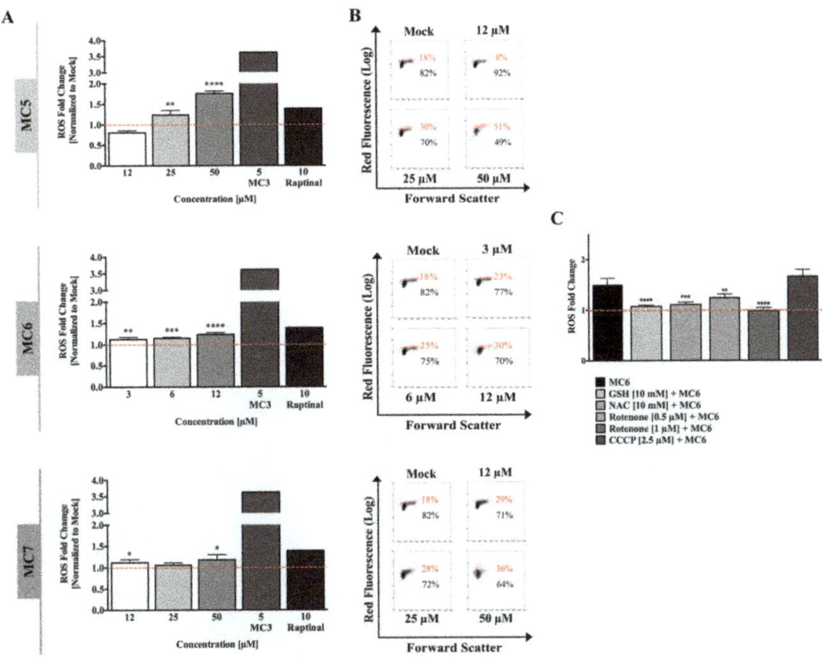

Figure 3. Total cellular reactive oxygen species (ROS) levels are moderately increased by naphthalimide-NHC analogues. (**A**) HCT116 cells were treated with various concentrations of the respective compound for 24 h, after which flow cytometric analysis of ROS generation was performed using the superoxide indicator, dihydroethidium (DHE). A 24 h treatment with the gold(I) NHC complex, MC3 [13] as well as the rapid apoptosis inducer, raptinal [16] was included as positive control. Cellular ROS levels were found to be concentration-dependently induced in response to all the three complexes, with the metal-free ligand (MC5) showing the highest induction. Data were normalized to mock (0.1% DMSO) treatment. Error bars ± SD; $n = 4$. Statistical significance between the respective treatment and mock was determined by two-tailed student's *t*-test. (**B**) Representative density plots of one out of four biological replicates shown in (**A**); (**C**) ROS levels induced by MC6 (6 µM, 24 h) were found to be significantly decreased in HCT116 cells pre-treated for 1 h with the anti-oxidants, *N*-acetyl-L-cysteine (NAC) and glutathione (GSH), as well as the mitochondrial complex I inhibitor, rotenone at the concentrations indicated. A 2 h co-treatment with the mitochondrial uncoupling reagent (CCCP), and Ru(II) complex (MC6) caused a mild increase in the latter's effects on ROS generation stained by DHE. Data were normalized to the values of mock (0.1% DMSO) as well as the corresponding single treatments. Statistical significance between MC6 in the absence/presence of each of the inhibitors was determined by two-tailed student's *t*-test. *, **, ***, and **** represent *p*-values less than or equal to 0.05, 0.01, 0.001, and 0.0001, respectively.

One of the main sources of intracellular ROS is mitochondria, known as mitochondrial ROS (mtROS), which are produced in the form of superoxide anions ($O_2{}^-$) as a by-product of oxidative metabolism [19]. To obtain further insight into the ROS inducing ability of naphthalimide-NHC conjugates, we performed a live cell analysis of MitoSox Red staining. In contrast to intracellular ROS levels, mtROS production was found to have the highest induction with the Ru(II) complex (MC6) at all of the tested time points, followed by MC5, and finally the Rh(I) analogue (Figures 4A,B and S3). As early as 3 h, mtROS were elevated up to 3.1-fold upon treatment with 12 µM of MC6 and reached the maximum after 12 h (5.4-fold), followed by a slight decrease at 24 h (Figure 4B). A similar decrease could be also observed with a concentration of 50 µM of MC5 (Figure 4B). Such a reduction after long-term treatments or higher concentrations might be due to the activation of anti-oxidant defense mechanisms by cancer cells, as previously reported for gold(I) NHC complexes [13].

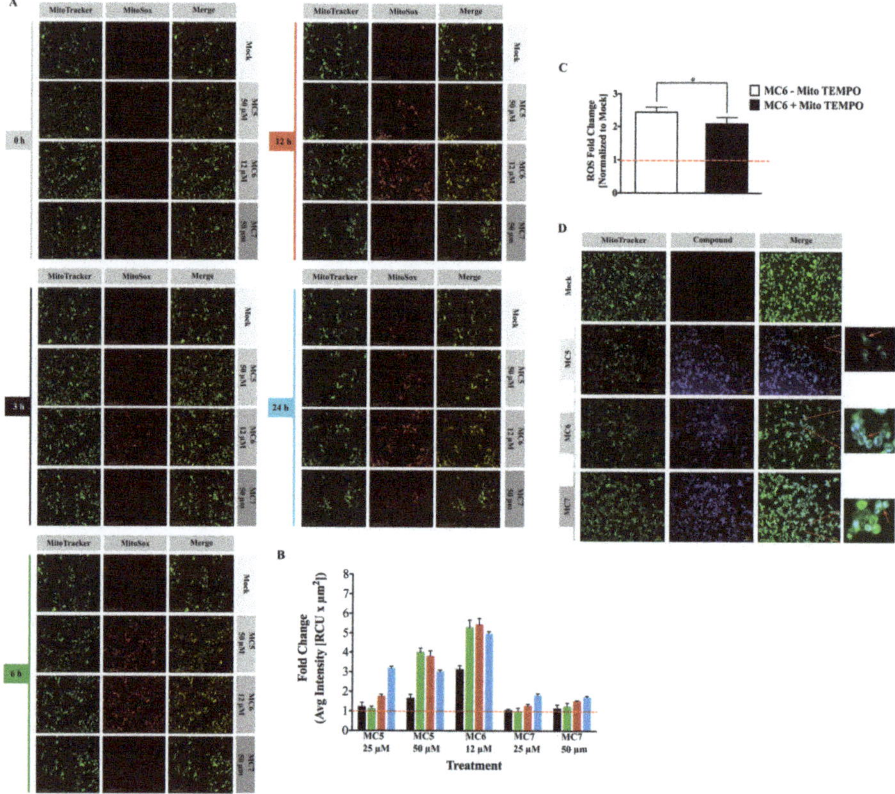

Figure 4. Mitochondrial ROS (mtROS) generation is strongly influenced by naphthalimide-NHC analogues. (**A**) Live cell imaging of mitochondrial superoxide generation stained with MitoSox Red and associated quantification (**B**) as described in the methods' section. MitoTracker Green was used to indicate mitochondria; (**C**) The mitochondria-targeted anti-oxidant, Mito TEMPO, attenuated the ROS induced by 12 µM of MC6, determined by flow cytometric analysis of MitoSox Red staining. Mito TEMPO (10 µM) was pre-incubated with HCT116 cells 2 h before the exposure to MC6 for 24 h. Data are shown as mean ± SD of three biological replicates. Comparison of ROS fold change between the two groups was performed by two-tailed student's *t*-test where a *p*-value less than or equal to 0.05 is denoted by *; (**D**) Fluorescence micrographs showing mitochondrial localization of the three complexes in HCT116 cells upon treatment with MC5 (50 µM), MC6 (12 µM), and MC7 (50 µM) for 4 h. Mitochondria were stained by MitoTracker Green. Scale bar: 40 µm. 0.1% DMSO was used as mock.

To further elucidate the source of ROS, Mito TEMPO, which is a mitochondria-specific ROS scavenger, was pre-incubated with the cells 2 h before MC6 treatment for 24 h. As shown in Figure 4C, anti-oxidant treatment was capable of reducing mitochondrial superoxide levels induced by 12 µM of MC6.

The rich photophysical properties of naphthalimides make them useful tools for monitoring their uptake and localization in living cells [2]. Using fluorescence microscopy, we detected a clear mitochondrial accumulation of all the three compounds in HCT116 cells after 4 h of treatment (Figure 4D). This may explain the stronger effect of the compounds on mtROS generation rather than that of cytosolic ROS.

2.4. Naphthalimide-NHC Derivatives Impact Mitochondrial Membrane Potential (MMP) in Different Ways

MtROS production is determined by a number of factors, one of which is MMP ($\Delta\psi$m) [20]. Flowcytometric analysis of $\Delta\psi$m in HCT116 cells revealed that, among the three complexes, MC5 and MC6 have the lowest and the highest potentials at all the tested time points, respectively (Figures S4 and 5A). It has been proposed by a "redox-optimized ROS balance hypothesis" that physiological ROS signaling occurs at optimized MMP levels, whereas oxidative stress can happen at either extreme (low or high) of $\Delta\psi$m [21]. This is in line with our results, showing the highest mtROS production in case of MC6 and MC5, which exhibit the most oxidized and reduced redox potentials, respectively. Accordingly, MC7, whose MMP does not move far from the basal levels (Figures S4 and 5A), shows only a moderate increase in mtROS generation, as compared to those of the other two compounds (Figures S4 and 5A).

Bcl-xL is known to govern the integrity of the mitochondrial outer membrane through protecting it from Bax-induced permeabilization, which leads to the release of cytochrome *c* and activation of caspases [22]. In this regard, we observed a concentration-dependent decrease in the protein expression of the pro-survival Bcl-2 member, Bcl-xL after 24 h of treatment, with MC6 and MC7 showing the highest and lowest reduction, respectively (Figure 5C,D). This was in parallel to the transcriptional activation of the apoptogenic factors, *Bax* and *Bad* (Figure 5B). The pro-apoptotic function of Bad is known to be mediated via its interaction with Bcl-2/Bcl-xL, which neutralizes the pro-survival activity of the latter proteins, thereby sensitizing cells to apoptosis [23]. In support of this, we observed elevated Bad protein levels upon 24 h of treatment with all three compounds at the indicated concentrations (Figure 5C,D). Of note, it has been repeatedly shown that only the unphosphorylated Bad is able to heterodimerize with Bcl2/Bcl-xL and that phosphorylation of the protein at either of the three serine residues, S112, S136, and S155 sequesters Bad away from mitochondrial membrane [23]. We therefore evaluated Bad phosphorylation status in response to the compounds and observed a reduction in two of the aforementioned phosphorylation sites (S112 and S136), suggesting the potential role of Bad in promoting cell death in response to the three analogues (Figure 5C,D). Taken together, all the above findings demonstrate the involvement of ROS and mitochondrial death pathway in the anti-tumor effects of the naphthalimide-NHC conjugates.

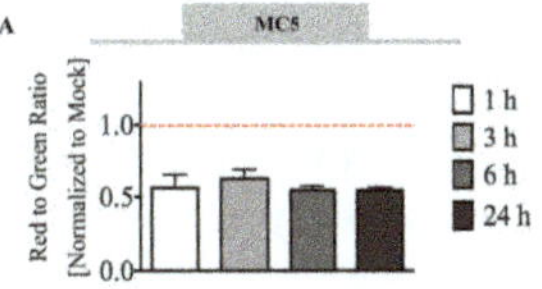
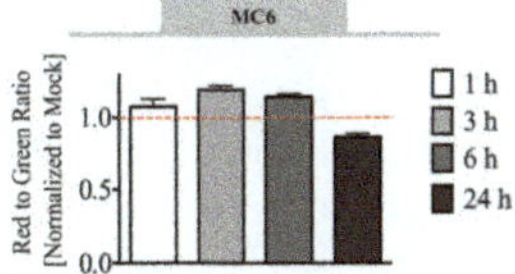
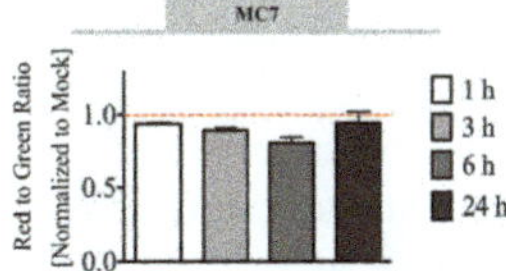

Figure 5. *Cont.*

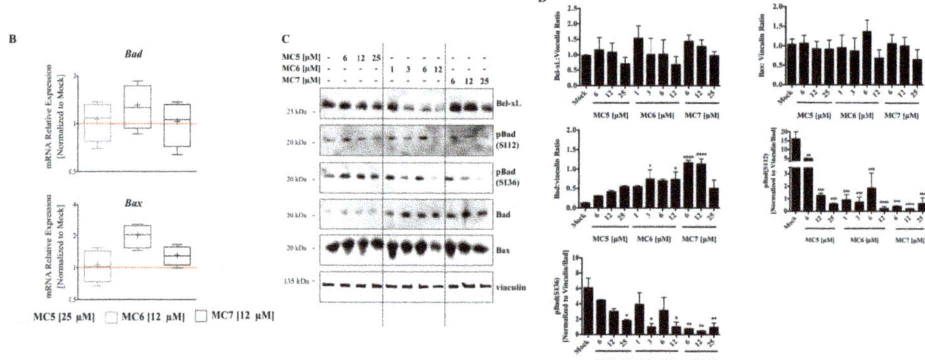

Figure 5. Naphthalimide-NHC analogues reduce, induce and have no clear effect on the Mitochondrial Membrane Potential (MMP) ($\Delta\psi$m) in HCT116 cells. (**A**) MMP was time-dependently decreased by MC5 (25 μM); it was increased in the earlier time points by MC6 (12 μM) followed by a decrease at 24 h; and was found to be mostly unaltered in response to MC7 (25 μM). Error bars are the SD of three biological replicates; (**B**) mRNA expression analysis of the pro-apoptotic Bcl-2 family members, *Bad* and *Bax*, after 24 h of treatment with the compounds at indicated concentrations. Relative expression was calculated by the $\Delta\Delta$ Ct method where the Ct values of the target genes were normalized to those of vinculin. Lower and upper ends of the bars denote the minimum and maximum values, respectively and "+" in the middle represents the mean. Error bars $\pm$ SD; n = 4; (**C**) 24 h of treatment with the respective compound led to a concentration-dependent decrease and increase in Bcl-xL and Bad protein levels, respectively, while it had no clear effect on Bax protein expression. Additionally, different phosphorylated forms of Bad were found to decrease in response to treatment; (**D**) Densitometric quantification of Bcl-2 family members normalized to the respective loading control (vinculin). The values of phosphorylated Bad were normalized to those of vinculin as well as total protein levels. Data are expressed as mean $\pm$ SEM of two independent experiments, one of which is presented in (**C**). Statistical comparisons were made between mock (0.1% DMSO), and the respective treatment using two-tailed student's *t*-test. *, **, ***, and **** represent *p*-values less than or equal to 0.05, 0.01, 0.001, and 0.0001, respectively.

2.5. p38 MAPK Signaling Is Activated in Response to Naphthalimide-NHC Complexes, with the Ru(II) Analogue Exhibiting the Strongest Effect

Several reports have demonstrated the activation of p38 signaling by a number of organometallic drugs, including gold(I) [13]- and Rh(I) [24] NHC complexes. MtROS generation, on the other hand, has been repeatedly associated with MAPK activation, leading to inflammatory responses, apoptosis, and autophagy [20]. We thus evaluated the involvement of this signaling molecule in the mode of action of naphthalimide-NHC analogues in HCT116 cells. After 24 h of treatments, the levels of phospho-p38 MAPK (pp38 MAPK; T180/Y182) were clearly increased across the three compounds, with the Ru(II) analogue (MC6) showing the highest fold change (~7) at a concentration of 12 μM (Figure 6A,B). The activation of p38 signaling by the Rh(I) and Ru(II) complexes was further confirmed as its downstream genes, *ATF2* and *Stat1* were found to be transcriptionally activated after 24 h of treatment with the respective concentrations of the compounds (Figure 6C). A time-dependent analysis of pp38 MAPK (T180/Y182) protein levels upon treatment with 12 μM of MC6 showed that p38 activation occurs as early as 1 h and it persists over the test period of 24 h (Figure 6D,E). Additionally, we observed a rapid and consistent activation of ATF2 in response to MC6, as determined by increased levels of its phosphorylated form (pATF2; T71) (Figure 6D,E).

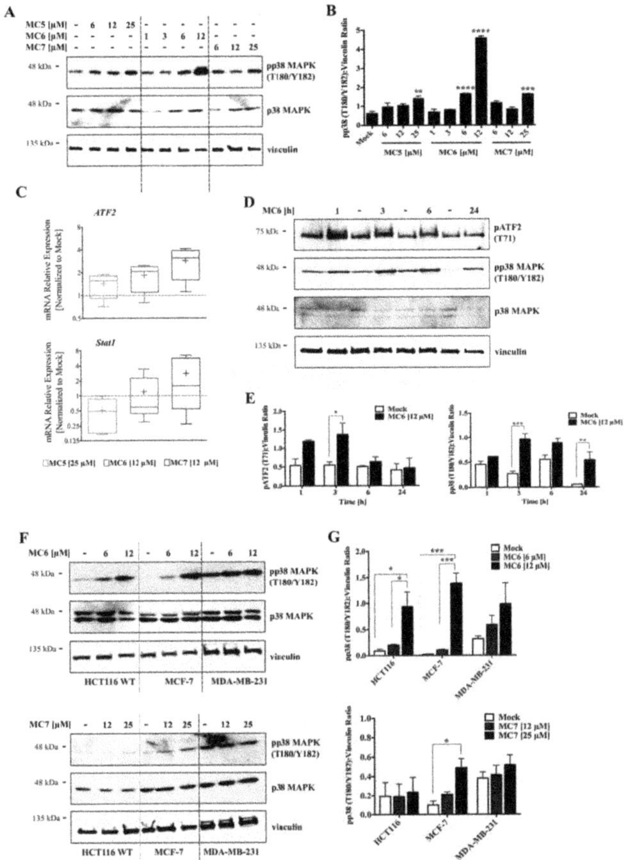

Figure 6. Naphthalimide-NHC derivatives activate the p38 pathway in human breast- and colon cancer cell lines. (**A**) Immunoblots as well as densitometric quantification (**B**) showing the accumulation of pp38 mitogen activated protein kinase (MAPK) (T180/Y182) protein levels by the three complexes in HCT116 CRC cells upon 24 h treatment with increasing concentrations of the respective compound as indicated. The induction appeared to be more profound in case of the Ru(II) analogue, MC6. Data in (**B**) are presented as mean ± SEM of three independent experiments, one of those is shown in (**A**); (**C**) qRT-PCR analysis of p38-associated signaling molecules, *ATF2* and *Stat1* in HCT116 cells treated with the compounds at indicated concentrations for 24 h. Lower and upper ends of the bars denote the minimum and maximum values, respectively, with the "+" sign representing the mean of four biological replicates. Error bars ± SD; (**D**) Time-course analyses of pp38 MAPK (T180/Y182) as well as its down-stream effector, pATF2 (T71) in HCT116 cells treated with 12 μM of MC6, determined by immunoblotting; (**E**) Densitometric analyses of pATF2 (T71) and pp38 (T180/Y182) bands obtained from three independent experiments, one of which is depicted in (**D**). Error bars ± SEM; (**F**) Regulation of p38 MAPK signaling was compared among HCT116, MCF-7 and MDA-MB-231 cancer cell lines treated for 24 h with the metal-containing analogues (MC6 and MC7). In case of MDA-MB-231 where the basal levels of pp38 (T180/Y182) are high, treatments did not profoundly impact the molecule's phosphorylation; (**G**) Densitometric quantifications illustrate no significant change in pp38 (T180/Y182) levels in MDA-MB-231 cells. Error bars ± SEM; n = 3. Statistical comparisons were made between mock (0.1% DMSO) and the respective treatment using two-tailed student's *t*-test. *p*-values less than or equal to 0.05, 0.01, 0.001, and 0.0001 are indicated as *, **, ***, and ****, respectively.

Next, we aimed to assess whether p38 activation is reproducible in cancer cell lines other than HCT116 CRC cells. As shown in Figure 6F,G, MC6 and MC7 were able to induce the levels of phospho-p38 MAPK (pp38 MAPK(T180/Y182)) in the breast cancer cell lines, MCF-7 and MDA-MB-231, however with a much lower efficiency in the case of the latter. With regards to this, it has been reported that breast cancer cell lines with a triple-negative (ER negative, PR negative and HER-2 negative) molecular profile exhibit enhanced expression and activity of p38 MAPK, which has been correlated with poor prognosis and survival in patients [25]. Thus, the minor effect on p38 in MDA-MB-231 is most likely due to the endogenous higher activity. In this context, p38 may act as a tumorigenic factor rather than a tumor suppressor. Therefore, its inhibition rather than its activation is shown to exert anti-proliferative effects in invasive breast cancer models [26].

p38 and other members of the MAPK family of proteins, ERK and JNK, have been shown to crosstalk at several levels, determining whether the cell survives or undergoes apoptosis in response to chemotherapeutic drugs [27,28]. We therefore evaluated the phosphorylation status of ERK and JNK using immunoblotting, as well as protein ELISA-microarray analysis [29]. We observed a mild inhibitory effect of the compounds, in particular MC6 and MC7, on the phospho-activation of ERK (Figure S5A–C), however the difference was not found to be statistically significant. With regards to JNK, 24 h treatment with increasing concentrations of the compounds did not show a clear alteration in the phosphorylated protein levels (Figure S5C), which most probably rules out the involvement of JNK signaling in the pro-apoptotic response of HCT116 cells to naphthalimide-NHC analogues.

2.6. Anti-Proliferative and Pro-Apoptotic Effects of the Ru(II) Naphthalimide-NHC Complex in HCT116 Cells Proceed via p38 MAPK Signaling, Involving ROS

To further elucidate the role of p38 in the anti-tumor properties of MC6, which triggered the strongest p38 activation across the three complexes (Figure 6A,B), we knocked-down the expression of *p38α* (*MAPK14*) in HCT116 cells while using siRNA (Figure 7D). As determined by SRB cytotoxicity assay, upon 24 h treatment with 12 µM of MC6, cellular viability was significantly increased in cells that were transfected with anti-*p38α* siRNA as compared to that of the negative control (siRNA NC) and non-transfected HCT116 cells (Figure 7A). We then aimed at investigating the influence of p38 knock-down on the MC6-mediated apoptosis using annexinV/propidium iodide (AV/PI) staining. As depicted in Figure 7B,C, 24 h treatment with 12 µM of MC6 resulted in the transition of cells through the early apoptotic (AV$^+$/PI$^-$) quadrant in HCT116 and HCT116 siRNA NC, as determined by ~1.8- and ~1.4-fold increase, respectively. This effect was found to be attenuated in response to the p38 knock-down, as the fold change of early apoptotic- as well as late apoptotic/necrotic (AV$^+$/PI$^+$) cells were significantly less than that of HCT116 and cells transfected with the non-targeting siRNA (Figure 7B). Similar results were obtained in the presence of two well-known chemical inhibitors of p38α, VX-702, and Ralimetinib. As shown in Figure 7E,F, the number of apoptotic cells upon 24 h of MC6 treatment was significantly decreased when combined with p38α inhibitors, as detected by AV/PI staining. Additionally, the pro-apoptotic response of HCT116 cells to MC6 was evaluated in the absence/presence of p38α activity using flow cytometric analysis of caspase 3 activation where significantly less cleaved caspase 3 levels were observed upon both knock-down of p38α (Figure 7G), as well as its pharmacological inhibition (Figure 7H,I). Moreover, we monitored the MC6-induced apoptosis using TUNEL assay and found significantly more apoptotic cells with 24 h of MC6 treatment, an effect that was clearly abrogated upon VX-702-mediated p38α inhibition (Figure 7J,K).

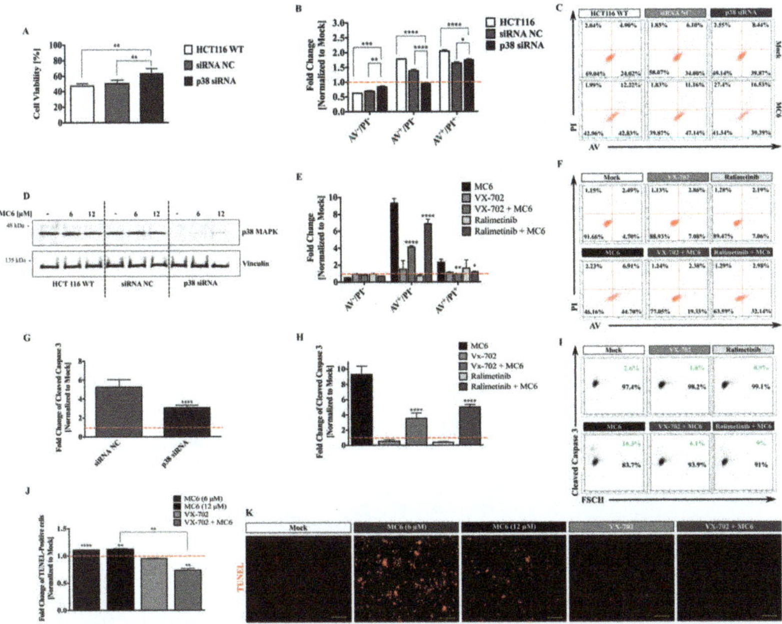

Figure 7. p38 signaling appears to be crucial for the MC6-mediated cytotoxic- and pro-apoptotic effects. (**A**) siRNA-mediated repression of *p38α* was found to hamper the growth inhibitory effects of MC6 in HCT116 cells treated with 12 µM of the compound for 24 h, measured by SRB assay. Percentage cell viability was calculated by normalizing the values of MC6-treated cells to those of the corresponding mock (0.1% DMSO) treatments. Error bars ± SD; *n* = 3; (**B**) Knock-down of *p38α* attenuates the pro-apoptotic response to MC6 (12 µM, 24 h), assessed by flow cytometric analysis of AV/PI staining. Percentage cell population in each quadrant was normalized to the respective mock (0.1% DMSO) treatment. Error bars ± SD. Multiple comparisons were performed using two-way ANOVA followed by a post-hoc Tukey test; (**C**) Density plots representative of three biological replicates illustrate increased population of AV$^+$/PI$^-$ and AV$^+$/PI$^+$ with treatment, however, to a lesser extent in case of cells transfected with anti-*p38α* siRNA; (**D**) Confirmation of knock-down efficiency, as determined by immunoblotting; (**E**) chemical inhibition of p38α abrogates the MC6-mediated apoptosis. HCT116 cells were treated with 12 µM of MC6 in the absence/presence of p38α inhibitors, VX-702 and Ralimetinib at a concentration of 0.5 µM for 24 h. Error bars ± SD, *n* = 3. Asterisks show significance in the amount of early- and late apoptotic population between cells treated with MC6 and each of the two inhibitors, and MC6 as a single agent, determined by two-tailed student's *t*-test; (**F**) Representative density plots of one out of three biological replicates demonstrate reduced number of AV$^+$/PI$^-$ and AV$^+$/PI$^+$ cells when p38α activity is inhibited; (**G**) Flow cytometric analysis of caspase 3 activation shows significantly less cleaved caspase 3 expression in cells transfected with anti-*p38α* siRNA as compared to that of the negative control. Error bars ± SD, *n* = 6. Statistical significance between the two groups was made using two-tailed student's *t*-test; (**H**) Chemical inhibition of p38α was found to decrease the levels of active caspase 3 in a similar manner to that of *p38α* knock-down. Error bars ± SD, *n* = 6. Statistical comparison was performed between combination treatments and MC6 using two-tailed student's *t*-test; (**I**) Representative density plots of one out of six biological replicates; (**J**) Detection of apoptotic cells using TUNEL assay. HCT116 cells were treated with the indicated concentrations of MC6 for 24 h in the absence/presence of VX-702 (0.5 µM). Statistical significance was calculated between mock and the respective treatment as well as MC6 as single agent and in combination with VX-702 using two-tailed student's *t*-test; (**K**) Representative fluorescence images of TUNEL reaction. Scale bar: 100 µm. *p*-values less than or equal to 0.05, 0.01, 0.001, and 0.0001 are denoted as *, **, ***, and ****, respectively.

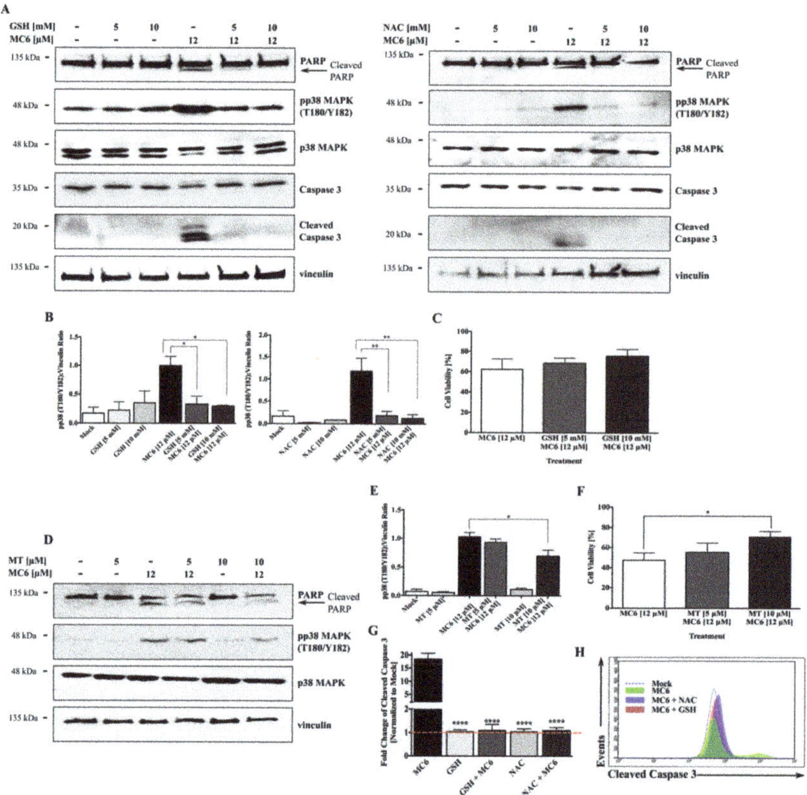

Figure 8. MC6-induced cytotoxicity and pro-apoptotic effects are mediated via the ROS-p38 signaling axis. (**A**) Treatment with GSH and NAC at the indicated concentrations 1 h prior to the addition of MC6 (12 μM) for 24 h blocked the activation of p38 as well as cleavages of caspase 3 and PARP, detected by immunoblotting; (**B**) Densitometric analyses show a significant reduction in the MC6-induced pp38 (T180/Y182) accumulation in the presence of anti-oxidants. Error bars indicate the SEM of three independent experiments, one of those is presented in (**A**); (**C**) Increased cellular survival of HCT116 cells pre-treated with either 5 or 10 mM of GSH 1 h before the addition of MC6 (12 μM) for 24 h, as determined by SRB assay. Data represent mean ± SD of three biological replicates, normalized to mock (0.1% DMSO) and the respective GSH treatment; (**D–F**) HCT116 cells pre-incubated with either 5 or 10 μM of Mito TEMPO (MT) for 2 h were treated with MC6 (12 μM) for 24 h. The mitochondria-targeted ROS scavenger was found to attenuate the MC6-mediated p38 activation as well as PARP cleavage at the highest used concentration (10 μM), as detected by immunoblotting (**D**) and the associated densitometric quantification (**E**), obtained from three independent experiments. Error bars ± SEM. Additionally, it rescued the MC6-mediated cytotoxic effects, as determined by SRB assay (**F**). Percentage cell viability of MC6-treated cells was normalized to mock (0.1% DMSO) and the respective MT treatment. Error bars ± SD; $n = 3$; (**G**) Flow cytometric analysis of caspase 3 activation illustrates significantly lower levels of the cleaved form of the protein in cells pre-treated for 1 h with either GSH (10 mM) or NAC (10 mM) as compared to that of MC6-treated cells (12 μM, 24 h). Error bars ± SD, $n = 3$; (**H**) Representative histogram of one out of three biological replicates presented in (**G**) demonstrates a left-ward shift in caspase 3 activity in the presence of ROS scavengers. Statistical significance between the MC6-treated cells in the absence/presence of anti-oxidants was calculated using two-tailed student's *t*-test. *, **, ***, and **** on the figures represent *p*-values that are less than or equal to 0.05, 0.01, 0.001, and 0.0001, respectively.

Activation of the pro-apoptotic p38 MAPK signaling can happen as result of ROS generation [19], and it has been previously implicated in the anti-cancer mechanism of metal NHC complexes [13]. We hence sought to investigate whether pp38 MAPK induction and subsequent growth inhibitory/pro-apoptotic effects of MC6 are due the elevated ROS levels. As shown in Figure 8A,B, pre-incubation with the ROS scavengers, reduced GSH as well as NAC, blocked the MC6-induced p38 activation as well as apoptosis, determined by the absence of PARP and caspase 3 cleavages, markers of apoptotic cell death. With respect to MC6-mediated cytotoxicity, we observed increased cellular survival percentages with GSH treatment, however, the difference failed to reach statistical significance (Figure 8C). Similarly, pre-treatment with a mitochondria-targeted scavenger, Mito TEMPO, hampered the effects of MC6 on p38 activation and PARP cleavage (Figure 8D,E), as well as its cytotoxicity (Figure 8F) in HCT116 cells. We further confirmed the protective effect of anti-oxidant treatment on MC6-mediated apoptosis by quantification of cleaved caspase 3 levels using flow cytometry. As shown in Figure 8G,H, pre-treatment of HCT116 cells with GSH and NAC rescued the MC6-mediated caspase 3 cleavage. In view of these findings, we propose that the in vitro anti-tumor activity of MC6 in HCT116 CRC cells may be regulated through mtROS-induced activation of the p38 MAPK pathway.

3. Discussion

Regulation of the redox state plays an important role in tumor cell survival. Although elevated ROS levels allow cancer cells to promote pro-tumorigenic signaling, excessive ROS production is usually associated with anti-tumorigenic pathways, which can trigger oxidative stress-induced cancer cell death [19]. The latter effects of ROS are mainly mediated through the ASK1/JNK and ASK1/p38 axes, leading to cell cycle arrest, growth inhibition, and apoptosis [19]. Here, we demonstrate that a novel Ru(II) naphthalimide-NHC complex is able of causing a remarkable increase in mtROS generation, which in turn activates the pro-apoptotic p38 signaling in HCT116 CRC cells.

The p38 pathway is a major regulator of stress responses, influencing various biological processes, including cellular proliferation and survival [8]. Studies have demonstrated that the role of p38 MAPK signaling in cancer therapy is contextual, depending on the nature of the stimuli, cancer type, as well as the status of other MAPKs (ERKs and JNKs) [30,31]. On the one hand, p38 MAPK activation mediates the sensitivity of tumor cells to a variety of chemotherapeutics, for example, cisplatin [9–11,32], oxaliplatin [33], and auranofin [12]. In particular, cisplatin has been reported to induce apoptosis in HCT116 CRC cells through the ROS-p38α axis downstream of p53 activation [10]. On the other hand, cancer cell lines with a high basal expression of pp38 MAPK tend to lose the tumor-suppressing functions of this molecule, possibly because of the inability to further activate p38 MAPK in response to anti-cancer treatments [26]. In light of this, we detected a clear induction of pp38 MAPK (T180/Y182) levels in response to all three naphthalimide-NHC derivatives in HCT116 and MCF-7 cells, with the Ru(II) analogue (MC6) exhibiting the strongest effect. However, MC6 was hardly able to promote p38 activation in the triple-negative MDA-MB-231 breast cancer cells in which the basal levels of phospho-p38 are elevated (Figure 6F,G). This is in line with a reduced cytotoxic response of MDA-MB-231 as compared with that observed for HCT116 and MCF-7 (Figure 1B), suggesting that the lack of p38 activation renders this cell line resistant to the anti-proliferative effects of naphthalimide-NHC conjugates. In this regard, several reports have shown the beneficial outcome of p38 MAPK inhibition rather than its activation for the treatment of invasive breast cancer models. For example, it has been proposed that p38 MAPK inhibition has synergistic effects with cisplatin for the treatment of breast cancer through the activation of ROS-mediated JNK signaling [27]. Our results suggest an anti-tumorigenic role for p38 signaling in response to MC6 as its genetic as well as chemical inhibition attenuates the cytotoxic- and pro-apoptotic effects of the compound in HCT116 cells (Figure 7). However, this requires further investigation in other cellular contexts, in particular, in cancer cells with enhanced basal p38 MAPK phosphorylation. To further address whether other MAPKs, including JNK and/or ERK signaling, are implicated in these processes, we also analyzed the regulatory state of these kinases in response to treatment with the three naphthalimide-NHC derivatives, as, for instance, JNK activation and/or ERK inhibition, which has been previously reported in the anti-cancer efficacy of several other organometallic drugs [27,34,35]. We, however, were not able to observe a profound regulation of MAPKs other than p38α in response to the naphthalimide-NHC analogues (Figure S5), further demonstrating the important role of p38α activation in sensitizing tumor cells to MC6-mediated apoptosis.

In this study, we show an essential role for mtROS generation in the cytotoxicity of naphthalimide-NHC analogues. There are two main sources of the signaling-associated ROS in cells: (i) membrane bound-NADPH oxidases (known as NOX enzymes) and (ii) mitochondrial electron transport chain (ETC) [19]. Our results demonstrate that intracellular ROS levels are minimally affected by the naphthalimide-NHC derivatives, whereas mitochondrial superoxide production is markedly increased in response to the molecules, with the Ru(II) complex showing the highest fold change. MtROS have been frequently mentioned in the regulation of pro-inflammatory/apoptotic responses via multiple mechanisms, among others, is the activation of MAPKs [20]. In view of this, we observed that treatment with a mitochondria-targeted ROS scavenger and MC6 blocks the latter's effect on p38 MAPK activity (Figure 8D,E), suggesting that mtROS generation may be acting upstream of MC6-mediated p38 induction. It is known that mitochondrial superoxide species sustain MAPK activity through the oxidation and inactivation of MAPK phosphatases (MKPs) [36]. In line with this, we found reduced MKP6 expression in response to the naphthalimide-NHC conjugates (data not shown), further supporting the involvement of mtROS in the induction of p38 signaling. Importantly, the cytotoxic and pro-apoptotic effects of MC6 were hampered upon general- and mitochondria-specific anti-oxidant treatments (Figure 8), implicating a ROS-mediated pathway underlying the in vitro anti-tumor efficacy of the complex.

One suggested mode of action of metal NHC complexes and naphthalimide derivatives is via mitochondrial accumulation, and perturbations in the MMP ($\Delta\psi$m) [13,37,38]. We here report that $\Delta\psi$m is differentially regulated by the three naphthalimide-NHC conjugates, showing all possible options; a timely decrease in case of MC5, a transient increase in case of MC6 followed by a decrease at 24 h, and no significant change in case of MC7. With regards to the relationship between mitochondria-driven ROS and $\Delta\psi$m, the general concept is that more polarized membrane (high $\Delta\psi$m) is associated with greater ROS production [20]. This is consistent with the observation of MC6 showing the highest MMP along with the highest mtROS generation as compared to the other two complexes. However, it fails to explain the influence of MC5 treatment on mitochondrial parameters, wherein, despite reduced $\Delta\psi$m, mtROS production increases. These disparate observations reconcile by a "redox-optimized ROS balance" model proposed by Aon et al. who demonstrated that oxidative stress can occur at either extreme of MMP (high $\Delta\psi$m or low $\Delta\psi$m), meaning that under oxidizing conditions mtROS can increase because of the compromised anti-oxidant defense mechanisms [21]. Taking all the mentioned observations into consideration, we suggest that the mode of anti-cancer activity of naphthalimide-NHC compounds is most likely through mitochondrial localization, leading to ROS-mediated activation of the pro-apoptotic p38 MAPK pathway.

4. Materials and Methods

4.1. Chemicals and Antibodies

1-(3'-(4"-ethylthio-1",8"-naphthalimid-*N*"-yl))-propyl-3-benzylimdazolium bromide (MC5), Dichloro[1-(3'-(4"-ethylthio-1",1"-naphthalimid-*N*"-yl))-propyl-3-benzyl-imidazol-2-ylidene] (η^6-*p*-cymene)ruthenium(II) (MC6), and Chlorido[(η^2, η^2-cycloocta-1,5-diene)-1-(3'-(4"-ethylthio-1", 8"-naphthalimid-*N*"-yl))-propyl-3-benzyl imidazol-2-ylidene] rhodium(I) (MC7) were synthesized and purified, as previously described [6,15]. Raptinal, antimycin, GSH, NAC, Mito TEMPO, rotenone, CCCP, and JC-1 were purchased from Sigma-Aldrich (Steinheim, Germany). Pharmacological inhibitors of p38α, VX-702 and Ralimetinib (LY2228820) were from selleckchem (Munich, Germany) and DHE was from Biomol GmbH (Hamburg, Germany). MitoTracker Green, MitoSox Red, and the transfection reagent, Lipofectamine 3000 were obtained from Thermo Fischer (Darmstadt, Germany). Primary antibodies against Bcl-xL (#2764), pATF2 (T71; #9221), pp38 MAPK (T180, Y182; #9211), PARP (#9542), Bax (#5023), phospho-Bad sampler kit (#9105), caspase 3 (#9662), cleaved caspase 3 (#9661), pERK (T202, Y204; #5683), pStat1 (Y701; #7649), pStat1 (S727; #8826), pStat3 (Y705; #9145), pStat3 (S727; #9134), Stat1 (#9176), Stat3 (#9139), as well as anti-mouse and rabbit IgG horseradish peroxidase (HRP)-linked antibodies were purchased form Cell Signaling Technologies. Anti-vinculin (SC-73614), anti-p38 (SC-7972), and anti-ERK (SC-135900) were from Santa Cruz Biotechnology.

4.2. Cell Culture

Human breast- (MCF-7, MDA-MB-231) and colon (HCT116 WT, HCT116 p53-/-, HT-29) cancer cell lines, as well as HFF were maintained in Dulbecco's modified Eagle medium (DMEM)-GlutaMax (Gibco, Darmstadt, Germany) supplemented with 10% (v/v) FCS (Gibco) and 1% penicillin/streptomycin (v/v) (Gibco) and they were incubated under 5% CO_2 and at 37 °C. All of the treatments were performed in the same media at the indicated conditions.

4.3. Cell Viability Assay

The SRB assay was employed to measure the inhibitory effects of MC5, MC6, and MC7 on the proliferation of HCT116 WT, HCT116 p53-/-, HT-29, MCF-7, and MDA-MB-231, as well as the influence of anti-oxidants (GSH and Mito TEMPO) and p38 knock-down on the toxicity of MC6-treated HCT116 CRC cells. Cells were seeded in either 96-well plates or 24-well plates at a density of 10,000- and 60,000 cells/well, respectively. At the end of treatments, plates were fixed with 10% ice-cold trichloroacetic acid (TCA), followed by incubation at 4 °C for at least one hour. Afterwards, well contents were washed three times with water and were dried at 60 °C. Next, 0.054% SRB sodium salt was added to the wells and incubated at room temperature for 30 mins. Plates were then washed three times with 1% acetic acid and were dried at 60 °C. Finally, the SRB dye was dissolved in 10 mM Tris (pH 10.5) and measurement was performed using the Tecan Ultra plate reader (Tecan, Crailsheim, Germany) at 535 nm absorbance wavelength.

4.4. Cell Cycle Analysis

For analysis of cell cycle phase distribution, 10^6 cells/well were harvested, washed with PBS, then fixed with 70% ice-cold ethanol, and incubated at -20 °C overnight. After twice washing with ice-cold PBS (Gibco), cells were treated with 200 µg/mL of RNase (Sigma-Aldrich) for 30 mins at 37 °C, followed by 15 min incubation with PI (50 µg/mL) (Sigma-Aldrich) in the dark for nucleic acid staining. Samples were analyzed by FACSCalibur (Becton Dickinson, Franklin Lakes, NJ, USA) and data analysis was performed using the software, CellQuest™ Pro (Becton Dickinson).

4.5. Intracellular ROS Measurement Using Flow Cytometry (FACS) and Live Cell Imaging

DHE and MitoSox Red were used for the detection of whole cell- and mitochondrial superoxide generation, respectively. For DHE staining, the HCT116 cells were seeded in 12-well plates at a density of 200,000 cells/well. After the indicated treatments, cells were incubated with phenol red-free DMEM containing 15 µM of DHE for 20 mins, harvested, and resuspended in fresh media. For MitoSox Red staining, 60,000 cells/well were seeded in 24-well plates. At the time points indicated, media were replaced with phenol red-free DMEM containing 5 µM of MitoSox Red for 15 mins. Cells were then harvested and resuspended in fresh media. FACS analysis was immediately performed using Guava easyCyte HT sampling flow cytometer (Guava Technologies, Hayward, CA, USA) and data were analyzed using GuavaSoft 3.1.1 software.

Live cell imaging of mitochondrial ROS production was performed using the Incucyte ZOOM live cell analysis system (Essen BioScience, Broadwater Road Welwyn Garden City, United Kingdom). HCT116 cells were seeded in a 24-well plate at a density of 60,000 cells/well. On the following day, the cells were first incubated with 2 nM of MitoTracker Green dissolved in FCS- and pheno red-free

DMEM for 20 mins, after which with fresh media containing 5 µM of MitoSox Red as well as the treatments. Time-lapse images were then taken every 30 mins for a period of 24 h. Data were analyzed using the Incucyte software 2016b. Briefly, the signal intensity was calculated based on a fluorescent area mask; with a top hat filter being used for excluding dead cells due to the higher auto-fluorescence. Nine pictures/well were used to determine the overall signal/well and each condition was performed in quadruplicates.

4.6. Fluorescence Microscopy

For assessing mitochondrial localization of naphthalimide-NHC analogues, HCT116 cells were cultured in 12-well plates at a density of 150,000 cells/well. After the indicated treatments, the cells were washed with PBS and were incubated with 2 nM of MitoTracker Green dissolved in FCS- and phenol red-free media for 20 mins in order to stain mitochondria. Cells were then washed with PBS and were imaged using the BIOREVO fluorescence microscope (BZ9000, KEYENCE; Neu-Isenburg, Germany).

4.7. Mitochondrial Membrane Potential Assessment

200,000 cells/well were seeded in 12-well plates. Treated cells were harvested and then incubated with phenol red-free DMEM containing 2 µM of the JC-1 dye for 15 mins in the dark. After resuspension in fresh media, FACS analysis was performed by Guava easyCyte HT sampling flow cytometer using GuavaSoft 3.1.1 software for data analysis.

4.8. RNA Isolation, Reverse Transcription, and Quantitative Real Time (qRT) PCR

At the end of treatments, total RNA was isolated using QIAzol lysis reagent (Qiagen, Hilden, Germany) and the quality of RNA samples was determined by NanoDrop 2000 UV-Vis Spectrophotometer (Thermo Scientific, Darmstadt, Germany). cDNA synthesis was performed from 500–1000 ng of total RNA using ProtoScript II first strand cDNA synthesis kit (New England Biolabs, Frankfurt am Main, Germany), according to the manufacturer's instructions. The thermal cycler q-Tower (Analytik Jena AG, Jena, Germany) was used to analyze gene expression levels. The amplification reaction solutions (5 µL) were prepared with 2.5× of ready to use master mix LightCycler® 480 SYBR Green I (Roche, Mannheim, Germany), 1× of nuclease-free H_2O and cDNA templates, as well as 1× of the following primer mixtures (Eurofins Genomics, Ebersberg, Germany): *CDKN1A* (5s: GACACCACTGGAGGGTGACT; 3as: CAGGTCCACATGGTCTTCCT), *Bax* (5s: GGG GACGAACTGGACAGTAA; 3as: CAGTTGAAGTTGCCGTCAGA), *Bad* (5s: GGTTCTGAGGGGAG ACTGAGG; 3as: GCTTCCTCTCCCACCGTAGC, *ATF2* (5 s: CAGCGTTTTACCAACGAGGA; 3as: GA ATCTTGTTGGTGTTGGGGTC), *Stat1* (5 s: GGAAAAGCAAGCGTAA TCTTCAGG; 3as: GAATATTCCCCGACTGAGCC), and *vinculin* (as reference gene) (5s-CAGTCAGACCCTTACTCA GTG-3′; 3as-CAGCCTCATCGAAGGTAAGGA).

4.9. Immunoblotting

After the respective treatments, cells were lysed using a urea-based lysis buffer supplemented with multiple phosphatase/protease inhibitors, namely sodium orthovanadate, aprotinin, PMSF, pepstatin, and sodium pyrophosphate. Protein concentration was determined with Bradford reagent (Sigma-Aldrich). 20–50 µg of total protein was separated on 10% SDS-PAGE, then transferred onto a PVDF membrane (GE Healthcare, Munich, Germany), after which it was blocked with 5% (*w/v*) milk in TBST (Tris-Buffered Saline Tween-20) for at least 1 h. Membranes were then incubated with primary antibody solutions (diluted following the manufacturer's instructions) overnight at 4 °C and were visualized by further incubation with the corresponding HRP-linked secondary antibodies (diluted at 1:5000 in 5% (*w/v*) milk TBST) for 1 h at room temperature, followed by three washing steps with TBST. Target proteins were finally detected by Western Lightning™ Plus ECl (Perkin Elmer, Waltham, USA) using the Fujifilm LAS-3000 imaging system and were quantified with ImageJ software.

4.10. Protein ELISA-Microarray Analysis

We have previously reported a detailed protocol on the assay [29]. Briefly, treated cells were lysed in a similar manner to that of immunoblotting. Proteins were then diluted 1:6 in a PBS-based buffer (containing 1 mM EDTA, 0.5% (*v/v*) Triton X-100, and 5 mM NaF). Protein concentration was assessed using the Pierce BCA Protein Assay kit (Thermo Fischer). The levels of phosphorylated ERK1/2 and JNK were quantified using sandwich ELISA microarrays that were based on the ArrayStrip platform (Alere Technologies GmbH, Jena, Germany). Signals were detected by the Arraymate reader (Alere Technology GmbH). Data were analyzed using the KOMA software [39] and were normalized to the amount of total proteins used.

4.11. siRNA-Mediated Knock Down

siRNA oligos against *p38α* (*MAPK14*) as well as the non-targeting siRNA were obtained from Thermo Fischer. HCT116 cells were transfected with 40 pmol of the respective siRNA diluted in 100 µL/well Opti-MEM Reduced Serum Medium (Gibco) and complexed with 1.5 µL/well of Lipofectamine 3000 in a 24-well plate and they were incubated for 20 mins at room temperature. Cell suspension was then added at a density of 60,000 cells/well and treatments were performed on the following day, as indicated.

4.12. Cell Death Analysis (AV/PI Staining)

AV- and PI staining were performed in parallel for the detection of early apoptotic and late apoptotic/necrotic cells, respectively. 250,000 cells/well of the non-transfected HCT116, HCT116 cells transfected with either a negative control- or anti-*p38α* siRNA, as well as HCT116 cells in the absence/presence of p38α pharmacological inhibitors were harvested, washed with AV binding buffer (BD Biosciences, Heidelberg, Germany), and then resuspended in 50 µL binding buffer containing 5 µL of both FITC-conjugated AV and PI (BD Biosciences). Samples were incubated for 15 mins in the dark, after which they were resuspended in binding buffer up to 500 µL for FACS analysis, which was done using Guava easyCyte HT sampling flow cytometer. Data were analyzed by GuavaSoft 3.1.1 software.

4.13. Caspase 3 Activation Assay

At the end of the treatments, 60,000 cells/wells were harvested, fixed, and permeabilized with 4% paraformaldehyde and 90% ice-cold methanol, respectively, as previously described [40]. Cells were then incubated overnight with cleaved caspase 3 antibody at a dilution of 1:800 and they were further incubated with secondary antibody (goat anti-rabbit Alexafluor 488-conjugated; Dianova, Hamburg, Germany) for 1 h at room temperature, thereafter analyzed using Guava easyCyte HT sampling flow cytometer and the GuavaSoft 3.1.1 software.

4.14. TUNEL Assay

TUNEL assay was performed for the detection of DNA fragmentation as a characteristic of late stage apoptosis using Cell Meter TUNEL Apoptosis Assay Kit (Biomol GmbH, Hamburg, Germany). Briefly, 60,000 cells/well were harvested, fixed, and permeabilized by 4% paraformaldehyde and 0.2% Triton X-100, respectively. After washing with PBS, cells were incubated with the tunnelyte reaction mixture for 1 h at 37 °C, thereafter the fluorescence intensity was measured by the Tecan Ultra plate reader (Tecan) at an excitation/emission wavelength of 550/650 nm, respectively. Additionally, the cells were analyzed using the BIOREVO Fluorescence microscope (BZ9000, KEYENCE).

4.15. Statistical Analyses

Data were analyzed using GraphPad Prism and Microsoft Excel. Densitometric analyses were performed by the ImageJ software. Statistical significance between the respective treatment and DMSO was determined by student's *t*-test. Multiple comparisons were performed using either one-way or two-way ANOVA, followed by a post-hoc Tukey test as indicated. *p*-values less than or equal to 0.05, 0.01, 0.001, and 0.0001 are presented as *, **, ***, and *** on the figures, respectively.

5. Conclusions

Several studies have supported the therapeutic use of high ROS levels present in tumor cell lines for the induction of pro-apoptotic pathways. In this study, we show that altered cellular redox balance induced by a novel Ru(II) naphthalimide-NHC complex—MC6—plays an important role in its in vitro anti-tumor efficacy, most likely via mediating the activation of p38 MAPK signaling. Nonetheless, several issues remain to be addressed with regards to the activation of ROS-p38 axis by naphthalimide-NHC analogues, such as: (i) if the pro-apoptotic functions of MC6-induced p38 MAPK could be extrapolated to other cellular contexts, in particular, the ones with enhanced p38 expression and (ii) whether the MC6-triggered elevation in mtROS levels is through an increase in ROS production at the mitochondrial respiratory chain complexes and/or an alteration in cellular/mitochondrial anti-oxidant systems, for instance, superoxide dismutases (SODs), glutathione peroxidases (GPXs), and thioredoxins (TRXs).

Supplementary Materials: The following are available online at http://www.mdpi.com/1422-0067/19/12/3964/s1. Figure S1: Human fibroblasts are less susceptible to the cytotoxic effects of naphthalimide-NHC analogues. Figure S2: Cytotoxic efficacy of naphthalimide-NHC analogues in CRC cells harboring deficient- or mutant p53. Figure S3: MtROS is strongly influenced by naphthalimide-NHC analogues. Figure S4: Naphthalimide-NHC analogues reduce, induce and have no clear effect on the MMP ($\Delta\psi$m) in HCT116 cells. Figure S5: Naphthalimide-NHC analogues do not appear to profoundly impact the activation of ERK and JNK MAPKs.

Author Contributions: Y.D. and X.C. designed the study. I.O. and W.S. designed and synthesized the complexes. Y.D., A.S., J.T., and B.B. performed experiments. Y.D., A.S., X.C. and S.W. analyzed and interpreted the data. Y.D. and X.C. wrote the manuscript. All authors revised and approved the final version of the manuscript prior to submission.

Funding: Y.D. is supported by the Landesgraduiertenförderung (LGF) fellowship program for individual doctoral training from Heidelberg University. B.B. is a recipient of a DAAD doctoral fellowship. S.W. is supported by BMBF (SysToxChip, FKZ 031A303E). X.C. is a recipient of the DFG grant program (CH 1690/2-1). We acknowledge financial support by Deutsche Forschungsgemeinschaft within the funding program Open Access Publishing, by the Baden-Württemberg Ministry of Science, Research and the Arts and by Ruprecht-Karls-Universität Heidelberg.

Conflicts of Interest: Authors declare no conflict of interest.

Abbreviations

NHC	*N*-heterocyclic carbene
CRC	Colorectal cancer
MAPK	Mitogen-activated protein kinase
JNK	c-Jun N-terminal kinase
ERK	Extracellular signal-related kinases
ROS	Reactive oxygen species
MtROS	Mitochondrial reactive oxygen species
MMP	Mitochondrial membrane potential
GSH	Glutathione
NAC	*N*-acetyl-L-cystein
NOX	NADPH oxidase
ETC	Electron transport chain
DMSO	Dimethyl sulfoxide

References

1. Tandon, R.; Luxami, V.; Kaur, H.; Tandon, N.; Paul, K. 1,8-Naphthalimide: A Potent DNA Intercalator and Target for Cancer Therapy. *Chem. Rec.* **2017**, *17*, 956–993. [CrossRef] [PubMed]

2.   Banerjee, S.; Veale, E.B.; Phelan, C.M.; Murphy, S.A.; Tocci, G.M.; Gillespie, L.J.; Frimannsson, D.O.; Kelly, J.M.; Gunnlaugsson, T. Recent advances in the development of 1,8-naphthalimide based DNA targeting binders, anticancer and fluorescent cellular imaging agents. *Chem. Soc. Rev.* **2013**, *42*, 1601–1618. [CrossRef] [PubMed]

3.   Bagowski, C.P.; You, Y.; Scheffler, H.; Vlecken, D.H.; Schmitza, D.J.; Ott, I. Naphthalimide gold(I) phosphine complexes as anticancer metallodrugs. *Dalton Trans.* **2009**, 10799–10805. [CrossRef] [PubMed]

4.   Kilpin, K.J.; Clavel, C.M.; Edafe, F.; Dyson, P.J. Naphthalimide-Tagged Ruthenium-Arene Anticancer Complexes: Combining Coordination with Intercalation. *Organometallics* **2012**, *31*, 7031–7039. [CrossRef]

5.   Meyer, A.; Oehninger, L.; Geldmacher, Y.; Alborzinia, H.; Wolfl, S.; Sheldrick, W.S.; Ott, I. Gold(I) N-Heterocyclic Carbene Complexes with Naphthalimide Ligands as Combined Thioredoxin Reductase Inhibitors and DNA Intercalators. *ChemMedChem* **2014**, *9*, 1794–1800. [CrossRef] [PubMed]

6.   Streciwilk, W.; Terenzi, A.; Misgeld, R.; Frias, C.; Jones, P.G.; Prokop, A.; Keppler, B.K.; Ott, I. Metal NHC Complexes with Naphthalimide Ligands as DNA-Interacting Antiproliferative Agents. *ChemMedChem* **2017**, *12*, 214–225. [CrossRef] [PubMed]

7.   Dhillon, A.S.; Hagan, S.; Rath, O.; Kolch, W. MAP kinase signalling pathways in cancer. *Oncogene* **2007**, *26*, 3279–3290. [CrossRef] [PubMed]

8.   Igea, A.; Nebreda, A.R. The Stress Kinase p38 alpha as a Target for Cancer Therapy. *Cancer Res.* **2015**, *75*, 3997–4002. [CrossRef]

9.   Hernandez Losa, J.; Parada Cobo, C.; Guinea Viniegra, J.; Sanchez-Arevalo Lobo, V.J.; Ramon y Cajal, S.; Sanchez-Prieto, R. Role of the p38 MAPK pathway in cisplatin-based therapy. *Oncogene* **2003**, *22*, 3998–4006. [CrossRef]

10.   Bragado, P.; Armesilla, A.; Silva, A.; Porras, A. Apoptosis by cisplatin requires p53 mediated p38alpha MAPK activation through ROS generation. *Apoptosis* **2007**, *12*, 1733–1742. [CrossRef]

11.   Mansouri, A.; Ridgway, L.D.; Korapati, A.L.; Zhang, Q.; Tian, L.; Wang, Y.; Siddik, Z.H.; Mills, G.B.; Claret, F.X. Sustained activation of JNK/p38 MAPK pathways in response to cisplatin leads to Fas ligand induction and cell death in ovarian carcinoma cells. *J. Biol. Chem.* **2003**, *278*, 19245–19256. [CrossRef] [PubMed]

12.   Park, S.J.; Kim, I.S. The role of p38 MAPK activation in auranofin-induced apoptosis of human promyelocytic leukaemia HL-60 cells. *Br. J. Pharmacol.* **2005**, *146*, 506–513. [CrossRef] [PubMed]

13.   Cheng, X.; Holenya, P.; Can, S.; Alborzinia, H.; Rubbiani, R.; Ott, I.; Wolfl, S. A TrxR inhibiting gold(I) NHC complex induces apoptosis through ASK1-p38-MAPK signaling in pancreatic cancer cells. *Mol. Cancer* **2014**, *13*, 221. [CrossRef] [PubMed]

14.   Shao, J.; Li, Y.; Wang, Z.; Xiao, M.; Yin, P.; Lu, Y.; Qian, X.; Xu, Y.; Liu, J. 7b, a novel naphthalimide derivative, exhibited anti-inflammatory effects via targeted-inhibiting TAK1 following down-regulation of ERK1/2- and p38 MAPK-mediated activation of NF-kappaB in LPS-stimulated RAW264.7 macrophages. *Int. Immunopharmacol.* **2013**, *17*, 216–228. [CrossRef] [PubMed]

15.   Streciwilk, W.; Terenzi, A.; Cheng, X.; Hager, L.; Dabiri, Y.; Prochnow, P.; Bandow, J.E.; Wolfl, S.; Keppler, B.K.; Ott, I. Fluorescent organometallic rhodium(I) and ruthenium(II) metallodrugs with 4-ethylthio-1,8-naphthalimide ligands: Antiproliferative effects, cellular uptake and DNA-interaction. *Eur. J. Med. Chem.* **2018**, *156*, 148–161. [CrossRef] [PubMed]

16.   Palchaudhuri, R.; Lambrecht, M.J.; Botham, R.C.; Partlow, K.C.; van Ham, T.J.; Putt, K.S.; Nguyen, L.T.; Kim, S.H.; Peterson, R.T.; Fan, T.M.; et al. A Small Molecule that Induces Intrinsic Pathway Apoptosis with Unparalleled Speed. *Cell Rep.* **2015**, *13*, 2027–2036. [CrossRef]

17.   Abbas, T.; Dutta, A. p21 in cancer: Intricate networks and multiple activities. *Nat. Rev. Cancer* **2009**, *9*, 400–414. [CrossRef]

18.   Zhang, P.Y.; Sadler, P.J. Advances in the design of organometallic anticancer complexes. *J. Organomet. Chem.* **2017**, *839*, 5–14. [CrossRef]

19.   Castro, J.P.; Grune, T.; Speckmann, B. The two faces of reactive oxygen species (ROS) in adipocyte function and dysfunction. *Biol. Chem.* **2016**, *397*, 709–724. [CrossRef]

20.   Li, X.Y.; Fang, P.; Mai, J.T.; Choi, E.T.; Wang, H.; Yang, X.F. Targeting mitochondrial reactive oxygen species as novel therapy for inflammatory diseases and cancers. *J. Hematol. Oncol.* **2013**, *6*, 19. [CrossRef]

21.   Aon, M.A.; Cortassa, S.; O'Rourke, B. Redox-optimized ROS balance: A unifying hypothesis. *Biochim. Biophys. Acta* **2010**, *1797*, 865–877. [CrossRef]

22.   Hata, A.N.; Engelman, J.A.; Faber, A.C. The BCL2 Family: Key Mediators of the Apoptotic Response to Targeted Anticancer Therapeutics. *Cancer Discov.* **2015**, *5*, 475–487. [CrossRef] [PubMed]

23. Danial, N.N. BAD: Undertaker by night, candyman by day. *Oncogene* **2008**, *27* (Suppl. 1), S53–S70. [CrossRef] [PubMed]

24. Oehninger, L.; Spreckelmeyer, S.; Holenya, P.; Meier, S.M.; Can, S.; Alborzinia, H.; Schur, J.; Keppler, B.K.; Wolfl, S.; Ott, I. Rhodium(I) N-Heterocyclic Carbene Bioorganometallics as in Vitro Antiproliferative Agents with Distinct Effects on Cellular Signaling. *J. Med. Chem.* **2015**, *58*, 9591–9600. [CrossRef] [PubMed]

25. Esteva, F.J.; Sahin, A.A.; Smith, T.L.; Yang, Y.; Pusztai, L.; Nahta, R.; Buchholz, T.A.; Buzdar, A.U.; Hortobagyi, G.N.; Bacus, S.S. Prognostic significance of phosphorylated P38 mitogen-activated protein kinase and HER-2 expression in lymph node-positive breast carcinoma. *Cancer* **2004**, *100*, 499–506. [CrossRef] [PubMed]

26. Chen, L.; Mayer, J.A.; Krisko, T.I.; Speers, C.W.; Wang, T.; Hilsenbeck, S.G.; Brown, P.H. Inhibition of the p38 kinase suppresses the proliferation of human ER-negative breast cancer cells. *Cancer Res.* **2009**, *69*, 8853–8861. [CrossRef] [PubMed]

27. Pereira, L.; Igea, A.; Canovas, B.; Dolado, I.; Nebreda, A.R. Inhibition of p38 MAPK sensitizes tumour cells to cisplatin-induced apoptosis mediated by reactive oxygen species and JNK. *EMBO Mol. Med.* **2013**, *5*, 1759–1774. [CrossRef]

28. Chiacchiera, F.; Grossi, V.; Cappellari, M.; Peserico, A.; Simonatto, M.; Germani, A.; Russo, S.; Moyer, M.P.; Resta, N.; Murzilli, S.; et al. Blocking p38/ERK crosstalk affects colorectal cancer growth by inducing apoptosis in vitro and in preclinical mouse models. *Cancer Lett.* **2012**, *324*, 98–108. [CrossRef]

29. Holenya, P.; Kitanovic, I.; Heigwer, F.; Wolfl, S. Microarray-based kinetic colorimetric detection for quantitative multiplex protein phosphorylation analysis. *Proteomics* **2011**, *11*, 2129–2133. [CrossRef]

30. Wagner, E.F.; Nebreda, A.R. Signal integration by JNK and p38 MAPK pathways in cancer development. *Nat. Rev. Cancer* **2009**, *9*, 537–549. [CrossRef]

31. Thornton, T.M.; Rincon, M. Non-classical p38 map kinase functions: Cell cycle checkpoints and survival. *Int. J. Biol. Sci.* **2009**, *5*, 44–51. [CrossRef] [PubMed]

32. St Germain, C.; Niknejad, N.; Ma, L.; Garbuio, K.; Hai, T.; Dimitroulakos, J. Cisplatin induces cytotoxicity through the mitogen-activated protein kinase pathways and activating transcription factor 3. *Neoplasia* **2010**, *12*, 527–538. [CrossRef]

33. Chiu, S.J.; Chao, J.I.; Lee, Y.J.; Hsu, T.S. Regulation of gamma-H2AX and securin contribute to apoptosis by oxaliplatin via a p38 mitogen-activated protein kinase-dependent pathway in human colorectal cancer cells. *Toxicol. Lett.* **2008**, *179*, 63–70. [CrossRef] [PubMed]

34. Guegan, J.P.; Ezan, F.; Theret, N.; Langouet, S.; Baffet, G. MAPK signaling in cisplatin-induced death: Predominant role of ERK1 over ERK2 in human hepatocellular carcinoma cells. *Carcinogenesis* **2013**, *34*, 38–47. [CrossRef] [PubMed]

35. Fan, C.; Zheng, W.; Fu, X.; Li, X.; Wong, Y.S.; Chen, T. Enhancement of auranofin-induced lung cancer cell apoptosis by selenocystine, a natural inhibitor of TrxR1 in vitro and in vivo. *Cell Death Dis.* **2014**, *5*, e1191. [CrossRef] [PubMed]

36. Kamata, H.; Honda, S.; Maeda, S.; Chang, L.; Hirata, H.; Karin, M. Reactive oxygen species promote TNFalpha-induced death and sustained JNK activation by inhibiting MAP kinase phosphatases. *Cell* **2005**, *120*, 649–661. [CrossRef]

37. Nandy, A.; Dey, S.K.; Das, S.; Munda, R.N.; Dinda, J.; Saha, K.D. Gold (I) N-heterocyclic carbene complex inhibits mouse melanoma growth by p53 upregulation. *Mol. Cancer* **2014**, *13*, 57. [CrossRef] [PubMed]

38. Zhang, J.J.; Muenzner, J.K.; Abu El Maaty, M.A.; Karge, B.; Schobert, R.; Wolfl, S.; Ott, I. A multi-target caffeine derived rhodium(i) N-heterocyclic carbene complex: Evaluation of the mechanism of action. *Dalton Trans.* **2016**, *45*, 13161–13168. [CrossRef]

39. Holenya, P.; Heigwer, F.; Wolfl, S. KOMA: ELISA-microarray calibration and data analysis based on kinetic signal amplification. *J. Immunol. Methods* **2012**, *380*, 10–15. [CrossRef]

40. Dabiri, Y.; Kalman, S.; Gürth, C.-M.; Kim, J.Y.; Mayer, V.; Cheng, X. The essential role of TAp73 in bortezomib-induced apoptosis in p53-deficient colorectal cancer cells. *Sci. Rep.* **2017**, *7*, 5423. [CrossRef]

MDPI

St. Alban-Anlage 66

4052 Basel

Switzerland

Tel. +41 61 683 77 34

Fax +41 61 302 89 18

www.mdpi.com

International Journal of Molecular Sciences Editorial Office

E-mail: ijms@mdpi.com

www.mdpi.com/journal/ijms